Dae Mann Kim

Introductory Quantum Mechanics for Semiconductor Nanotechnology

Related Titles

Waser, R. (ed.)

Nanoelectronics and Information Technology

Advanced Electronic Materials and Novel Devices

2010

ISBN: 978-3-527-40927-3

Wolf, E. L.

Quantum Nanoelectronics

An Introduction to Electronic Nanotechnology and Quantum Computing

2009

ISBN: 978-3-527-40749-1

Reiher, Markus / Wolf, Alexander

Relativistic Quantum Chemistry

The Fundamental Theory of Molecular Science

2009

ISBN: 978-3-527-31292-4

Köhler, Michael / Fritzsche, Wolfgang

Nanotechnology

An Introduction to Nanostructuring Techniques
Second, Completely Revised Edition

2007

ISBN: 978-3-527-31871-1

Coleman, Charles C.

Modern Physics for Semiconductor Science

2007

ISBN: 978-3-527-40701-9

Dae Mann Kim

Introductory Quantum Mechanics for Semiconductor Nanotechnology

WILEY-VCH Verlag GmbH & Co. KGaA

The Author

Prof. Dae Mann Kim
Seoul National University
Korea Inst. f. Advanced Study
Cheongnyangni 2-dong
Seoul 130-722
Republik Korea

Library of Congress Card No.: applied for

British Library Cataloguing-in-Publication Data
A catalogue record for this book is available from the British Library.

Bibliographic information published by the Deutsche Nationalbibliothek
The Deutsche Nationalbibliothek lists this publication in the Deutsche Nationalbibliografie; detailed bibliographic data are available on the Internet at http://dnb.d-nb.de.

Composition Thomson Digital, Noida, India
Printing and Bookbinding T.J. International Ltd., Padstow, Cornwall
Cover Design Formgeber, Eppelheim

Printed in Great Britain
Printed on acid-free paper

ISBN: 978-3-527-40975-4

To my grandma and my family

Contents

Introductory Quantum Mechanics for Semiconductor Nanotechnology. Dae Mann Kim

ISBN: 978-3-527-40975-4

Preface

This is an introductory textbook on quantum mechanics, intended for students majoring in nanotechnology-related engineering or applied sciences. The physics and chemistry majors who are already well versed in quantum mechanics could also benefit from the book by acquiring a general feel for the real-life impacts made by quantum mechanics. The level of discussion has been geared toward undergraduate seniors and graduate students in their early phases of graduate studies.

The book is focused on application-specific fundamentals, briefly on formulations but comprehensively on applications. The students could thus get the essential workings of quantum mechanics without spending too much time on covering wide realms of physics proper. The applications are directed toward time-tested main-stream technologies to render the discussions practical and useful for students to pursue their professional careers later on.

The book is built on (i) a brief review of classical and statistical mechanics and electromagnetism as a general background for understanding quantum mechanics in proper perspective, and (ii) compact discussions of atoms and molecules, the interaction of radiation with matter, and quantum statistics and semiconductor devices.

The quantum mechanics as a natural tool for bridging various disciplines in science and engineering was pioneered by Mckelvey in 1970s and by Yariv in mid-1980s and this book is in line with those interdisciplinary approaches for teaching. An effort has been made to fuse the topics covered as smoothly as possible. Also, the semiconductor devices are treated from a simple and unified standpoint of interface physics and equilibrium and nonequilibrium statistics.

The devices singled out for discussion include among others the p–n junctions as diode, photo and laser diodes, solar cell, and so on; the bipolar junction transistors; the metal oxide silicon field effect transistors (MOSFETs); and Schottky ohmic contacts and diodes. The ubiquitous roles of tunneling in a variety of devices are highlighted.

As well known, the quantum mechanics has been instrumental for inducing epoch-making technological breakthroughs. If history is any guide, it will continue to exert similar roles in years to come for both enhancing the performance of

Introductory Quantum Mechanics for Semiconductor Nanotechnology. Dae Mann Kim

ISBN: 978-3-527-40975-4

existing devices and overcoming the limitations of such devices. A goal of this book is to provide the technical backgrounds and motivations for the students to carry on further studies on quantum mechanics and to put the quantum insights into useful applications.

The organization of the book is summarized as follows.

First, the essentials of classical mechanics are reviewed by considering a simple example of harmonic oscillator. The statistical physics and electromagnetic waves are reviewed, based on the simplified version of Boltzmann transport equation and Maxwell's equations. The milestone discoveries leading to the advent of quantum mechanics are briefly described, and Bohr's theory of hydrogen atom is presented as the culmination of the old quantum theory. The Schrödinger equation is then introduced heuristically and essential features of operator algebra are discussed, together with the basic postulates of quantum mechanics.

The Schrödinger equation is then applied to problems of practical interest, involving quantum wells, and such key concepts as subbands and one-, two-, and three-dimensional densities of states are highlighted. Also, the scattering of particles from potential wells or barriers are considered, with a strong emphasis on tunneling and its applications. The simple band theory in one-dimensional crystal is presented in conjunction with Bloch wavefunctions and in correlation with the resonant tunneling.

Two key bound systems are treated in detail. First, the harmonic oscillator is revisited and treated quantum mechanically and its operator formulation is also introduced to pave the way for field quantization. The Schrödinger treatment of hydrogen atom as the simplest atomic system is next taken up as the second example of key bound system. Here, the atomic wavefunctions or orbitals are discussed in detail and utilized for describing multielectron atoms and molecules. In so doing, the Pauli exclusion principle is brought in and covalent and ionic chemical bonds and van der Waals attraction are briefly touched upon.

The perturbation theory as a necessity for dealing with complicated problems of practical interest is discussed and the time-dependent perturbation scheme is utilized to account for the important interaction of light with matter. Here, the discussion begins with the field interacting with a single atom and is extended to the general case of ensemble of atoms, constituting the optical medium. A brief discussion on the novel features and operating principles of laser devices follows.

The quantum statistics of fermions, bosons, and distinguishable particles are discussed, starting from the first principles. The semiconductor statistics and the transport of electrons and holes are considered in detail as two key factors for charge control. The drift–diffusion and generation–recombination currents are analyzed under both equilibrium and nonequilibrium conditions.

Finally, the basic concepts of quantum mechanics and device physics underlying the semiconductor device operation are highlighted. For this purpose, the p–n junction is first singled out for consideration and the general background issues of equilibrium versus nonequilibrium, Fermi level versus quasi-Fermi level, and interface band bending are addressed. The p–n junctions used and operated in a variety of active devices are highlighted in light of the quantum mechanical principles involved. The bipolar junction transistor working on the basis of two interacting junction diodes is discussed.

The metal oxide silicon field effect transistor is discussed as one of the mainstream technology drivers and also one of the first devices based on quantum wells, induced and controlled by the gate voltage. The simple MOSFET structure serving as a platform for high-density memory cells and image sensors is highlighted. The relentless drive for downsizing the device for higher integration and higher performance is considered, together with accompanying physical issues. Finally, Schottky contact as a key component for nano-CMOS and molecular devices is discussed both as diodes and ohmic contacts.

The contents of this book have evolved from the courses offered in parts in the departments of electrical and computer engineering at Rice University, Houston, TX, United States and also at POSTECH and Sungkyunkwan university, Korea. The active participation and hard works put in on the part of attending students made it a joyful experience to teach the courses. I would like to express my sincere thanks to those students. I would also like to express special thanks to my students Kang Whee Kim, Bomsoo Kim, and Chang-Ki Baek for their tireless devotions for preparing the figures and graphs used in the book. Finally I would like to express my thanks to Korean Ministry of Education, Science and Technology for the support provided through Korean Nano Research Society.

Seoul, Korea *Dae Mann Kim*

1
Brief Review of Classical Theories

Science as a quantitative tool for understanding and describing natural phenomena in a causal context was initiated by Newton, who introduced the basic postulate of the equation of motion. The framework of classical mechanics provides the general background that facilitates understanding of quantum mechanics in a proper perspective. Also, the basic concepts of statistical mechanics and electromagnetic waves are inseparably intermingled with quantum mechanics. Hence, a brief review of these classical theories is a convenient starting point for discussing quantum mechanics.

1.1
Harmonic Oscillator

A particle attached to a spring is called the harmonic oscillator (HO), as sketched in Figure 1.1. When the spring is stretched or compressed from its equilibrium position by pulling or pushing the particle, it provides a restoring force, pulling or pushing the particle toward the equilibrium position. As a result, the particle executes an oscillatory motion, the understanding of which is essential for considering a wide range of phenomena, for example, molecular vibrations, electromagnetic waves, and so on.

The Oscillator Equation The Newton's equation of motion of the oscillator is then given by

$$m\ddot{x} = -kx, \tag{1.1}$$

where m is the mass, x is the displacement from the equilibrium position, and k is the spring constant. The double dots denote the second-order differentiation with respect to time. The constant k representing Hook's restoring force generally depends on x, especially for large x, but is taken constant for simplicity. Dividing both sides of (1.1) by m and transferring the force term to the left results in

$$\ddot{x} + \omega^2 x = 0, \quad \omega^2 \equiv (k/m) \tag{1.2}$$

with ω representing the characteristic frequency.

Introductory Quantum Mechanics for Semiconductor Nanotechnology. Dae Mann Kim

ISBN: 978-3-527-40975-4

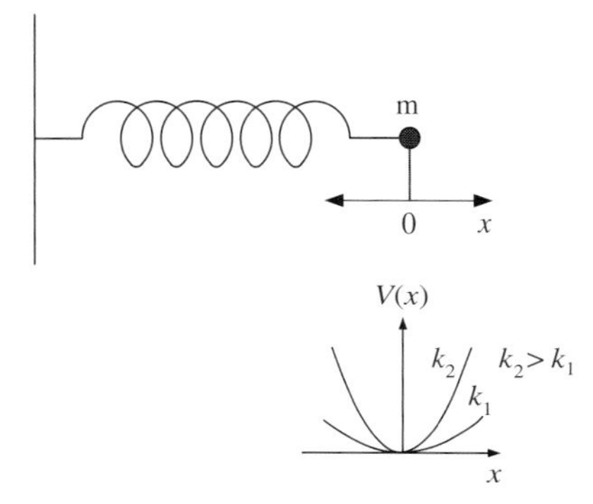

Figure 1.1 A harmonic oscillator: particle of mass m attached to a spring with constant k and oscillating with frequency $\omega = \sqrt{k/m}$.

The solution of this harmonic oscillator equation can be chosen from a rich repertoire of functions such as $\sin \omega t$, $\cos \omega t$, and $\exp \pm i\omega t$, with two constants of integration available for fitting the boundary or the initial conditions. To be specific, consider the case in which the oscillator is pulled by x_0 and gently released, that is, $x(t=0) = x_0$ and $\dot{x}(t=0) \equiv v(t=0) = 0$. The solution satisfying the initial condition is then given by

$$x(t) = x_0 \cos \omega t, \tag{1.3}$$

in which case the velocity reads as

$$v(t) \equiv \dot{x}(t) = -\omega x_0 \sin \omega t. \tag{1.4}$$

Thus, both $x(t)$ and $v(t)$ are shown to oscillate in quadrature, as shown in Figure 1.2. The period of oscillation is defined in terms of the angular frequency ω and the frequency ν in Hz as

$$T \equiv 2\pi/\omega \equiv 1/\nu. \tag{1.5}$$

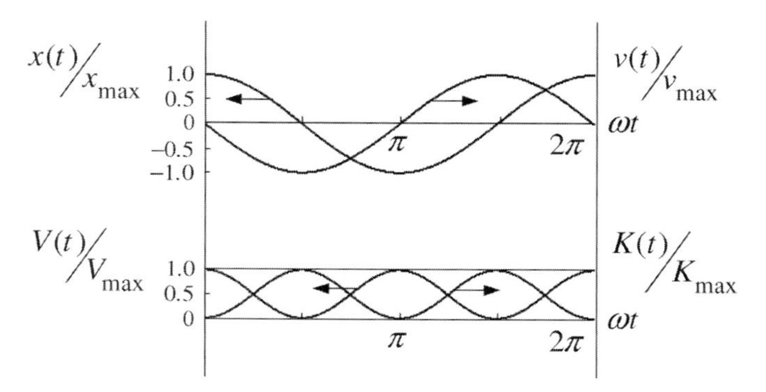

Figure 1.2 The displacement (x), velocity (v), kinetic (K), and potential (V) energies of the harmonic oscillator versus time ωt. The plots are scaled with respective maximum values. The total energy $K + V$ is constant in time.

The Energies The kinetic energy of the oscillator is given in this case by

$$K \equiv \frac{1}{2}m\dot{x}^2 = \frac{1}{2}m\omega^2 x_0^2 \sin^2 \omega t = \frac{1}{2}kx_0^2 \sin^2 \omega t, \tag{1.6}$$

with $k = m\omega^2$ (see (1.2)). The potential energy is defined as the work done against the force and is given by

$$V \equiv -\int_0^x dx(-kx) = \frac{1}{2}kx^2 = \frac{1}{2}kx_0^2 \cos^2 \omega t. \tag{1.7}$$

Here, $V(x)$ is taken zero at the reference point at $x = 0$. The total energy of the oscillator,

$$E = K + V = \frac{1}{2}kx_0^2(\sin^2 \omega t + \cos^2 \omega t) = \frac{1}{2}kx_0^2, \tag{1.8}$$

is therefore constant in time.

A system whose total energy is time invariant is called the conservative system, and the harmonic oscillator is a typical example. As shown in Figure 1.2, the kinetic and potential energies are oscillating in time with different phases but their sum is always maintained at a constant level, specified by the initial condition. It is important to point out that in the classical mechanics, the total energy E is an analog quantity, varying by any infinitesimal quantity. That is, the oscillator can execute an oscillation with any value of E chosen by the initial condition.

1.1.1
The Hamiltonian

The Hamiltonian H of a system is the total energy consisting of the kinetic and potential energies. For the harmonic oscillator under consideration, one can thus express H in terms of the linear momentum p_x and the displacement x as

$$H \equiv K + V = \frac{p_x^2}{2m} + \frac{1}{2}kx^2. \tag{1.9}$$

The equation of motion can also be expressed in terms of the Hamiltonian as

$$\dot{x} = \frac{\partial H}{\partial p_x} = \frac{p_x}{m}, \quad \dot{p}_x = -\frac{\partial H}{\partial x} = -kx. \tag{1.10}$$

The pair of equations in (1.10) is known as the Hamilton's equation of motion and when combined reduces to Newton's equation of motion, which can be readily verified. The variables x and p_x appearing in the Hamilton's equation of motion are called the canonically conjugate variables.

Also, p_x appearing in (1.10) is called the generalized momentum for x. A dynamic system having a set of coordinates $\{q_j\}$ and corresponding momenta $\{p_j\}$ is characterized by the Hamiltonian given by

$$H = \sum_{i,j} \frac{p_{ij}^2}{2m} + V[\{q_{ij}, p_{ij}\}], \tag{1.11}$$

where $i = x,\ y,\ z$ and j denotes the number of particles comprising the system. The equations of motion for the jth conjugate variables read as

$$\dot{q}_{ij} = \frac{\partial H}{\partial p_{ij}} = \frac{p_{ij}}{m}, \quad \dot{p}_{ij} = -\frac{\partial H}{\partial q_{ij}}. \tag{1.12}$$

Clearly, Newton's and Hamilton's equations of motion are identical. By solving these equations, one can precisely predict or follow the position and momentum of the particle in a deterministic manner. That is, once x and p_x are known, say at $t = 0$, it is possible to exactly follow these dynamic variables at all times. In classical mechanics, the dynamic variables such as energy, position, momentum, and so on are all analog quantities varying continuously by any infinitesimal amount.

1.2
Boltzmann Transport Equation

Macroscopic quantities are ultimately determined by the cumulative effects of a large number of microscopic objects such as electrons, holes, atoms, and molecules. Statistical mechanics describes such cumulative effects via the distribution function $f(\boldsymbol{r}, \boldsymbol{v}, t)$. Here, $\boldsymbol{r}$ and $\boldsymbol{v}$ are the position and velocity of the particle at time t and $f(\boldsymbol{r}, \boldsymbol{v}, t)d\boldsymbol{r}d\boldsymbol{v}$ represents the probability of finding the particle in the phase-space volume element $d\boldsymbol{r} \times d\boldsymbol{v}$ centered at $\boldsymbol{r},\ \boldsymbol{v}$. Thus, when multiplied by the number density n of the particles, $f(\boldsymbol{r}, \boldsymbol{v}, t)d\boldsymbol{r}d\boldsymbol{v}$ represents the number of particles in the phase-space volume element under consideration.

The change in time of $f(\boldsymbol{r}, \boldsymbol{v}, t)$ is then simply given by the total differentiation of $f(\boldsymbol{r}, \boldsymbol{v}, t)$ with respect to time, that is,

$$\begin{aligned} \frac{df(\boldsymbol{r}, \boldsymbol{v}, t)}{dt} &= \frac{\partial f}{\partial t} + \frac{\partial f}{\partial x}\frac{\partial x}{\partial t} + \cdots + \frac{\partial f}{\partial v_x}\frac{\partial v_x}{\partial t} + \cdots \\ &= \frac{\partial f}{\partial t} + \boldsymbol{v} \cdot \nabla f + (\boldsymbol{F}/m) \cdot \nabla_v f, \end{aligned} \tag{1.13}$$

where

$$\nabla = \hat{x}(\partial/\partial x) + \hat{y}(\partial/\partial y) + \hat{z}(\partial/\partial z) \quad \text{and} \quad \nabla_v = \hat{x}(\partial/\partial v_x) + \hat{y}(\partial/\partial v_y) + \hat{z}(\partial/\partial v_z)$$

are the gradient operators with respect to $\boldsymbol{r}$ and $\boldsymbol{v}$, respectively, and the acceleration $\boldsymbol{a} = \dot{\boldsymbol{v}}$ is expressed via the force, that is, $\boldsymbol{a} = \boldsymbol{F}/m$.

The change in time of the distribution function $f(\boldsymbol{r}, \boldsymbol{v}, t)$ is caused by the collisions the particles encounter. That is, the particles may be pushed out of the phase-space volume element centered at $\boldsymbol{r}$, $\boldsymbol{v}$ or pushed into it due to collisions. Hence, one can write the total time rate of change of f as

$$\frac{\partial f}{\partial t} + \boldsymbol{v} \cdot \nabla f + (\boldsymbol{F}/m) \cdot \nabla_v f = \left.\frac{\delta f}{\delta t}\right|_{\text{coll}}, \tag{1.14}$$

where the collision term on the right-hand side represents the collision-induced changes of the distribution function. This change is generally specified by an integral in which $f(\boldsymbol{r}, \boldsymbol{v}, t)$ appears as a part of the integrand. Equation 1.14 is thus an integro-differential equation and is known as the Boltzmann transport equation.

1.2.1
Equilibrium Distribution Function and Equipartition Theorem

In the thermodynamic equilibrium, the physical quantities are invariant in time; hence, the equilibrium distribution function f_0 should also be time independent, that is, $(\partial/\partial t)f_0 = 0$. Furthermore, the collision term in (1.14) should be put to zero because of the detailed balancing of the two opposing collision processes operative in equilibrium, one pushing the particle out of and the other pushing the particle into the volume element under consideration. Thus, the transport equation (1.14) reduces in equilibrium to

$$\boldsymbol{v} \cdot \nabla f_0 - \frac{1}{m} \nabla \varphi \cdot \nabla_v f_0 = 0, \quad \boldsymbol{F} \equiv -\nabla \varphi, \tag{1.15}$$

where the force is expressed via the potential φ.

One may look for the solution of (1.15) in the form

$$f_0(\boldsymbol{r}, \boldsymbol{v}) \propto R(r) e^{-mv^2/2k_B T}, \tag{1.16}$$

where k_B is the Boltzmann constant, namely, $k_B = 1.381 \times 10^{-23}$ J/K or 8.617×10^{-5} eV/K. Inserting (1.16) into (1.15) and performing the operations therein and dividing both sides of (1.15) by (1.16) results in

$$\boldsymbol{v} \cdot \left[\frac{\nabla R(\boldsymbol{r})}{R(\boldsymbol{r})} + \frac{\nabla \varphi(\boldsymbol{r})}{k_B T} \right] = 0. \tag{1.17}$$

The solution is thus given by

$$R(\boldsymbol{r}) \propto e^{-\varphi(\boldsymbol{r})/k_B T}, \tag{1.18}$$

which can be readily verified by direct substitution. Hence, by combining (1.16) and (1.18), f_0 is given by

$$f_0(\boldsymbol{r}, \boldsymbol{v}) = N e^{-E(\boldsymbol{r})/k_B T}, \tag{1.19}$$

where

$$E(\boldsymbol{r}) = \frac{1}{2} m v^2 + \varphi(\boldsymbol{r}) \tag{1.20}$$

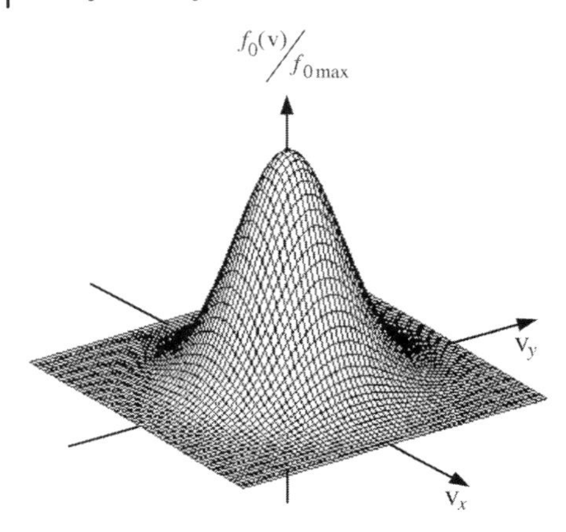

Figure 1.3 The plot of equilibrium distribution function of a system of free particles, $f_0(\boldsymbol{v})/f_0(0)$. $f_0(\boldsymbol{v})$ is symmetric with respect to v_x, v_y, and v_z.

is the total energy at $\boldsymbol{r}$ and N is the normalization constant. For the special case of a free particle system in which $\varphi = 0$, $f_0(\boldsymbol{r}, \boldsymbol{v})$ is reduced to the well-known Maxwell–Boltzmann distribution function

$$f_0(\boldsymbol{v}) = \left(\frac{m}{2\pi k_B T}\right)^{3/2} e^{-mv^2/2k_B T}, \quad v^2 = v_x^2 + v_y^2 + v_z^2. \tag{1.21}$$

The exponential factor in (1.21) is often called the Boltzmann probability factor.

The distribution function (1.21) well describes the equilibrium properties. Specifically, $f_0(\boldsymbol{v})$ is symmetric in $\boldsymbol{v}$ and there is no preferred direction in $\boldsymbol{v}$, as clearly shown in Figure 1.3. A particle moving from left to right, for instance, is balanced by its inverse process of moving from right to left. This behavior is explicitly built into $f_0(\boldsymbol{v})$ in (1.21). Therefore, the average velocity in equilibrium is zero as evident from the parity considerations, that is,

$$\langle v \rangle = \int_{-\infty}^{\infty} dv_x \int_{-\infty}^{\infty} dv_y \int_{-\infty}^{\infty} dv_z \boldsymbol{v} f_0(\boldsymbol{v}) = 0 \tag{1.22}$$

(see Figure 1.3). However, $\langle v_x^2 \rangle$ is not zero and is found by using $f_0(\boldsymbol{v})$ in (1.21) as

$$\begin{aligned} \langle v_x^2 \rangle &= \left(\frac{m}{2\pi k_B T}\right)^{3/2} \int_{-\infty}^{\infty} dv_x v_x^2 e^{-mv_x^2/2k_B T} \int_{-\infty}^{\infty} dv_y e^{-mv_y^2/2k_B T} \int_{-\infty}^{\infty} dv_z e^{-mv_z^2/2k_B T} \\ &= \left(\frac{m}{2\pi k_B T}\right)^{1/2} \left(-\frac{\partial}{\partial \beta}\right) \int_{-\infty}^{\infty} dv_x e^{-\beta v_x^2}, \quad \beta = m/2k_B T, \\ &= k_B T/m. \end{aligned} \tag{1.23}$$

In evaluating (1.23), the v_y and v_z integrations have been performed, using the formula

$$\int_{-\infty}^{\infty} dx e^{-ax^2+bx} = \sqrt{\frac{\pi}{a}} e^{b^2/4a}. \tag{1.24}$$

Also, the v_x integral was first reduced to the same format as (1.24) and the integrated result was differentiated with respect to β to obtain the final result. One can likewise write by inspection

$$\langle v_x^2 \rangle = \langle v_y^2 \rangle = \langle v_z^2 \rangle = k_B T/m \tag{1.25}$$

and obtain the celebrated equipartition theorem

$$\frac{1}{2} m \langle v^2 \rangle = \frac{1}{2} m \left(\langle v_x^2 \rangle + \langle v_y^2 \rangle + \langle v_z^2 \rangle \right) = \frac{3}{2} k_B T. \tag{1.26}$$

1.2.2
Nonequilibrium Distribution Function and Relaxation Approach

When a system is pushed away from the equilibrium, for example, by irradiation or by applying a voltage, the system relaxes back to equilibrium by collisions. This process is often described by a simple phenomenology in which the collision term in the transport equation is replaced by

$$\left. \frac{\delta f}{\delta t} \right|_{\text{coll}} = -\frac{f - f_0}{\tau}, \tag{1.27}$$

where τ denotes the relaxation time and f and f_0 are the nonequilibrium and equilibrium distribution functions, respectively.

The role of the relaxation term is best illustrated by considering a free and homogeneous system of electrons, which is pushed away from the equilibrium instantaneously and left alone all by itself. In this case, in view of the spatial homogeneity, $\nabla f = 0$ and there is no force acting on the system, that is, $\boldsymbol{F} = 0$. Thus, the transport equation (1.14) is simplified as

$$\frac{\partial f}{\partial t} = -\frac{f - f_0}{\tau}. \tag{1.28}$$

By grouping together the two terms containing f to the left-hand side and introducing an integration factor, one can rewrite (1.28) as

$$\frac{\partial}{\partial t}(e^{t/\tau} f) = \frac{f_0}{\tau} e^{t/\tau}. \tag{1.29}$$

Equation 1.29 can be readily verified to be identical to (1.28) by performing the time differentiation. One can easily integrate (1.29), obtaining

$$f(t) = f(0) e^{-t/\tau} + f_0 (1 - e^{-t/\tau}), \tag{1.30}$$

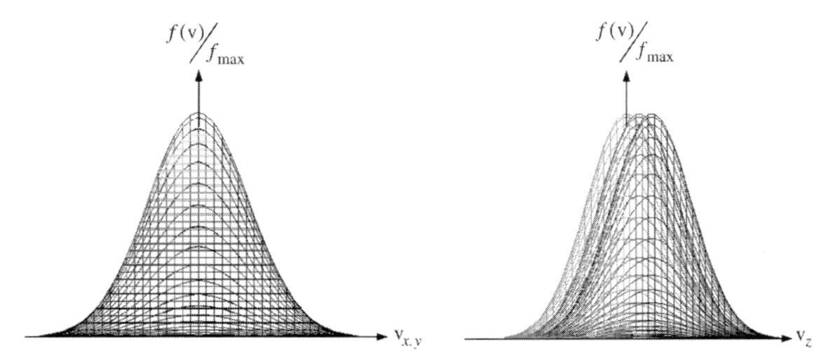

Figure 1.4 The plot of the nonequilibrium electron distribution function $f(\boldsymbol{v})/f(0)$ brought about by the electric field along the z-direction. $f(\boldsymbol{v})$ is symmetric in x and y directions, but the symmetry is broken progressively with increasing field strength in the z-direction.

where $f(0)$ is the initial nonequilibrium distribution function. Thus, $f(t)$ is shown to relax back to f_0 in a few relaxation times τ, while $f(0)$ decays to essentially zero level in the same time span.

With the role of the relaxation term thus clearly demonstrated, consider next the effect of applying a constant electric field to the same system of free and spatially homogenous electrons. At steady state to which the discussion is confined to, f is independent of time, that is, $\partial f/\partial t = 0$, and the transport equation (1.14) is given in relaxation approximation by

$$-\frac{q\mathrm{E}_0}{m}\frac{\partial f}{\partial v_z} = -\frac{f - f_0}{\tau}, \tag{1.31}$$

where the force acting on the electron $-q\mathrm{E}_0$ is specified via the electronic charge $-q$ and the electric field E_0 applied in the z-direction. For the simplicity of discussion, f is taken not to depart too much from f_0, so that $f - f_0 \ll f, f_0$. Then, one may replace f appearing on the left-hand side of (1.31) by f_0 to the first-order iteration, obtaining

$$f \approx f_0 + \frac{q\mathrm{E}_0\tau}{m}\frac{\partial f_0}{\partial v_z} = f_0\left(1 - \frac{q\mathrm{E}_0\,\tau v_z}{k_B T}\right). \tag{1.32}$$

Here the Boltzmann distribution function f_0 derived in (1.21) has been used.

It is therefore clear from (1.32) that the nonequilibrium distribution function differs from f_0. Specifically, f is asymmetric in v_z, along which the electric field E_0 has been applied, and this is clearly illustrated in Figure 1.4.

1.2.3
Mobility and Conductivity

The nonequilibrium distribution function at steady state carries an interesting consequence. First, f is still symmetric with respect to v_x and v_y; hence, $\langle v_x \rangle = \langle v_y \rangle = 0$, as evident from the parity consideration. However, the average velocity along the direction of the field is not zero, but is given by

$$\langle v_z \rangle = \frac{\int\limits_{-\infty}^{\infty} dv_x \int\limits_{-\infty}^{\infty} dv_y \int\limits_{-\infty}^{\infty} dv_z v_z f_0 \left(1 - \frac{q\text{E}_0 \tau v_z}{k_B T}\right)}{\int\limits_{-\infty}^{\infty} dv_x \int\limits_{-\infty}^{\infty} dv_y \int\limits_{-\infty}^{\infty} dv_z f_0 \left(1 - \frac{q\text{E}_0 \tau v_z}{k_B T}\right)}. \tag{1.33}$$

In evaluating (1.33), one can again invoke the parity arguments to eliminate the first integral in the numerator. By the same token, the second integral in the denominator goes to zero, in which case the denominator becomes unity (see (1.21)). Thus, (1.33) can be formally expressed as

$$\langle v_z \rangle = -\frac{q\text{E}_0}{m} \langle \tau \rangle, \tag{1.34}$$

where

$$\langle \tau \rangle \equiv \frac{m}{k_B T} \int\limits_{-\infty}^{\infty} dv_x \int\limits_{-\infty}^{\infty} dv_y \int\limits_{-\infty}^{\infty} dv_z v_z^2 \tau(v) f_0 \tag{1.35}$$

is the average relaxation time that depends, in general, on v. For the special case where τ is constant and independent of v, that is, $\tau(v) = \tau_0$, it follows from (1.35) that $\langle \tau \rangle = \tau_0$, as expected.

The average velocity found in (1.34) represents the drift velocity with which all electrons in the system move uniformly on top of their random thermal motion. This drift velocity is driven by the external electric field by exerting force or acceleration on these electrons and can therefore be viewed as the output of E_0, that is,

$$v_{dn} \equiv \langle v_z \rangle = -\frac{q\text{E}_0 \langle \tau \rangle}{m} \equiv -\mu_n \text{E}_0. \tag{1.36}$$

The response function

$$\mu_n = \frac{q\langle \tau \rangle}{m}, \tag{1.37}$$

connecting the input field E_0 to the output drift velocity v_{dn}, is called the electron mobility. The drift velocity is thus to be viewed approximately as the uniform velocity the electrons acquire in between collisions on top of their random thermal motions.

The drift current density contributed by electrons is therefore given by

$$\begin{aligned} \boldsymbol{J}_n &\equiv -q \sum_{j=1}^{n} (\boldsymbol{v}_{j\text{th}} + \boldsymbol{v}_{dn}) \\ &= -qn\boldsymbol{v}_{dn} = qn\mu_n \mathbf{E}_0 \equiv \sigma_n \mathbf{E}_0, \end{aligned} \tag{1.38}$$

where n is the electron density and the thermal velocities $\boldsymbol{v}_{j\text{th}}$ of the electrons sum up to be zero due to their random nature. Here again, the drift current density can be viewed as the output driven by the input field and the quantity

$$\sigma_n = qn\mu_n \tag{1.39}$$

connecting J_n to E_0 is called the electrical conductivity of the system of electrons. The mobility and conductivity are key transport coefficients.

1.3
Maxwell's Equations

The Maxwell's equations provide the basis for describing the dynamics of electromagnetic fields. When the charge and current densities ϱ and $\boldsymbol{J}$ are spatially distributed and vary in time, the electric ($\mathbf{E}(\boldsymbol{r}, t)$) and magnetic ($\boldsymbol{B}(\boldsymbol{r}, t)$) fields are generated. These fields are dynamically coupled in a manner specified by the Maxwell's equations. The equations read in differential form as

$$\nabla \times \mathbf{E} = -\frac{\partial \boldsymbol{B}}{\partial t}, \tag{1.40}$$

$$\nabla \times \boldsymbol{H} = \boldsymbol{J} + \frac{\partial \boldsymbol{D}}{\partial t}, \tag{1.41}$$

$$\nabla \cdot \mathbf{E} = \varrho/\varepsilon, \tag{1.42}$$

$$\nabla \cdot \boldsymbol{B} = 0, \tag{1.43}$$

where the displacement vector $\boldsymbol{D}(\boldsymbol{r}, t)$ and the magnetic field intensity $\boldsymbol{H}(\boldsymbol{r}, t)$ are related to the electric and magnetic fields by

$$\boldsymbol{D} = \varepsilon \mathbf{E}, \tag{1.44}$$

$$\boldsymbol{B} = \mu \boldsymbol{H}, \tag{1.45}$$

and ε and μ are the permittivity and the permeability of the medium. An additional auxiliary equation is the charge conservation equation

$$\frac{\partial \varrho}{\partial t} + \nabla \cdot \boldsymbol{J} = 0. \tag{1.46}$$

The first Maxwell's equation (1.40) is the mathematical statement of the law of induction, found empirically by Faraday. It quantifies the time-varying flux of $\boldsymbol{B}(\boldsymbol{r}, t)$ inducing the electromotive force $\mathbf{E}$ around the $\boldsymbol{B}$ field lines. The second equation (1.41) represents Ampere's circuital law and describes the current acting as the source for generating $\boldsymbol{B}(\boldsymbol{r}, t)$ at every point in space. The third equation (1.42) is the usual Gauss flux theorem representing the well-known Coulomb's law, and the fourth equation (1.43) simply points to the fact that no magnetic monopole has been observed thus far.

It is important to note that the circuital law of (1.41), as found empirically by Ampere, was modified by Maxwell, who introduced the additional source term $\partial \boldsymbol{D}/\partial t$, called the displacement current. This was necessitated by the fact that the curl of any vector, $\nabla \times \boldsymbol{A}$, should be solenoidal, that is, $\nabla \cdot \nabla \times \boldsymbol{A} \equiv 0$, which can be readily verified by carrying out the operation on any vector. With the displacement current term built into the circuital law, the divergence operation on the right-hand side of (1.41) simply yields the charge conservation equation (1.46) via (1.42) and (1.44), and therefore becomes zero, thereby making $\boldsymbol{H}$ solenoidal. Thus, the Maxwell's equations are shown to stand firmly on the observed empirical laws of nature and also to satisfy the consistency condition of $\boldsymbol{H}$ being solenoidal. The equations have successfully undergone the test of time and have been the source of unceasing applications.

1.3.1 Wave Equation

The coupled electric and magnetic fields in (1.40) and (1.41) can be decoupled from each other and singled out for examination as a function of space and time. Thus, consider the electric and magnetic fields in vacuum, in which there is no current or the charge, that is, $\boldsymbol{J} = \varrho = 0$. Performing the curl operation on both sides of (1.40), one obtains from the left-hand side

$$\nabla \times \nabla \times \mathbf{E} \equiv \nabla \nabla \cdot \mathbf{E} - \nabla^2 \mathbf{E} = -\nabla^2 \mathbf{E}, \tag{1.47}$$

where a well-known vector identity has been used, together with (1.42) for $\varrho = 0$. Similarly, with the use of (1.41), (1.44), and (1.45) and for $\boldsymbol{J} = 0$, the right-hand side yields

$$\nabla \times \left(\frac{-\partial \boldsymbol{B}}{\partial t} \right) = -\mu_0 \frac{\partial}{\partial t} \left(\boldsymbol{J} + \varepsilon_0 \frac{\partial}{\partial t} \mathbf{E} \right) = -\mu_0 \varepsilon_0 \frac{\partial^2}{\partial t^2} \mathbf{E}. \tag{1.48}$$

Hence, equating (1.47) and (1.48) yields the wave equation

$$\nabla^2 \mathbf{E} - \frac{1}{c^2} \frac{\partial^2}{\partial t^2} \mathbf{E} = 0, \quad \mu_0 \varepsilon_0 \equiv 1/c^2, \tag{1.49}$$

where c is the velocity of light in vacuum and is specified in terms of the vacuum permittivity ε_0 and permeability μ_0. One can similarly derive the identical wave equation for $\boldsymbol{H}$.

The wave equation derived in vacuum can be generalized to an optical medium in which there is no charge or current source. In this case, the only modification required is to replace μ_0 and ε_0 by μ and ε, respectively. Thus, one can write

$$\nabla^2 \mathbf{E} - \frac{1}{v^2} \frac{\partial^2}{\partial t^2} \mathbf{E} = 0, \quad \mu\varepsilon = \mu_0 \mu_r \varepsilon_0 \varepsilon_r \equiv 1/(c/n)^2 = 1/v^2. \tag{1.50}$$

Here, the permeability and permittivity of the medium have been scaled with those of the vacuum and the analysis has been confined to nonmagnetic materials in which $\mu_r = 1$. The dielectric constant ε_r of the medium is related to the index of refraction n via $\varepsilon_r = n^2$, and $v < c$ since $n > 1$.

It should be pointed out that the Maxwell's displacement current term is crucial in bringing out the wave nature of electromagnetic fields, as explicitly demonstrated in the derivation of the wave equation.

1.3.2 Plane Waves and Wave Packets

The electric field satisfying the wave equation (1.50) can be expressed in the form of a traveling plane wave,

$$\mathbf{E}(\boldsymbol{r}, t) = \text{Re}\, \mathbf{E}_0 e^{-i(\omega t - \boldsymbol{k}\cdot\boldsymbol{r})}. \tag{1.51}$$

Here, $\mathbf{E}_0$ is the amplitude of the field and is, in general, a complex quantity consisting of the magnitude and the phase, both of which are the information carriers. The exponential mode function is characterized by the angular frequency ω and the wave vector $\boldsymbol{k}$, which should satisfy the dispersion relation to make (1.51) the solution of the wave equation. Inserting (1.51) into (1.50) results in the dispersion relation

$$k^2 = \omega^2/v^2, \quad v = c/n, \tag{1.52a}$$

or equivalently

$$\omega = vk. \tag{1.52b}$$

The dispersion relation (1.52), specifying ω as a function of k or vice versa, plays a key role in characterizing the electric field.

Now, the mode function in (1.51) generally represents a signal propagating in space with a certain velocity. This can be explicitly illustrated by taking $\mathbf{E}_0$ real and the wave vector $\boldsymbol{k}$ in the z-direction, that is, $\boldsymbol{k} = \hat{z}k$. In this case, (1.51) reduces to

$$\mathbf{E}(z, t) = \mathbf{E}_0 \cos(\omega t - kz). \tag{1.53}$$

The spatial profiles of (1.53) are plotted in Figure 1.5 at different times. As is clear from the figure, at any fixed time, the spatial pattern of $\mathbf{E}(z, t)$ repeats itself over an interval λ, called the wavelength. This distance λ over which the pattern repeats itself should obviously be determined by the condition $k\lambda = 2\pi$, namely,

$$\lambda \equiv 2\pi/k = 2\pi/(\omega/v) = v/f, \tag{1.54}$$

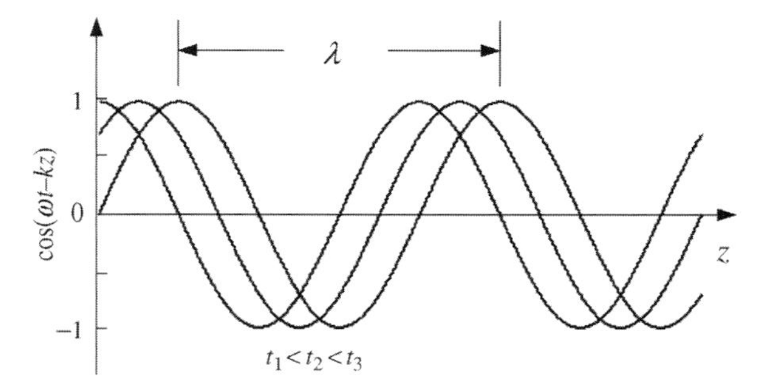

Figure 1.5 The plot of the plane wave mode function $\cos(\omega t - kz)$ versus z at different times t. The wave propagates in time and the wavelength is determined by $k\lambda = 2\pi$, namely, $\lambda \equiv 2\pi/k$.

where the dispersion relation (1.52) has been used and the angular frequency is related to the frequency in Hz as $\omega = 2\pi f$. Also shown in the figure is the spatial profile of $\mathbf{E}(z,t)$ propagating in time with a velocity. This velocity can be found by following in time a constant phase front, say the peak of the profile, for example, in which $\omega t - kz = 0$. Differentiating the constant phase with respect to time yields the velocity with which the signal propagates, hence the name phase velocity, that is,

$$v_p \equiv \partial z/\partial t = \omega/k. \tag{1.55}$$

Superposed Waves Inasmuch as the wave equation (1.50) is linear, a linear superposition of two plane waves, each satisfying the equation, is also the solution. Thus, consider two such plane waves, propagating in the z-direction with equal amplitude, namely,

$$\mathbf{E}(z,t) = \mathbf{E}_0[\cos(\omega_1 t - k_1 z) + \cos(\omega_2 t - k_2 z)]. \tag{1.56}$$

The resulting electric field can be expressed in a more compact and transparent form by using the identity $\cos\alpha + \cos\beta \equiv 2\cos[(\alpha-\beta)/2]\cos[(\alpha+\beta)/2]$ as

$$\mathbf{E}(z,t) = 2\mathbf{E}_0 \cos\left[\frac{(\omega_2-\omega_1)t}{2} - \frac{(k_2-k_1)z}{2}\right]\cos\left[\frac{(\omega_2+\omega_1)t}{2} - \frac{(k_2+k_1)z}{2}\right]. \tag{1.57}$$

The superposed electric field is thus shown to consist of two components, one oscillating at the average carrier frequency $(\omega_1+\omega_2)/2$ and the other at the difference frequency $(\omega_1-\omega_2)/2$. Clearly, the former represents the effective mode function, while the latter describes the modulated envelope of the field amplitude, arising from the beating of two plane waves. These two components of the wave propagate in time with phase and group velocities specified in the same way as in (1.55) by

$$v_p = \frac{(\omega_2+\omega_1)/2}{(k_2+k_1)/2} = \frac{\langle\omega\rangle}{\langle k\rangle}, \tag{1.58}$$

$$v_g = \frac{(\omega_2-\omega_1)/2}{(k_2-k_1)/2} = \frac{\Delta\omega}{\Delta k}. \tag{1.59}$$

The mode function and the modulated envelope of the superposed wave are shown in Figure 1.6. The interference pattern of the two plane waves can be either viewed in space z at fixed time or detected in time t at fixed position.

Wave Packets The linear superposition of plane waves can obviously include any number of waves and the wave packet typifies such an example. The wave packet generally consists of a large number of plane waves and is represented by

$$\mathrm{E}(z,t) = \mathrm{Re}\sum_n \mathbf{E}_n e^{-i(\omega_n t - k_n z)} = \mathrm{Re}\int_{-\infty}^{\infty} dk\mathbf{E}(k)e^{-i(\omega t - kz)}, \tag{1.60}$$

where the amplitude $\mathbf{E}_n$ or $\mathbf{E}(k)$ represents the spectral profile of the packet.

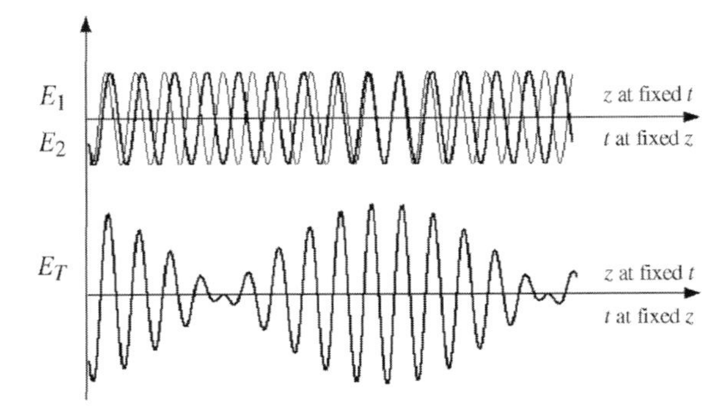

Figure 1.6 The plot of two traveling plane waves $\cos(\omega_j t - k_j z)$, $j = 1, 2$, versus z at fixed time or versus t at fixed z. Also shown is the resulting envelope of two superposed waves.

The dispersion relation governing ω as a function of k understandably plays a central role in determining the characteristics of the wave packet in (1.60). This can be shown by Taylor expanding ω at k_0 up to the second power in k for simplicity:

$$\omega(k) = \omega(k_0) + v_g(k - k_0) + \alpha(k - k_0)^2 + \cdots, \tag{1.61}$$

where the expansion coefficient linear in k,

$$v_g \equiv \partial\omega(k_0)/\partial k, \tag{1.62a}$$

represents the group velocity as in (1.59) and

$$\alpha \equiv \frac{1}{2}\frac{\partial^2\omega(k_0)}{\partial k^2} \tag{1.62b}$$

is the first nonlinear coefficient called the frequency chirping term.

In a linear optical medium in which $\omega = (c/n)k$ (see (1.52)) and α and all higher order terms in k are zero, one may insert (1.61) into (1.60), rearrange the terms, and cast the wave packet into a form

$$\mathbf{E}(z, t) = \mathrm{Re}\, e^{-i(\omega_0 t - k_0 z)} \int_{-\infty}^{\infty} dk \mathbf{E}(k) e^{i(z - v_g t)(k - k_0)}. \tag{1.63}$$

In this case, the mode function oscillating with the carrier frequency ω_0 propagates with the phase velocity v_p, while the envelope propagates with the group velocity v_g.

As a specific example, consider a Gaussian spectral density centered at k_0, that is,

$$\mathbf{E}(k) = \mathbf{E}_0 e^{-(k - k_0)^2/2\sigma^2}. \tag{1.64}$$

Inserting (1.64) in (1.63) and performing the integration using the formula given in (1.24) results in

$$\mathbf{E}(z, t) = \mathrm{Re}\, e^{-i(\omega_0 t - k_0 z)} \mathbf{E}_0 \sqrt{2\pi}\sigma e^{-\sigma^2(z - v_g t)^2/2}. \tag{1.65}$$

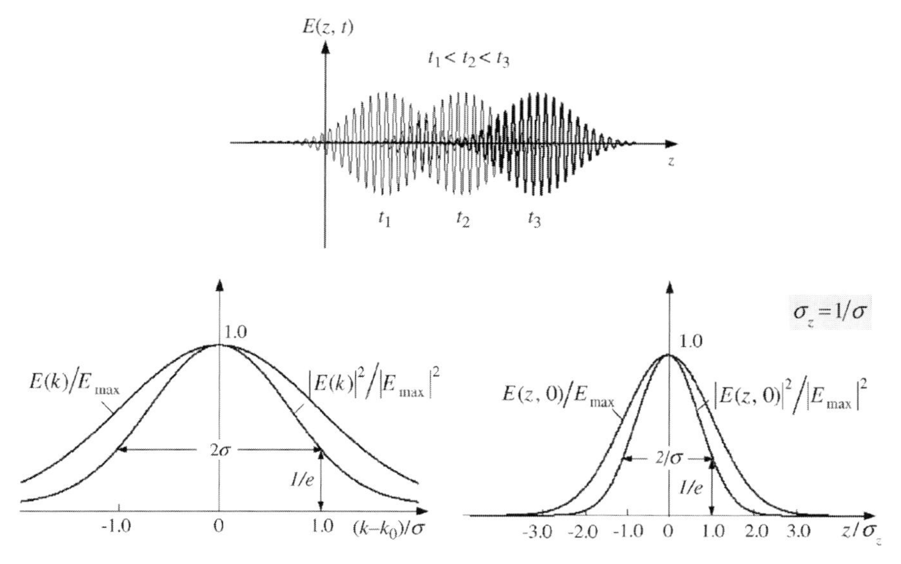

Figure 1.7 The plot of spatial profiles of wave packet at different times. The mode function fast oscillating within the envelope propagates with the phase velocity, while the packet travels with the group velocity. Also shown are the power spectrum $|\mathbf{E}(\omega)|^2$ and the intensity profile $|\mathbf{E}(z, t = 0)|^2$.

Thus, the resulting wave packet consists of a Gaussian envelope propagating with v_g, while the mode function oscillating rapidly within the envelope propagates with v_p.

Figure 1.7 shows the spatial profile of the wave packet at different times together with the power spectrum $|\mathbf{E}(k)|^2$ in (1.64) and the intensity $|\mathbf{E}(z, t = 0)|^2 \propto \exp{-\sigma^2 z^2}$ obtained from (1.65). The bandwidth of k, as defined, for example, by $1/e$ points from the peak, is given by $\Delta k = 2\sigma$. The bandwidth Δk determines in turn the bandwidth in ω via the dispersion relation $\Delta\omega = v_g \Delta k = 2v_g\sigma$. Also, the spatial extent of the wave packet, as determined by $1/e$ points from the peak of the envelope at say $t = 0$, is given by $\Delta z = 2/\sigma = 4/\Delta k$. Then, $\Delta z/v_g$ clearly represents the time duration of the wave packet measured at a fixed point in space, namely, $\Delta t = \Delta z/v_g = 2/\sigma v_g = 4/\Delta\omega$. Hence, an important conclusion can be drawn that Δz and Δt of the wave packet are characterized by the relation

$$\Delta z \propto 1/\Delta k, \quad \Delta t \propto 1/\Delta\omega, \tag{1.66}$$

where the proportionality constants are of the order of unity and depend on the dispersion relation operative within its power spectrum. The relationship in (1.66) represents one of the fundamental properties of the waves and further carry important implications in quantum mechanics, as will become clear subsequently.

It is interesting to note that given the same Gaussian spectral profile (1.64), the ω–k dispersion curve generally plays a crucial role in shaping the wave packet. For instance, in the presence of the linear chirping term α in (1.61), one can obtain using the formula (1.24)

$$\mathbf{E}(z,t) \propto \exp - \frac{\sigma^2 (z - v_g t)^2}{2(1 - 2i\alpha\sigma^2 t)}.$$

Hence, the spatial profile of the intensity of the wave packet,

$$|\mathbf{E}(z,t)|^2 \propto \exp - \frac{\sigma^2 (z - v_g t)^2}{[1 + (2\alpha\sigma^2 t)^2]},$$

broadens progressively in time. Also, the phase and group velocities are generally different in such media.

1.3.3 Interference Effects

Another fundamental property of the wave is its ability to interfere. The interference is nearly synonymous with the wave and is in fact a fundamental footprint of the electromagnetic waves. Such an effect was first demonstrated by Young with his double slit experiment, as illustrated in Figure 1.8. Here, two light beams emanating from a source are incident on two slits. In the far-field limit, the light beams passing through the slits can be approximated by two plane waves:

$$\mathbf{E}_j(\boldsymbol{r}_j, t) = \mathrm{Re}\, \mathbf{E}_0 e^{-i(\omega t - \boldsymbol{k}_j \bullet \boldsymbol{r}_j)}, \quad j = 1, 2. \tag{1.67}$$

These two beams are detected on an observation plane, L distance away from the slit. At an observation point P, the total field is given by the sum of these two plane waves,

$$\mathbf{E}(\boldsymbol{r}, t) = \mathbf{E}_1(\boldsymbol{r}_1, t) + \mathbf{E}_2(\boldsymbol{r}_2, t), \tag{1.68}$$

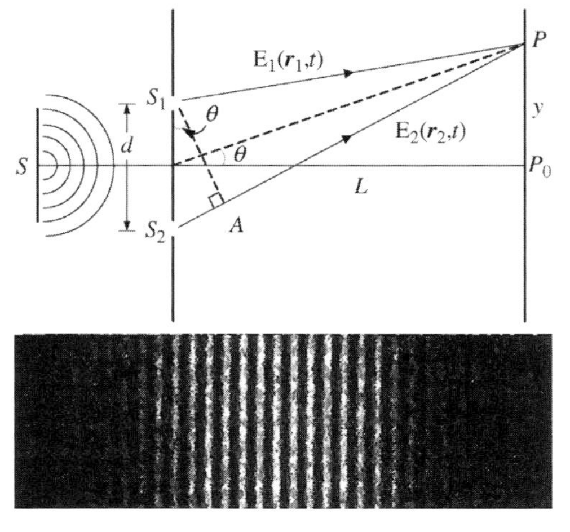

Figure 1.8 The schematics of Young's double slit experiment and observed fringe pattern.

and registers itself via the time-averaged intensity given by

$$I = \langle (\mathbf{E}_1 + \mathbf{E}_2) \cdot (\mathbf{E}_1^* + \mathbf{E}_2^*) \rangle_t = |\mathbf{E}_1|^2 + |\mathbf{E}_2|^2 + (\mathbf{E}_1 \cdot \mathbf{E}_2^* + \mathbf{E}_2 \cdot \mathbf{E}_1^*), \tag{1.69}$$

where only DC components contribute. The two interfering terms in the parentheses are affected by the difference in optical paths the two beams have traversed before reaching P. The resulting phase difference is given in the far-field approximation by $kd \sin \theta$ (see Figure 1.8). Hence, with the use of (1.67) one can write

$$I = 2|\mathbf{E}_0|^2 (1 + \cos \varphi), \quad \varphi = kd \sin \theta \simeq (2\pi/\lambda) d (y/L), \tag{1.70}$$

where d is the slit distance and y is the position of P on the observation plane, and one can put $\sin \theta \simeq \tan \theta \simeq y/L$ for $L \gg y$.

Hence, the intensity at P consists of the background term arising from the self-beating of the electric field and the interference term brought about by beating of two waves. This latter term adds to or subtracts from the background, depending on the relative phase between the two beams. Specifically, the maximum intensity I_{max} ensues for $\varphi = 2n\pi$ with n an integer, that is, $(2\pi/\lambda) d (y/L) = 2n\pi$. That is, a bright strip should appear at $y_n = (\lambda L/d) n$. By the same token, dark strips should appear at $y_n = (\lambda L/d)(n + 1/2)$ (Figure 1.8).

1.4 Summary of Classical Theories

The key features of classical theories are summarized as follows. The classical mechanics is based upon the hypothesis that is embodied in the Newton's equation of motion. It describes the motion of a particle or a system of particles in a deterministic manner. That is, if the position and momentum of these particles are known at t, it is possible to precisely follow the particles at all times. This deterministic description presupposes that these dynamic variables can be measured to an arbitrary order of accuracy. That is to say, the dynamic variables, for example, energy, momentum, position, and so on, of the system are taken as analog quantities varying in continuous fashion by any infinitesimal amount.

The statistical description of the motion of the ensemble of particles is based on the use of the distribution function f and the function evolves in time according to the Boltzmann transport equation. The distribution function operative in nonequilibrium conditions differs from the equilibrium distribution function f_0 and quantifies the transport coefficients of the particle.

The Maxwell's equations are firmly rooted in the empirical laws such as Faraday's law of induction, Ampere's circuital law, and Coulomb's law and the fact no magnetic monopoles are observed. The equations provide the basic and general frameworks for describing the electromagnetic field and also the interaction of radiation with matter.

1.5 Problems

1.1. The Lagrangian of a dynamic system is defined as the difference between its kinetic (K) and potential (V) energies, that is, $L \equiv K - V$, and the Lagrange equation of motion reads as

$$\frac{d}{dt}\frac{\partial L}{\partial \dot{q}_j} - \frac{\partial L}{\partial q_j} = 0,$$

where q_j denotes its jth coordinate and $\dot{q}_j$ the time derivative.

(a) Write down the Lagrangian of the one-dimensional harmonic oscillator with k_x spring constant and the equation of motion. Compare the result with Hamilton's and Newton's equations of motion.

(b) A particle of mass m is attached to two springs with spring constants k_x and k_y along x and y directions, respectively. Write down both the Hamiltonian and Lagrangian of the particle and the respective equations of motion and compare the results (Figure 1.9).

(c) The particle is initially displaced by x_0 and y_0 distances from the equilibrium position and gently released. Describe the motion of the particle.

1.2. Verify and not prove the following vector identities:

$$\nabla \times \nabla \times \boldsymbol{A} = \nabla\nabla \cdot \boldsymbol{A} - \nabla^2 \boldsymbol{A},$$

$$\nabla \cdot (f^* \nabla f - f \nabla f^*) = f^* \nabla^2 f - f \nabla^2 f^*,$$

where f is an arbitrary function.

1.3. A molecule can be roughly modeled by two atoms with masses m_1 and m_2, respectively, and coupled with each other via a spring with constant k (Figure 1.10). The Hamiltonian thus reads as

$$H = \frac{1}{2} m_1 \dot{x}_1^2 + \frac{1}{2} m_2 \dot{x}_2^2 + \frac{1}{2} k (x_1 - x_2)^2.$$

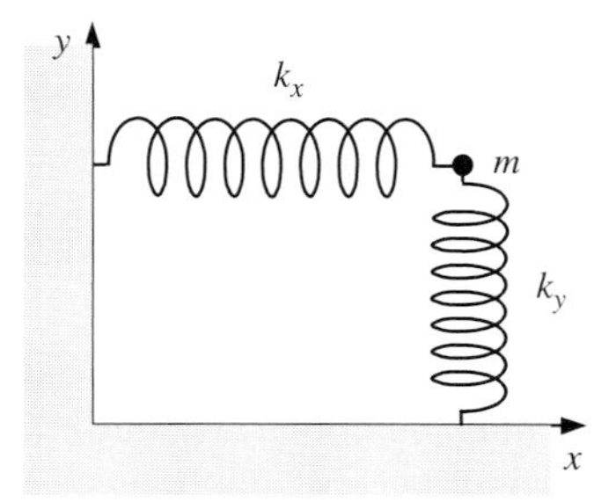

Figure 1.9 2-D harmonic oscillator: a particle of mass m attached to two springs with constants k_x and k_y and movable in x, y directions, respectively.

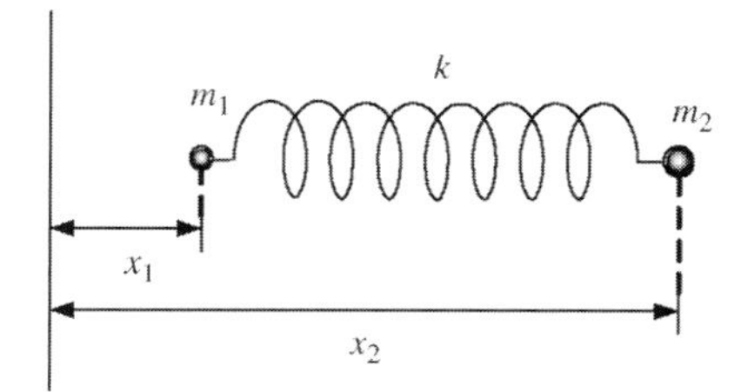

Figure 1.10 A molecule as modeled by two atoms of masses m_1, m_2 and coupled via spring with constant k.

(a) Introduce the center of mass and relative coordinates as

$$X = x_1 + x_2, \quad x = x_1 - x_2$$

and express the Hamiltonian in terms of these new coordinates.

(b) Write down the Hamilton's equations of motion for X and x and interpret the results.

(c) If the frequency of oscillation for x is $3 \times 10^{13}\,\mathrm{s}^{-1}$, find the corresponding spring constant, k by taking $m_1 = m_2 = m_H$, m_H denoting the mass of hydrogen atom.

1.4. Using the equipartition theorem, find the thermal speed of (a) electron, (b) proton, (c) CO_2 molecule, and (d) particle of mass 1 g at $T = 10$, 100, 300, and 1000 K, respectively.

1.5. (a) If an electric field is given by

$$\mathbf{E} = \hat{x}E_0 e^{-i(\omega t - kz)},$$

find $\boldsymbol{H}$ and the Poynting vector $\boldsymbol{P} = \mathbf{E} \times \boldsymbol{H}$ by using the Maxwell's equations.

(b) A helium–neon laser emits light at a wavelength 6.328×10^{-7} m in air. Find the frequency, period, and wave number ($\equiv 1/\lambda$). If the light enters a dielectric medium with μ_0 and $\varepsilon = 4\varepsilon_0$, find the frequency, wave number, and propagation velocity.

Suggested Reading

1 Serway, R.A., Moses, C.J., and Moyer, C.A. (2004) *Modern Physics*, 3rd edn, Brooks Cole.
2 Halliday, D., Resnick, R., and Walker, J. (2007) *Fundamentals of Physics Extended*, 8th edn, John Wiley & Sons, Inc.
3 Shen, L.C. and Kong, J.A. (1987) *Applied Electromagnetism*, 2nd edn, PWS Publishing Company.

2
Milestone Discoveries and Old Quantum Theory

Science progresses with fundamental discoveries and theorizations. In this chapter, such key discoveries leading to quantum theory are briefly discussed. The basic concepts underpinning the quantum mechanics are highlighted and an overall view is presented for quantum mechanics evolving from classical theories and successfully accounting for the microscopic phenomena. The theory of hydrogen atom by Bohr is presented as the culmination of the old quantum theory and also as a bridge to the full-fledged wave mechanics.

2.1
Blackbody Radiation and Quantum of Energy

A puzzling problem confronting theoretical physicists at the turn of the twentieth century was the spectral intensity emanating from the blackbody in thermodynamic equilibrium. A blackbody is a material that absorbs all radiations incident on its surface. A cavity with a small hole, for example, is an approximate implementation of the blackbody, as sketched in Figure 2.1. Once the light passes through the hole into the cavity, it undergoes multiple reflections off the inner surface of the cavity and eventually ends up being absorbed by the material. The thermodynamic equilibrium is then established between the radiation field and the atoms on the wall, which constantly emit and absorb the same amount of radiation.

The observed spectral energy density, $\varrho(\nu)$ of the blackbody radiation is shown in Figure 2.2 versus frequency. As shown in this figure, $\varrho(\nu)$ rises and falls with increasing frequency ν at the given temperature T. Also, the height of the peak and the frequency ν_m at which the peak occurs increase with T. These $\varrho(\nu)$ profiles are universal, independent of the material and determined solely by the temperature.

Rayleigh and Jeans modeled $\varrho(\nu)$ in terms of the number of standing wave modes in the cavity in the frequency range ν and $\nu + d\nu$, and the average energy of the radiation field therein. The former is to be readily found from the usual boundary condition of the standing wave modes, namely, $8\pi n^3\nu^2/c^3$ where c, n are the velocity of light and index of refraction, respectively (as will be derived later). For the latter, they used the Boltzmann distribution function at equilibrium f_0 and obtained the

Introductory Quantum Mechanics for Semiconductor Nanotechnology. Dae Mann Kim

ISBN: 978-3-527-40975-4

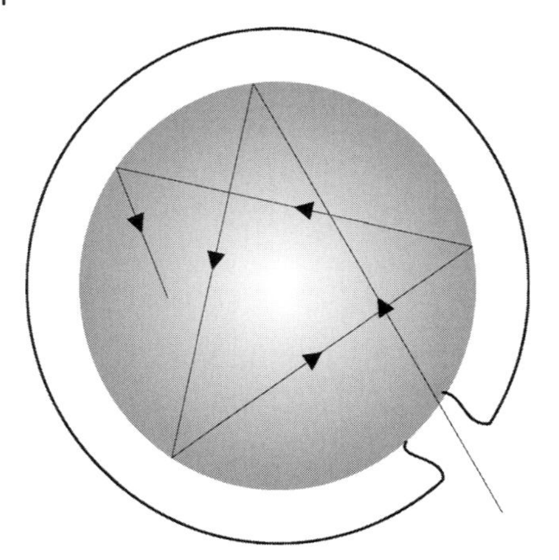

Figure 2.1 A hollow sphere as a blackbody. The light entering through a small hole is repeatedly bounced off the inner surface until absorbed by one of the surface atoms.

average energy:

$$\langle \varepsilon \rangle = \frac{\int_0^\infty d\varepsilon \varepsilon f_0(\varepsilon)}{\int_0^\infty d\varepsilon f_0(\varepsilon)}, \quad f_0(\varepsilon) \propto e^{-\varepsilon/k_B T}. \tag{2.1}$$

Now, (2.1) can be compacted as

$$\langle \varepsilon \rangle = -\frac{\partial}{\partial \beta} \ln \int_0^\infty d\varepsilon e^{-\beta\varepsilon}, \quad \beta \equiv 1/k_B T. \tag{2.2}$$

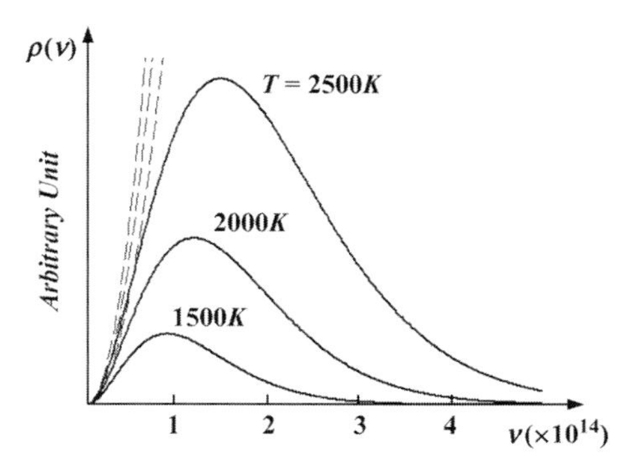

Figure 2.2 The spectral intensity or field energy density of the blackbody radiation versus frequency at different temperatures. Also shown are corresponding Rayleigh–Jean's theory (broken lines).

Recasting (2.1) into (2.2) is a convenient technique, often used in statistical mechanics for finding average values (see (1.23). The equivalence between (2.1) and (2.2) can be easily verified in a straightforward manner by differentiating (2.2) with respect to β and comparing the result with (2.1). Thus, one can find from (2.2) that

$$\langle\varepsilon\rangle = -\frac{\partial}{\partial\beta}\ln\left(\frac{1}{\beta}\right) = \frac{1}{\beta} = k_B T. \tag{2.3}$$

Hence, the Rayleigh–Jean's theory is given by

$$\varrho(\nu) = (8\pi n^3 \nu^2/c^3)k_B T. \tag{2.4}$$

The theory agrees with the data in the small frequency regime, as clearly shown in Figure 2.2. However, at high frequencies, the data rapidly fall down to be negligible, while the theory predicts a continued increase of $\varrho(\nu)$, following the power law ν^2. This drastic disagreement between the theory and experiment is known as the ultraviolet catastrophe. More seriously, the integration of $\varrho(\nu)$ with respect to ν diverges, while the measured total energy density at a given temperature is finite.

Quantum of Energy The ultraviolet catastrophe was resolved by Planck, who introduced the concept of the quantum of energy. Planck postulated that (i) a system oscillating with frequency ν is associated with the basic quantum of energy, not to be divided, (ii) the quantum of energy is specified by $\varepsilon = h\nu$ with h denoting the universal Planck constant, having the value 6.626×10^{-34} J s or 4.136×10^{-15}eV s, and (iii) the energy of the system thus varies digitally in steps of $h\nu$, that is, $\varepsilon_n = nh\nu$, with $n = 1, 2, 3, \ldots$.

With these basic assumptions, the average energy of the radiation mode oscillating at ν can be found again by resorting to the same Boltzmann probability factor and the same mathematical technique as was used in the Rayleigh–Jean's theory. But the summations involved are discrete due to the energy contents varying digitally. Thus, one may write

$$\langle\varepsilon\rangle = \frac{\sum_{n=0}^{\infty}\varepsilon_n e^{-\beta\varepsilon_n}}{\sum_{n=0}^{\infty} e^{-\beta\varepsilon_n}} = -\frac{\partial}{\partial\beta}\ln\sum_{n=0}^{\infty} e^{-\beta\varepsilon_n}, \quad \varepsilon_n = nh\nu,\ \beta \equiv 1/k_B T. \tag{2.5}$$

The discrete infinite geometric series in (2.5) is to be summed as

$$\sum_{n=0}^{\infty} e^{-\beta\varepsilon_n} = 1 + e^{-\beta h\nu} + e^{-2\beta h\nu} + \cdots = \frac{1}{1-e^{-\beta h\nu}}. \tag{2.6}$$

Hence,

$$\langle\varepsilon\rangle = -\frac{\partial}{\partial\beta}\ln\left[\frac{1}{1-e^{-\beta h\nu}}\right] = \frac{h\nu e^{-\beta h\nu}}{1-e^{-\beta h\nu}} = \frac{h\nu}{e^{h\nu/k_B T}-1}. \tag{2.7}$$

Therefore, the spectral energy density $\varrho(\nu)$ is obtained by replacing k_BT in Rayleigh–Jean's formula in (2.4) by the following new expression of $\langle\varepsilon\rangle$:

$$\varrho(\nu) = \frac{8\pi n^3 \nu^2}{c^3} \frac{h\nu}{(e^{h\nu/k_BT} - 1)}. \quad (2.8)$$

Equation 2.8 is the celebrated Planck's theory and the theory explained quantitatively the observed data. In the small frequency regime where $h\nu \ll k_BT$, the denominator of (2.8) can be Taylor expanded to yield

$$e^{(h\nu/k_BT)} - 1 = 1 + h\nu/k_BT + \cdots - 1 \approx h\nu/k_BT.$$

In this limit, the Planck's $\varrho(\nu)$ expression reduces to the Rayleigh–Jean's formula and agrees with the data, as pointed out. Furthermore, in the large frequency range, in which $h\nu \gg k_BT$ and $\varrho(\nu) \propto \exp-(h\nu/k_BT)$, it falls down exponentially with increasing frequency, again in agreement with the data.

The basic difference between these two theories is to be traced to the different procedure for finding the average energy, and this can in turn be attributed to the quantum of energy introduced in an oscillating system. Because the quantum of energy is not to be divided in Planck's theory, the energy has to vary digitally in steps of $h\nu$ and, therefore, the Boltzmann factor $\exp-h\nu/k_BT$ renders it prohibitively small for atoms in the cavity wall to emit radiation modes, whose quantum of energy $h\nu$ is much larger than the thermal energy, that is, $h\nu \gg k_BT$. This is clearly shown in Figure 2.3, where the average energy $\langle\varepsilon\rangle$ obtained classically as well as by Planck's formula is plotted versus the frequency of the radiation. In the case of the former theory, the average energy is pinned at k_BT, while in the latter case, it decreases monotonously with ν to essentially vanish for large ν.

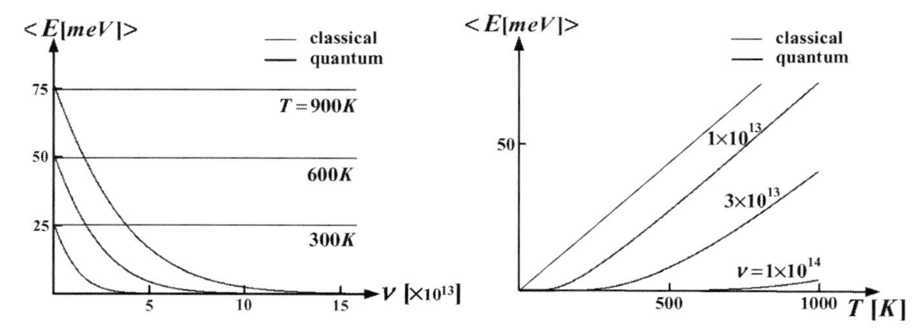

Figure 2.3 The average field energy $\langle E\rangle$ versus ν at different T and $\langle E\rangle$ versus T at different ν. Classically, $\langle E\rangle$ linearly increases with T regardless of ν. Quantum mechanically, $\langle E\rangle$ exponentially decreases with ν at given T and the rate of increase with T drastically slows down with ν.

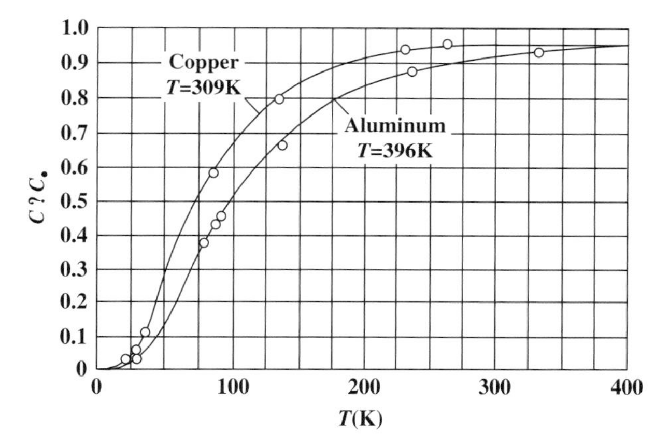

Figure 2.4 The specific heat of solid versus temperature, measured data (circles), and theoretical curves (solid line) (after Debye, P. (1912) *Ann. Phys.*, **39**, 789).

2.2 Specific Heat of Solids

Another observation that defied the explanation by the classical theory and pointed to the quantum of energy is the specific heat C of solids. The observed behavior of C as a function of temperature T is shown in Figure 2.4. At low T, C increases following the power law T^3, while at high T, C is pinned at a constant value.

The specific heat is generally attributed to the vibrational motion of atoms, centered around the equilibrium position in the solid. The average energy of such atomic vibration in each direction is again specified in the classical theory by (2.3) and is equal to $k_B T$ and independent of the frequency of oscillation. The atoms vibrating in x, y, and z directions can be accounted for via the equipartition theorem, which has been discussed in Chapter 1, and one can write the internal energy of the solid as

$$E = 3Nk_B T, \tag{2.9}$$

with N denoting the number of atoms per unit volume and the factor 3 accounting for the 3D vibrational motion. Thus, the heat capacity is given according to the classical theory by

$$C \equiv \partial E/\partial T = 3Nk_B \tag{2.10}$$

and is independent of temperature, in clear disagreement with observed data at low T.

Energy Quanta and Phonons The disagreement between the theory and experiment was resolved by Einstein, who applied Planck's concept of quantum of energy to the vibrational motion of the atoms in the solids. When the concept of the energy quantum is introduced, the procedures for finding the average energy are identical to those leading to (2.7). Hence, the internal energy is specified as

$$E = 3N\frac{h\nu}{(e^{h\nu/k_BT}-1)}, \tag{2.11}$$

where ν stands for the characteristic frequency of the oscillation and the quantum of energy $h\nu$ in this case is called phonon. At high temperatures at which $k_BT \gg h\nu$ and $\exp(h\nu/k_BT)-1 \simeq h\nu/k_BT$, E is reduced to the classical result of (2.9). For low temperature, however, for which $k_BT \ll h\nu$, $E \propto \exp-h\nu/k_BT$ and decreases rapidly with decreasing temperature in general agreement with the data. The satisfactory analysis of C exhibiting the power law $C \propto T^3$ at low temperatures was carried out by Debye who further refined Einstein's theory.

2.3 Photoelectric Effect

The conduction of electric current in the cathode ray tube attracted the attention of many physicists in the latter half of the nineteenth century. The cathode ray tube consists of a glass tube filled with rarefied gas and two metallic electrodes, anode and cathode, inserted therein (see Figure 2.5). The tube was instrumental in bringing out key discoveries and concepts in the history of physics and the photoelectric effect is just one of such discoveries. The effect consists of an output current induced by an input light incident on the cathode (see Figure 2.5).

The general features of the photocurrent are shown in Figure 2.6. The current is induced instantaneously by an incident light and grows and saturates with increasing positive bias at the anode. Also, the current flows only when the frequency of the incident light ν exceeds a critical value for the given metal at the cathode and the current level is commensurate with the light intensity. Finally, the photocurrent

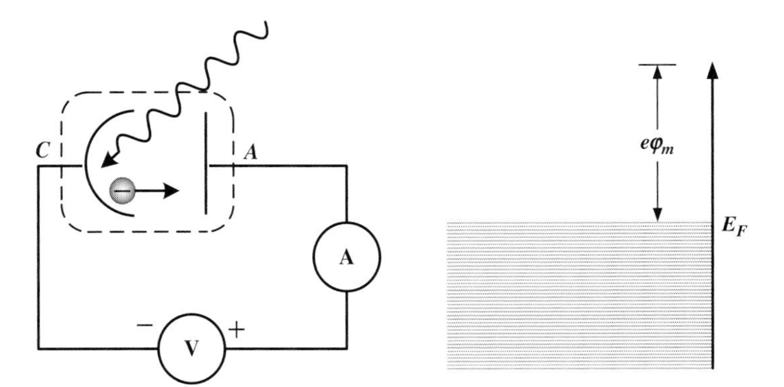

Figure 2.5 The schematics of the photoelectric effect, as occurring in a vacuum tube. Photoemitted electrons from the cathode are guided to anode to close the current loop. A metal is often modeled by sea of electrons filled up to Fermi level and confined by the potential barrier at the surface, called the work function.

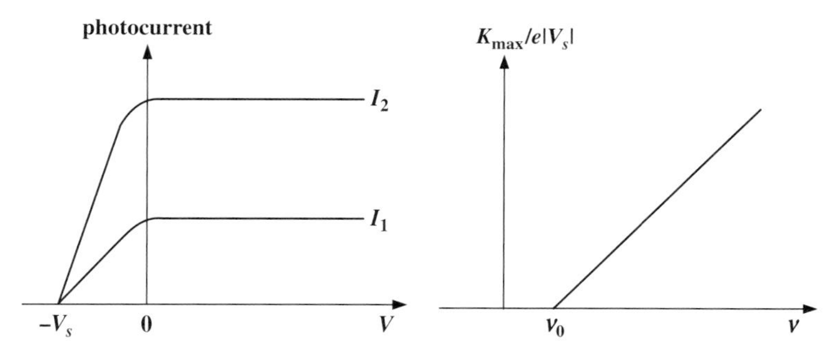

Figure 2.6 The photocurrent versus anode voltage at different light intensities (left) and the stopping power or the maximum kinetic energy of photoemitted electrons versus frequency (right).

persists even under a negative bias and ceases to flow at a given negative voltage $-V_S$. The stopping power eV_S depends only on the frequency of light and increases linearly with ν, as shown in the figure.

Evidently, the photocurrent is due to the electrons emitted from the cathode after absorbing sufficient energy from the incident light. After emission the electrons are attracted to the anode by the positive voltage applied therein, thereby closing the current loop. A few key features of the photocurrent defied the explanation by the prevailing theories. In the classical theory, the energy gained by the electrons is proportional to the light intensity. This implies that the photocurrent should flow as long as the light is incident on the cathode, regardless of its frequency. However, the current is observed only when the light frequency is greater than a threshold value for the given cathode material.

Also, the stopping power can be viewed as the potential barrier for the emitted electrons to overcome before reaching the anode. Inasmuch as the kinetic energy of the emitted electrons is commensurate with the light intensity, the stopping power should depend on the light intensity. However, the stopping power was observed to depend only on the frequency and not on the intensity. Finally, it should take some time for the electrons in the cathode to gain sufficient kinetic energy to overcome the barrier at the surface of the metal, called the work function (see Figure 2.5). However, the current instantaneously flowed the moment the light is incident on the cathode.

Energy Quanta and Photons Einstein resolved these puzzling features of the data again by invoking Planck's concept of quantum of energy. Specifically, Einstein observed that the light of frequency ν consists of photons that carry the basic undividable quantum of energy $h\nu$ and travel with the velocity of light c. In this corpuscular picture of the light, the intensity I is comprised of the flux of photons $I/h\nu$ crossing unit area per unit time.

Because the energy $h\nu$ of these photons is not dividable, the photons should interact digitally with electrons in the cathode. That is, a photon imparts all of its energy to an electron or none at all. The absorbed photon energy should then be consumed in part for the electron to overcome the metal work function, while the remainder is converted to the kinetic energy of the emitted electrons. Thus, using this conservation of energy, one can write

$$h\nu = e\varphi + K, \quad K = mv^2/2, \tag{2.12}$$

where the first term on the right-hand side is the potential barrier provided by the work function and the second term is the kinetic energy of emitted electrons.

Equation 2.12 single-handedly explains the entire characteristics of the photocurrent: (i) higher light intensity with larger photon flux should emit more electrons from the cathode, thereby increasing the photocurrent. With increasing anode voltage, the efficiency of guiding the emitted electrons to the anode reaches unity, thereby leading to the current saturation; (ii) given the work function of the cathode, the photoemission of electrons will occur only when $h\nu \geq e\varphi$, thereby accounting for the critical frequency of the radiation inducing the current; (iii) the stopping power eV_S, as simply determined by K in (2.12), depends only on and increases linearly with ν, in agreement with the observed data.

Moreover, Einstein's theory as compacted by (2.12) also provides a simple method by which to determine the universal Planck's constant and the metal work function. If the stopping power $eV_S(= K)$ is measured for the given metal versus the light frequency ν, both h and φ can be extracted from the slope and intercept of $K-\nu$ curve (see Figure 2.6). This simple measurement scheme has become a standard means of extracting the fundamental physical constant and parameter. Thus, Einstein's theory has explained the photoelectric effect and in doing so revived the corpuscular nature of light.

2.4
Compton Scattering

The electromagnetic radiation composed of photons was also unambiguously demonstrated by Compton, who performed the X-ray scattering experiments. The experimental scheme together with the data obtained is sketched in Figure 2.7. As shown in the figure, the wavelength of the scattered X-ray is shifted by an amount depending solely on the angle of scattering. This observation was again in contradiction with the classical theory, which predicted that the shift in wavelength should depend on the intensity and illumination time of the radiation field.

Compton interpreted his observed data by modeling the X-ray to consist of photons, streaming with the velocity of light c and having the quantum of energy unit $h\nu$. The schematics of the scattering are shown in Figure 2.8. As shown in the figure, a photon is scattered off a free electron. Although the target electron is bound to the atom and is in constant motion, its kinetic and binding energies can be neglected, compared with the energy of the X-ray photon or the amount of energy exchanged during the scattering.

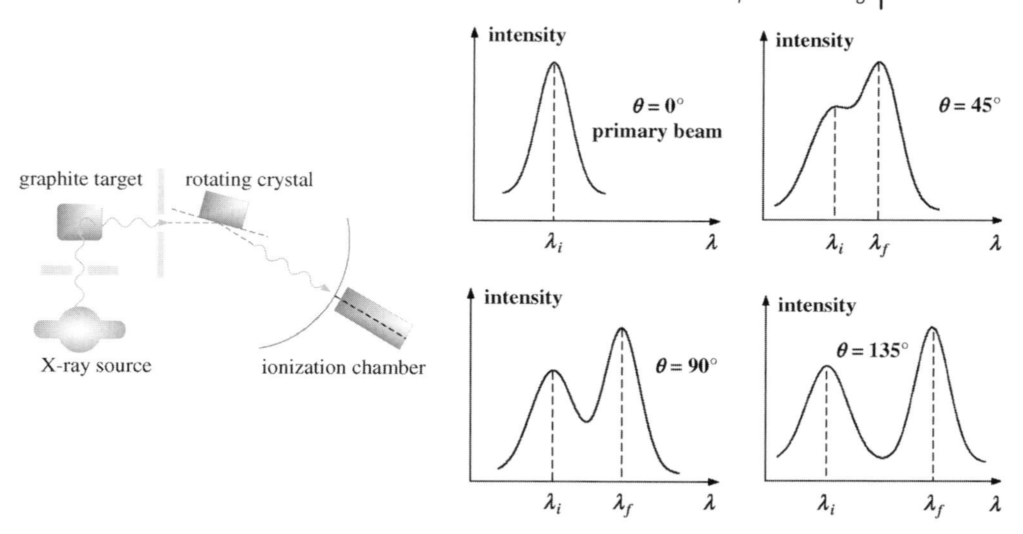

Figure 2.7 The experimental set up for X-ray scattering and observed spectra of scattered X-rays showing the shifted wavelengths.

A photon moves with the velocity of light c and has zero rest mass, while possessing the energy $h\nu$. Hence, it follows from the special theory of relativity that

$$\varepsilon = [m_{\text{ph}}^2 c^4 + c^2 p^2]^{1/2} = cp, \tag{2.13}$$

while

$$\varepsilon = h\nu. \tag{2.14}$$

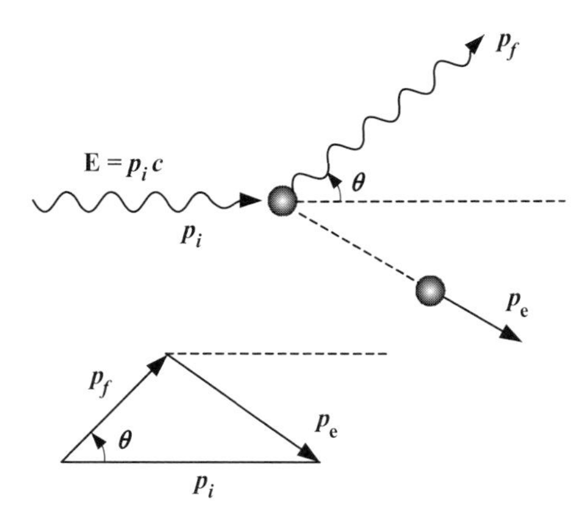

Figure 2.8 The X-ray scattering off a free electron. Both momentum and energy are conserved during the scattering.

Thus, combining (2.13) and (2.14), one can write

$$p = \frac{h\nu}{c} = \frac{h}{\lambda}, \tag{2.15}$$

where λ is the wavelength satisfying the relation $\lambda\nu = c$.

Now, during the collision, both the energy and momentum are conserved:

$$p_i c + mc^2 = p_f c + (m^2c^4 + p_e{}^2 c^2)^{1/2}, \tag{2.16}$$

$$\boldsymbol{p}_i = \boldsymbol{p}_f + \boldsymbol{p}_e, \tag{2.17}$$

where $\boldsymbol{p}_i$ and $\boldsymbol{p}_f$ are the photon momenta before and after the scattering, $\boldsymbol{p}_e$ is the momentum of the electron after the scattering due to recoil, and m is the rest mass of electron. One can express p_e^2 from (2.16) and (2.17), respectively, as

$$p_e^2 = (p_i - p_f + mc)^2 - m^2c^2 = (p_i - p_f)^2 + 2mc(p_i - p_f), \tag{2.18}$$

$$p_e^2 = \boldsymbol{p}_e \cdot \boldsymbol{p}_e = (\boldsymbol{p}_i - \boldsymbol{p}_f) \cdot (\boldsymbol{p}_i - \boldsymbol{p}_f) = p_i^2 + p_f^2 - 2p_i p_f \cos\theta, \tag{2.19}$$

with θ denoting the scattering angle between $\boldsymbol{p}_i$ and $\boldsymbol{p}_f$. Equating the right-hand sides of (2.18) and (2.19), one obtains

$$2p_i p_f (1 - \cos\theta) = 2(p_i - p_f)mc. \tag{2.20}$$

When both sides of (2.20) are divided by $p_i p_f$ and the trigonometric identity $(1 - \cos\theta) = 2\sin^2(\theta/2)$ is used, one obtains

$$\frac{1}{p_f} - \frac{1}{p_i} = \frac{2}{mc}\sin^2\left(\frac{\theta}{2}\right). \tag{2.21}$$

The shift in wavelength due to scattering can, therefore, be found from (2.21) by expressing p_i and p_f in terms of the corresponding wavelengths from (2.15), namely,

$$\Delta\lambda \equiv \lambda_f - \lambda_i = \frac{2h}{mc}\sin^2\left(\frac{\theta}{2}\right) \equiv 4\pi\lambda_e \sin^2\left(\frac{\theta}{2}\right), \tag{2.22}$$

where

$$\lambda_e = \frac{\hbar}{mc}, \quad \hbar \equiv h/2\pi, \tag{2.23}$$

is the Compton wavelength $\simeq 4 \times 10^{-13}$ m or 4×10^{-4} nm. The final result of (2.22) is in agreement with the observed data and the particle-like nature of the electromagnetic radiation was once again confirmed. Since the Compton wavelength is independent of the wavelength, the relative shift $\Delta\lambda/\lambda_i$ of the scattered radiation should depend sensitively on the incident wavelength λ_i. For instance, the maximum shift $\Delta\lambda/\lambda$ at $\theta = \pi$ is as much as $\simeq 5 \times 10^{-2}$ for $\lambda = 0.1$ nm, indicating that the

Compton effect is more readily observed at shorter wavelength regime such as X-ray regime.

2.5 Duality of Matter

The light exhibiting the interference or diffraction was firmly established in the classical optics over a long period of time. Young's simple double slit experiment, for example, provides a clear evidence for the wave nature of light, as discussed in Chapter 1. In addition, light was shown as an integral part of electromagnetic waves obeying Maxwell's equations. At the same time, the corpuscular or particle-like nature of light was also unambiguously demonstrated by the photoelectric effect and X-ray scattering. Thus, the light composed of photons carrying undividable quantum of energy $h\nu$ and traveling with the velocity of light was also firmly established. These two different tracks of discoveries pointed to the duality of light – it exhibits both the wave- and particle-like natures.

On the material side, the particle nature of the fundamental building blocks such as electrons, atoms, and molecules have long been taken for granted. However, the most daring postulate in the development of the quantum theory was proposed by de Broglie, who brought forth the concept of the matter waves. Specifically, de Broglie postulated that a particle also exhibits the wave-like nature and is inherently associated with the wavelength given by

$$\lambda = \frac{h}{p}, \tag{2.24}$$

where h is the Planck's constant and p is the magnitude of the linear momentum of the particle (see Figure 2.9).

The hypothesis introduced by de Broglie was subsequently confirmed by Davisson and Germer at Bell Laboratories. They successfully obtained the diffraction pattern of electrons from a crystal in a manner identical to the X-ray diffraction experiment, satisfying the Bragg diffraction condition

$$2d\sin\theta = l\lambda, \quad l = 1, 2, 3, \ldots, \tag{2.25}$$

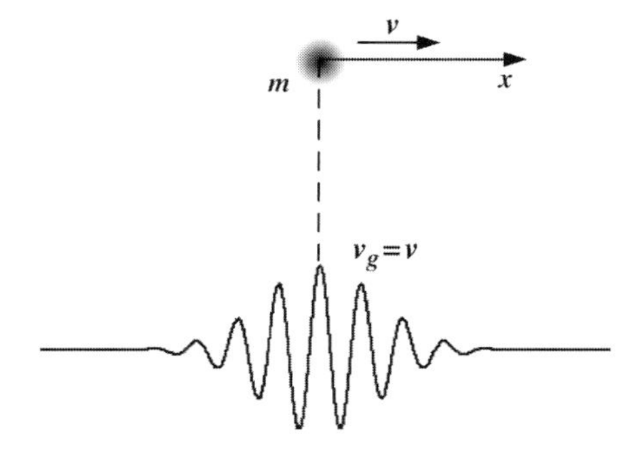

Figure 2.9 The de Broglie matter wave: the particle as a wave is to be represented by a spatially localized wave packet, traveling with group velocity.

Figure 2.10 The diffraction pattern of 50 keV electrons from Cu_3Au film (Courtesy of the late Dr. L.H. Germer).

where d is the lattice spacing and λ is the wavelength as specified in (2.24) (see Figure 2.10).

In this manner, the duality of matter was also firmly established. That is, microscopic particles such as atoms, molecules, electrons, holes, and so on also exhibit both the particle-and wave-like nature. This abstract concept that a particle behaves like a wave was not only demonstrated in the laboratory but has also become an integral part of everyday life. For instance, the electron microscope is based on utilizing the electrons as waves just as the optical microscope uses visible light sources for imaging an object. For the case of electron microscope, the wavelength is to be tuned by varying the kinetic energy of electrons $E = p^2/2m$, and de Broglie wavelength is, therefore, given by $\lambda = h/[2mE]^{1/2}$. For example, the wavelength of the electron accelerated by a voltage V is given in terms of nm by

$$\lambda = \frac{1.226}{\sqrt{E(\mathrm{eV})}}\ \mathrm{nm}. \tag{2.26}$$

The wavelength of other particles such as thermal neutrons can likewise be tuned by varying the energy of the particle.

2.6
Bohr's H-Atom Theory

The theory of the hydrogen atom proposed by Bohr is a landmark achievement in theoretical physics and it culminates in the old quantum theory. The theory revealed

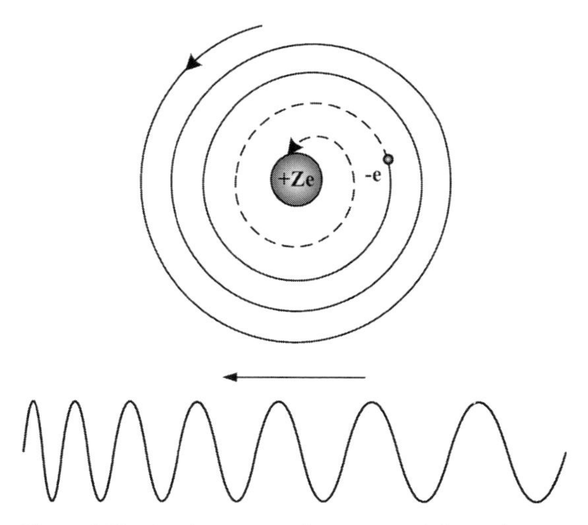

Figure 2.11 An electron spiraling in toward the nucleus emits radiation to conserve energy, the frequency of which increases monotonously with decreasing radius.

for the first time that the structure of atom involves quantized electron orbits having discrete energy levels. It also explained a vast amount of spectral data accumulated over a long period of time.

Before Bohr's theory, a few puzzling theoretical issues called for explanation. First, the α-particle scattering experiments by Rutherford pointed to the fact that the atom consists of a nucleus at a center and electrons revolving around it. An electron in a circular orbit, for example, is under a constant acceleration and should, therefore, emit radiation, according to the theory of electrodynamics. This unavoidably suggests that the electron revolving around the nucleus loses its energy constantly and should, therefore, spiral into the nucleus, as illustrated in Figure 2.11. This raised the basic issue of the stability of atom and/or matter. In reality, however, atoms are stable.

In addition, the emitted radiation was observed to consist of several sets of infinite number of discrete lines, rather than the continuous spectra, as predicted by the electrodynamics for charged particles spiraling in. In fact, the observed spectral lines could be accurately fitted by the empirical formula given by

$$\frac{1}{\lambda} = R\left(\frac{1}{n^2} - \frac{1}{m^2}\right), \tag{2.27}$$

where λ is the wavelength, n and m are positive integers, and R is the Rydberg constant with the value $R = 10,973,732\ \text{m}^{-1}$ or $R = 0.010973732\ \text{nm}^{-1}$. The infinite series of discrete lines was accurately fitted by fixing n and varying m in (2.27): Lyman series for $n = 1$ and $m \geq 2$; Balmer series for $n = 2$ and $m \geq 3$; Paschen series for $n = 3$ and $m \geq 4$; Bracket series for $n = 4$ and $m \geq 5$; and Pfund series for $n = 5$ and $m \geq 6$. Figure 2.12 shows the Balmer series as an example.

These fundamental theoretical issues were resolved by Bohr, who came up with a new theory based on a few assumptions:

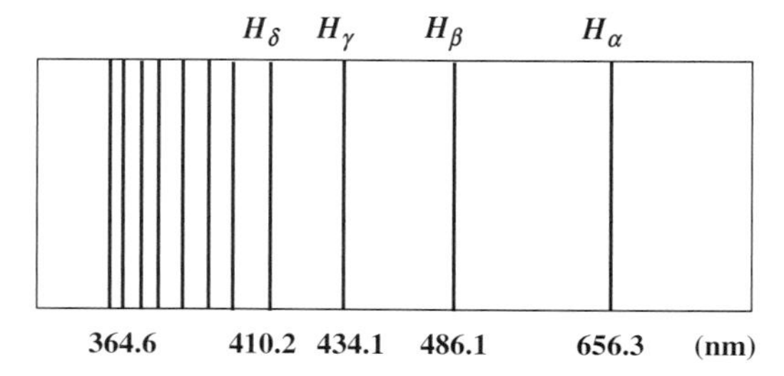

Figure 2.12 The Balmer emission spectral lines from the hydrogen atom.

1) **Quantized Orbits**: The electrons in atom reside in stable, nonradiating orbit whose angular momentum L with respect to the nucleus is characterized by a discrete value, obeying the quantization rule,

$$L_n = n\hbar, \quad \hbar \equiv h/2\pi, \tag{2.28}$$

with n representing a positive integer.

2) **Quantum Transition**: The electron can make a transition from a higher lying orbit to a lower lying orbit or vice versa, emitting or absorbing radiation to conserve energy.

With these postulates Bohr quantified the observed spectral data in a simple way. First, the quantized angular momentum in (2.28) is specified as

$$L_n \equiv mv_n r_n = n\hbar, \quad v_n \equiv \omega_n r_n, \tag{2.29}$$

with m, v_n, ω_n, and r_n denoting, respectively, the electron mass, velocity, angular frequency, and displacement from the nucleus in the nth circular orbit.

Now, the circular orbit of the electron is well known to be maintained by the balance of two forces acting in opposite directions, namely,

$$\frac{e_M^2}{r_n^2} = \frac{mv_n^2}{r_n}, \quad e_M^2 \equiv \frac{e^2}{4\pi\varepsilon_0}, \tag{2.30}$$

with ε_0 denoting the vacuum permittivity. The left-hand side represents the centripetal force provided by the attractive Coulomb force between the proton and the electron. On the other hand, the right-hand side represents the centrifugal force associated with the circular motion (Figure 2.13). Hence, combining (2.29) and (2.30) and eliminating v_n, one finds r_n as

$$r_n = r_B n^2, \quad r_B \equiv \frac{\hbar^2}{me_M^2}, \tag{2.31}$$

where r_B is known as the Bohr radius and naturally scales the radius r_n in terms of the atomic dimension with $r_B = 0.053\,\text{nm}$. Once r_n is specified by (2.31), v_n is found from (2.29) as

$$v_n = \frac{n\hbar}{mr_n}. \tag{2.32}$$

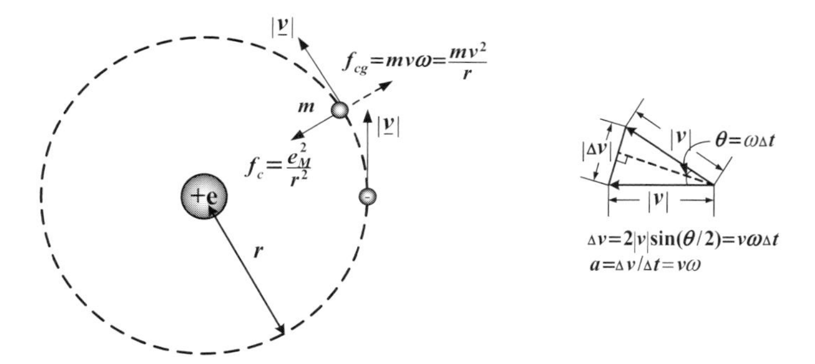

Figure 2.13 An electron maintained in circular orbit via the balanced centripetal and centrifugal forces. Also shown is the specification of the centrifugal force.

The kinetic energy K_n of the electron in the nth orbit can then be correlated to the potential energy therein by using (2.32) and (2.31):

$$K_n \equiv \frac{1}{2}mv_n^2 = \frac{e_M^2}{2r_n} = -\frac{1}{2}V_n, \quad V_n \equiv -\frac{e_M^2}{r_n}. \tag{2.33}$$

Therefore, the total energy of the electron in the nth orbit is given by

$$E_n = \frac{1}{2}mv_n^2 - \frac{e_M^2}{r_n} = -\frac{1}{2}\frac{e_M^2}{r_n}. \tag{2.34}$$

Or, with the use of (2.31), one can also express the total energy as

$$E_n = -E_0\frac{1}{n^2}, \quad E_0 = \frac{e_M^4 m}{2\hbar^2} \equiv \frac{e^4 m}{2(4\pi\varepsilon_0)^2\hbar^2}, \tag{2.35}$$

with $E_0 = 13.64\,\text{eV}$ denoting the well-known ionization energy of the hydrogen atom. The energy levels of the stable orbits bound to the nucleus are thus explicitly quantized and constitute a key feature of Bohr's theory. The positive integer n entering in the quantized energy expression is called the principal quantum number.

As stated in the basic assumption, the electron can make a transition from a higher lying orbit with the quantum number n_i to a lower lying orbit with n_f, thereby emitting a photon of frequency ν or wavelength λ to conserve the energy, that is,

$$h\nu = \frac{hc}{\lambda} = E_0\left(\frac{1}{n_f^2} - \frac{1}{n_i^2}\right). \tag{2.36}$$

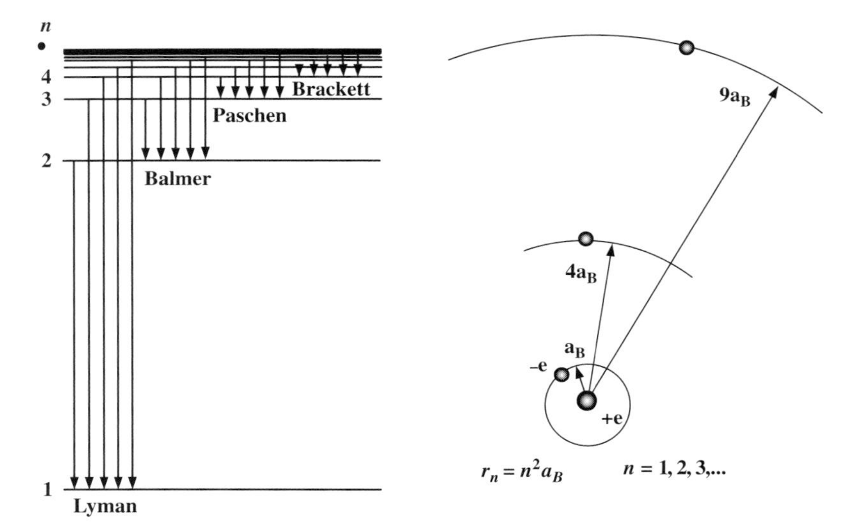

Figure 2.14 The quantized energy level diagram of hydrogen atom. The electron transitions between energy levels give rise to several sets of infinite line spectra. Also shown are the corresponding stable electron orbits.

Comparing (2.36) with the empirical formula (2.27), the Rydberg constant can be theoretically identified as

$$R = \frac{E_0}{hc} = \frac{e^4 m}{4\pi(4\pi\varepsilon_0)^2 \hbar^3 c}. \tag{2.37}$$

The precise quantification of the Rydberg constant in terms of the basic physical constants is one of the highlights of the theoretical physics. The lowest energy level corresponding to $n = 1$ is called the ground state of the atom and the energy E_0 is then the energy required to ionize the atom by knocking out the electron from the ground state to the vacuum level. The discrete electron orbits and the corresponding energy levels are shown in Figure 2.14.

Stable Orbits and de Broglie Wavelength The key point of Bohr's theory consists of quantizing the angular momentum, which characterizes the stable electron orbits. These orbits can be interpreted in terms of the standing wave condition for de Broglie wavelength of the electron circling in the orbit. Specifically, the de Broglie wavelength of the electron in the nth orbit is given from (2.24) by

$$\lambda_n = \frac{h}{p_n} = \frac{h}{m v_n}. \tag{2.38}$$

Thus, (2.38) when combined with the quantized angular momentum (2.29) yields

$$r_n = \frac{n\hbar}{m v_n} = n\frac{\lambda_n}{2\pi}. \tag{2.39}$$

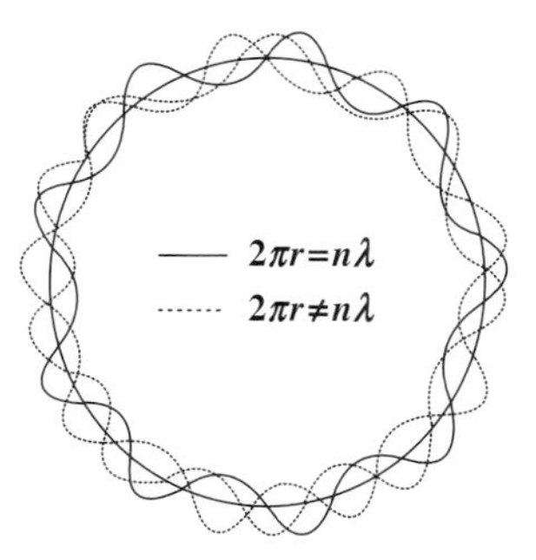

Figure 2.15 The stable orbits and de Broglie wavelengths. The circumference of stable orbits are integer multiples of de Broglie wavelength of the electron circling in the orbits, the condition for standing waves.

Equivalently, (2.39) can be rewritten as

$$2\pi r_n = n\lambda_n. \quad (2.40)$$

Thus, Equation 2.40 states that the circumference of the stable orbit is an integer multiple of de Broglie wavelength of the electron therein. That is, the optical path of the stable orbit satisfies the standing wave condition of the de Broglie wavelength of orbiting electron. If this condition is not met, the wave interferes with itself destructively and cannot exist in the orbit, as illustrated in Figure 2.15.

2.7 Problems

2.1. Calculate the energy (in eV unit) and momentum of a photon with the wavelengths 10 m (radiowave), 1 m (microwave), 10,000 nm (infrared), 600 nm (visible), 200 nm (ultraviolet), 50 nm (EUV), and 1 nm (X-ray).

2.2. The average power generated by the sun is roughly estimated as 3.7×10^{26} W. (a) Take the average frequency of the sun's radiation at 500 nm and estimate the photon flux (number of photons per area per time) at the surface of the sun. (b) Estimate the total flux reaching the earth.

2.3. Calculate the de Broglie wavelengths for (a) electron, proton, neutron, and hydrogen atoms and CO_2 molecule at room temperature, (b) electron with kinetic energy 200 eV, 50 keV, and 100 keV, (c) proton with the same kinetic energies as the electron in (b), and (d) electron in the Bohr orbits with $n = 1$ and $n = 100$.

2.4. Calculate the de Broglie wavelength of electron in the ground state of donor atom in silicon. The donor atom has the hydrogen-like structure consisting of a nucleus with effective charge of one proton and an electron having an effective mass m_n. The medium in between proton and electron has the permittivity $\varepsilon = \varepsilon_r\varepsilon_0$, with the dielectric constant $\varepsilon_r = 11.9$. Take $m_n = 0.9m_0$ and $m_n = 0.1m_0$, with m_0 denoting the rest mass of electron.

2.5. Estimate the error introduced by neglecting the effect of the special relativity in calculating the de Broglie wavelengths of (i) 10 keV electron, (ii) 100 keV proton, and (iii) the electron in the ground state of the hydrogen atom.

2.6. (a) Lithium, beryllium, and mercury have the work function of 2.3, 3.9, and 4.5 eV, respectively. When the light of wavelength 300 nm is incident on these metals, determine which will exhibit the photoelectric effect and find the stopping power in such metals.

(b) When the surface of the copper is irradiated with light with wavelength 253.7 nm, the stopping power is −0.24 V. Determine the longest wavelength capable of inducing the photoelectric effect.

(c) The stopping power of photoelectrons from aluminum is −2.3 V for radiation of 200 nm and −0.9 V for radiation of 313 nm. From these data, find the Planck's constant and the work function of the aluminum.

2.7. Consider the hydrogen atom in its ground ($n = 1$) and first excited ($n = 2$) states. Compare (a) the radius, (b) magnitude of linear and angular momenta, and (c) the kinetic, potential, and total energy of the electron in these two states.

2.8. X-ray with energy $h\nu = 2\times 10^5$ eV is scattered off an electron at rest. If the scattered beam is detected at 90° with respect to the incident direction, find (a) the shift in wavelength, (b) the energy of the scattered X-ray, and (c) the energy of the recoiling electron.

2.9. The ionized helium atom He^+ consists of two protons at the nucleus and one electron revolving around the nucleus. Use the Bohr theory and discuss the quantized energy levels and compare them with those of the hydrogen atom. Find the shortest atomic radius and compare it with that of the hydrogen atom. Compare the shortest and longest wavelengths corresponding to Balmer series.

2.10. Consider a hydrogen atom in its ground state with $n = 1$. According to the electrodynamics, a charged particle under acceleration emits radiation. Estimate the time in which the electron hits the nucleus from its ground state orbit by spiraling in while radiating away its energy. Refer to any textbook on electromagnetic theory for the radiative power emitted by an accelerated charged particle.

Suggested Reading

1 Serway, R.A., Moses, C.J., and Moyer, C.A. (2004) *Modern Physics*, 3rd edn, Brooks Cole.

2 Halliday, D., Resnick, R., and Walker, J. (2007) *Fundamentals of Physics Extended*, 8th edn, John Wiley & Sons, Inc.

3 Singh, J. (1996) *Quantum Mechanics, Fundamentals & Applications to Technology*, John Wiley & Sons, Inc.

4 Liboff, R.L. (2002) *Introductory Quantum Mechanics*, 4th edn, Addison-Wesley Publishing Company, Reading, MA.

5 Gasiorowicz, S. (2003) *Quantum Physics*, 3rd edn, John Wiley & Sons, Inc.

3
Schrödinger Equation and Operator Algebra

The Schrödinger equation to quantum mechanics is what the Newton's equation of motion is to classical mechanics. The equation represents the basic hypothesis, not to be derived but the starting point for the wave mechanical description, understanding, and control of the microscopic natural phenomena. Its validity should ultimately rest on the agreement between the theory and experiment. The versatility of the equation was first amply demonstrated, for example, by reproducing and at the same time providing much more information than Bohr's H-atom theory. This was achieved without imposing the hypothesis of the angular momentum quantization. Rather, the quantization is a natural consequence of wavefunctions physically well behaving.

The essential features of the operator algebra underlying the Schrödinger equation are compactly discussed together with the postulates of quantum mechanics.

3.1
Schrödinger Equation

As discussed in Chapter 1, a wave packet is formed by linearly superposing plane waves

$$\psi(\boldsymbol{r}, t) = \int d\boldsymbol{k} f(\boldsymbol{k}) e^{-i(\omega t - \boldsymbol{k} \cdot \boldsymbol{r})}. \tag{3.1}$$

For the case of electromagnetic waves, the frequency ω and the wave vector $\boldsymbol{k}$ are related by the dispersion relation, thereby making these plane waves the solutions of the wave equation. As discussed, the spatially localized envelope of the packet propagates with the group velocity.

In Chapter 2, a particle was shown to exhibit both the particle-and wave-like behaviors. This duality points to the possible representation of the particle by means of the spatially localized wave packet. To this end, however, ω and k should be appropriately correlated in terms of the dynamic parameters of the particle.

Introductory Quantum Mechanics for Semiconductor Nanotechnology. Dae Mann Kim

ISBN: 978-3-527-40975-4

Particle Dispersion Relation Consider first a free particle moving with the linear momentum $\boldsymbol{p}$. The energy of the particle is to be transcribed into the frequency in analogy with the concept of the photon,

$$E = h\nu \equiv \hbar\omega, \quad \hbar \equiv h/2\pi \tag{3.2}$$

while the momentum is to be related to k via the de Broglie wavelength

$$p = \frac{h}{\lambda} \equiv \frac{\hbar 2\pi}{\lambda} = \hbar k. \tag{3.3}$$

Thus, with the use of (3.2) and (3.3), one can express the kinetic energy as

$$E = \hbar\omega = \frac{p^2}{2m} = \frac{\hbar^2 k^2}{2m} \tag{3.4}$$

and view (3.4) as the dispersion relation of the free particle. Incorporating this dispersion relation in (3.1) and identifying the group velocity as the particle velocity $\hbar k/m$, the free particle can be represented by the wave packet.

Given such a wave packet, differentiate (3.1) with respect to time, obtaining

$$i\hbar\frac{\partial}{\partial t}\psi(\boldsymbol{r},t) = i\hbar\int d\boldsymbol{k} f(\boldsymbol{k})\frac{\partial}{\partial t}e^{-i(\omega t-\boldsymbol{k}\cdot\boldsymbol{r})} = \int d\boldsymbol{k} f(\boldsymbol{k})\hbar\omega e^{-i(\omega t-\boldsymbol{k}\cdot\boldsymbol{r})}. \tag{3.5}$$

Also, the Laplacian operator

$$\nabla^2 \equiv \partial^2/\partial x^2 + \partial^2/\partial y^2 + \partial^2/\partial z^2$$

acting on (3.1) leads to

$$-\frac{\hbar^2}{2m}\nabla^2\psi(\boldsymbol{r},t) = -\frac{\hbar^2}{2m}\int d\boldsymbol{k} f(\boldsymbol{k})\nabla^2 e^{-i(\omega t-\boldsymbol{k}\cdot\boldsymbol{r})} = \int d\boldsymbol{k} f(\boldsymbol{k})\frac{\hbar^2 k^2}{2m}e^{-i(\omega t-\boldsymbol{k}\cdot\boldsymbol{r})}. \tag{3.6}$$

Clearly, (3.5) and (3.6) are identical in content in view of the dispersion relation (3.4). Hence, one can write

$$i\hbar\frac{\partial}{\partial t}\psi(\boldsymbol{r},t) = -\frac{\hbar^2}{2m}\nabla^2\psi(\boldsymbol{r},t) \tag{3.7}$$

and (3.7) is the Schrödinger equation of a free particle. Implicit in (3.7) is the assumption that physical quantities are represented by operators. For instance,

$$\hat{p}_x \rightarrow -i\hbar\frac{\partial}{\partial x}, \tag{3.8a}$$

$$p \rightarrow -i\hbar\nabla, \quad \nabla \equiv \hat{x}\frac{\partial}{\partial x} + \hat{y}\frac{\partial}{\partial y} + \hat{z}\frac{\partial}{\partial z}, \tag{3.8b}$$

$$E \rightarrow i\hbar\frac{\partial}{\partial t}. \tag{3.8c}$$

The Schrödinger equation of (3.7) can be generalized to the case of a particle moving in the presence of a potential by introducing the Hamiltonian as

$$\hat{H} = \frac{p_x^2 + p_y^2 + p_z^2}{2m} + V(\boldsymbol{r}) = -\frac{\hbar^2}{2m}\nabla^2 + V(\boldsymbol{r}) \tag{3.9}$$

and writing

$$i\hbar\frac{\partial}{\partial t}\psi(\boldsymbol{r}, t) = \hat{H}\psi(\boldsymbol{r}, t). \tag{3.10}$$

Equation 3.10 is the celebrated Schrödinger equation that governs the evolution in time of dynamic systems. The equation is a linear, second-order, partial differential equation and the wavefunction ψ carries all possible dynamical information of the system.

3.1.1 Energy Eigenequation

The time-dependent Schrödinger equation (3.10) can be generally recast into time independent one by using the separation of variable technique, that is, by looking for the solution of (3.10) in the form

$$\psi(\boldsymbol{r}, t) = T(t)u(\boldsymbol{r}) = e^{-i(Et/\hbar)}u(\boldsymbol{r}). \tag{3.11}$$

Inserting (3.11) into (3.10) and dividing both sides by (3.11) results in

$$E = \frac{\hat{H}u(\boldsymbol{r})}{u(\boldsymbol{r})}. \tag{3.12}$$

Therefore, one can write

$$\hat{H}u(\boldsymbol{r}) = Eu(\boldsymbol{r}). \tag{3.13}$$

This time-independent Schrödinger equation represents the typical eigenequation in which an operator, in this case $\hat{H}$, acting on u reproduces the same function u multiplied by a constant E. Here, u and E are called the (energy) eigenfunction and (energy) eigenvalue, respectively. Solving the energy eigenequation constitutes the bulk of the wave mechanical analysis.

The eigenequation generates a set of eigenfunctions with eigenvalues, and the wavefunction can be generally expanded in terms of these eigenfunctions:

$$\psi(\boldsymbol{r}, t) = \sum_n a_n e^{-i\omega_n t}u_n(\boldsymbol{r}), \quad \omega_n \equiv E_n/\hbar, \tag{3.14}$$

where a_n is the expansion coefficient.

3.2
Momentum Eigenfunction and Fourier Series

One-Dimensional Momentum Eigenfunction Consider one-dimensional momentum eigenequation:

$$-i\hbar\frac{\partial}{\partial x}u(x) = p_x u(x), \quad \hat{p}_x \equiv -i\hbar\frac{\partial}{\partial x}, \tag{3.15}$$

where $u(x)$ and p_x are the eigenfunction and eigenvalue, respectively. Rearranging (3.15) as

$$\frac{\partial u(x)}{u(x)} = \frac{ip_x}{\hbar}\partial x \equiv ik_x \partial x \tag{3.16}$$

and performing the integration on both sides, one obtains

$$u(x) = N \exp ik_x x, \quad p_x/\hbar = k_x. \tag{3.17}$$

Here, N is the constant of integration and the eigenvalue p_x and/or k_x is determined by the boundary conditions imposed. For example, when a periodic boundary condition is imposed in the interval $0 \le x \le L$, that is, $u(x=0) = u(x=L)$, then p_x or k_x takes up a discrete set of eigenvalues, satisfying the condition

$$(p_{xn}/\hbar)L \equiv k_{xn}L = 2\pi n, \quad n = 0, \pm 1, \pm 2, \ldots. \tag{3.18}$$

Hence, one can write the normalized eigenfunction as

$$u_n(x) = Ne^{in(2\pi/L)x}, \quad N = 1/\sqrt{L}. \tag{3.19}$$

It is important to note that the normalization condition is an integral part of the eigenfunction u_n together with the condition for the orthogonality,

$$\int_0^L dx u_m^*(x) u_n(x) = \frac{1}{L}\int_0^L e^{-im(2\pi/L)x} e^{in(2\pi/L)x} dx = \delta_{\mathrm{nm}}, \tag{3.20}$$

where the Kronecker delta function δ_{nm} is defined as

$$\delta \equiv \begin{cases} 0, & n \ne m, \\ 1, & n = m. \end{cases} \tag{3.21}$$

The functions satisfying (3.20) are said to be orthonormal.

It is interesting to note that the momentum eigenfunctions given in (3.19) are identical to the complex Fourier series and any function in the interval $0 \le x \le L$ can, therefore, be represented in terms of this set of eigenfunctions, that is,

$$f(x) = \frac{1}{\sqrt{L}}\sum_{n=-\infty}^{\infty} c_n e^{in(2\pi/L)x}, \tag{3.22}$$

where the expansion coefficients are specified by the operation,

$$\begin{aligned}\frac{1}{\sqrt{L}}\int_0^L dx e^{-im(2\pi/L)x} f(x) &= \frac{1}{L}\sum_{n=-\infty}^{\infty} c_n \int_0^L dx e^{-im(2\pi/L)x} e^{in(2\pi/L)x} \\ &= \sum_{n=-\infty}^{\infty} c_n \delta_{nm} = c_m.\end{aligned} \tag{3.23}$$

Therefore, one can represent an arbitrary function in terms of these eigenfunctions:

$$f(x) = \frac{1}{\sqrt{L}}\sum_{n=-\infty}^{\infty} c_n e^{in(2\pi/L)x}, \quad c_n \equiv \frac{1}{\sqrt{L}}\int_0^L dx e^{-in(2\pi/L)x} f(x). \tag{3.24}$$

Three-Dimensional Momentum Eigenfunction The generalization from 1D to 3D analysis is straightforward. The eigenequation of the momentum operator now reads as

$$-i\hbar\nabla u(\boldsymbol{r}) = \boldsymbol{p}u(\boldsymbol{r}), \quad \boldsymbol{p} = \hat{x}p_x + \hat{y}p_y + \hat{z}p_z. \tag{3.25}$$

The solution of the eigenfunction can be sought by putting

$$u(r) \propto f(x)g(y)h(z) \tag{3.26}$$

and inserting (3.26) into (3.25), and again dividing both sides with (3.26) after performing the differentiations, obtaining

$$\hat{x}\frac{\partial f(x)/\partial x}{f(x)} + \hat{y}\frac{\partial g(y)/\partial y}{g(y)} + \hat{z}\frac{\partial h(z)/\partial z}{h(z)} = \frac{i}{\hbar}(\hat{x}p_x + \hat{y}p_y + \hat{z}p_z). \tag{3.27}$$

Thus, the equations for x, y, and z components are identical to the 1D momentum eigenequation (3.15) and one can use, for example, the same periodic boundary conditions in intervals from 0 to L_x, L_y, and L_z, respectively, and write the eigenfunctions as

$$u(r) = \frac{1}{(L_x L_y L_z)^{1/2}} e^{i\boldsymbol{k}_{lmn}\cdot\boldsymbol{r}}, \quad \boldsymbol{k}_{lmn} = \hat{x}\frac{2\pi}{L_x}l + \hat{y}\frac{2\pi}{L_y}m + \hat{z}\frac{2\pi}{L_z}n, \tag{3.28}$$

with l, m, and n denoting integers. Again, any function can be expanded in terms of these eigenfunctions as

$$f(x,y,z) = \frac{1}{(L_x L_y L_z)^{1/2}}\sum_{n,l,m=-\infty}^{\infty} c_{nlm} e^{il(2\pi/L_x)x} e^{im(2\pi/L_y)y} e^{in(2\pi/L_z)z} \tag{3.29}$$

with the expansion coefficient given by

$$c_{lmn} \equiv \frac{1}{(L_x L_y L_z)^{1/2}}\int_0^{L_x} dx e^{-il(2\pi/L_x)x}\int_0^{L_y} dy e^{-im(2\pi/L_y)y}\int_0^{L_z} dz e^{-il(2\pi/L)z} f(x,y,z). \tag{3.30}$$

3.3 Hermitian Operator and Bra–Ket Notations

As noted earlier, a physical observable, such as position, momentum, and energy, is associated with its own operator. Any such physical operator $\hat{A}$ should satisfy the condition,

$$\int d\boldsymbol{r} f^*(\boldsymbol{r})\hat{A}g(\boldsymbol{r}) = \int d\boldsymbol{r}[\hat{A}f(\boldsymbol{r})]^* g(\boldsymbol{r}), \tag{3.31}$$

where f and g are arbitrary but well-behaving functions, namely, differentiable and vanishing at infinity. An operator satisfying (3.31) is called the Hermitian operator. The momentum operator $-i\hbar\partial/\partial x$, for instance, is a Hermitian operator. This can be readily seen to be the case by performing the integration by parts:

$$\int_{-\infty}^{\infty} dx f^*\left(-i\hbar\frac{\partial g}{\partial x}\right) = -i\hbar\left[f^*g|_{-\infty}^{\infty} - \int_{-\infty}^{\infty} dx g\frac{\partial f^*}{\partial x}\right] = \int_{-\infty}^{\infty} dx\left(-i\hbar\frac{\partial}{\partial x}f\right)^* g \quad \text{q.e.d.}$$

The spatial integral involving the product of two functions, f^*g, is called the inner product and compactly denoted by

$$\int d\boldsymbol{r} f^*(\boldsymbol{r})g(\boldsymbol{r}) \equiv \langle f|g\rangle, \tag{3.32a}$$

where the functions f^* and g are denoted by bra and ket vectors, respectively, as

$$f^* \rightarrow \langle f|, \tag{3.32b}$$

$$g \rightarrow |g\rangle. \tag{3.32c}$$

By the same token, an operator acting on a function and transforming it is compactly denoted by

$$\hat{A}g \rightarrow \hat{A}|g\rangle = |\hat{A}g\rangle. \tag{3.33}$$

With this bra–ket notations devised by Dirac, the Hermitian condition (3.31) is compacted as

$$\langle f|\hat{A}g\rangle = \langle\hat{A}f|g\rangle. \tag{3.34}$$

3.4 The Orthonogonality and Completeness of Eigenfunctions

The eigenfunctions of a Hermitian operator are orthogonal and the eigenvalues are real. In proving this important theorem, both the conventional and Dirac notations are used in parallel to get familiar with bra–ket notations. Thus, consider two eigenfunctions of Hermitian operator $\hat{A}$, that is,

$$\hat{A}u_j(\mathbf{r}) = a_j u_j(\mathbf{r}), \quad j = m, n, \tag{3.35a}$$

or equivalently,

$$\hat{A}_j|u_j\rangle = a_j|u_j\rangle, \quad j = m, n. \tag{3.35b}$$

Performing the inner product on (3.35) with respect to u_m, one can write

$$\int_{-\infty}^{\infty} dr u_m^* \hat{A} u_n = a_n \int_{-\infty}^{\infty} dr u_m^* u_n, \tag{3.36a}$$

$$\langle u_m|\hat{A}u_n\rangle = a_n\langle u_m|u_n\rangle. \tag{3.36b}$$

Since u_j is an eigenfunction of $\hat{A}$ and $\hat{A}$ is a Hermitian operator, one can write (3.36) using (3.31), (3.34), and (3.35) as

$$\int_{-\infty}^{\infty} dr u_n^* \hat{A} u_m = a_m \int_{-\infty}^{\infty} dr u_n^* u_m = \int_{-\infty}^{\infty} dr (\hat{A}u_n)^* u_m, \tag{3.37a}$$

$$\langle u_n|\hat{A}|u_m\rangle = a_m\langle u_n|u_m\rangle = \langle \hat{A}u_n|u_m\rangle. \tag{3.37b}$$

Thus, by taking the complex conjugate of the second identities in (3.37), one can write

$$\int_{-\infty}^{\infty} dr u_m^* (\hat{A}u_n) = a_m^* \int_{-\infty}^{\infty} dr u_m^* u_n, \tag{3.38a}$$

$$\langle u_m|\hat{A}u_n\rangle = a_m^*\langle u_m|u_n\rangle. \tag{3.38b}$$

Subtracting (3.8) from (3.36) results in

$$(a_n - a_m^*) \int_{-\infty}^{\infty} dr u_m^* u_n = 0, \tag{3.39a}$$

$$(a_n - a_m^*)\langle u_m|u_n\rangle = 0. \tag{3.39b}$$

One can therefore draw a few conclusions from (3.39). For the nondegenerate case, in which the eigenvalues are not the same, that is, $a_n \neq a_m$ if $n \neq m$, the eigenfunctions are orthogonal, namely, $\langle u_m|u_n\rangle = 0$. In addition, for $n = m$, $u_n^* u_n$ is positive definite and therefore $\langle u_n|u_n\rangle \neq 0$, hence $a_n = a_n^*$. That is, the eigenvalues are real. For the degenerate case in which the eigenvalues can be the same even if $n \neq m$, the present proof does not apply. However, the degenerate eigenfunctions can be made orthogonal by forming appropriate liner combinations of these eigenfunctions.

Closure Property With the orthonormality of the eigenfunctions thus established, one can expand the wavefunction $\psi(\boldsymbol{r})$ in terms of these eigenfunctions as

$$|\psi(\boldsymbol{r})\rangle = \sum_{n=0}^{\infty} c_n |u_n(\boldsymbol{r})\rangle = \sum_{n=0}^{\infty} c_n |n\rangle, \tag{3.40}$$

where $|u_n\rangle$ is further compacted by $|n\rangle$. The expansion coefficient c_n is specified via the inner product:

$$\langle k|\psi(\boldsymbol{r})\rangle = \sum_{n=0}^{\infty} c_n \langle k|n\rangle = \sum_{n=0}^{\infty} c_n \delta_{kn} = c_k. \tag{3.41}$$

Inserting this coefficient in (3.40) results in

$$|\psi(\boldsymbol{r})\rangle = \sum_{n=0}^{\infty} \langle n|\psi(\boldsymbol{r})\rangle |n\rangle = \sum_{n=0}^{\infty} |n\rangle\langle n|\psi(\boldsymbol{r})\rangle. \tag{3.42}$$

In (3.42), the constant expansion coefficient c_n has been slipped to the right of the ket vector. It is therefore evident from (3.42) that

$$\sum_{n=0}^{\infty} |n\rangle\langle n| = I. \tag{3.43}$$

The identity (3.43) is called the closure property and represents the completeness of the eigenfunctions of Hermitian operators.

It is interesting to notice a similarity existing between the expansion scheme of (3.42) and the representation of a 3D vector in terms of unit vectors:

$$\boldsymbol{A} = \hat{x}A_x + \hat{y}A_y + \hat{z}A_z.$$

Here, because of the orthonormality of $\hat{x}$, $\hat{y}$, and $\hat{z}$, that is, $\hat{x}\cdot\hat{x} = 1$, $\hat{x}\cdot\hat{y} = 0$, and so on, the three components of the vector are obtained by performing the scalar products, that is, $A_x = \boldsymbol{A}\cdot\hat{x}$, $A_y = \boldsymbol{A}\cdot\hat{y}$, and $A_z = \boldsymbol{A}\cdot\hat{z}$. In the same contexts, (3.42) can be viewed as representing a state vector in infinite Hilbert space with $|n\rangle, c_n$ playing the role of unit basis vector and projected component.

3.5
Basic Postulates of Quantum Mechanics

Now that the Schrödinger equation has been discussed together with some elementary operator algebra, it is an opportune time to summarize the basic postulates of quantum mechanics:

1) A dynamical system, for example, particle, atom, molecule, and so on, is associated with a wavefunction $\psi(\boldsymbol{r}, t)$ that contains all possible information regarding the system.

2) The wavefunction evolves in time according to the Schrödinger equation,

$$i\hbar \frac{\partial}{\partial t}\psi(\boldsymbol{r},t) = \hat{H}\psi(\boldsymbol{r},t). \tag{3.44}$$

3) The quantity $\psi^*(r,t)\psi(r,t)d\boldsymbol{r}$ represents the probability of finding the system between $\boldsymbol{r}$ and $\boldsymbol{r}+d\boldsymbol{r}$ at time t, with $\psi^*(r,t)\psi(r,t)$ representing the probability density (Born interpretation).
4) Physical observables have their own Hermitian operators and the expectation or average values are theoretically described by

$$\langle\hat{A}\rangle = \frac{\int_{-\infty}^{\infty} d\boldsymbol{r}\psi^*(\boldsymbol{r},t)\hat{A}\psi(\boldsymbol{r},t)}{\int_{-\infty}^{\infty} d\boldsymbol{r}\psi^*(\boldsymbol{r},t)\psi(\boldsymbol{r},t)} = \frac{\langle\psi(\boldsymbol{r},t)|\hat{A}|\psi(\boldsymbol{r},t)\rangle}{\langle\psi(\boldsymbol{r},t)|\psi(\boldsymbol{r},t)\rangle}. \tag{3.45}$$

   The probability density $\psi^*(r,t)\psi(r,t)$ plays a role similar to a distribution function, except that the operator is inserted in between the wavefunctions.
5) Physical operator $\hat{A}$ generates a complete set of eigenfunctions, that is,

$$\hat{A}|u_n\rangle = a_n|u_n\rangle, \quad n = 1,2,3,\ldots, \tag{3.46}$$

   and the wavefunction at fixed time, say at $t=0$, can be expanded in terms of these eigenfunctions:

$$|\psi(\boldsymbol{r},t=0)\rangle = \sum_{n=0}^{\infty} c_n|u_n\rangle, \quad c_n = \langle u_n|\psi(\boldsymbol{r},t=0)\rangle, \tag{3.47}$$

   with c_n denoting the expansion coefficient. The normalization condition of the wavefunction is then given by

$$1 = \langle\psi|\psi\rangle = \sum_m c_m^* \sum_n c_n\langle u_m|u_n\rangle = \sum_m c_m^* \sum_n c_n\delta_{mn} = \sum_n |c_n|^2 \tag{3.48}$$

   and $|c_n|^2$ therefore naturally represents the probability of the system being in the nth eigenstate, and such probability should be summed up to be unity. Likewise, the expectation value of $\hat{A}$ is given by

$$\langle\hat{A}\rangle = \sum_m c_m^* \sum_n c_n\langle u_m|\hat{A}|u_n\rangle = \sum_m c_m^* \sum_n c_n a_n\langle u_m|u_n\rangle = \sum_n |c_n|^2 a_n \tag{3.49}$$

   and is described by the weighted average of the eigenvalues.

3.6 Commutation Relation

As noted in Chapter 1, the motion of a particle is deterministically described in classical mechanics. That is, its position and momentum can be followed exactly, once the initial values are known. This deterministic description presupposes the fact that the act of measuring the position and the momentum does not perturb the motion itself. In contrast, the quantum description is probabilistic in nature. The probabilistic description is necessitated by the fundamental fact that the act of measurement could disturb and modify the system. This observation-induced perturbation is particularly pronounced in microsystems such as atoms, molecules, electron, holes, and so on.

Commutation Relation and Measurement A thought experiment of taking a snapshot of the hydrogen atom could make this point clear. To resolve the atomic diameter of about 0.1 nm (see the Bohr radius, (2.31)), the wavelength of the probing light should obviously be equal to or less than 0.1 nm. Equivalently, the frequency $\nu(= c/\lambda)$ of the light should be equal to or greater than 3×10^{18} Hz. This means that the photons used for taking the snapshot should have the energy $h\nu$ greater than $\sim 1.23 \times 10^4$ eV, which is greater than the binding energy of the atom by orders of magnitude and the measurement could end up ionizing the atom.

The fact that the measurement process itself perturbs or modifies the system suggests that the consecutive measurements of two physical observables, when performed in reverse orders, do not necessarily yield the same results. Theoretically, this corresponds to

$$\langle\psi|\hat{A}\hat{B}|\psi\rangle \neq \langle\psi|\hat{B}\hat{A}|\psi\rangle. \tag{3.50}$$

Equivalently, in terms of the commutator,

$$[\hat{A}, \hat{B}] \equiv \hat{A}\hat{B} - \hat{B}\hat{A} \neq 0. \tag{3.51}$$

Thus, the commutation relation of the two physical observables carries important consequences. For example, the necessary and sufficient condition of two operators sharing common eigenfunction is that the two operators commute. This is proven as follows:

First, introduce the eigenfunction of $\hat{B}$:

$$\hat{B}|u_n\rangle = b_n|u_n\rangle. \tag{3.52}$$

If $\hat{A}$ and $\hat{B}$ commute, that is, $\hat{A}\hat{B} = \hat{B}\hat{A}$, then it follows from (3.52) that

$$\hat{A}\hat{B}|u_n\rangle = b_n\hat{A}|u_n\rangle \equiv \hat{B}\hat{A}|u_n\rangle. \tag{3.53}$$

It is therefore clear from (3.53) that the new function $|v_n\rangle \equiv \hat{A}|u_n\rangle$ is also an eigenfunction of $\hat{B}$. Since an eigenfunction is determined to within a constant, one can put

$$|v_n\rangle \equiv \hat{A}|u_n\rangle \propto |u_n\rangle = a_n|u_n\rangle, \tag{3.54}$$

thereby proving that $|u_n\rangle$ is also the eigenfunction of $\hat{A}$.

Also, if $\hat{A}$ and $\hat{B}$ share a common eigenfunction, one can write by definition,

$$\langle u_n|\hat{A}\hat{B}|u_n\rangle = \langle u_n|\hat{B}\hat{A}|u_n\rangle = a_n b_n\langle u_n|u_n\rangle = a_n b_n. \tag{3.55}$$

Hence, $\hat{A}$ and $\hat{B}$ commute. An additional implication of (3.55) is that it is possible to measure two commuting observables simultaneously.

On the other hand, if $\hat{A}$ and $\hat{B}$ do not commute, it is not generally possible to measure both $\hat{A}$ and $\hat{B}$ simultaneously. To show this, let $|u_n\rangle$ be the eigenfunction of $\hat{B}$. Then the observable $\hat{B}$ is theoretically described as

$$\langle u_n|\hat{B}|u_n\rangle = b_n\langle u_n|u_n\rangle = b_n. \tag{3.56}$$

Now, since $\hat{A}$ and $\hat{B}$ do not commute, $|u_n\rangle$ is not the eigenfunction of $\hat{A}$ but it can be expanded in terms of the eigenfunctions of $\hat{A}$, that is,

$$|u_n\rangle = \sum_m c_m|v_m\rangle, \quad \hat{A}|v_m\rangle = a_m|v_m\rangle. \tag{3.57}$$

The theoretical description of $\hat{A}$, when carried out in terms of the eigenfunctions of $\hat{B}$, $|u_n\rangle$, namely,

$$\langle u_n|\hat{A}|u_n\rangle = \sum_m c_m^* \sum_n c_n\langle v_m|\hat{A}|v_n\rangle = \sum_m c_m^* \sum_n c_n a_n \delta_{mn} = \sum_n |c_n|^2 a_n, \tag{3.58}$$

explicitly indicates the statistical nature of observing $\hat{A}$ in terms of the weighted average of the eigenvalues. Hence, while the observable $\hat{B}$ can be measured and described precisely by $|u_n\rangle$, the noncommuting observable $\hat{A}$ cannot be precisely measured and described simultaneously.

3.7 Conjugate Variables and Uncertainty Relation

In Chapter 1, the canonically conjugate variables were introduced in conjunction with the Hamilton's equation of motion. The conjugate pairs such as position and momentum do not commute and the commutation relation is given by

$$[x, p_x] = [y, p_y] = [z, p_z] = i\hbar. \tag{3.59}$$

These commutation relations can be proven in a single step as follows:

$$[x, p_x]f(x) \equiv -i\hbar\left\{x\frac{\partial}{\partial x}f(x) - \frac{\partial}{\partial x}[xf(x)]\right\} = i\hbar f(x) \quad \text{q.e.d.}$$

However, all other combinations of the position and momentum commute.

Since x and p_x do not commute, the implication is that it is not possible to measure the position and momentum simultaneously. In fact, the uncertainty in the position and the momentum is given by

$$\Delta x \Delta p_x \approx \hbar, \quad \Delta y \Delta p_y \approx \hbar, \quad \Delta z \Delta p_z \approx \hbar. \tag{3.60}$$

This relation is the crux of Heisenberg's uncertainty principle. The same kind of uncertainty relation holds true for the canonically conjugate variables such as energy and time, namely,

$$\Delta E \Delta t \approx \hbar. \tag{3.61}$$

Uncertainty in Position and Momentum The uncertainty relation is an essential ingredient of quantum mechanics and is rooted in the wave nature of particles. These relations are heuristically discussed by taking a free particle as represented by a Gaussian wave packet. The packet is taken to propagate in the x-direction with velocity v, or the group velocity v_g, while oscillating at the carrier frequency ω_0 (see (1.65), for example):

$$\psi(x,t) = Ne^{-(i\omega_0 t - k_0 x)} e^{-(x-v_g t)^2/2\sigma^2}, \tag{3.62}$$

where N is the normalization constant. At a fixed time, say at $t = 0$, the spatial profile of the probability density is given by (see Figure 3.1)

$$|\psi(x, t=0)|^2 \propto e^{-(x^2/\sigma^2)}. \tag{3.63}$$

Then, the variance Δx of the position is given by

$$(\Delta x)^2 \equiv \left\langle (x - \langle x \rangle^2 \right\rangle = \left\langle x^2 - 2x\langle x \rangle + \langle x \rangle^2 \right\rangle = \langle x^2 \rangle - \langle x \rangle^2. \tag{3.64}$$

Here, one can put $\langle x \rangle = 0$ from the parity consideration and write

$$\langle x^2 \rangle \equiv \frac{\int_{-\infty}^{\infty} dx x^2 e^{-(x^2/\sigma^2)}}{\int_{-\infty}^{\infty} dx e^{-(x^2/\sigma^2)}} = -\frac{\partial}{\partial(1/\sigma^2)} \ln \int_{-\infty}^{\infty} dx e^{-(x^2/\sigma^2)} = \frac{\sigma^2}{2}. \tag{3.65}$$

Thus, the uncertainty in position reads from (3.64) and (3.65) as

$$\Delta x = \frac{\sigma}{\sqrt{2}}. \tag{3.66}$$

In (3.65), the simple technique for finding the average values has been used (see (1.23)), together with the integral formula of (1.24).

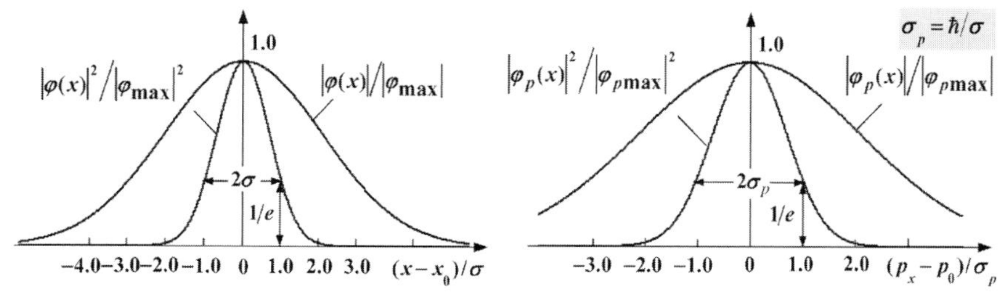

Figure 3.1 The envelope and intensity of Gaussian wave packet versus position x/σ and the corresponding spectrum versus momentum p_x/σ_p, $\sigma_p = \hbar/\sigma$.

At fixed position, say at $x = 0$, the temporal profile of the probability density (3.62) is given by

$$|\psi(t, x = 0)|^2 \propto e^{-(t^2/\sigma_t^2)}, \quad \sigma_t^2 \equiv \sigma^2/v_g^2, \tag{3.67}$$

and the variance in time can be directly transcribed from (3.66) as

$$\Delta t = \frac{1}{\sqrt{2}}\sigma_t \equiv \frac{1}{\sqrt{2}}\frac{\sigma}{v_g} \tag{3.68}$$

and the result indicates that the uncertainty in time of detecting the wave packet at fixed position is given approximately by its transit time through the detecting plane.

Now, as discussed in Section 3.2, the wave packet introduced in (3.62) can be expanded in terms of the momentum eigenfunctions, namely,

$$\psi(x, t = 0) = \int_{-\infty}^{\infty} dp\varphi_p e^{ikx}, \quad k \equiv p_x/\hbar, \tag{3.69}$$

where the expansion coefficient can be obtained by using again the formula (1.24) as

$$\varphi_p \equiv \int_{-\infty}^{\infty} dx e^{-ikx}\psi(x, t = 0) \propto \int_{-\infty}^{\infty} dx e^{-ikx} e^{k_0 x} e^{-x^2/2\sigma^2} \propto e^{-\sigma^2(k-k_0)^2/2}. \tag{3.70}$$

It is therefore clear from (3.69) and (3.70) that the Gaussian wave packet in space is composed of the Gaussian momentum spectrum, as shown in Figure 3.1:

$$\left|\varphi_p\right|^2 \propto e^{-[(p-p_0)^2/(\hbar^2/\sigma^2)]}. \tag{3.71}$$

Hence, the variance of p_x, centered around the average value $\langle p_x \rangle = p_0$ can be found exactly in the same way as the variance in position, Δx, and the result can be directly transcribed from (3.66) as

$$\Delta p_x = \frac{1}{\sqrt{2}}\frac{\hbar}{\sigma}. \tag{3.72}$$

Therefore, it follows from (3.66) and (3.72) that if σ is small, for example, the uncertainty in position is reduced, while the uncertainty in momentum is increased. Naturally, the reverse conclusions can also be drawn for large σ. Hence, by combining (3.66) and (3.72), one can write the fundamental relation

$$\Delta x \Delta p_x \approx \hbar, \tag{3.73}$$

which explicitly states that the product of the uncertainties in position and momentum is not zero but is finite, approximately equal to $\hbar$. In this manner, the Heisenberg's uncertainty principle has been explicitly shown for the position and the momentum.

Uncertainty in Energy and Time The uncertainty in energy and time can also be shown from the dispersion relation of the same free particle, namely, $E = p_x^2/2m$. By differentiating E with respect to p_x, one can write

$$\Delta E = \frac{p_x}{m}\Delta p_x = v_g \Delta p_x, \tag{3.74}$$

where v_g represents the particle velocity. Thus, by combining (3.66), (3.68), and (3.74), the uncertainty relation between the time and the energy is obtained as

$$\Delta E \Delta t \approx \Delta p_x \Delta x \approx \hbar. \tag{3.75}$$

The implication of (3.75) is that the accuracy of measuring the energy of a system depends on the measurement time. Moreover, in view of $\Delta E \approx \hbar\Delta\omega$, (3.75) is also in line with the basic relationship existing between the time duration and frequency bandwidth in pulsed electromagnetic waves, as detailed in connection with (1.66).

3.8
Operator Equation of Motion and Ehrenfest Theorem

As pointed out, a physical observable is associated with an operator, the expectation value of which is given by

$$\langle \hat{A} \rangle = \langle \psi | \hat{A} | \psi \rangle. \tag{3.76}$$

The change in time of this expectation value, therefore, reads as

$$\frac{d}{dt}\langle \hat{A} \rangle = \langle \dot{\psi} | \hat{A} | \psi \rangle + \langle \psi | \hat{A} | \dot{\psi} \rangle + \langle \psi | \dot{\hat{A}} | \psi \rangle. \tag{3.77}$$

Now, the wavefunction evolves in time according to the Schrödinger equation,

$$i\hbar|\dot{\psi}\rangle = \hat{H}|\psi\rangle, \quad \hat{H} = -\frac{\hbar^2}{2m}\nabla^2 + V(\boldsymbol{r}), \tag{3.78a}$$

and by the same token, one may write the complex conjugate of the Schrödinger equation in the bra–ket notation as

$$-i\hbar\langle\dot{\psi}| = \langle\psi|\hat{H}. \tag{3.78b}$$

Thus, inserting (3.78) into (3.77) results in the equation of motion of the expectation value of the operator:

$$\begin{aligned}\frac{d}{dt}\langle \hat{A} \rangle &= \frac{1}{-i\hbar}\langle\psi|\hat{H}\hat{A}|\psi\rangle + \frac{1}{i\hbar}\langle\psi|\hat{A}\hat{H}|\psi\rangle + \langle\psi|\dot{\hat{A}}|\psi\rangle \\ &= \frac{i}{\hbar}\langle[\hat{H},\hat{A}]\rangle + \left\langle\frac{\partial\hat{A}}{\partial t}\right\rangle, \quad [\hat{H},\hat{A}] \equiv \hat{H}\hat{A}-\hat{A}\hat{H}.\end{aligned} \tag{3.79}$$

Ehrenfest Theorem By applying the operator equation (3.79) to the position and momentum operators, the Ehrenfest theorem can be proven, thereby bridging the quantum and classical descriptions. Thus, consider the time rate of change of $\langle x \rangle$, which reads from (3.79) as

$$\frac{d}{dt}\langle x \rangle = \frac{i}{\hbar}\langle \psi | [\hat{H}, x] | \psi \rangle, \tag{3.80}$$

where the variable x is independent of time and $\dot{x} \equiv 0$. Inserting the Hamiltonian in (3.78a) and using the fact that $V(\boldsymbol{r})$ and x commute, one can simplify (3.80) as

$$\frac{d}{dt}\langle x \rangle = \frac{i}{\hbar}\langle \psi | \left[\frac{-\hbar^2}{2m}\frac{\partial^2}{\partial x^2}, x\right] | \psi \rangle. \tag{3.81}$$

Since it follows from the chain rule that

$$\frac{\partial^2}{\partial x^2}[x\psi(x)] - x\frac{\partial^2}{\partial x^2}\psi(x) = 2\frac{\partial}{\partial x}\psi(x), \tag{3.82}$$

(3.81) reduces with the use of (3.82) to

$$\frac{d}{dt}\langle x \rangle = \frac{1}{m}\langle \psi | -i\hbar\frac{\partial}{\partial x} | \psi \rangle = \frac{\langle p_x \rangle}{m}. \tag{3.83}$$

Evidently, (3.83) specifies the average value of the velocity in terms of the average value of momentum, just as in Hamilton's equation of motion.

One can likewise consider the time rate of the change in momentum, obtaining

$$\begin{aligned}\frac{d}{dt}\langle p_x \rangle &= \frac{i}{\hbar}\langle \psi | [\hat{H}, \hat{p}_x] | \psi \rangle = \frac{i}{\hbar}\left\langle \left[\frac{-\hbar^2}{2m}\nabla^2 + V(r), -i\hbar\frac{\partial}{\partial x}\right] \right\rangle \\ &= \left\langle \left[V, \frac{\partial}{\partial x}\right] \right\rangle = \langle \psi | V\frac{\partial}{\partial x} - \frac{\partial}{\partial x}V | \psi \rangle \\ &= \langle \psi | -\frac{\partial V}{\partial x} | \psi \rangle = \left\langle -\frac{\partial V}{\partial x} \right\rangle \end{aligned} \tag{3.84}$$

in agreement with the Newton's or Hamilton's equation of motion. It is therefore clear from (3.83) and (3.84) that in quantum mechanics, the expectation values of the position and momentum evolve in time in the same way as in classical mechanics.

3.9 Problems

3.1. The Schrödinger equation is the first-order differential equation in time.

(a) Verify by direct substitution that given an initial wavefunction $\psi(\boldsymbol{r}, t = 0)$, the function evolves in time according to

$$\psi(\boldsymbol{r}, t) = \exp\left[-\frac{i}{\hbar}\int_0^t dt' \hat{H}(t')\right]\psi(\boldsymbol{r}, t=0). \tag{3.85}$$

(b) If the Hamiltonian is time independent, how does the solution in (3.85) compare with the solution obtained in the text, using the separation of variable technique.

3.2. Show by using the integration by parts that the Hamiltonian

$$\hat{H} = -\frac{\hbar^2}{2m}\nabla^2 + V(r)$$

is a Hermitian operator.

3.3. Given an operator $\hat{A}$, the operator $\hat{B}$ satisfying the relation

$$\int f^*(\boldsymbol{r})\hat{A}g(\boldsymbol{r})d\boldsymbol{r} = \int [\hat{B}f(\boldsymbol{r})]^* g(\boldsymbol{r})d\boldsymbol{r},$$

with f and g being arbitrary but physically well-behaving functions, is called the Hermitian adjoint and denoted by $\hat{B} = \hat{A}^+$.

(a) Show that a Hermitian operator is its own Hermitian adjoint, namely,

$$\hat{A}^+ = \hat{A} \tag{3.86}$$

(b) Show that $i(\hat{A}^+ - \hat{A})$ is a Hermitian operator for any $\hat{A}$.
(c) Write your answer in (b) by using bra–ket notations.

3.4. Prove the following operator relations:

(a) $[x, p_x^n] = i\hbar n p_x^{n-1}$.
(b) $[p_x, x^n] = -i\hbar n x^{n-1}$.

3.5. Use the uncertainty relation to estimate the minimum energy of an electron, spatially confined in a cube of edge, 10, 1, and 0.5 nm, respectively.

3.6. Take the diameter of an atomic nucleus to be 1×10^{-5} nm. Use the uncertainty relation to estimate the minimum kinetic energy for (a) an electron and (b) a proton to have within the nucleus. If the binding energy per nucleon is 5×10^6 eV, what is the physical implication of your results obtained in (a) and (b)?

3.7. When the electron in hydrogen atom is excited from the ground state with $n = 1$ to the first excited state with $n = 2$, for example, its lifetime therein is short, typically of the order of 10 ns before returning to the ground state. Find (i) the wavelength of the radiation when the electron makes the transition from $n = 2$ state to $n = 1$ state and (ii) the spread of wavelength arising from the finite lifetime of the electron in the excited state.

Suggested Reading

1 Singh, J. (1996) *Quantum Mechanics, Fundamentals & Applications to Technology*, John Wiley & Sons, Inc.
2 Yariv, A. (1982) *An Introduction to Theory and Applications of Quantum Mechanics*, John Wiley & Sons, Inc.
3 Liboff, R.L. (2002) *Introductory Quantum Mechanics*, 4th edn, Addison-Wesley Publishing Company, Reading, MA.
4 Gasiorowicz, S. (2003) *Quantum Physics*, 3rd edn, John Wiley & Sons, Inc.
5 Robinett, R.W. (2006) *Quantum Mechanics, Classical Results, Modern Systems and Visualized Examples*, Oxford University Press.
6 Kroemer, H. (1994) *Quantum Mechanics for Engineering, Materials Science, and Applied Physics*, International edn, Prentice Hall.

4
Particle in Potential Well

The problem of a particle in a simple potential well is interesting and useful, providing valuable insights into the bound states. It also offers concrete example, in which Schrödinger equation is analytically solved with simple algebra and at the same time the novel features of quantum mechanics are explicitly brought out. The energy quantization rooted deeply in the wave nature of particle and naturally resulting from the self-evident fact of the wavefunction physically well behaving is clearly demonstrated. Moreover, the results obtained from the simple potential wells are directly applicable to the problems of practical interest and play key roles in both designing and understanding the semiconductor devices, for example.

4.1
Infinite Square Well Potential

The ideal case of a particle confined in an infinite one-dimensional (1D) square potential well is a useful example to consider. It provides insights into the bound states and clearly brings out how the quantized discrete energy levels come about. In addition, it enables a quick and accurate estimation of the energy levels in practical problems of interest. Thus, consider a particle of mass m in an infinite 1D potential well of width L, as shown in Figure 4.1:

$$V(x) = \begin{cases} 0, & |x| \leq L/2, \\ \infty, & \text{otherwise}. \end{cases} \tag{4.1}$$

The Hamiltonian of the particle in the well is thus given by

$$\hat{H} = \frac{\hat{p}^2}{2m} + V(x) = -\frac{\hbar^2}{2m}\frac{\partial^2}{\partial x^2} \tag{4.2}$$

and the Schrödinger equation reads as

$$i\hbar\frac{\partial}{\partial t}\psi(x,t) = \hat{H}\psi(x,t). \tag{4.3}$$

Introductory Quantum Mechanics for Semiconductor Nanotechnology. Dae Mann Kim

ISBN: 978-3-527-40975-4

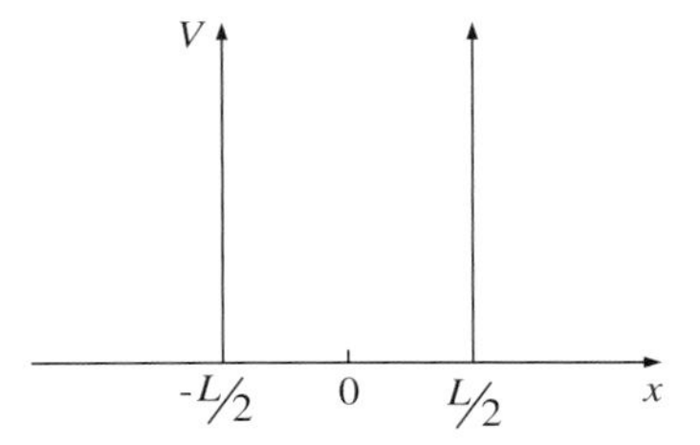

Figure 4.1 A square well potential with infinite barrier height and width L.

The equation can be solved by using the separation of variable technique, which has already been discussed in Chapter 3. Thus, look for the solution in the form,

$$\psi(x,t) = e^{-iEt/\hbar}u(x). \tag{4.4}$$

Inserting (4.4) into (4.3) and canceling out the exponential factor from both sides, one obtains the energy eigenequation as

$$-\frac{\hbar^2}{2m}\frac{\partial^2}{\partial x^2}u(x) = Eu(x). \tag{4.5}$$

The equation can be further compacted as

$$u'' + k^2 u = 0, \quad k^2 \equiv 2mE/\hbar^2. \tag{4.6}$$

Here, k is the wave vector and the primes denote differentiation with respect to x.

Equation 4.6 is identical in form to the harmonic oscillator equation, with t replaced by x. Hence, $\cos kx$ and $\sin kx$, for example, are the possible solutions, possessing even and odd parities. In addition, the infinite barrier height requires that $u(x)$ vanish at the edges of the well, so that the probability of finding the particle outside the well can be kept zero. This boundary condition requires for the case of the even solution $\cos kx$, $k_n(L/2) = (n+1/2)\pi$, $n = 0, 1, 2, \ldots$, and for the odd solution $\sin kx$, $k_n(L/2) = n\pi$, $n = 1, 2, \ldots$,. Also, the eigenfunctions of both parities can be readily normalized and one can write

$$u_n(x) = (2/L)^{1/2}\begin{cases}\cos k_n x, & k_n = (2n+1)\pi/L,\\ \sin k_n x, & k_n = 2n\pi/L,\end{cases} \tag{4.7}$$

so that the probability density $u_n^*(x)u(x)$ when integrated over the entire well yields unity, as it should.

The corresponding energy levels $E_n = \hbar^2 k_n^2/2m$ as specified in (4.6) for even and odd parity eigenfunctions are given, respectively, by

$$E_n = \begin{cases}\dfrac{\hbar^2}{2m}(2n+1)^2\dfrac{\pi^2}{L^2}, & n = 0, 1, 2, \ldots,\\[2ex] \dfrac{\hbar^2}{2m}(2n)^2\dfrac{\pi^2}{L^2}, & n = 1, 2, 3, \ldots.\end{cases} \tag{4.8}$$

The two sets of energy levels in (4.8) can be combined into one as

$$E_n = \frac{\hbar^2 \pi^2}{2mL^2} n^2, \quad n = 1, 2, 3, \ldots, \tag{4.9}$$

where odd and even integers correspond to the even and odd parity wavefunctions. In this manner, the energy levels are naturally quantized from the boundary conditions of the energy eigenfunction. Also, the lowest ground state is associated with the even parity eigenfunction ($n = 1$) and the even and odd parity eigenfunctions alternate for describing higher lying energy eigenstates. The integer n appearing in the energy eigenvalues is called the quantum number.

It is interesting and important to notice that the lowest ground-state energy E_1 is not zero but is finite, and in fact this finite value increases as the well width L gets smaller (see (4.9)). This is in marked contrast with the classical theory but is consistent with the uncertainty principle. Classically, a particle can be completely at rest in the potential well in which case $E_1 = 0$. Quantum mechanically, however, it is not possible for a particle to be at complete rest. Otherwise, it would be possible to measure simultaneously both the position and momentum of the particle in contradiction with the uncertainty principle.

Also, the more narrowly the particle is confined in smaller well width L, the more there will be uncertainty in the momentum or equivalently the more kinetic energy of the particle, thereby raising the ground-state energy E_1. The typical wavefunctions, probability densities are sketched in Figure 4.2.

Alternatively, one can shift the origin of x by $L/2$ and model the potential as

$$V(x) = \begin{cases} 0, & 0 \leq x \leq L, \\ \infty, & \text{otherwise}. \end{cases} \tag{4.10}$$

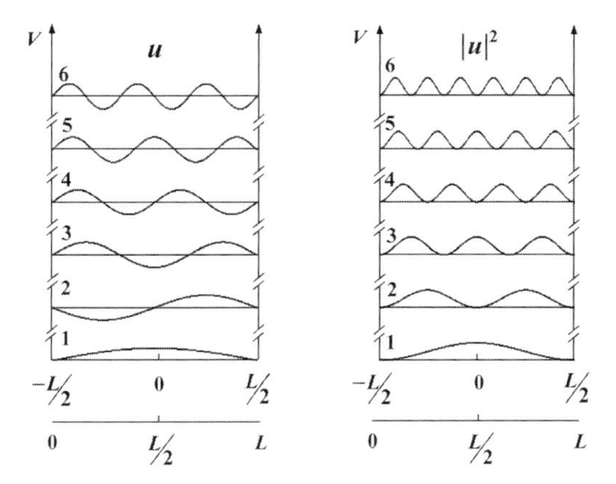

Figure 4.2 Typical energy eigenfunctions (u_n) and corresponding probability densities in infinite square well potential.

In this case, one can write

$$u_n(x) = (2/L)^{1/2} \sin k_n x, \quad k_n = n\pi/L, \; n = 1, 2, \ldots, \tag{4.11a}$$

and satisfy the required boundary conditions $u_n(x=0) = u_n(x=L) = 0$. The resulting energy levels

$$E_n = \frac{\hbar^2 k_n^2}{2m} = \frac{\hbar^2 \pi^2}{2mL^2} n^2, \quad n = 1, 2, 3, \ldots \tag{4.11b}$$

are identical to (4.9), as it should. With $u_n(x)$ and E_n thus found, the total time-dependent wavefunction in (4.4) is now completely specified.

4.2 Particle in 3D Box

An electron in bulk solids is often modeled as a free particle in three-dimensional box. The 3D box is in turn often approximated by 3D infinite square well potential, and it is therefore interesting and important to consider the motion of a particle in such a potential. The 3D infinite potential well can thus be modeled by

$$V(\mathbf{r}) = \begin{cases} 0, & 0 \le x, y, z \le L, \\ \infty, & \text{otherwise}, \end{cases} \tag{4.12a}$$

and the energy eigenequation of the particle in the box reads as

$$-\frac{\hbar^2}{2m}\left(\frac{\partial^2}{\partial x^2} + \frac{\partial^2}{\partial y^2} + \frac{\partial^2}{\partial z^2}\right) u(x, y, z) = Eu(x, y, z). \tag{4.12b}$$

Equation 4.12b can again be solved by using the separation of variable technique, namely, by looking for the solution in the form

$$u(x, y, z) = X(x)Y(y)Z(z). \tag{4.13}$$

Inserting (4.13) into (4.12), performing the differentiations entailed in (4.12b), and dividing both sides by (4.13) itself results in

$$-\frac{\hbar^2}{2m}\left(\frac{X''}{X} + \frac{Y''}{Y} + \frac{Z''}{Z}\right) = E. \tag{4.14}$$

An inspection of (4.14) readily reveals that the only way to satisfy (4.14) is to have the three terms on the left-hand side, depending solely on x, y, and z, respectively, should yield constants, so that these constants sum up to provide the total energy E. Thus, one can write

$$X'' + \frac{2mE_x}{\hbar^2} X = 0, \quad Y'' + \frac{2mE_y}{\hbar^2} Y = 0, \quad Z'' + \frac{2mE_z}{\hbar^2} Z = 0, \tag{4.15}$$

with

$$E_x + E_y + E_z = E. \tag{4.16}$$

The three equations in (4.15) are identical to the 1D energy eigenequation considered in Section 4.1 and the results obtained therein can be directly transcribed.

Thus, using (4.4) and (4.11), one can write

$$\psi(\mathbf{r}, t) = \exp{-i(E_n t/\hbar)} u_n(x, y, z), \tag{4.17a}$$

with

$$u_n(x, y, z) = \left(\frac{2}{L}\right)^{3/2} \sin\left(\frac{n_x \pi}{L} x\right) \sin\left(\frac{n_y \pi}{L} y\right) \sin\left(\frac{n_z \pi}{L} z\right), \tag{4.17b}$$

$$E_n = \frac{\hbar^2 \pi^2}{2mL^2}(n_x^2 + n_y^2 + n_z^2). \tag{4.17c}$$

Here, the three quantum numbers n_x, n_y, and n_z are positive integers. The ground state is specified by $n_x = n_y = n_z = 1$, while the first excited state is characterized by three different combinations of the quantum numbers, namely, $n_x = 2, n_y = n_z = 1$, $n_y = 2, n_x = n_z = 1$, and $n_z = 2, n_x = n_y = 1$. This indicates that there exists three-fold degeneracy in the first excited state and the degree of degeneracy increases for higher lying energy levels.

4.3 Density of States: 1D, 2D, and 3D

As noted earlier and will further be elaborated in due course, an electron in solids can be taken as a free particle, confined therein by the potential barrier at the surface. This means that the electron can be modeled as a free particle in 3D box. The stationary boundary condition of the 3D box requiring the wavefunction to vanish at the edge of the potential well further ensures that electrons are well confined in solids in one of those bound states in the box, as discussed in Section 4.2.

An additional boundary condition, called the periodic boundary condition, is often used to focus on the motion of a particle freely propagating and spatially unconfined. The condition can thus be utilized for describing the motion of electrons in the bulk of a solid. The periodic boundary condition is tantamount to stating that a particle exiting at $x + L$ reenters the box at x, as schematically illustrated in Figure 4.3.

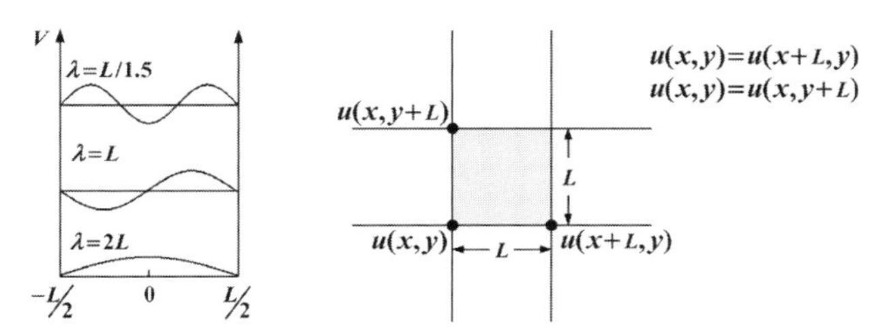

Figure 4.3 Graphical sketch of stationary (left) and periodic (right) boundary conditions.

Now, the wavefunction of the particle satisfying the energy eigenequation (4.12b) can also be expressed in the traveling waveform as

$$u(\boldsymbol{r}) = \frac{1}{\sqrt{V}} e^{\pm i\boldsymbol{k}\cdot\boldsymbol{r}}, \quad V = L^3. \tag{4.18}$$

Here, V is the volume of the box and the positive and negative branches in the exponent, when combined with the time component of the wavefunction in (4.4), respectively represent the forward, namely, traveling in the direction of $\boldsymbol{k}$, and backward components. The periodic boundary conditions are specified by

$$\begin{aligned} u(x, y, z) &= u(x + L, y, z), \\ u(x, y, z) &= u(x, y + L, z), \\ u(x, y, z) &= u(x, y, z + L), \end{aligned} \tag{4.19}$$

so that the wave vector $\boldsymbol{k}$ is constrained by

$$k_x L = 2\pi n_x; \quad k_y L = 2\pi n_y; \quad k_z L = 2\pi n_z, \tag{4.20}$$

with n_x, n_y, and n_z denoting integers, both positive and negative. The total energy of the particle is then given by

$$E_{\boldsymbol{n}} = \frac{\hbar^2 k^2}{2m} = \frac{\hbar^2}{2m}\left(\frac{2\pi}{L}\right)^2 (n_x^2 + n_y^2 + n_z^2) = \frac{\hbar^2}{2m}\left(\frac{2\pi}{L}\right)^2 \boldsymbol{n}\cdot\boldsymbol{n}. \tag{4.21}$$

The $\boldsymbol{k}$ vector specified by (4.20) or the corresponding momentum $\boldsymbol{p} = \hbar\boldsymbol{k}$ represents a single quantum state with the eigenfunction (4.18) and eigenenergy (4.21) and corresponding momentum. It is important to note that there is one-to-one correspondence between $\boldsymbol{k}$ and the integer $\boldsymbol{n}(n_x, n_y, n_z)$.

A key quantity of interest in the semiconductor physics, for example, is the number of quantum states in the interval k and $k + dk$ or equivalently E and $E + dE$ of a freely propagating particle in 1D, 2D, or 3D environment. These quantities can be analyzed by introducing the $\boldsymbol{k}$ space in corresponding dimensions, as shown in Figure 4.4. As is clear from the figure, in the $\boldsymbol{k}$ space scaled with $2\pi/L$, the basic volume elements containing a single dot, that is, one quantum state in 3D, 2D, and 1D cases are given, respectively, by

$$(2\pi/L)^3, (2\pi/L)^2, (2\pi/L)^1. \tag{4.22}$$

Also, it is clear from Figure 4.4 that the respective differential volume elements between k and $k + dk$ are specified by

$$4\pi k^2 dk, 2\pi k dk, 2dk. \tag{4.23}$$

Thus, for a particle in 3D space, the number of quantum states, namely, the number of the lattice points in $\boldsymbol{k}$ space differential volume element, is obtained by dividing the differential volume in (4.23) by the basic unit volume element in (4.22):

$$\frac{4\pi k^2 dk}{(2\pi/L)^3}. \tag{4.24}$$

Now, for each quantum state for given $\boldsymbol{k}$, there are two independent quantum states, corresponding to the spin of the electron, namely, the spin up and spin down states, as

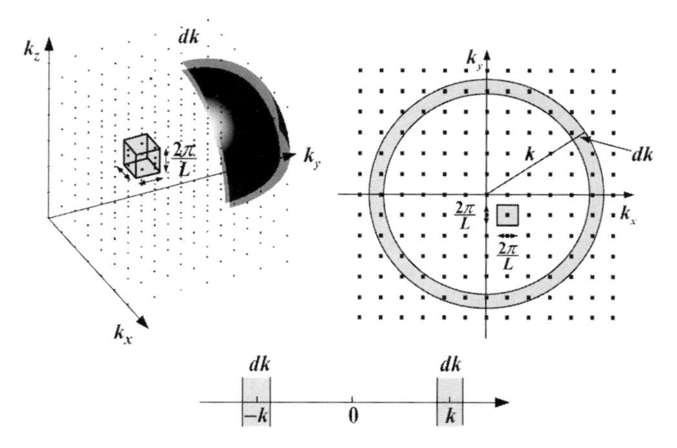

Figure 4.4 The 3D, 2D, and 1D k spaces and the corresponding volume elements between k and $k + dk$. Each dot represents one quantum state.

will be detailed later. Thus, the number of the quantum states between k and $k + dk$ per volume is given by

$$g_{3D}(k)dk = 2\frac{4\pi k^2 dk}{(2\pi/L)^3}\frac{1}{L^3} = \frac{k^2 dk}{\pi^2}. \tag{4.25}$$

The quantity $g_{3D}(k)$ introduced in this manner in (4.25) is called the 3D density of states in k space. One can likewise express the same number of quantum states in energy space by transcribing k into E, using the dispersion relation of the free particle,

$$E = \frac{\hbar^2 k^2}{2m}. \tag{4.26}$$

With the use of (4.26), one can express k and dk in terms of E and dE, obtaining

$$g_{3D}(k)dk \equiv g_{3D}(E)dE = \frac{\sqrt{2}m^{3/2}E^{1/2}}{\pi^2\hbar^3}dE. \tag{4.27}$$

The quantity $g_{3D}(E)$ thus defined in (4.27) is called the density of states in E space and is one of the most extensively used quantities in semiconductor physics and devices. Note that $g_{3D}(E)$ increases with energy following the power law $g(E) \propto E^{1/2}$.

For the case of the 2D motion, one can obtain the number of states between k and $k + dk$ by following exactly the same procedure as in 3D motion. Thus, one can write

$$g_{2D}(k)dk = 2\frac{2\pi k dk}{(2\pi/L)^2}\frac{1}{L^2} = \frac{k dk}{\pi}. \tag{4.28}$$

The only modification required in this case consists of using the 2D differential volume element, together with the basic 2D volume element containing one quantum state (see (4.22) and (4.23)). In terms of E, (4.28) can likewise be expressed as

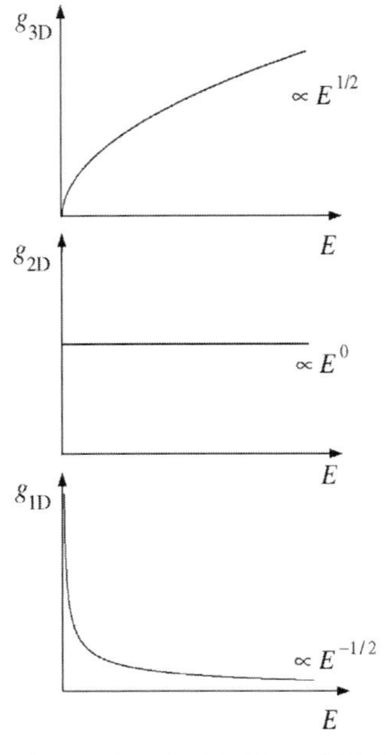

Figure 4.5 The 3D, 2D, and 1D densities of states versus energy.

$$g_{2D}(E)dE = \frac{m}{\pi\hbar^2}dE \tag{4.29}$$

and the 2D density of states, $g_{2D}(E)$, is thus shown to be independent of E.

The 1D density of states can be obtained in a similar way and one can write

$$g_{1D}(k)dk = 2\frac{2dk}{(2\pi/L)}\frac{1}{L} = \frac{2dk}{\pi} \tag{4.30}$$

or in terms of E

$$g_{1D}(E)dE = \frac{\sqrt{2}m^{1/2}}{\pi\hbar}\frac{1}{E^{1/2}}dE. \tag{4.31}$$

The 3D density of states is a key parameter used for modeling the bulk semiconductor devices such as MOSFET. The 2D and 1D densities of states also play the same roles in modeling quantum well and nanowire devices, respectively. Figure 4.5 shows the density of states versus energy in 3D, 2D, and 1D systems, respectively.

4.4
Particle in Quantum Well

A potential well with a finite barrier height is called the quantum well and it has emerged as an important element in semiconductor and optoelectronic device

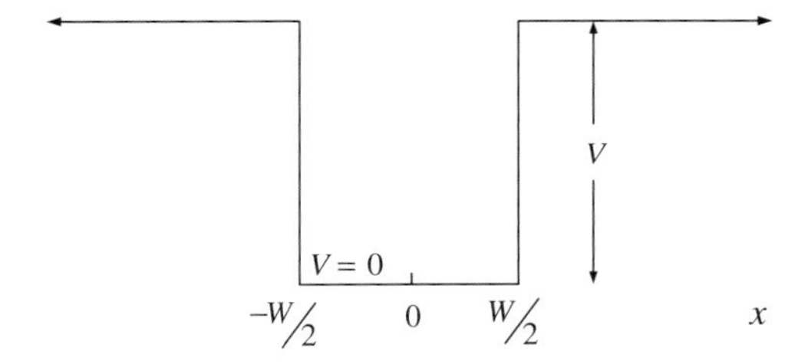

Figure 4.6 A quantum well with depth V and width W.

structures, for example, laser diodes, resonant tunneling devices, and so on. Thus, the analysis of the quantum well has come to carry an additional practical importance. Consider a square potential well of width W and height V and located symmetrically around the origin, as shown in Figure 4.6, namely,

$$V(x) = \begin{cases} 0, & |x| \le W/2, \\ V, & |x| \ge W/2. \end{cases} \tag{4.32}$$

The Hamiltonian of the particle is then given by

$$\hat{H}(x) = -\frac{\hbar^2}{2m}\frac{\partial^2}{\partial x^2} + V(x) \tag{4.33}$$

and the energy eigenequation reads as

$$\hat{H}(x)u(x) = Eu(x). \tag{4.34}$$

The analysis of (4.34) is focused on the bound states, that is, for $E < V$.

To facilitate the analysis, a simple and useful theorem is first considered. If the Hamiltonian is even in x, that is,

$$\hat{H}(x) = \hat{H}(-x), \tag{4.35}$$

the eigenfunction is either even or odd in x. To show this, let $x \to -x$ in (4.34), obtaining

$$\hat{H}(-x)u(-x) = \hat{H}(x)u(-x) = Eu(-x), \tag{4.36}$$

where (4.35) has been used. Clearly, (4.36) is identical to the original eigenequation (4.34). Therefore, both $u(x)$ and $u(-x)$ must be the same eigenfunction within a constant K, namely,

$$u(-x) \propto u(x) = Ku(x). \tag{4.37}$$

Next, to determine the proportionality constant, K introduce the inversion operator $\hat{P}$ and apply it to $u(x)$, obtaining

$$\hat{P}u(x) \equiv u(-x) = Ku(x), \tag{4.38}$$

where (4.37) has been used. When the inversion operator is again applied to (4.38), there results

$$\hat{P}\hat{P}u(x) = K\hat{P}u(x) = K^2u(x), \tag{4.39a}$$

while by definition,

$$\hat{P}\hat{P}u(x) = \hat{P}u(-x) = u(x). \tag{4.39b}$$

It is therefore clear from (4.39a) and (4.39b) that $K^2 = 1$, that is, $K = \pm 1$, and this result when combined with (4.37) shows unambiguously that $u(x)$ is either even or odd in x.

Now, since the Hamiltonian of the quantum well given in (4.33) is even in x, the eigenfunction $u(x)$ should therefore be either even or odd in x. Also, for the bound state, for which $E < V$, the eigenequation (4.34) is divided into two regions, namely, inside and outside the well:

$$u'' + k^2 u = 0, \quad k^2 \equiv 2mE/\hbar^2, \ |x| \leq W/2, \tag{4.40a}$$

$$u'' - \varkappa^2 u = 0, \quad \varkappa^2 \equiv 2m(V - E)/\hbar^2, \ |x| > W/2. \tag{4.40b}$$

It is evident from (4.40) that $u(x)$ is sinusoidal inside the well, while it varies exponentially outside the well, namely, $u(x) \sim \exp \pm \varkappa x$. One can therefore combine these solutions into even and odd parity eigenfunctions, namely,

$$u_e(x) = N \begin{cases} Ae^{\varkappa x}, & x < -W/2, \\ \cos kx, & |x| \leq W/2, \\ Ae^{-\varkappa x}, & x > W/2, \end{cases} \tag{4.41}$$

$$u_o(x) = N \begin{cases} -Ae^{\varkappa x}, & x < -W/2, \\ \sin kx, & |x| \leq W/2, \\ Ae^{-\varkappa x}, & x > W/2. \end{cases} \tag{4.42}$$

Here, only those exponential functions are chosen, so that $u(x)$ vanishes at infinity. Also, the constants of integration, A and N, are introduced to meet the necessary boundary conditions and to normalize $u(x)$, if necessary.

Boundary Conditions There are two boundary conditions, namely, that both $u(x)$ and its derivative $\partial u(x)/\partial x$ be continuous everywhere. The first requirement is needed since $u^*(x)u(x)$ representing the probability density should vary continuously versus x. The second condition is required since the first derivative is commensurate with the momentum or the velocity of the particle and should also be continuous everywhere.

It is clear from (4.41) and (4.42) that these two boundary conditions are automatically met within the quantum well and outside the well where $u_e(x)$ and $u_o(x)$ are described by analytical functions. This leaves only the two edges of the well on which the conditions should be imposed. But because of the definite parities the eigenfunctions possess, the boundary conditions need to be imposed only on one of these two edges, say, at $x = W/2$. For the case of $u_e(x)$, the two boundary conditions are specified by

$$\cos \xi = Ae^{-\eta}, \quad \xi \equiv \frac{kW}{2}, \ \eta \equiv \frac{\varkappa W}{2}, \tag{4.43}$$

$$-k \sin \xi = -\varkappa A e^{-\eta}. \tag{4.44}$$

Multiplying both sides of (4.44) by $W/2$ and dividing it with (4.43), the two conditions are compacted into one:

$$\xi \tan \xi = \eta. \tag{4.45}$$

One can likewise specify the boundary conditions for $u_o(x)$ at $x = W/2$:

$$\sin \xi = A e^{-\eta}, \quad \xi \equiv \frac{kW}{2}, \; \eta \equiv \frac{\varkappa W}{2}, \tag{4.46}$$

$$k \cos \xi = -\varkappa A e^{-\eta}. \tag{4.47}$$

These two conditions can again be compacted to yield

$$-\xi \cot \xi = \eta. \tag{4.48}$$

Also, the dimensionless parameters ξ and η are related with each other via (4.40) as

$$\xi^2 + \eta^2 \equiv \left(\frac{kW}{2}\right)^2 + \left(\frac{\varkappa W}{2}\right)^2 = \frac{mVW^2}{2\hbar^2}. \tag{4.49}$$

The problem has thus been reduced to determining k and $\varkappa$ that are embedded in $u_e(x)$ and $u_o(x)$. Once k and $\varkappa$ are determined, $u_e(x)$ and $u_o(x)$ can satisfy the required boundary conditions, and at the same time the energy eigenvalue follows explicitly and naturally from (4.40). Now, the values of ξ and η satisfying (4.45) and (4.49) for the case of $u_e(x)$ and (4.48) and (4.49) for the case of $u_o(x)$, respectively, can be found by a simple numerical or graphical means. Here, the latter approach is chosen to illustrate the general features of the bound state in quantum wells.

Figure 4.7 plots η as a function of ξ as specified by (4.45) and (4.48). Also plotted in the figure are the circular trajectories (4.49) for different radii, corresponding to

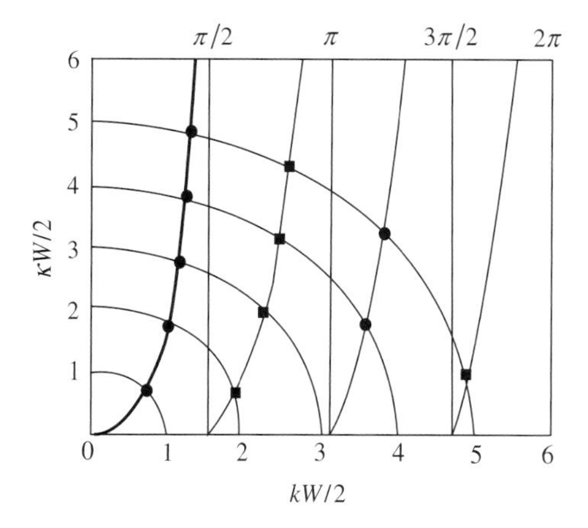

Figure 4.7 The graphical analysis of $E-k$ dispersion, occurring in quantum well. Plotted in the figure are $\eta = \xi \tan \xi$ (thicker lines), $\eta = -\xi \cot \xi$ (thinner lines), and family of circles for $\xi^2 + \eta^2 = mVW^2/2\hbar^2$. The intersection points between two curves involved (filled circles for $u_e(x)$ and filled squares for $u_o(x)$) determine the bound-state energy levels.

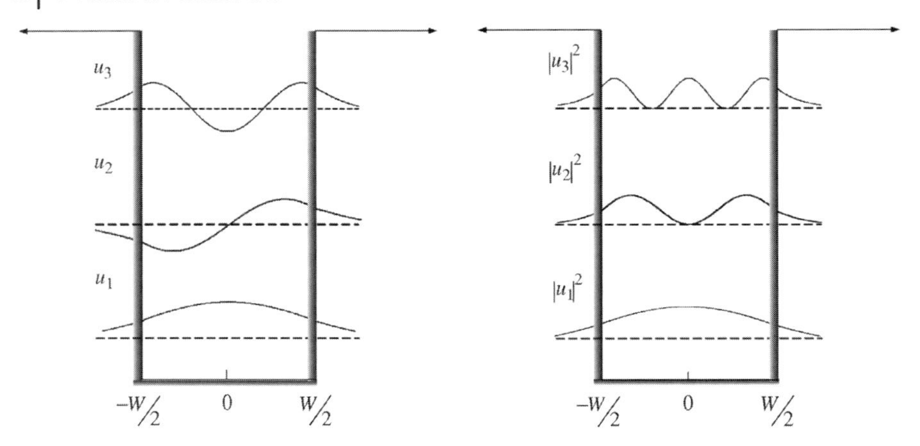

Figure 4.8 Typical even and odd parity wavefunctions and corresponding probability densities in quantum well. These wavefunctions penetrate into the classically forbidden region.

various potential depths V or the width W of the quantum well. Thus, finding the values of ξ and η or equivalently k and $\varkappa$ has been reduced to finding the coordinates of the crossing points of the two curves (4.45) and (4.49) for $u_e(x)$ and (4.48) and (4.49) for $u_o(x)$, respectively.

A few typical eigenfunctions and eigenvalues found in this way are shown in Figure 4.8, together with the corresponding probability densities. The general features of the bound states in the quantum well naturally emerge from these graphical solutions. When the radius of the circular trajectory is increased or equivalently when the potential depth V of the quantum well is increased, there are more crossing points on a given circular curve, that is, more bound states exist in the well. Also, the lowest ground state is always associated with $u_e(x)$ and higher lying excited states alternate between even and odd parity quantum states.

Moreover, no matter how shallow the quantum well is, there always exists at least one bound state. In the other limit of the infinite well depth, it is clear from Figure 4.7 that the intersection points should now occur at

$$\xi_n \equiv \frac{k_n W}{2} = \frac{\pi}{2}(2n+1), \quad n = 0, 1, 2, \ldots, \tag{4.50a}$$

for $u_e(x)$, while for $u_o(x)$,

$$\xi_n \equiv \frac{k_n W}{2} = n\pi, \quad n = 1, 2, \ldots, \tag{4.50b}$$

so that these two conditions (4.50a) and (4.50b) when combined with (4.40) yield

$$E_n = \frac{\hbar^2 \pi^2}{2mW^2} n^2, \quad n = 1, 2, 3, \ldots, \tag{4.51}$$

in complete agreement with the result of the infinite potential well, as it should (see (4.9) or (4.11b)).

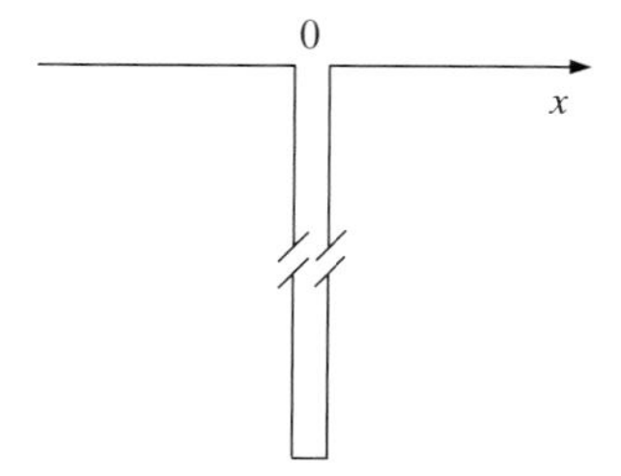

Figure 4.9 The single delta function potential well at $x = 0$.

4.5
Particle in Delta Function Potential Well

A particle in a δ-function potential well has been analyzed primarily for modeling molecules. However, with the rapid advances in nanotechnology, it has gained added significance. Specifically, the δ-function potential well is often used for representing a quantum dot, which is fast becoming an important element in optoelectronic, nanoelectronic, and quantum computing devices. As schematically shown in Figure 4.9, the delta function potential well

$$V(x) = -\frac{\hbar^2\lambda}{2ma}\delta(x) \tag{4.52}$$

is specified by the mass of the particle, m, confined therein and two parameters, a carrying the dimension of the length and λ characterizing the depth.

As well known, the delta function $\delta(x)$ is nonanalytical and is zero everywhere except at $x = 0$, where it goes to infinity in such a manner that the x-integration yields unity, namely,

$$\int_{-\infty}^{\infty} dx\delta(x) = 1. \tag{4.53}$$

As a consequence, given a function $f(x)$, one can write

$$\int_{-\infty}^{\infty} dxf(x)\delta(x) = f(0). \tag{4.54}$$

In addition, (4.53) implies that $\delta(x)$ has the dimension of the inverse length, so that $V(x)$ as represented by (4.52) has the dimension of energy, as it should.

Single Delta Function Given the single delta function potential well, (4.52), the energy eigenequation for bound state reads as

$$-\frac{\hbar^2}{2m}\frac{\partial^2}{\partial x^2}u(x) - \frac{\hbar^2\lambda}{2ma}\delta(x)u(x) = -|E|u(x), \tag{4.55}$$

where $E < 0$. Because of the nonanalytic behavior of $\delta(x)$, the bound-state wavefunction should satisfy the boundary conditions that are slightly different from the usual boundary conditions used thus far. To specify this condition, integrate (4.55) over an infinitesimal interval $-\varepsilon \leq x \leq \varepsilon$ across the potential well, obtaining

$$u'(x \to 0^+) - u'(x \to 0^-) = -\frac{\lambda}{a} u(0). \tag{4.56}$$

In (4.56), use has been made of (4.54) and also the fact that $|E|u(0)2\varepsilon \to 0$ for $\varepsilon \to 0$.

Since $V(x) = 0$ everywhere except at the origin, the energy eigenequation outside the potential well is given from (4.55) by

$$u(x)'' - \varkappa^2 u(x) = 0, \quad \varkappa^2 \equiv \frac{2m|E|}{\hbar^2}, \tag{4.57}$$

and the eigenfunction is readily found as

$$u(x) = A \begin{cases} e^{\varkappa x}, & x < 0, \\ e^{-\varkappa x}, & x > 0. \end{cases} \tag{4.58}$$

Here, the two exponential functions, $\exp \pm \varkappa x$, as the solution of the eigenequation have been assigned to the two regions in such a way that $u(x)$ vanishes at $x \to \pm\infty$, respectively.

Inserting (4.58) into the boundary condition (4.56) and using the definition of $\varkappa$ given in (4.57) results in

$$2\varkappa \equiv 2\sqrt{\frac{2m|E|}{\hbar^2}} = \frac{\lambda}{a}. \tag{4.59}$$

Therefore, the energy eigenvalue of the bound state is obtained from (4.59), namely,

$$|E| = \frac{\hbar^2 \lambda^2}{8ma^2}. \tag{4.60}$$

It is interesting to notice that there exists only one bound state in the single δ-function potential well, and this bound state is associated with the symmetric wavefunction, as detailed in Section 4.4.

Double Delta Function Next, consider a particle bound by and oscillating between the two δ-function potential wells, which are separated by $2a$ and centered around the origin in symmetric fashion (Figure 4.10). The eigenequation is still given by (4.57) everywhere, except at $x = \pm a$ where the delta function potential wells are located. With $V(x)$ set even in x, the Hamiltonian is centrosymmetric and the eigenfunctions should therefore be either even or odd. For the even parity eigenfunction, one can write

$$u_e(x) = \begin{cases} e^{\varkappa x}, & x < -a, \\ A \cosh \varkappa x, & |x| \leq a, \\ e^{-\varkappa x}, & x > a. \end{cases} \tag{4.61}$$

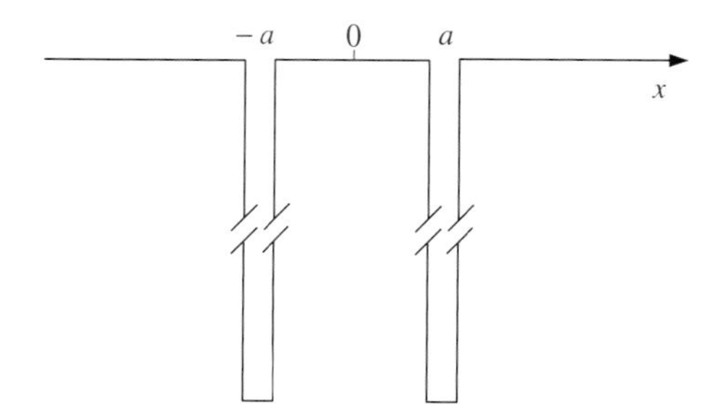

Figure 4.10 The double delta function potential well separated by $2a$ and centered at $x = 0$.

Here the exponential solutions $\exp \pm \varkappa x$ have been symmetrically combined with the proviso that $u_e(x)$ vanishes at $x \to \pm\infty$, respectively.

Again, $u_e(x)$ has to be continuous everywhere, in particular at $x = \pm a$, and has also to satisfy the boundary condition (4.56) with x shifted from the origin to $x = \pm a$. However, as already discussed, once the boundary conditions are met at $x = a$, the conditions are automatically satisfied at $x = -a$ due to $u_e(x)$ possessing definite parity. With this fact in mind, one can first specify the condition, namely, that $u_e(x)$ is continuous at a, by

$$e^{-\varkappa a} = A \cosh \varkappa a. \tag{4.62}$$

Also, the boundary condition (4.56) when shifted to $x = a$ yields

$$u'(x \to a^+) - u'(x \to a^-) = -\frac{\lambda}{a} u(a). \tag{4.63}$$

Thus, inserting (4.61) into (4.63) results in

$$-\varkappa e^{-\varkappa a} - \varkappa A \sinh \varkappa a = -\frac{\lambda}{a} e^{-\varkappa a}. \tag{4.64}$$

The two boundary conditions (4.62) and (4.64) are combined into one by eliminating A from both equations, namely,

$$\tanh \varkappa a = \frac{\lambda}{\varkappa a} - 1. \tag{4.65}$$

The bound-state energy can therefore be obtained from this transcendental dispersion relation. The transcendental relation (4.65) can again be analyzed by graphical means, as illustrated in Figure 4.11. In this figure are plotted both sides of (4.65) versus $\varkappa a$. As usual, the points of intersection of these two curves determine the value of $\varkappa$, hence E. It is clear from the figure that there exists only one intersection point, that is, only one bound state for $u_e(x)$. The magnitude of the associated binding energy can be examined as follows. Since $\tanh x \leq 1$, the real value of $\varkappa$ can be found from (4.65) only for $\varkappa$ values such that $\lambda/\varkappa a \leq 2$. That is,

$$\sqrt{\frac{2m|E|}{\hbar^2}} \equiv \varkappa \geq \frac{\lambda}{2a} \tag{4.66}$$

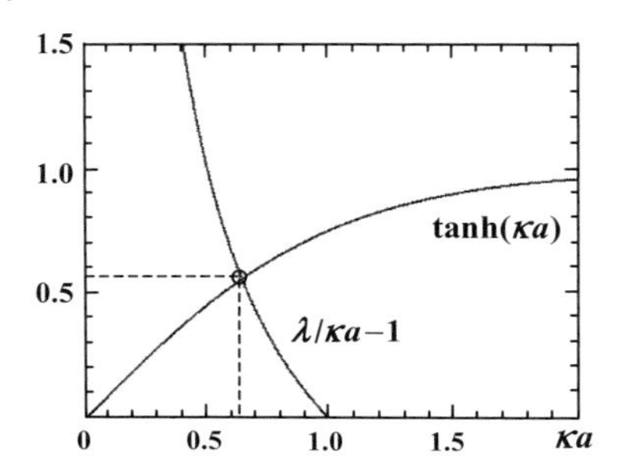

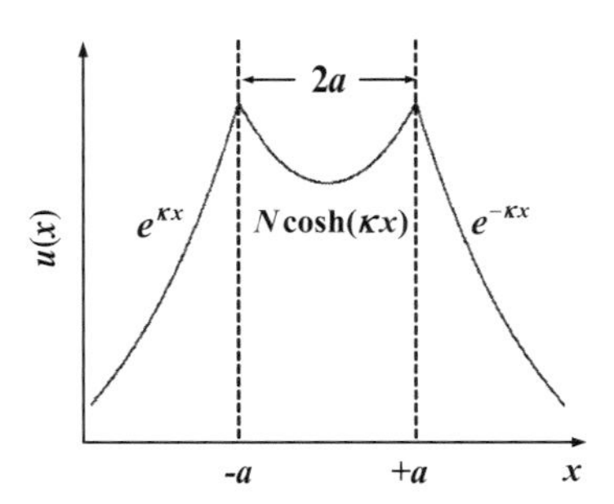

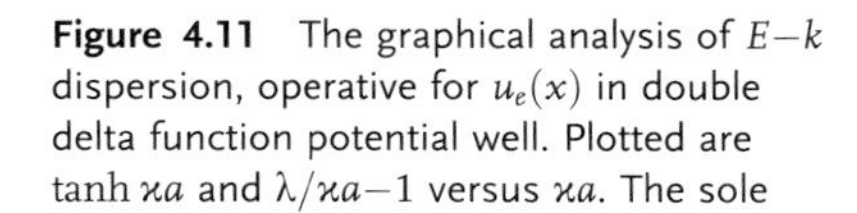

Figure 4.11 The graphical analysis of $E-k$ dispersion, operative for $u_e(x)$ in double delta function potential well. Plotted are $\tanh \varkappa a$ and $\lambda/\varkappa a-1$ versus $\varkappa a$. The sole intersection point determines the single energy eigenvalue of $u_e(x)$. The corresponding spatially symmetric wavefunction is also plotted.

so that

$$|E| \geq \frac{\hbar^2 \lambda^2}{8ma^2}. \tag{4.67}$$

When (4.67) is compared with (4.60), it is clear that the bound state corresponding to the symmetric wavefunction in double delta function potential well is more tightly bound than the bound state in the single delta function potential well. The associated energy eigenfunction $u_e(x)$ is also shown in Figure 4.11.

The energy eigenfunction possessing the odd parity can likewise be analyzed. The eigenfunction is given in this case by

$$u_o(x) = \begin{cases} -e^{\varkappa x}, & x < -a, \\ A \sinh \varkappa x, & |x| \leq a, \\ e^{-\varkappa x}, & x > a. \end{cases} \tag{4.68}$$

The boundary conditions respectively corresponding to (4.62) and (4.64) read as

$$e^{-\varkappa a} = A \sinh \varkappa a, \tag{4.69}$$

$$-\varkappa e^{-\varkappa a} - \varkappa A \cosh \varkappa a = -\frac{\lambda}{a} e^{-\varkappa a}. \tag{4.70}$$

Again, eliminating A in (4.69) and (4.70) results in

$$\tanh \varkappa a = \frac{1}{\frac{\lambda}{\varkappa a} - 1}. \tag{4.71}$$

In Figure 4.12 are plotted both sides of (4.71) for different values of λ and clearly there is only one point of intersection between these two curves, that is, there exists only one bound state for $u_o(x)$, provided $\lambda > 1$. Also, since $\tanh x \leq 1$, a real value of $\varkappa$ can be found from (4.71), provided $\lambda/\varkappa a \geq 2$. That is,

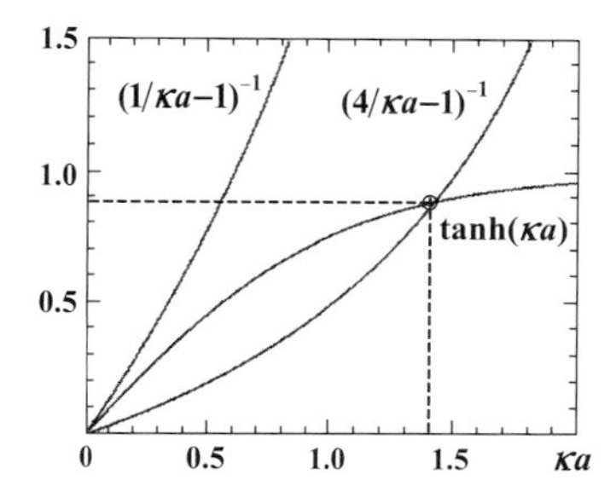

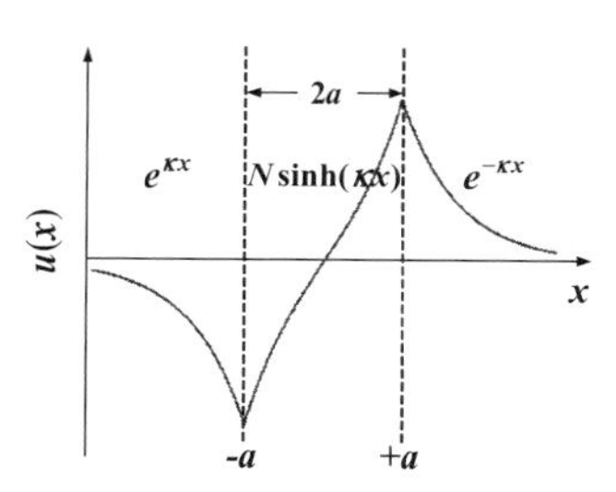

Figure 4.12 The graphical analysis of $E-k$ dispersion, operative for $u_o(x)$ in the double delta function potential well. Plotted are $\tanh \varkappa a$ and $(\lambda/\varkappa a-1)^{-1}$ versus $\varkappa a$. The sole intersection point occurring only for $\lambda > 1$ determines the single energy eigenvalue of $u_o(x)$. The corresponding spatially antisymmetric wavefunction is also plotted.

$$\sqrt{\frac{2m|E|}{\hbar^2}} \equiv \varkappa \leq \frac{\lambda}{2a} \tag{4.72}$$

so that

$$|E| \leq \frac{\hbar^2\lambda^2}{8ma^2}. \tag{4.73}$$

It is therefore clear from (4.73) that the bound state associated with $u_o(x)$ is less tightly bound than the bound state in the single potential well, if it exists at all. Figure 4.12 also shows the odd parity energy eigenfunction in the double δ-function potential well. The symmetric and antisymmetric wavefunctions play an important role in discussing molecules and quantum dots. It is interesting to notice that for the case of u_e, the probability of the bound particle being located in between the two potential wells is high, while for u_o, it is practically zero in the same spatial region.

4.6 Quantum Well and Wire

The bound states in simple potential wells are interesting and instructive by themselves but these states have come to carry practical importance in nanoscience and technology. An essential part of this branch of science is to design and implement novel nanostructures. For instance, it has become a routine procedure to grow atomic layers in controlled manner and to realize the desired bandgap engineering. This has been made possible with the development of molecular beam epitaxy (MBE) or metal organic chemical vapor deposition (MOCVD) techniques. By using such deposition techniques, it is possible to fabricate desired quantum wells, as shown in Figure 4.13.

As pointed out earlier, an electron in a semiconducting material behaves as a free particle in certain energy ranges, called the conduction and valence bands. These bands are separated by the bandgap and the electrons are forbidden to reside and

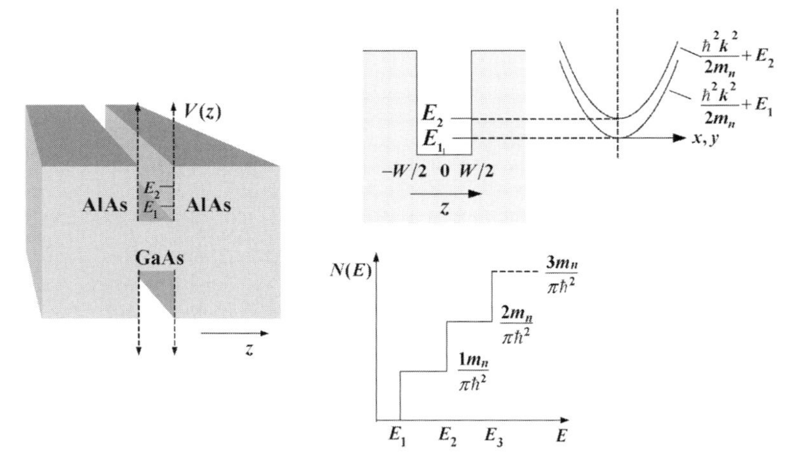

Figure 4.13 A quantum well, the subbands, dispersion curves of 2D electrons, and density of states versus E.

propagate in such forbidden energy gaps. With the use of epitaxial deposition techniques, the atomic layers of varying thicknesses and bandgaps can be grown and interfaced, thereby forming a quantum well for the electrons to reside in (Figure 4.13). By the same token, the mirror image of the quantum well formed in the valence band is capable of confining holes therein. These energy bands will be discussed subsequently.

Quantum Well Consider electrons confined in a quantum well. These electrons are confined in one direction, say in the z-direction, while propagating freely in x and y directions, thereby forming the 2D electron gas. The potential energy is then given simply by

$$V(x, y, z) = \begin{cases} 0, & |z| \leq W/2, \\ V, & |z| \geq W/2. \end{cases} \tag{4.74}$$

The 2D electron gas system can be readily analyzed, based on the results obtained from simple potential wells considered in this chapter. The energy eigenequation of the 2D electron gas system is given by

$$\left[-\frac{\hbar^2}{2m_x}\frac{\partial^2}{\partial x^2} - \frac{\hbar^2}{2m_y}\frac{\partial^2}{\partial y^2} - \frac{\hbar^2}{2m_z}\frac{\partial^2}{\partial z^2} + V(x, y, z)\right] u(x, y, z) = Eu(x, y, z). \tag{4.75}$$

Naturally, Equation 4.74 can be reduced to three 1D equations in a simple manner by putting

$$u(x, y, z) = u(x)u(y)u(z) \tag{4.76}$$

and performing the usual procedure for separating the variables. In x and y directions, one obtains the 1D free particle eigenequations, while in the z-direction, there ensues the energy eigenequation in a quantum well, as discussed in Section 4.4. Hence, the total electron energy consists of the respective kinetic energies in x and y directions, and the quantized energy formed along the z-direction, namely,

$$E_n = \frac{\hbar^2 k_x^2}{2m_x} + \frac{\hbar^2 k_y^2}{2m_y} + E_{zn}. \tag{4.77}$$

In (4.77), the effective mass of electron in each direction can differ from each other, as will become clear subsequently. Figure 4.13 also shows the energy level of the 2D electron gas system, together with the density of states. The discrete energy levels arising from the spatial confinement in the z-direction are called the subbands. Since the 2D density of states is constant and is independent of energy, as discussed in Section 4.3, the number of the quantum states increases stepwise whenever E crosses those discrete subband energy levels. Also, the subband states with E_{zn} levels are inseparably associated with quantum states representing two degrees of freedom with which electrons move around freely.

Quantum Wire A particle in a quantum well enjoys two degrees of freedom. In quantum wire structure, however, as sketched in Figure 4.14, the degree of freedom is reduced to one. The energy spectrum then consists of two sets of discrete energy levels corresponding to the spatial confinements in, say, y and z directions and the free propagation in x-direction. The total energy is, therefore, given by

$$E_{n,m} = \frac{\hbar^2 k_x^2}{2m_x} + E_{ym} + E_{zn}, \tag{4.78}$$

where the integers m and n denote the discrete energy levels. Figure 4.14 shows the quantum wire energy level, together with the corresponding density of states. Since

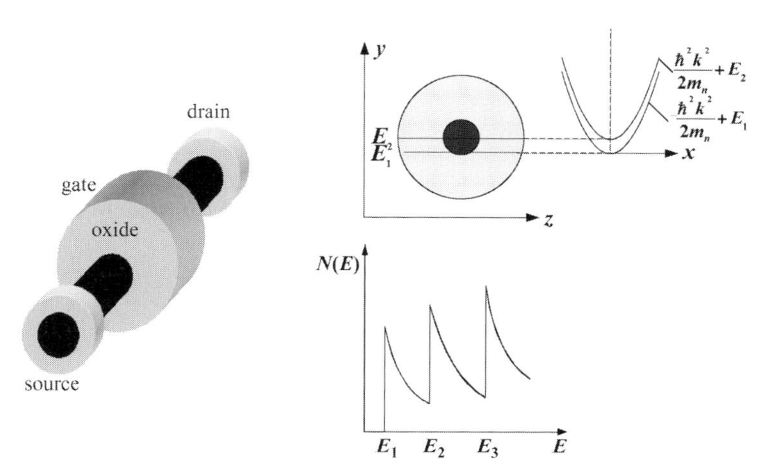

Figure 4.14 The quantum wire constituting the channel in field effect transistor, for example the subbands, dispersion curves of 1D electrons, and density of states versus E.

the 1D density of states varies according to the power law, $E^{-1/2}$, the total number of states exhibits a sawtooth-like characteristics versus E.

As well known, the quantum wells have become an integral part of semiconductor devices. For example, in high-efficiency laser diodes, electrons and holes are injected into quantum wells, formed in the junction region, and are allowed to have longer interaction time, while confined in the well for radiative recombinations. In addition, the metal oxide semiconductor field effect transistors (MOSFET) fabricated in bulk silicon operates, based entirely upon inducing 2D electrons or holes in the channel and transporting them in a controlled manner to the drain for generating current. Moreover, one possible and promising way to scale the bulk FETs to nanoscale regime is to utilize the 1D electrons or holes in a quantum wire. The quantum mechanical principles underpinning such devices will be discussed subsequently.

4.7 Problems

4.1. An electron is contained in one-dimensional box of cross-sectional area $1\text{nm} \times 1\text{nm}$.

(a) Find the lowest four energy levels and calculate the wavelengths of photons that will be emitted when the electron cascades down the energy levels from top to bottom.

(b) Repeat the calculation for the quantum well with width 1 nm and compare the two results.

4.2. The three-dimensional density of states has been found in the text by using the periodic boundary condition in the cubic box of volume $V = L^3$. Derive the 3D density of states in the same box by using instead the stationary boundary conditions, in which the wavefunctions vanish at the edges of the box. Show that both results are the same.

4.3. A particle of mass m is confined in one-dimensional box of cross-sectional lengths L_x and L_y, respectively.

(a) Find its eigenfunctions and eigenenergies.

(b) If $L_x = L_y$, some of the eigenenergies become degenerate. Discuss the degeneracy of the eigenfunctions.

4.4. An electron is confined in a potential well given by

$$V(x) = \begin{cases} \infty, & x < \dfrac{-d}{2}, \\ -V, & \dfrac{-d}{2} < x < \dfrac{d}{2}, \\ 0, & x > \dfrac{d}{2}. \end{cases}$$

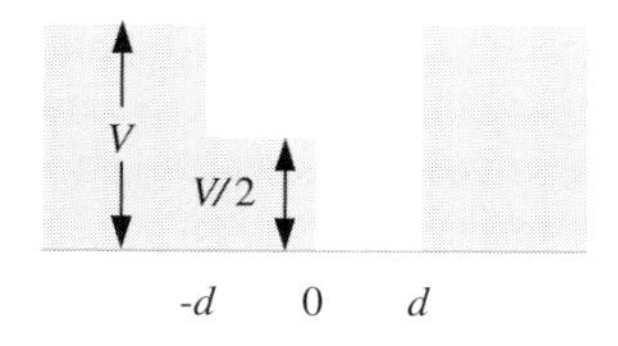

Figure 4.15 Quantum well formed by two potential barriers of different shape.

(a) Calculate the bound-state energy in eV unit by taking $V = 5\,\text{eV}$ and $d = 2$ nm and the effective mass of electron $m_n = 0.1\,m_0$ with m_0 denoting the rest mass.
(b) Compare the result with those corresponding to infinite potential well of the same width.

4.5. A quantum well is formed by a potential given by

$$V(z) = \begin{cases} 3.1\,\text{eV} & z \leq -W/2, z \geq W/2, \\ 0\,\text{eV} & -W/2 \leq z \leq W/2, \end{cases}$$

with the well width $W = 1.5$ nm.

(a) Find the energy eigenfunctions and eigenvalues of the two-dimensional electrons, moving freely on x, y plane with an effective mass $m_n = 0.1\,m_0$ with m_0 denoting the rest mass.
(b) Plot the density of states in the well.
(c) Write a short program enabling the analysis of bound states for varying well width W and depth V. Plot the ground and first excited sublevels versus W for $V = 3.1$ eV.
(d) Find W at which the ground-state sublevel is comparable to the thermal energy k_BT at room temperature.

4.6. Consider a quantum well given by

$$V(x) = \begin{cases} V & \text{for } x < -d, \\ V/2 & \text{for } -d < x < 0, \\ 0 & \text{for } 0 < x < d, \\ V & \text{for } x > d, \end{cases}$$

as sketched in Figure 4.15.

(a) Write down the energy eigenfunctions in each regions for $E > V$, $V > E > V/2$, and $E < V/2$, respectively.
(b) Write down the boundary conditions of these eigenfunctions.
(c) Find the eigenenergies in the ranges $V > E > V/2$ and $E < V/2$.

Suggested Reading

1 Singh, J. (1996) *Quantum Mechanics, Fundamentals & Applications to Technology*, John Wiley & Sons, Inc.

2 Yariv, A. (1982) *An Introduction to Theory and Applications of Quantum Mechanics*, John Wiley & Sons, Inc.

3 Liboff, R.L. (2002) *Introductory Quantum Mechanics*, 4th edn, Addison-Wesley Publishing Company, Reading, MA.

4 Gasiorowicz, S. (2003) *Quantum Physics*, 3rd edn, John Wiley & Sons, Inc.

5 Robinett, R.W. (2006) *Quantum Mechanics, Classical Results, Modern Systems and Visualized Examples*, Oxford University Press.

6 Kroemer, H. (1994) *Quantum Mechanics for Engineering, Materials Science, and Applied Physics*, International edn, Prentice Hall.

5
Scattering of a Particle at 1D Potentials

The scattering constitutes a key process from which to extract important physical information and, therefore, a great deal of attention has been devoted to its analysis. In this chapter, the discussion is mainly confined to the scattering of 1D particle from 1D potential well or barrier. The emphasis is placed on discussing the probability current density and the reflection and transmission coefficients. The basic difference between the classical and quantum mechanical descriptions of the scattering processes is pointed out and the analogy existing between the reflection and transmission of the light and those of particles is highlighted. The physical significance of a particle penetrating into the classically forbidden region is stressed, together with the resonant transmission across the quantum well to pave the way for discussing the tunneling and the energy bands in solids.

5.1
Scattering at Step Potential

The step potential is the simplest potential from which the essential features of scattering processes can be brought out. The step potential, as shown in Figure 5.1, is an idealized version of a barrier potential. Classically, a particle incident on a potential barrier either flies over it energetically or is bounced back. More specifically, when the kinetic energy E of the particle is greater than the barrier height V, it slows down upon hitting the barrier to conserve the energy but it is transmitted over the barrier with 100% certainty. When the incident energy E is less than the barrier V, it comes to a complete rest at the barrier and is bounced back again with 100% certainty. Quantum mechanically, however, the particle can be both transmitted across or reflected from the barrier with certain probabilities, depending on E and V.

Thus, consider a particle incident on the step potential with energy greater than the potential barrier, $E > V$ (Figure 5.1). The step potential is given by

$$V(x) = \begin{cases} 0, & x \leq 0, \\ V, & x > 0, \end{cases} \tag{5.1}$$

Introductory Quantum Mechanics for Semiconductor Nanotechnology. Dae Mann Kim

ISBN: 978-3-527-40975-4

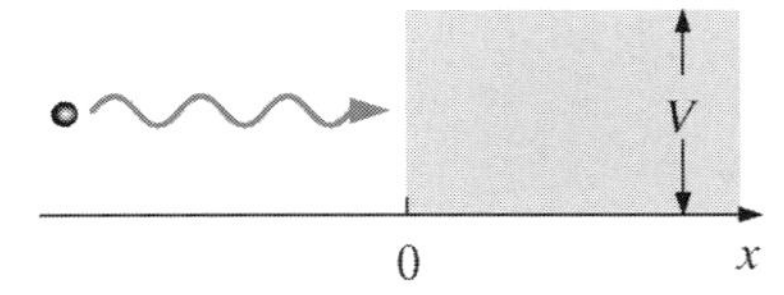

Figure 5.1 A particle incident on a step potential barrier with height V.

and the energy eigenequation of the particle thus reads as

$$u(x)'' + \alpha^2 u(x) = 0, \tag{5.2}$$

where the wave vector is specified by

$$\alpha^2 = \begin{cases} k_0^2, \quad k_0^2 = \dfrac{2mE}{\hbar^2}, & x \leq 0, \\ k^2 \quad k^2 = \dfrac{2m(E-V)}{\hbar^2}, & x > 0. \end{cases} \tag{5.3}$$

As discussed, the eigenfunction in this case is given by the exponential functions $\exp \pm i\alpha x$ and when these functions are incorporated into the total wavefunction (see (4.4)),

$$\psi(x,t) \propto e^{-i(\omega t \mp \alpha x)}, \quad \omega = E/\hbar, \tag{5.4}$$

the eigenfunctions describe a particle propagating in the positive or negative x directions, respectively. With this fact in mind, one can write the solution for the eigenfunctions as

$$u(x) = \begin{cases} i_0 e^{ik_0 x} + re^{-ik_0 x}, & x \leq 0, \\ te^{ikx}, & x > 0. \end{cases} \tag{5.5}$$

Here the forward and backward components are retained in the region where $x \leq 0$, in which the incident as well as reflected beams have to be taken into account. In the region $x > 0$, only the forward component is needed, since once the particle overcomes the barrier and is transmitted, there is no potential to reflect it back. Thus, the i_0 term represents the incident beam, while r and t terms denote the reflected and transmitted components, respectively. The constants of integration, r and t, are to be determined from the boundary conditions.

As discussed, the boundary conditions are (i) $u(x)$ and (ii) its derivative $u'(x)$ should be continuous at $x = 0$ where $V(x)$ undergoes a sudden change from zero to V. These requirements when imposed on (5.5) lead to

$$i_0 + r = t, \tag{5.6}$$

$$k_0(i_0 - r) = kt. \tag{5.7}$$

Given these two constraints, r and t can be found in terms of i_0 as

$$\frac{r}{i_0} = \frac{k_0 - k}{k_0 + k}, \tag{5.8a}$$

$$\frac{t}{i_0} = \frac{2k_0}{k_0 + k}. \tag{5.8b}$$

5.1.1
The Probability Current Density

The physical significance of (5.8) can be explicitly brought out by introducing the probability current density, which naturally enters into the consideration of the time rate of change of the probability density, namely,

$$\frac{\partial}{\partial t}\psi^*\psi = \left(\frac{\partial}{\partial t}\psi^*\right)\psi + \psi^*\left(\frac{\partial}{\partial r}\psi\right). \tag{5.9}$$

Now, (5.9) can be put into a more transparent form by using the Schrödinger equation,

$$i\hbar\frac{\partial\psi(\boldsymbol{r},t)}{\partial t} = -\frac{\hbar^2}{2m}\nabla^2\psi(\boldsymbol{r},t) + V(\boldsymbol{r})\psi(\boldsymbol{r},t). \tag{5.10}$$

Inserting (5.10) and its complex conjugate into (5.9) and using the well-known vector identity

$$\psi^*\nabla^2\psi - \psi\nabla^2\psi^* \equiv \nabla\cdot(\psi^*\nabla\psi - \psi\nabla\psi^*) \tag{5.11}$$

results in

$$\frac{\partial}{\partial t}\psi^*\psi = -\nabla\cdot\boldsymbol{S}, \tag{5.12}$$

with

$$\boldsymbol{S} \equiv \frac{\hbar}{2mi}(\psi^*\nabla\psi - \psi\nabla\psi^*) \tag{5.13}$$

representing the probability current density. Evidently, (5.12) is analogous to the charge conservation equation and represents the conservation of the probability density.

5.1.2
Reflection and Transmission Coefficients

The time components of the wavefunction and its complex conjugate appearing in the probability current density (5.13) cancel each other out, so that S can also be expressed in terms of the energy eigenfunction as

$$S \equiv \frac{\hbar}{2mi}\left(u^*(x)\frac{\partial}{\partial x}u(x) - \text{c.c.}\right), \tag{5.14}$$

where c.c. denotes the complex conjugate.

Next, the probability current density is applied to the energy eigenfunctions that have been found for the step potential. Thus, inserting (5.5) into (5.14), one finds the probability current densities in the two regions:

$$S_I(x) = \frac{\hbar k_0}{m}|i_0|^2 - \frac{\hbar k_0}{m}|r|^2, \quad x \leq 0, \tag{5.15a}$$

and

$$S_{II}(x) = \frac{\hbar k}{m}|t|^2, \quad x > 0. \tag{5.15b}$$

Evidently, the first term in S_I represents the incident flux, while the second term accounts for the reflected flux. By the same token, S_{II} describes the transmitted flux. It is important to note that each flux is composed of the probability density of the eigenfunction, namely, $|i_0|^2$, $|r|^2$, and $|t|^2$, and the velocity with which it is propagating, $\hbar k_0/m$ and $\hbar k/m$, as it should.

Thus, one can quantify the reflection and transmission coefficients by combining (5.8) and (5.15):

$$R \equiv \frac{(\hbar k_0/m)|r|^2}{(\hbar k_0/m)|i_0|^2} = \frac{(k_0 - k)^2}{(k_0 + k)^2} \tag{5.16}$$

and

$$T \equiv \frac{(\hbar k/m)|t|^2}{(\hbar k_0/m)|i_0|^2} = \frac{4kk_0}{(k_0 + k)^2}. \tag{5.17}$$

It is interesting that both R and T are generally nonzero, as clearly shown in Figure 5.2. This is in contrast with the classical description, in which $R = 0$ and $T = 1$ for $E > V$. Only in the limit $E \gg V$, $k_0 \approx k$ (see (5.3)) and $R \rightarrow 0$, while $T \rightarrow 1$. The difference between the quantum and classical descriptions of the scattering

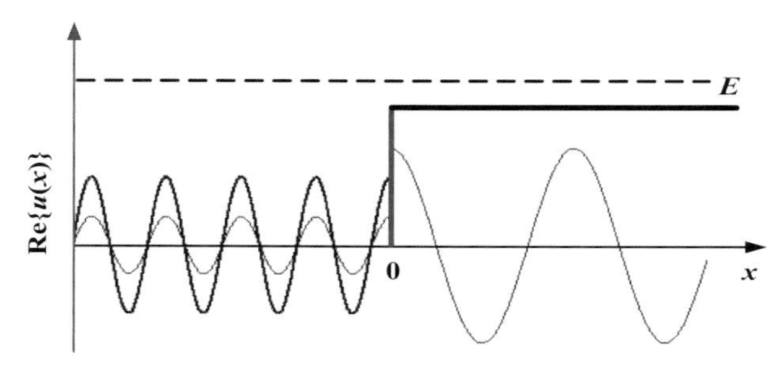

Figure 5.2 The incident (thicker line), reflected (thinner line), and transmitted (thinner line) wavefunctions of the electron, incident with kinetic energy $E = 12\,\text{eV}$ on the step barrier with $V = 10\,\text{eV}$.

is again rooted in the wave nature of the particle. Note that the reflection and transmission coefficients add up to be unity, as they should in view of the conservation of the probability density:

$$R + T = 1. \tag{5.18}$$

The behaviors of R and T are in close analogy with the reflection and transmission of a light beam, incident on a dielectric interface. Also, the entire analysis leading to the results of R and T in (5.16) and (5.17) is directly applicable to the case of the negative barrier in which $V < 0$. The only modification needed is to redefine k as $k^2 = 2m(E + V)/\hbar^2$. Here again, R and T are generally nonzero in contrast with the classical description, in which $R = 0$ and $T = 1$.

When the energy of the incident particle is less than the potential barrier, that is, $E < V$, the analysis can be carried out in similar fashion. The main modification required in this case consists of accounting for the fact that $k^2 < 0$ in (5.3), which obviously necessitates the introduction of an imaginary wave vector,

$$k \to i\varkappa, \quad \varkappa^2 \equiv \frac{2m(V - E)}{\hbar^2}. \tag{5.19}$$

The energy eigenfunction is then given by

$$u(x) = \begin{cases} i_0 e^{ik_0 x} + re^{-ik_0 x}, & x \leq 0, \\ te^{-\varkappa x}, & x > 0, \end{cases} \tag{5.20}$$

where the physically well-behaving exponential term has only been retained in the region $x > 0$. Then, the expression for r can be transcribed from (5.8) as

$$\frac{r}{i_0} = \frac{k_0 - i\varkappa}{k_0 + i\varkappa}$$

and the reflection coefficient (5.16) now reads as

$$R \equiv \left|\frac{r}{i_0}\right|^2 = \frac{k_0 - i\varkappa}{k_0 + i\varkappa} \cdot \frac{k_0 + i\varkappa}{k_0 - i\varkappa} = 1, \tag{5.21}$$

while $T = 0$, as can be readily verified by inserting the corresponding wavefunction into the probability current density (5.14). Waves become evanescent when the wave vector turns purely imaginary. This result is consistent with the conservation of the probability density (5.18) and further agrees with the classical picture of 100% reflection for $E < V$.

However, there is an interesting and profound difference between the quantum and classical descriptions, namely, that the particle penetrates into the classically forbidden barrier region by an amount $\delta \approx 1/(2\varkappa)$ before undergoing the total reflection. The penetration depth increases with increasing E for the given V, as shown in Figure 5.3. The finite penetration of the particle before the total reflection leads to novel effects when the width of the step potential becomes finite, and this will be taken up in Chapter 6.

Thus far the analysis has been made by representing the particle via the spatially unconfined plane waves. The general case of a particle as represented by wave packet

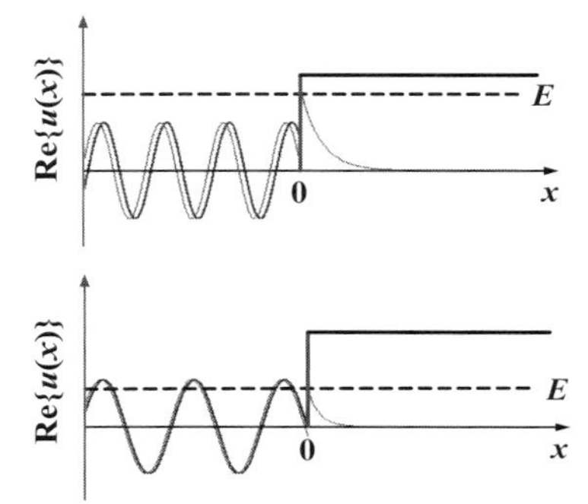

Figure 5.3 The incident (thicker line) and totally reflected (thinner line) wavefunctions of the electron incident with kinetic energies $E = 8$ eV (top) and $E = 4$ eV (bottom) on step potential with $V = 10$ eV. Note increasing penetration depth with increasing kinetic energy of incident electron.

can be analyzed by decomposing the packet into plane waves and following each plane wave in the manner discussed in this section. The reflected and transmitted output plane waves can then be combined into the reflected and transmitted wave packets.

5.2 Scattering at Quantum Well

When a particle is incident on a potential well rather than a barrier, as sketched in Figure 5.4, the incident flux again divides into reflected and transmitted fluxes, both of which are generally nonzero. This is again in stark contrast with the classical description in which the particle is bound to proceed in the same direction and with added velocity over the quantum well. This difference is again to be attributed to the wave nature of the particle. The energy eigenequation in this case is split into two regimes, inside and outside the well:

$$u(x)'' + \alpha^2 u(x) = 0, \tag{5.22}$$

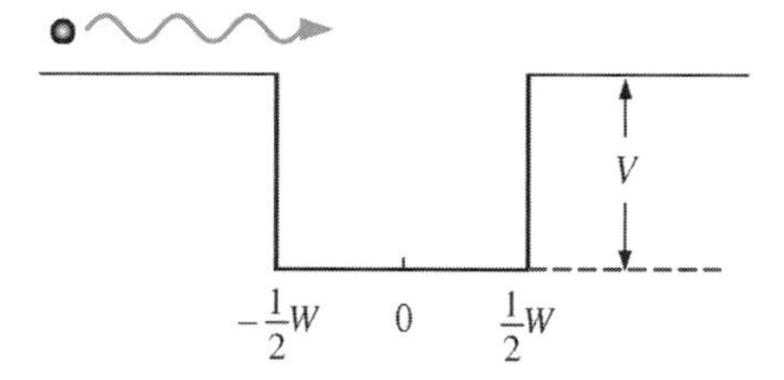

Figure 5.4 A particle incident on a quantum well with depth V and width W.

where

$$\alpha^2 = \begin{cases} k_0^2, \quad k_0^2 = \dfrac{2mE}{\hbar^2}, & |x| \geq \dfrac{W}{2}, \\ k^2, \quad k^2 = \dfrac{2m(E+V)}{\hbar^2}, & |x| \leq \dfrac{W}{2}, \end{cases} \tag{5.23}$$

and the energy eigenfunction can thus be represented by

$$u(x) = \begin{cases} i_0 e^{ik_0 x} + r e^{-ik_0 x}, & x < -W/2, \\ A e^{ikx} + B e^{-ikx}, & |x| \leq W/2, \\ t e^{ik_0 x}, & x > W/2. \end{cases} \tag{5.24}$$

As before, $u(x)$ contains the incident and reflected components to the left of the well, two counterpropagating components inside the well and one forward component to the right of the well.

The constants appearing in (5.24) are determined as usual from the boundary conditions, applied in this case at both edges of the well, namely, $x = \mp W/2$. The condition that $u(x)$ is continuous leads to

$$i_0 e^{-ik_0 W/2} + r e^{ik_0 W/2} = A e^{-ikW/2} + B e^{ikW/2}, \tag{5.25}$$

$$A e^{ikW/2} + B e^{-ikW/2} = t e^{ik_0 W/2}, \tag{5.26}$$

while $u'(x)$ being continuous yields

$$i_0 k_0 e^{-ik_0 W/2} - r k_0 e^{ik_0 W/2} = A k e^{-ikW/2} - B k e^{ikW/2}, \tag{5.27}$$

$$A k e^{ikW/2} - B k e^{-ikW/2} = t k_0 e^{ik_0 W/2}. \tag{5.28}$$

Thus, there are four conditions to be satisfied and five unknowns and one can take i_0 as the input parameter and determine the remaining constants in terms of i_0, obtaining

$$\frac{t}{i_0} = e^{-ik_0 W} \frac{2k_0 k}{2k_0 k \cos(kW) - i(k_0^2 + k^2)\sin kW}, \tag{5.29}$$

$$\frac{r}{i_0} = i e^{-ik_0 W} \frac{(k^2 - k_0^2)\sin(kW)}{2k_0 k \cos(kW) - i(k_0^2 + k^2)\sin kW}. \tag{5.30}$$

The resulting R and T are therefore found from (5.29) and (5.30) as

$$T \equiv |t/i_0|^2 = \frac{1}{1 + \Lambda(k_0, k)}, \tag{5.31a}$$

with

$$\Lambda(k_0, k) \equiv [(k^2 - k_0^2)/2k_0 k]^2 \sin^2(kW) \tag{5.31b}$$

and

$$R \equiv |r/i_0|^2 = \frac{\Lambda(k_0, k)}{1 + \Lambda(k_0, k)}. \tag{5.32}$$

It is noted here that the velocities with which the incident, reflected, and transmitted fluxes are propagating are the same, namely, $\hbar k_0/m$. By expressing k_0 and k in terms of the quantum well parameters and the incident energy E (see (5.23)), one can specify the reflection and transmission coefficients as

$$T = \frac{1}{1 + \Lambda(E, V, W)} \tag{5.33a}$$

with

$$\Lambda(E, V, W) \equiv \frac{V^2}{4E(E+V)} \sin^2\left[W\sqrt{\frac{2m}{\hbar^2}(E+V)}\right] \tag{5.33b}$$

and

$$R = \frac{\Lambda(E, V, W)}{1 + \Lambda(E, V, W)}. \tag{5.34}$$

Obviously, R and T thus found add up to unity, as they should.

5.2.1 Resonant Transmission

As indicated earlier, both the reflection and transmission coefficients are generally nonzero in distinct contrast with the classical description, as specifically illustrated in Figure 5.5. However, for $E \gg V$, T approaches unity, as is clear from (5.33), in agreement with the classical description. More important, even for E comparable with V (5.31) indicates that 100% transmission could ensue, provided the incident energy satisfies the condition

$$kW = n\pi, \tag{5.35}$$

where n is a positive integer. The physical significance entailed in this condition can be understood in view of the de Broglie wavelength λ. Since $k = 2\pi/\lambda$, (5.35) is equivalent to the condition $2W = n\lambda$, thereby indicating that the round-trip distance of the quantum well is an integer multiple of de Broglie wavelength of the particle. This is exactly the same condition by which to induce 100% transmission of light in Fabry Perot etalon or Bragg diffraction. Under this condition, the reflected and transmitted light beams interfere destructively and constructively to yield $R = 0$ and $T = 1$. This total transmission of a particle across the potential well is known as the resonant transmission. Figure 5.5 shows the typical reflection and transmission coefficients of a particle incident on the quantum well with varying energies.

The resonant transmission lends to an interesting interpretation when the condition (5.35) is transcribed in terms of energy using (5.23), namely,

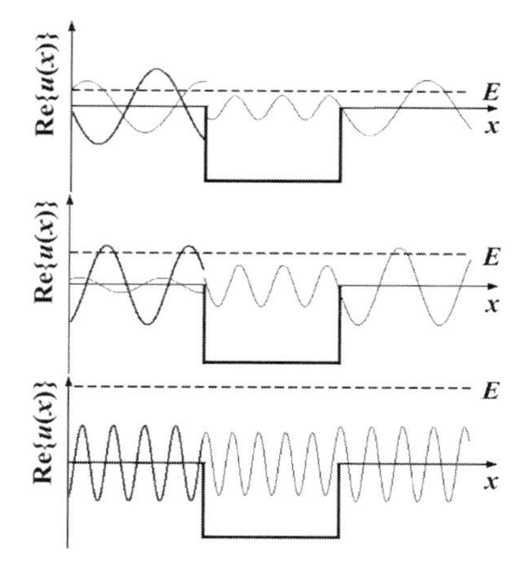

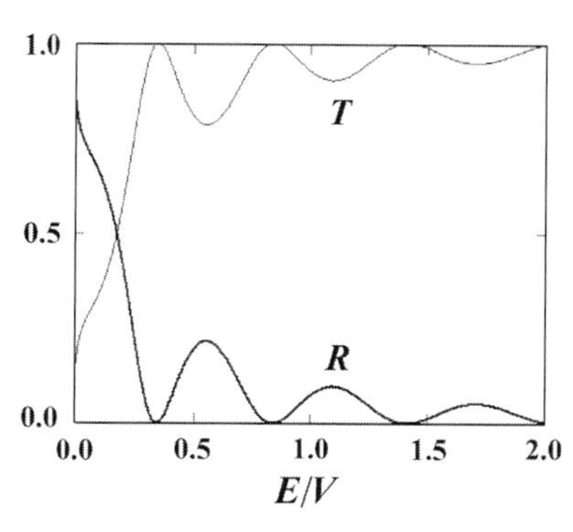

Figure 5.5 The incident (thicker line), reflected (thinner line), and transmitted (thinner line) wavefunctions of an electron. The electron is incident on the quantum well with depth $V = 10\,\text{eV}$ and width $W = 1\,\text{nm}$ with varying incident energies E. Also shown are R and T versus incident energy E/V. The resonant transmission peaks are clearly noticeable.

$$W\sqrt{\frac{2m}{\hbar^2}(E_n + V)} = n\pi. \tag{5.36}$$

From (5.36), the incident energy inducing the resonant transmission is specified by

$$E_n + V = \frac{\pi^2\hbar^2}{2mW^2}n^2. \tag{5.37}$$

Clearly, (5.37) indicates that if the incident energy as viewed from the bottom of the square potential well corresponds to one of the possible eigenstates of the infinite square well potential (see (4.9)), the total transmission of the particle ensues. This carries an important bearing on the band theory of solids, as will later become clear.

5.3 Problems

5.1. (a) Starting from the boundary conditions for the energy eigenfunctions given in (5.25)–(5.28) in the text, fill in the algebraic steps for finding t and r in terms of i_0 and verify (5.29) and (5.30).

(b) With the use of the solutions thus found in (a), derive the reflection and transmission coefficients in (5.33) and (5.34).

5.2. A particle is incident on a two-step potential barrier with an energy E greater than the potential height V_2, as sketched in Figure 5.6.

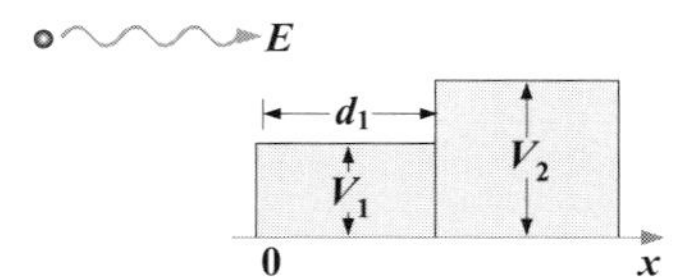

Figure 5.6 A particle incident on two step barrier potential with $E > V_2$.

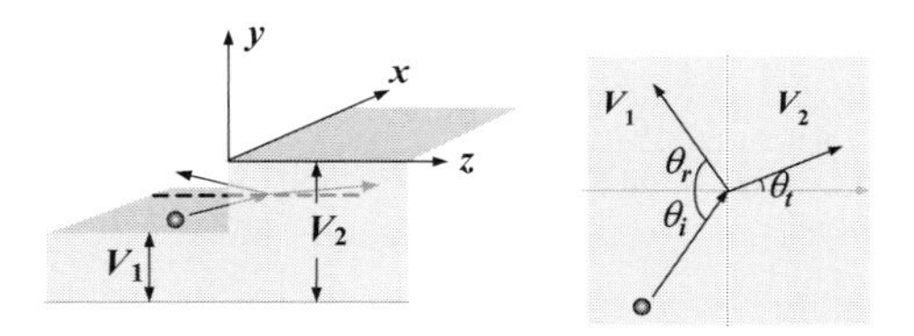

Figure 5.7 A particle incident on step barrier potential $V_2 - V_1$ with $E > V_2$ at an angle θ_i with respect to the normal. The particle is reflected off at θ_r or transmitted at θ_t.

(a) Write down the energy eigenfunctions for $x < 0$ and $x > 0$.
(b) By imposing the boundary conditions at $x = 0,\ d_1$, find the reflection and transmission coefficients.
(c) For the given E, determine the first step barrier height V_1 and width d_1 such that there is no reflection and discuss the result.

5.3. A particle is incident on a potential barrier from the region $V = V_1$ to the region $V = V_2$ at an angle θ_i with respect to the normal direction, as sketched in Figure 5.7. The energy E of the particle is greater than the barrier height V_2.

(a) Write down the wavefunctions in the two regions.
(b) Find the angle of reflection and transmission of the particle by imposing the boundary conditions of the wavefunction at the potential boundary.
(c) Find reflection and transmission coefficients, R and T, of the particle.
(d) Discuss the results in analogy with the reflection and refraction of light at a dielectric interface.

5.4. A particle is incident on two square potential barriers with height V and width d and separated from each other by W, as sketched in Figure 5.8. The energy of the particle is greater than the barrier height V.

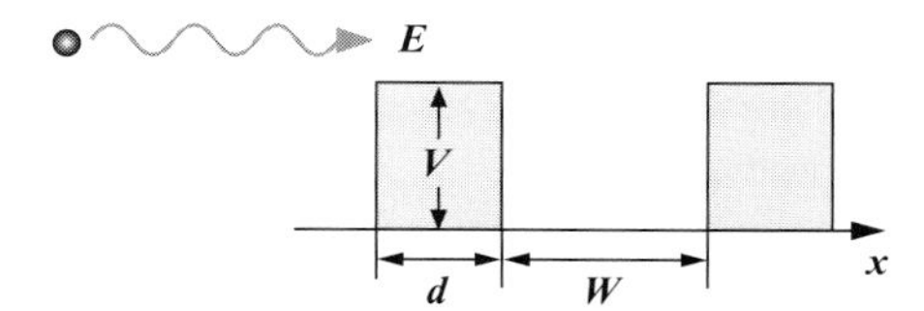

Figure 5.8 A particle incident on two barrier potentials of height V, width d and W distance apart with $E > V$.

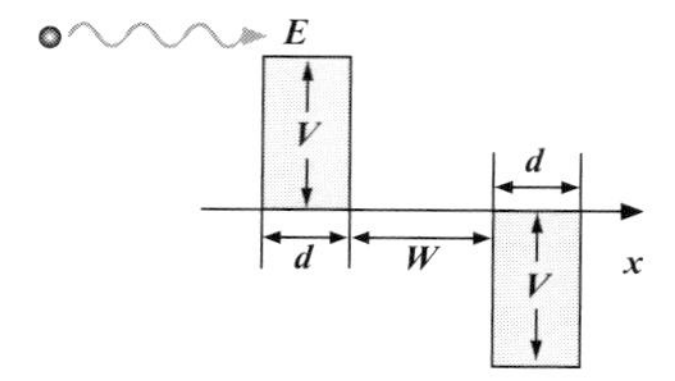

Figure 5.9 A particle incident on a composite potential consisting of barrier and well of width d, height or depth V, and W distance apart with $E > V$.

(a) Write down the energy eigenfunctions in each region.
(b) Write down the boundary conditions for these eigenfunctions.
(c) Find the reflection and transmission coefficients.
(d) What is the energy of the incoming particle at which $T = 1$?
(e) Generalize the analysis to a string of identical barrier potentials separated by quantum wells of the same width and discuss the results obtained.

5.5. A particle is incident on 1D potential consisting of a square barrier potential and a quantum well, as sketched in Figure 5.9. Repeat the analysis of problem 4 for this potential and compare the results.

Suggested Reading

1 Singh, J. (1996) *Quantum Mechanics, Fundamentals & Applications to Technology*, John Wiley & Sons, Inc.
2 Yariv, A. (1982) *An Introduction to Theory and Applications of Quantum Mechanics*, John Wiley & Sons, Inc.
3 Liboff, R.L. (2002) *Introductory Quantum Mechanics*, 4th edn, Addison-Wesley Publishing Company, Reading, MA.
4 Gasiorowicz, S. (2003) *Quantum Physics*, 3rd edn, John Wiley & Sons, Inc.
5 Robinett, R.W. (2006) *Quantum Mechanics, Classical Results, Modern Systems and Visualized Examples*, Oxford University Press.
6 Kroemer, H. (1994) *Quantum Mechanics for Engineering, Materials Science and Applied Physics*, International edn, Prentice Hall.

6
Tunneling and Its Applications

A particle incident on a potential barrier higher than its own kinetic energy has a finite probability of transmitting through the barrier. The transport mechanism for such a process is known as tunneling. The tunneling has been singled out from the 1D scattering problems in Chapter 5 and is treated more comprehensively in this chapter. The tunneling is one of the unique features of quantum mechanics, deeply rooted in the wave nature of the particle and plays crucial roles in semiconductor and nanodevices, flat panel displays, nanometrology, nonvolatile memory cells, and so on. More important, the resonant tunneling underlies the band theory of solids, thereby providing the basic foundations of semiconductor physics and devices.

6.1
Tunneling Across Square Potential Barrier

When a particle is incident on a square potential barrier with height V and width d, as shown in Figure 6.1, the wavefunction can be expressed in a manner similar to (5.24):

$$u(x) = \begin{cases} i_0 \exp ik_0 x + r \exp -ik_0 x, & x < -d/2, \\ A \exp i\alpha x + B \exp -i\alpha x, & |x| \leq 2/d, \\ t \exp ik_0 x, & x \geq d/2, \end{cases} \tag{6.1}$$

where $k_0^2 = 2mE/\hbar^2$ and α is either k or $\varkappa$, depending on the value of E, that is,

$$\alpha = \begin{cases} k, \quad k = \sqrt{\dfrac{2m(E-V)}{\hbar^2}}, & E > V, \\ i\varkappa, \quad \varkappa = \sqrt{\dfrac{2m(V-E)}{\hbar^2}}, & E < V. \end{cases} \tag{6.2}$$

For $E > V$, the entire results obtained in Section 5.2 for R and T can be directly transcribed, provided k as defined in (6.2) is used and the well width W is replaced by the width of the barrier d.

Introductory Quantum Mechanics for Semiconductor Nanotechnology. Dae Mann Kim

ISBN: 978-3-527-40975-4

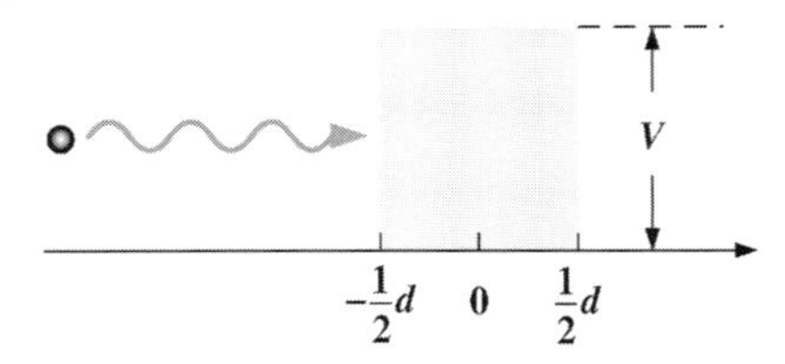

Figure 6.1 A particle incident on a potential barrier with height V and thickness d.

For $E < V$, most of the results obtained in Section 5.2 can still be applied with the use of identities $\sin ix = i \sinh x$ and $\cos ix = \cosh x$. Thus, one can write from (5.29) and (5.30)

$$\frac{t}{i_0} = \exp(-ik_0 d)\frac{2k_0\varkappa}{2k_0\varkappa\cosh(\varkappa d) - i(k_0^2 - \varkappa^2)\sinh\varkappa d}, \tag{6.3}$$

$$\frac{r}{i_0} = -i\exp(-ik_0 d)\frac{(\varkappa^2 + k_0^2)\sinh(\varkappa d)}{2k_0\varkappa\cosh(\varkappa d) - i(k_0^2 - \varkappa^2)\sinh\varkappa d}. \tag{6.4}$$

The transmission and reflection coefficients can therefore be expressed as in (5.33) and (5.34) in the form

$$T = \left|\frac{t}{i_0}\right|^2 = \frac{1}{1 + \Lambda(E, V, d)} \tag{6.5}$$

with

$$\Lambda(E, V, d) \equiv \frac{V^2}{4E(V - E)}\sinh^2\left[d\sqrt{\frac{2m}{\hbar^2}(V - E)}\right] \tag{6.6}$$

and

$$R = \left|\frac{r}{i_0}\right|^2 = \frac{\Lambda(E, V, d)}{1 + \Lambda(E, V, d)}. \tag{6.7}$$

In (6.5) and (6.7), $\cosh^2 x = 1 + \sinh^2 x$ has been used and the fact is that the incident, reflected, and transmitted velocities are same. Obviously, $R + T = 1$, as evident from (6.5) and (6.7).

Figure 6.2 shows the transmission coefficient versus the energy of the incident particle for both $E < V$ and $E > V$. For the former, (6.5) is used while for the latter (5.33a) is used with k defined in (6.2). Also plotted in the figure are typical eigenfunctions of the incident, reflected, and transmitted components. Note that the two expressions of T for $E < V$ and $E > V$ give identical results at $E = V$, namely,

$$T = \frac{1}{1 + (mVd^2/2\hbar^2)},$$

as they should. It is interesting that the resonant peaks at which $T = 1$ also appear for $E > V$, indicating that the resonant transmission is also operative for potential

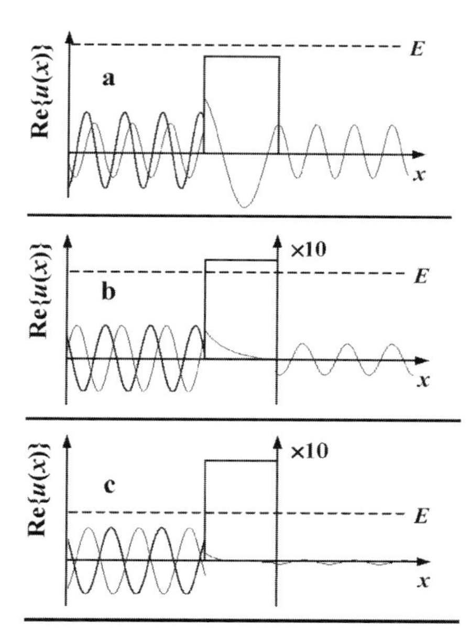

Figure 6.2 The incident (thicker line), reflected (thinner line), and transmitted (thinner line) wavefunctions of an electron incident on a potential barrier $V = 1$ eV with different energies E. The transmission coefficient is also plotted versus E. For $E > V$, the resonant transmission peaks are noticeable, while for $E < V$, T decreases exponentially with decreasing E.

barriers as well as quantum wells. For $E < V$, T decreases exponentially with decreasing E but there is a finite probability for the particle to transmit across the barrier via tunneling.

Tunneling and Penetration Depth As mentioned at the outset of this chapter, the tunneling is a process unique in quantum mechanics and is firmly rooted in the wave nature of the particle. The tunneling can be understood in light of the finite penetration the particle makes into the classically forbidden barrier region in the step potential before it undergoes the total reflection. Obviously, this total reflection at the step potential barrier is due to the infinite width of the barrier. When the barrier width is cut to a finite value, the probability density of the wavefunction of the incident particle becomes finite beyond the barrier width, as apparent in Figure 6.2. This indicates that the particle has a finite probability of penetrating beyond the width of the barrier. That is to say, the particle has a finite probability of tunneling through the barrier.

As noted, the penetration depth is analogous to the skin depth of the light at the surface of a metal and the tunneling has optical analog as well, for example, the directional coupler as shown in Figure 6.3. As well known, the light propagates in a waveguide or an optical fiber, based on the total internal reflection. But in the presence of another waveguide in proximity, the light confined in one waveguide leaks into the other, thereby inducing power coupling and this phenomenon is used

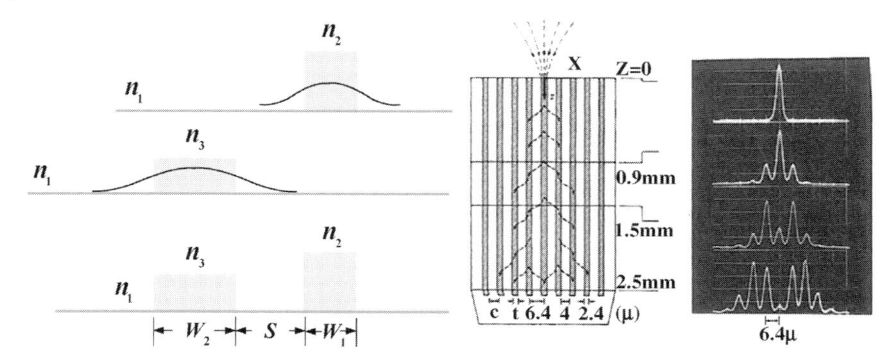

Figure 6.3 Schematics for an observed directional coupling of light. The coupling requires the waveguides to be placed within the exponential tailing of the field amplitude (taken from Yariv, A. (1985) *Optical Electronics*, Holt, Rinehart and Winston).

for the modulation or switching of light. This coupling of power is due to the guided electromagnetic waves tailing out of the waveguide film or the optical fiber and this tailing of the field amplitude out of the confined spatial region underpins such coupling processes. By the same token, the finite penetration of the wavefunction into the classically forbidden region underpins the tunneling.

6.2
Fowler–Nordheim and Direct Tunneling

The tunneling analysis that has been carried out for the simple square potential barrier can be extended to an arbitrary shape potential barrier $V(x)$. Given a barrier $V(x)$, as sketched in Figure 6.4, one can decompose it into a juxtaposition of square barriers with thickness Δx and height $V(n\Delta x)$, as indicated in the figure. One can then take the tunneling through each square barrier element as statistically independent and multiply the differential tunneling probabilities for obtaining the net tunneling probability. In so doing, the probability of tunneling through the jth barrier element can further be simplified from (6.5) as

$$T_j \approx \exp -2\sqrt{\frac{2m}{\hbar^2}(V_j - E)}dx, \tag{6.8}$$

where approximations were made of $\sinh x \approx (\exp x)/2$ for $x \gg 1$. Thus, the total probability of tunneling is approximately given by

$$T = \prod_j T_j = \exp -\frac{2\sqrt{2m}}{\hbar}\int_{x_1}^{x_2} dx[(V(x) - E)]^{1/2}, \tag{6.9}$$

where the upper and lower limits of the integration is determined by the condition $V = E$ (see Figure 6.4).

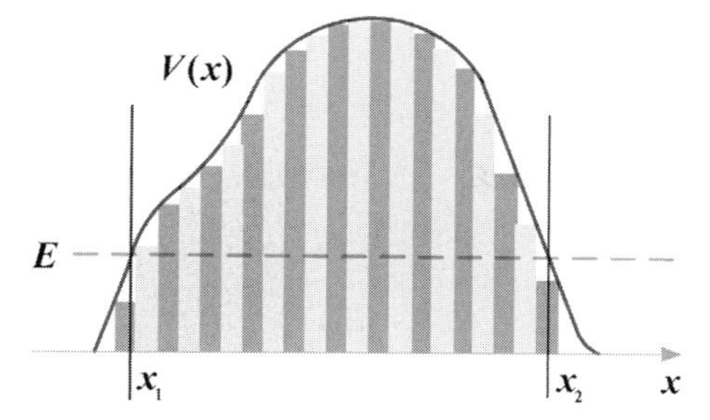

Figure 6.4 An arbitrary shape potential barrier and its juxtaposition via square potential barrier elements.

The tunneling probability obtained for an arbitrary potential barrier $V(x)$ in (6.9) is next applied to the trapezoidal potential barrier, as shown in Figure 6.5. This kind of potential barrier is often encountered by an electron or a hole incident on a dielectric layer in the presence of an external field **E**. In this case $V(x)$ is given by

$$V(x) = V - q|\mathbf{E}|x, \tag{6.10}$$

where V is the barrier height provided at the dielectric material, q is the magnitude of the electronic charge, and **E** is the applied electric field. Inserting (6.10) into (6.9) and performing the integration, one finds

$$T = \exp-\left\{\frac{4(2m)^{1/2}}{3q|\mathbf{E}|\hbar}\left[(V-E)^{3/2}-(V-E-q|\mathbf{E}|d)^{3/2}\right]\right\}, \tag{6.11}$$

where d is the barrier width. As is clear from (6.11), the tunneling probability in this case is primarily determined by three factors, namely, the applied electric field **E**, the width of the barrier d, and the incident energy E. Any of these parameters affects T exponentially. The tunneling through the trapezoidal barrier is known as the direct tunneling. It is this direct tunneling of electrons that imposes fundamental

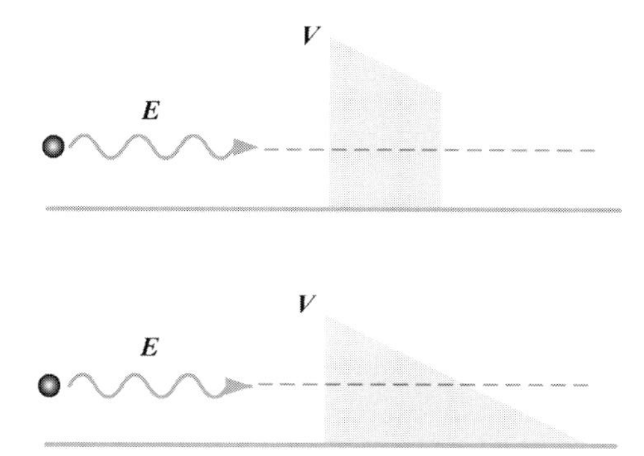

Figure 6.5 The direct tunneling through trapezoidal potential barrier and Fowler–Nordheim tunneling through triangular potential barrier.

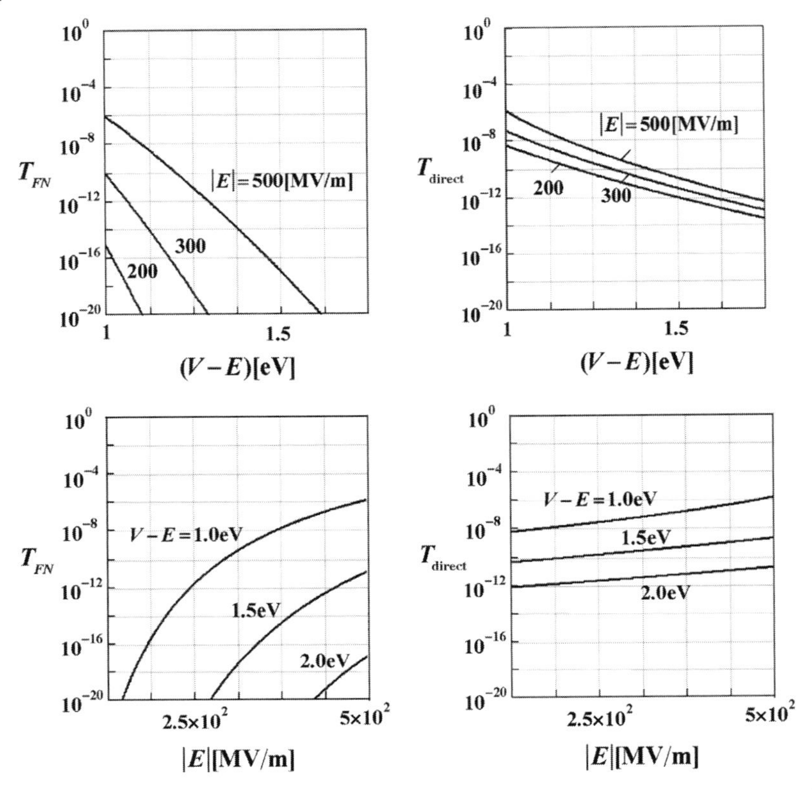

Figure 6.6 The F–N and direct tunneling probabilities of an electron versus effective barrier height $V-E$ for fixed V of 2 eV and for different electric field E. The same tunneling probabilities versus E for different effective barrier height $V-E$.

limitations on downscaling the semiconductor devices into nanoregimes, as will be further elaborated subsequently.

When the potential barrier is of a triangular shape instead, the second term in (6.11) automatically drops out (see Figure 6.5) and the tunneling probability reduces to

$$T = \exp - \frac{4(2m)^{1/2}}{3q|\mathbf{E}|\hbar}(V-E)^{3/2} \tag{6.12}$$

and is known as the Fowler–Nordheim (F–N) tunneling. Figure 6.6 shows T versus the incident electron energy for different electric field for both direct and F–N tunnelings. In all of these curves, the tunneling probabilities are shown critically dependent on E and the incident energy. The F–N tunneling also fundamentally limits the device operation and downscaling because it enhances the gate leakage in logic devices. However, it also provides the driving force for the operation of nonvolatile memory cells, as will be further elaborated subsequently.

6.3
Resonant Tunneling

The multiple quantum wells formed by a string of barrier potentials are routinely fabricated by means of the bandgap engineering, based on metal organic chemical vapor deposition (MOCVD) or molecular beam epitaxy (MBE) techniques. These structures have become an integral part of semiconductor and optoelectronic devices. The particle therein may tunnel through these barriers. In such event, the resonant tunneling becomes operative and important just like the resonant transmission across the quantum well.

Thus, consider an electron incident on two square potential barriers with height V and W distance apart, as sketched in Figure 6.7. An electron incident on the input plane at z_j with $E < V$ and transmitting at the output plane at z_{j+3} is required to undergo two successive tunneling through each barrier, combined with any number of internal reflections in the quantum well in between. The net tunneling, therefore, is accompanied by the interference effect of the matter wave of the electron in the quantum well.

The tunneling analysis in this case is lengthy but straightforward and can be carried out by utilizing the results already obtained. First, one can write the wavefunction in the region to the left of the first barrier at z_j (jth region) and within the barrier (($j+1$)th region) as

$$u_j(z) = A_j e^{ikz} + B_j e^{-ikz}, \quad k \equiv (2mE/\hbar^2)^{1/2}, \tag{6.13}$$

$$u_{j+1}(z) = A_{j+1} e^{-\varkappa z} + B_{j+1} e^{\varkappa z}, \quad \varkappa \equiv [2m(V-E)/\hbar^2]^{1/2}. \tag{6.14}$$

The boundary conditions of these two wavefunctions at the interface z_j are specified by

$$u_j(z_j) = u_{j+1}(z_j), \tag{6.15}$$

$$\frac{1}{m_j} u_j'(z_j) = \frac{1}{m_{j+1}} u_{j+1}'(z_j). \tag{6.16}$$

In (6.16), the continuity of the electron velocity rather than momentum has been imposed by introducing the effective mass of the electron in two regions, so that the continuity of the current is ensured. The effective mass of electron is in general

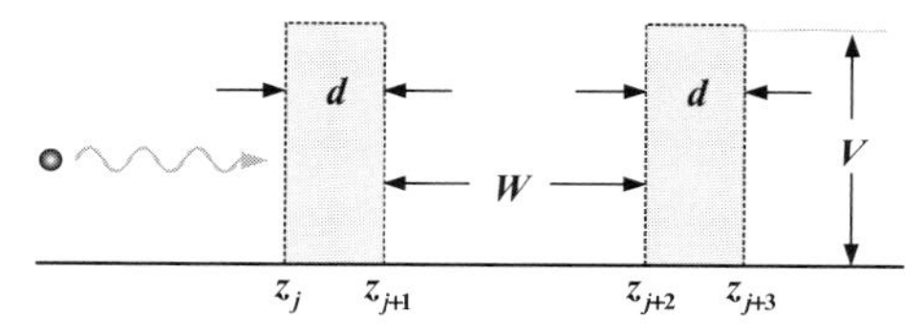

Figure 6.7 A particle incident on two potential barriers of height V and thickness d and W distance apart, thereby forming a quantum well between the two barriers.

different in the well and the barrier regions. For simplicity of analysis, however, the two effective masses are taken to be the same, that is, $m_j = m_{j+1}$.

The two boundary conditions (6.15) and (6.16), when applied to the eigenfunctions (6.13) and (6.14), relate the constants A_j and B_j to A_{j+1} and B_{j+1} and one can write after solving for A_j, B_j in terms of A_{j+1}, B_{j+1}.

$$\begin{pmatrix} A_j \\ B_j \end{pmatrix} = M(z_j; i\varkappa, k)\begin{pmatrix} A_{j+1} \\ B_{j+1} \end{pmatrix}, \tag{6.17}$$

where the 2×2 transfer matrix is given by

$$M(z_j; i\varkappa, k) = \frac{1}{2}\begin{pmatrix} \left(1+\dfrac{i\varkappa}{k}\right)e^{i(i\varkappa-k)z_j} & \left(1-\dfrac{i\varkappa}{k}\right)e^{-i(i\varkappa+k)z_j} \\ \left(1-\dfrac{i\varkappa}{k}\right)e^{i(i\varkappa+k)z_j} & \left(1+\dfrac{i\varkappa}{k}\right)e^{-i(i\varkappa-k)z_j} \end{pmatrix}. \tag{6.18}$$

Similarly, the constants A_{j+1} and B_{j+1} can be correlated to A_{j+2} and B_{j+2}, which specify the wavefunction in the well to the right of the first barrier by

$$\begin{pmatrix} A_{j+1} \\ B_{j+1} \end{pmatrix} = M(z_j + d; k, i\varkappa)\begin{pmatrix} A_{j+2} \\ B_{j+2} \end{pmatrix}. \tag{6.19}$$

Here, the transfer matrix in (6.19) is to be directly transcribed from (6.18) by simply interchanging $i\varkappa$ with k and replacing z_j by $z_j + d$.

In this manner, the constants A_j and B_j at z_j can be correlated to A_{j+2} and B_{j+2} at z_{j+1} $(= z_j + d)$ by the straightforward multiplication of two transfer matrices involved, namely,

$$\begin{pmatrix} A_j \\ B_j \end{pmatrix} = M(z_j; d)\begin{pmatrix} A_{j+2} \\ B_{j+2} \end{pmatrix}, \tag{6.20}$$

where

$$\begin{aligned} M(z_j; d) &= M(z_j; i\varkappa, k)M(z_j + d; k, i\varkappa) \\ &= \begin{pmatrix} m_{11}(d) & m_{12}(z_j, d) \\ m_{12}^*(z_j, d) & m_{11}^* \end{pmatrix} \end{aligned} \tag{6.21a}$$

and

$$m_{11}(d) = e^{ikd}\left(\cosh \varkappa d - i\frac{k^2 - \varkappa^2}{2k\varkappa}\sinh \varkappa d\right), \tag{6.21b}$$

$$m_{12}(z_j, d) = ie^{-ik(2z_j + d)}\frac{k^2 + \varkappa^2}{2k\varkappa}\sinh \varkappa d. \tag{6.21c}$$

The matrix thus derived in (6.21) provides the basic unit by which to analyze the tunneling through multiple barriers.

Tunneling Through Single Barrier The transfer matrix given in (6.21) accounts for the tunneling probability through a single square potential barrier considered in Section 6.1. To show this, A_j and B_j in (6.20) can be taken to be the incident and reflected components, while A_{j+2} is to be viewed as the transmitted component for the single barrier. Once the particle is transmitted across the barrier, there is no reflection, hence one should put $B_{j+2}=0$. Therefore, the tunneling probability through a single barrier is given from (6.20) and (6.21) by

$$T=\left|\frac{A_{j+2}}{A_j}\right|^2=\left|\frac{1}{m_{11}}\right|^2=\frac{1}{\left|\cosh \varkappa d-i\frac{(k^2-\varkappa^2)}{2k\varkappa}\sinh \varkappa d\right|^2} \tag{6.22}$$

and (6.22) is in complete agreement with the expression for T obtained in (6.3).

Tunneling Through Double Barriers Next, the analysis of the tunneling through two successive potential barriers can likewise be carried out by a straightforward extension of (6.20), that is,

$$\begin{aligned}\begin{pmatrix}A_j\\B_j\end{pmatrix}&=\prod_{j=1}^{2}M(z_j;d)M(z_{j+2};d)\begin{pmatrix}A_{j+4}\\0\end{pmatrix},\quad z_{j+2}=z_j+W+d\\&=\begin{pmatrix}m_{11}(d)&m_{12}(z_j,d)\\m_{12}^*(z_j,d)&m_{11}^*(d)\end{pmatrix}\begin{pmatrix}m_{11}(d)&m_{12}(z_{j+2},d)\\m_{12}^*(z_{j+2},d)&m_{11}^*(d)\end{pmatrix}\begin{pmatrix}A_{j+4}\\0\end{pmatrix}\end{aligned} \tag{6.23}$$

where B_{j+4} has again been put to zero, since there is no reflection after the second barrier. Here, the two barriers are located at z_j and z_j+d+W with d and W denoting the widths of the barrier and the well, respectively. Performing the matrix multiplication with (6.21) in (6.23), one finds

$$\begin{aligned}\frac{A_{j+4}}{A_j}&=\frac{1}{m_{11}(d)m_{11}(d)+m_{12}(z_j,d)m_{12}^*(z_{j+2},d)},\quad z_{j+2}=z_j+W+d\\&=\frac{-e^{-2ikd}4k^2\varkappa^2}{[(k^2-\varkappa^2)\sinh \varkappa d+2ik\varkappa\cosh \varkappa d]^2-e^{2ikW}(k^2+\varkappa^2)^2\sinh^2\varkappa d.}\end{aligned} \tag{6.24}$$

Next, (6.24) can be put into a more transparent form. For this purpose, note that the first bracket in the denominator when put into the phasor notation and combined with the numerator leads to the tunneling probability for a single potential barrier, T_{1B} (see (6.3), (6.5) or (6.22)). Thus, (6.24) can be compacted as

$$\frac{A_{j+4}}{A_j}=\frac{-e^{-2ikd-2i\theta}T_{1B}}{1-e^{2i(kW-\theta)}R_{1B}}, \tag{6.25}$$

where $R_{1B}=1-T_{1B}$ and

$$\tan\theta = \frac{2k\varkappa\cosh\varkappa d}{(k^2-\varkappa^2)\sinh\varkappa d} \tag{6.26}$$

Hence, the probability of tunneling the two successive barriers is obtained from (6.25) as

$$T_{2B} = \left|\frac{A_{j+4}}{A_j}\right|^2 = \frac{1}{1+4\frac{R_{1B}}{T_{1B}^2}\sin^2(kW-\theta)}, \tag{6.27}$$

where the identities $T_{1B} = 1-R_{1B}$, $|1-f\exp i\chi|^2 = 1+f^2-2f\cos\chi$, and $\cos x = 1-2\sin^2(x/2)$ have been used.

Thus, it naturally follows from (6.27) that the resonant tunneling, that is, 100% transmission ensues, provided $kW \approx n\pi$. This condition in turn can be expressed in terms of E via (6.2) as

$$E_n = \frac{\hbar^2\pi^2 n^2}{2\,mW^2}. \tag{6.28}$$

It is, therefore, clear from (6.28) that the resonant tunneling occurs when the energy of the incident electron approximately coincides with one of the possible bound-state energies of quantum well located in between the two barriers. This carries an important bearing for understanding the band theory of solids, as will become clear in Chapter 7. The general features of the transmission characteristics of the electron across two barriers are shown in Figure 6.8. Here, T is plotted versus the incident electron energy for different barrier heights V and widths d. Indeed, the tunneling probability is drastically reduced with increasing V and d, as expected, but the general features of the resonant tunneling are preserved. It is particularly interesting that the resonant peaks are considerably broadened with reduced potential barrier height or width.

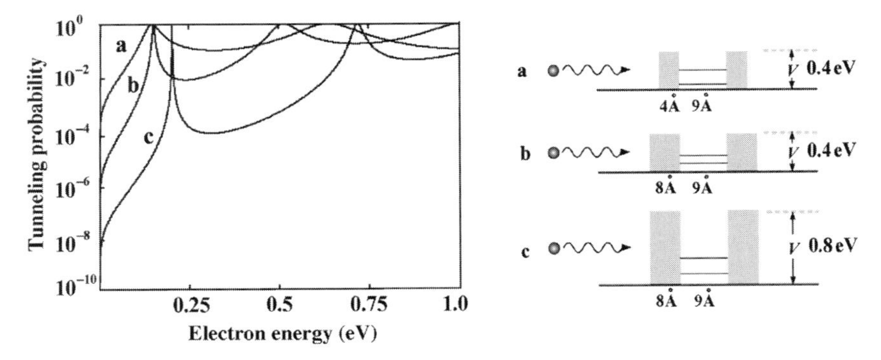

Figure 6.8 The tunneling probabilities through the double potential barriers versus incident electron energy for different barrier height V and thickness d.

6.4
The Applications of Tunneling

The tunneling commands a wide range of applications in semiconductor and optoelectronic devices and the list of applications is fast increasing. For example, the tunneling underpins the operation of tunnel diodes, Schottky ohmic contacts, resonant tunneling devices, single-electron transistors, nonvolatile memory cells, and so on. In addition, the tunneling offers an efficient means for nanometrology, such as scanning tunneling microscopy. More important, the resonant tunneling provides the physical basis where electrons or holes can be taken as free particles in solids. Some of these applications are briefly described.

Direct and F–N Tunnelings The direct or Fowler–Nordheim tunneling discussed in Section 6.2 has played interesting and important roles. Figure 6.9 describes the cold emission of electrons from the surface of metal. As noted earlier, a metal is often modeled by a sea of free electrons filled up to a certain energy level, called the Fermi energy, and these electrons are confined in the metal by a potential barrier at the surface, called the work function. Thus, when two metals having the same work function, for example, are brought together, the electrons remain confined in each metal by the square potential barrier. When a voltage is applied between the two metals, however, the barrier is reduced from the square shape to the trapezoidal or triangular shape, depending on the voltages applied and the distance between the two metals. In this case, either direct or F–N tunneling becomes operative and the tunneling probability is enhanced exponentially, as detailed in Figure 6.6. The resulting emission of electrons from the metallic surface is known as the cold or field emission.

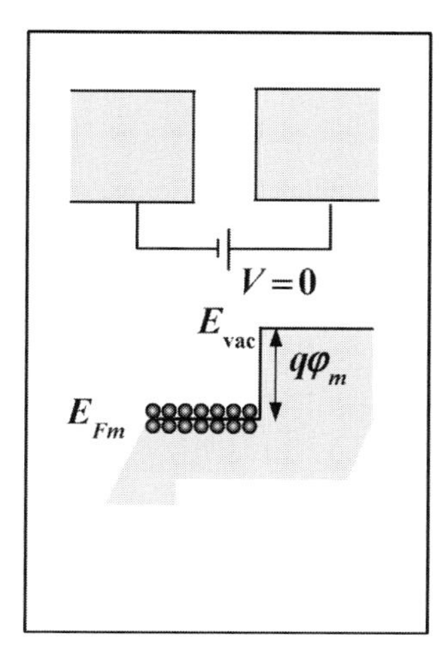

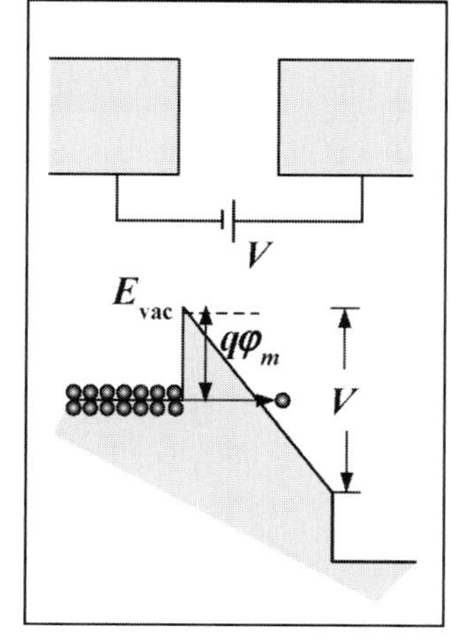

Figure 6.9 The schematics of cold and/or field emission of electrons from the surface of metal.

The same kinds of trapezoidal or triangular potential barriers come into play in metal-oxide–silicon field effect transistors (MOSFETs). Here the barrier is provided by the dielectric layer such as the silicon dioxide sandwiched between the gate electrode and the silicon substrate. Without bias between the gate and the substrate, the insulator can effectively block the tunneling of electrons or holes from the gate to the semiconductor or vice versa. However, when a bias is applied between the metal gate and the silicon substrate, the rectangular barrier is again transformed into a trapezoidal or triangular shape and the tunneling probability of electrons is greatly enhanced. As a consequence, there ensues a gate current, contributed by these tunneling electrons. If the tunneling occurs in a logic device, it leads to the gate leakage, degrading the device performance and also fundamentally limiting the device scaling. If the tunneling occurs in nonvolatile memory cells instead, the tunneling provides the driving force for the cell operation. This will be further amplified upon later.

Nanometrology Figure 6.10 shows the schematics for the scanning tunneling microscope, a key tool for nanometrology. Here, the tunneling provides a convenient mechanism for probing the surface of material with atomic scale resolution. The working principle of the microscope simply consists of keeping the tunnel current

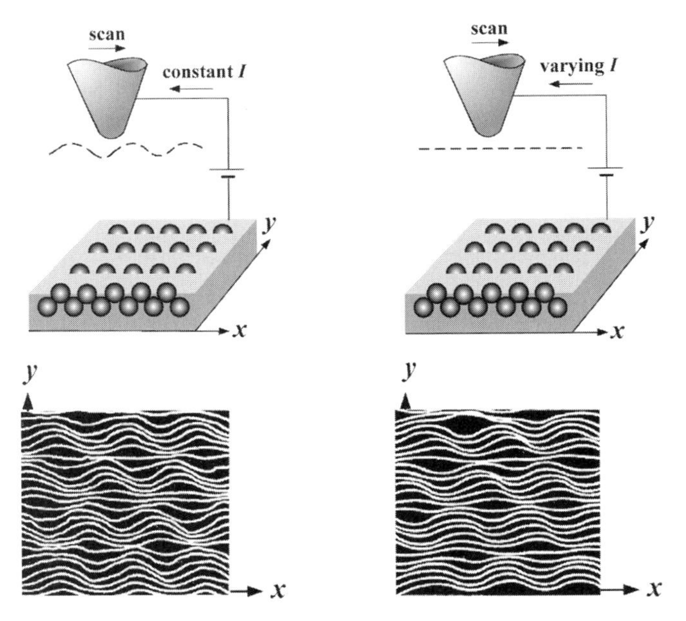

Figure 6.10 The schematic illustration of scanning tunneling microscopy. The current is kept fixed at given bias by adjusting the tip distance from the sample surface and converting the difference in distance into the surface structure (left). Alternatively, the tip distance with respect to sample is fixed, while the resulting change in tunneling current at given bias is translated into surface morphology.

between the probe tip and the surface atoms fixed at a given bias or the height of the tip fixed with respect to the sample surface. As detailed, the F–N or direct tunneling probability is mainly dictated by the distances involved and the voltages or electric fields applied. Hence, for the case of the former scheme, to keep the tunnel current constant at the given voltage, the distance of the probe tip with respect to the surface atom has to be maintained at fixed distance. This requires an adjustment of the height of the tip as it scans the surface. The adjusted height of the tip versus the x–y scanning reveals the surface morphology with about 0.1 nm in accuracy. For the latter case, the varying tunneling current at fixed voltage is translated into the surface morphology.

Field Emission Display The schematics for the field emission display are shown in Figure 6.11. The image information to be displayed is transferred in the form of applied voltages from the driver circuitry to the array of the metallic tips, forming the pixels. These voltages transferred in turn give rise to the field crowding at the metallic tip, thereby enhancing the tunneling probability. Those electrons thus tunneling out from the metallic tips carry the image information to the screen for display.

Ohmic Contact Another useful application of the tunneling consists of the ohmic contact, which is an essential element of any semiconductor device, connecting it to the outside world with negligible resistance. Such a contact can be fabricated using the interface between the metal and the semiconductor, as sketched in Figure 6.12. At the interface, a potential barrier is generally formed resulting from the difference between two work functions involved, as will be further discussed in chapters to follow. However, the width of this potential barrier can be made very thin by heavily doping the semiconductor, in which case the potential barrier becomes nearly transparent to tunneling electrons or holes. Thus, under bias, these charge carriers can be readily injected from the metal into the semiconductor or vice versa with negligible voltage drop at the contact interface.

Resonant Tunneling Device The resonant tunneling device is based entirely on the resonant tunneling, as the device nomenclature suggests. The device consists of two potential barriers, processed in semiconductor and in contact with metal electrodes

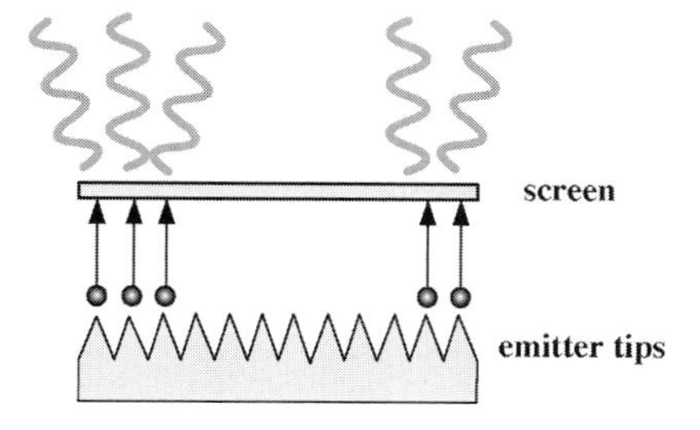

Figure 6.11 The schematics of the field emission based flat panel display. Electrons are emitted from pixel tips to transfer the image information transmitted from the driver circuitry to the display screen.

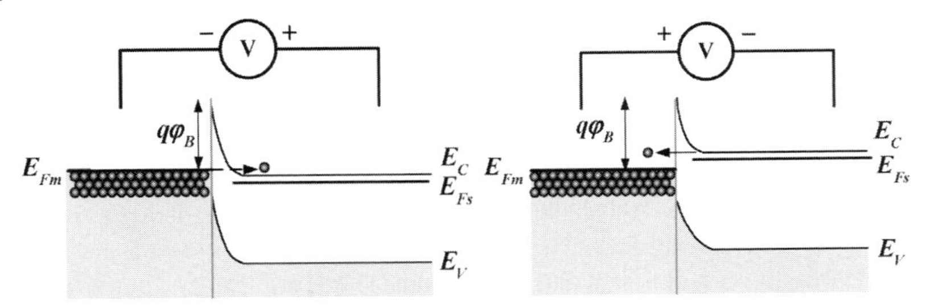

Figure 6.12 The tunnel-based ohmic contact. The potential barriers existing between metal electrode and semiconductor are made sufficiently thin to be nearly transparent to tunneling electrons.

while forming a quantum well between the two, as sketched in Figure 6.13. Without bias and at equilibrium, the energy level of the bound state in the quantum well does not line up with the metal Fermi level E_F, as indicated in the figure. In this configuration, the probability of an electron on top of the Fermi level tunneling from metal L to metal R is evidently small. However, when a positive voltage V is applied at metal R, E_F therein is lowered with respect to that in metal L and drags down the energy level of the bound state in the quantum well. This is because the positive bias V lowers the electronic energy level by the amount $-eV$. As a consequence, the transmission probability starts increasing with V, peaking at the bias at which the bound-state energy lines up with E_F in the metal L. At this bias point, the condition for the resonant tunneling is realized, namely, the energy level of the incident electron matches exactly with the bound-state energy level of the quantum well (see Figure 6.8).

With further increase in V, the bound-state energy level slides further down, the resonant tunneling condition is destroyed, and the tunneling current decreases. As a result, a negative resistance regime is formed (Figure 6.13). With further increase in

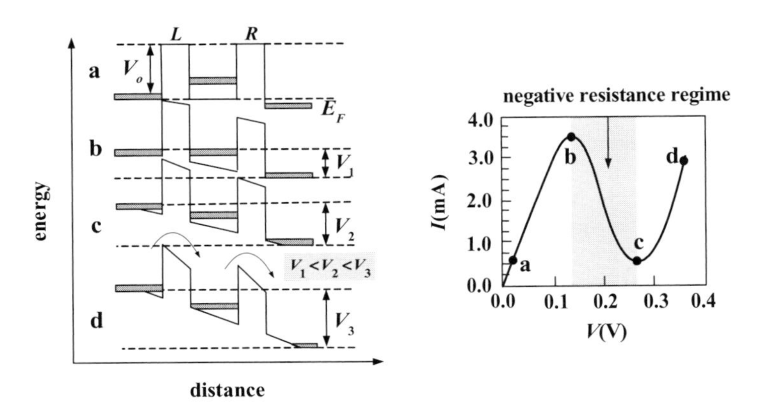

Figure 6.13 The resonant tunneling device consisting of two metal contacted potential barriers forming a quantum well between the two. Also shown is the energy level diagram corresponding to each regime in static I–V behavior. The tunnel current is modulated via voltage controlled overlap of subband with E_F.

V, the effective barrier height as encountered by electrons in metal L is lowered, and electrons can then fly over the barrier energetically, contributing to the current. The resulting I–V behavior is also sketched in the figure. The resonant tunneling device holds up the promise as a high-speed device, due to the extremely short tunneling time, that is, the transit time of electrons between the two electrodes.

Single-Electron Transistor and Coulomb Blockade With shrinking device geometry, it has become possible to control the tunneling of a single electron. All the devices discussed in the preceding paragraphs share a common feature namely that the device operation requires a large number of tunneling electrons. However, with the device rapidly downsized to the nanoscale geometry, controlling a single-electron transfer is fast becoming a reality even at the room temperature. In fact, the single-electron transistor is entirely based on such controlled single-electron transfer. The device consists of an island made of a metal or semiconductor, sandwiched between the "source" and "drain" electrodes via the thin dielectric layer, as sketched in Figure 6.14.

If the size of the island is sufficiently small, the electronic potential energy $-eV$ therein can vary due to the presence or absence of a single electron to such an extent that it profoundly affects the tunneling process itself. This size effect can be explicitly illustrated by taking the island junction as a simple parallel plate capacitor, for simplicity of discussion. The capacitance is then given by the area A and thickness d of the junction by

$$C = \frac{\varepsilon A}{d}, \quad \varepsilon = \varepsilon_r \varepsilon_0, \tag{6.29}$$

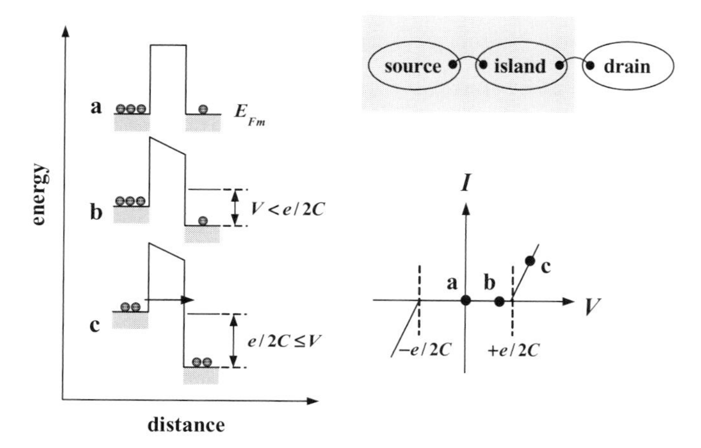

Figure 6.14 The single-electron transistor comprised of a quantum dot, isolated from the source and drain by thin dielectric layers. Also shown are the energy level diagrams corresponding to static I–V operation regimes, including the Coulomb blockade.

with ε denoting the permittivity of the medium specified in terms of the vacuum permittivity ε_0 and the dielectric constant ε_r. Also, the energy required to charge the capacitor with a charge Q is well known to be given by $Q^2/2C$. Hence, the basic charging energy of a single electron is given by

$$E_c = \frac{e^2}{2C}. \tag{6.30}$$

This charging energy, although small, can affect the tunneling process, provided it far exceeds the thermal energy, namely,

$$\frac{e^2}{2C} \gg k_B T. \tag{6.31}$$

Otherwise, the charging effect could be buried by the thermal fluctuation. For $A \simeq 1 \times 1\,\mu\text{m}^2$, $d \approx 10\,\text{nm}$, and $\varepsilon_r \approx 10$, for example, E_c is much smaller than the thermal energy even at extremely low temperatures. However, with shrinking device geometry, for example, at $A \approx 10 \times 10\,\text{nm}^2$ and $d \approx 2\,\text{nm}$, E_c is of the order of 15 meV and becomes comparable with the thermal energy at room temperature of approximately 25 meV. This clearly indicates the possibility of controlling the single electron effects by reducing the island size deep into nano dimensions.

An additional constraint required is that E_c should exceed the uncertainty in energy of the single electron, inherently associated with its finite lifetime. The lifetime of an excess electron in the island can be roughly estimated by $\tau = R_T C$, where R_T represents the effective tunneling resistance, which is inversely proportional to the tunneling probability. Evidently, this lifetime is analogous to the RC time constant of a capacitor, connected in series with resistance. The charging energy E_c that is much greater than the uncertainty in energy of the excess electron can be expressed from the uncertainty relation by

$$\Delta E \approx \frac{h}{\tau} = \frac{h}{R_T C} \ll \frac{e^2}{2C}. \tag{6.32a}$$

Equivalently,

$$R_K \ll R_T, \quad R_K \equiv \frac{h}{e^2}. \tag{6.32b}$$

The resistance $R_K \simeq 25.8\,\text{k}\Omega$ thus introduced is one of the basic physical parameters, known as the quantum resistance. The condition (6.32) is essentially equivalent to ensuring that there is a quantum state for the excess electron to jump in and be localized in the island.

When the two basic constraints (6.31) and (6.32) are met, the charging energy of a single electron (6.30) can play a crucial role in the operation of the single-electron transistor. This is illustrated in Figure 6.14, where the island of quantum dot is surrounded by two potential barriers set up by dielectric layers placed in between the electrodes and the island. Thus, the electron may tunnel through the barrier from the source to the island or vice versa. In the process of an excess electron being tunneled into the island, the amount of charge Q therein changes by a discrete amount $-e$ and the resulting change in the electrostatic energy of the island is given by

$$\Delta E = \frac{Q^2}{2C} - \frac{(Q-e)^2}{2C} = \frac{e(Q-e/2)}{C}. \tag{6.33}$$

Since the tunneling is an elastic process, it cannot occur unless this change in charging energy is positive, that is,

$$\Delta E \geq 0. \tag{6.34}$$

Thus, for $Q = 0$ before the tunneling in of an excess electron, the condition (6.34) requires an equivalent charge to compensate for $-e/2$. This requirement can be met by applying a positive voltage V at the drain, such that $CV \geq e/2$. By the same token, the same magnitude of negative voltage at the drain could cause the tunneling in of an electron from the drain to the quantum dot. It is therefore clear that in the range of drain bias,

$$-\frac{e}{2C} < V < \frac{e}{2C}, \tag{6.35}$$

the condition (6.34) is not satisfied and there is no tunneling current. This is known as the Coulomb blockade.

Figure 6.14 shows the resulting current–voltage characteristics, including the Coulomb blockade. When the single electron tunnels into the quantum dot from the source, the electronic energy level in the dot is raised by $e^2/2C$, thereby blocking the tunneling process itself. The only way to suppress this self-blocking is to apply a positive voltage at the drain and bring down the energy of the quantum dot by $V = e/2C$, so that the electronic energy levels at the source and the quantum dot are kept lined up during the tunneling (Figure 6.14). With further increase of V beyond this blocking voltage, the condition (6.34) is automatically satisfied and electrons can undergo the usual tunneling to give rise to current. By the same token, if a negative bias is applied to the drain beyond $-e/2C$, the tunneling current ensues flowing in the opposite direction.

6.5 Problems

6.1. Consider a quantum well formed by two potential barriers with a common height V but with different thicknesses d_L and d_R, and W distance apart, as sketched in Figure 6.15.

(a) First, approximate the quantum well by an infinite square well potential by taking $V \to \infty$. Find the ground-state energy E_0 of an electron in the well as a function of W in nanometer. Take the effective mass of the electron in the well as $m_n \simeq m_0$ and $m_n \simeq 0.2\, m_0$ with m_0 denoting the rest mass.

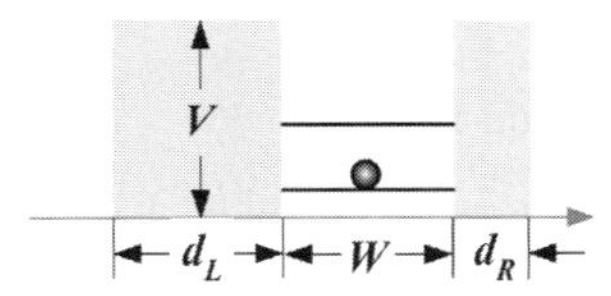

Figure 6.15 Quantum well formed by two potential barriers with different thicknesses.

(b) Take $V = 3.1$ eV and find a few electron eigenfunctions and eigenenergies therein and compare with the result obtained in (a).

(c) At what value of W is the ground state energy E_0 equal to the thermal energy $k_BT/2$ at room temperature?

(d) The electrons residing in the well form two-dimensional gas system, confined via two potential barriers in one direction but propagating freely on a plane parallel to the potential surfaces. Plot the density of states over the energy interval $0 < E < 2.5$ eV as a function of W for $2\text{nm} \leq W \leq 200$ nm.

6.2. Consider the same quantum well as considered in (6.1).

(a) Find the tunneling probabilities of electron through the two barriers with the common height $V = 3.1$ eV but with thickness $d_L = 50$ nm and $d_R = 10$ nm, respectively, and compare the results.

(b) Find the lifetime of the electron in the ground state subband for $W = 200$, 10, and 1 nm, respectively. The lifetime, τ is defined by the tunneling probability, T times the number of encountering the barrier N being equal to unity, namely,

$$TN = 1, \quad \tau = N/\nu$$

where the frequency of encounter at each barrier, ν is defined by the inverse round-trip time

$$\nu = \frac{1}{2W/v}; \quad \frac{mv^2}{2} = E.$$

6.3. Consider the same quantum well as considered in (6.1) and (6.2).

(a) If an electric field E is applied between the quantum well and the barrier on the right-hand side, the potential shape is transformed from square to trapezoidal or triangular shape, as shown in Figure 6.16. Find the electron lifetime versus the electric field applied for $2\text{ nm} \leq W \leq 200$ nm.

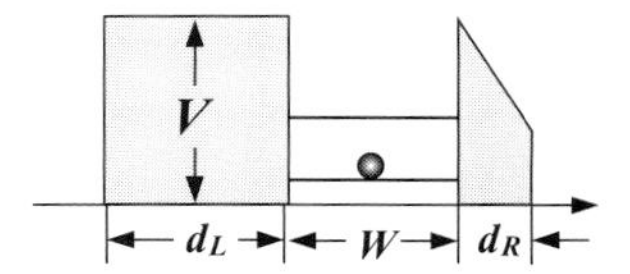

Figure 6.16 Quantum well in which the potential barrier on the right hand side is modified by applying electric field.

(b) At what value of E does the trapezoidal shape reduces to the triangular shape for the electron in the ground state subband for $W = 50$, 200 nm? Calculate the field amplitude E necessary to shorten the electron lifetime to 1 μs.

6.4. In scanning tunneling microscope, a metal tip has the work function of 4.5 eV.

(a) Find the electric field at which the electron tunneling probability is 10^{-4} if the distance between the tip and sample is 1 nm.
(b) If 100 mV is applied between the tip and sample, estimate the distance between the tip and sample to attain the same tunneling probability of 10^{-4}.

6.5. (a) Apply the boundary conditions (6.15) and (6.16) with the assumption $m_j = m_{j+1}$ on the wavefunctions (6.13) and (6.14) and obtain the transfer matrix (6.18) in the text.
(b) Fill in the algebraic steps and derive the transfer matrix in (6.21).
(c) Show that (6.22) is identical to (6.5).
(d) Show that $|B_j/A_j|^2$ in (6.20) leads to the same reflection coefficient R as in (6.7).

6.6. Reproduce the tunneling probabilities through the two potential barriers considered in Figure 6.8. Next, fix the barrier height and thickness and vary the width of the quantum well and plot the tunneling probability versus incident electron energy. Discuss the emerging general features of the resonant tunneling.

Suggested Reading

1 Singh, J. (1996) *Quantum Mechanics, Fundamentals & Applications to Technology*, John Wiley & Sons, Inc.

2 Kroemer, H. (1994) *Quantum Mechanics for Engineering, Materials Science, and Applied Physics*, International edn, Prentice Hall.

7
Periodic Potential and Energy Bands

The band theory of solids is one of the most important theories in quantum mechanics and it provides the bases, for example, upon which the semiconductor devices operate. The basic concept of the energy band is discussed, using a simple one-dimensional crystal in conjunction with Kronig–Penny model and resonant tunneling. Both the role and physical significance of the dispersion relations in energy bands are highlighted and the concept of the effective mass of electrons in solids is discussed.

7.1
One-Dimensional Crystal and Kronig–Penny Model

The one-dimensional (1D) crystal is often modeled by a linear array of positive ions, located periodically, as sketched in Figure 7.1. The potential encountered by an electron in such a crystal is then provided by the attractive Coulomb potential between the electron and ions. The resulting periodic potential could in turn be simplified by a string of square barrier potentials, separated by quantum wells of identical width, as sketched in the same figure. This idealized version of the periodic 1D potential is known as Kronig–Penny (K–P) potential and it brings out the concept of the energy bands in a simple and elegant manner. With quantum wires rapidly gaining practical importance, the K–P model has acquired an additional significance.

7.1.1
The Bloch Wavefunction

Consider the K–P potential with its unit cell shown in Figure 7.2. Here, the widths of the well and barrier are denoted by a and b, respectively, and the total length of the cell is $d\ (= a + b)$. A central theme of the discussion is the electron wavefunction in the presence of a periodic potential. Inasmuch as the potential is periodic, that is,

$$V(x) = V(x + d), \tag{7.1}$$

the Hamiltonian

Introductory Quantum Mechanics for Semiconductor Nanotechnology. Dae Mann Kim

ISBN: 978-3-527-40975-4

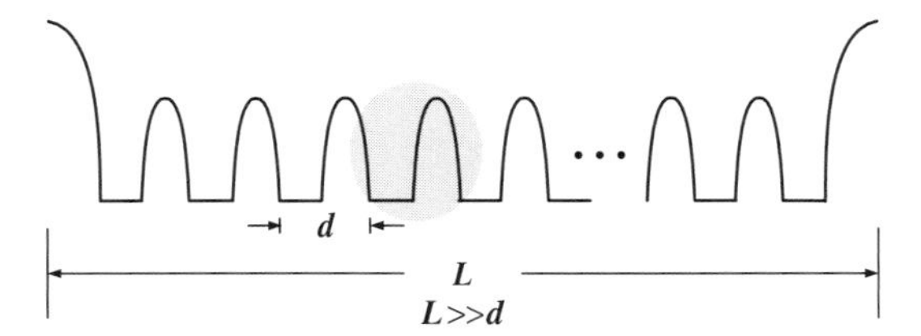

Figure 7.1 The periodic potential in 1D crystal.

$$\hat{H}(x) = -\frac{\hbar^2}{2m}\frac{\partial^2}{\partial x^2} + V(x) \tag{7.2}$$

is also periodic, that is,

$$\hat{H}(x) = \hat{H}(x+d). \tag{7.3}$$

In this case, the eigenfunction of $\hat{H}$ can be generally represented by a modulated plane wave

$$\varphi(x) = e^{ikx}u(x), \tag{7.4}$$

with the modulation function $u(x)$ periodic over the same distance d, namely,

$$u(x) = u(x+d). \tag{7.5}$$

The wavefunction (7.4) is known as the Bloch wavefunction and it accurately describes the behavior of electrons in the crystal for arbitrary potential shape $V(x)$ within the unit cell.

The validity of the Bloch theorem can be shown simply by considering the displacement operator

$$\hat{D}f(x) = f(x+d), \tag{7.6}$$

where $f(x)$ is an arbitrary function. Then, it follows from (7.4)–(7.6) that

$$\hat{D}\varphi(x) = \hat{D}e^{ikx}u(x) \equiv e^{ik(x+d)}u(x+d) = e^{ikd}\varphi(x). \tag{7.7}$$

Hence, the Bloch wavefunction is an eigenfunction of $\hat{D}$ with the eigenvalue exp ikd.

It is also evident from (7.3) and (7.6) that the two operators $\hat{H}$ and $\hat{D}$ commute with each other. Since two commuting operators can share the common eigenfunction

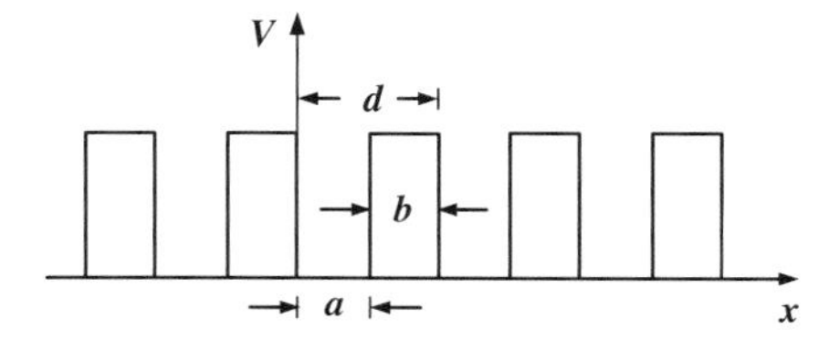

Figure 7.2 The Kronig–Penny periodic potential. The unit cell consists of square well of width a and potential barrier of thickness b, so that the total length is $d = a + b$.

and since the Bloch wavefunction is the eigenfunction of $\hat{D}$, it is also the eigenfunction of $\hat{H}$ as long as $V(x)$ is periodic. It is essential to notice that the probability density of the electron as described by the Bloch function is generally characterized by

$$|\varphi(x+nd)|^2 = \left|e^{ik(x+nd)}u(x+nd)\right|^2 = \left|e^{iknd}\varphi(x)\right|^2 = |\varphi(x)|^2, \quad n = 1, 2, \ldots \tag{7.8}$$

and is consistent with the basic premise of a periodic system, namely, that there is no preferred ionic site and the electron is therefore to be found in all unit cells with equal probability.

7.1.2
Bloch Wavefunction in K–P Potential

Next, the specific Bloch wavefunction in K–P potential is analyzed and discussed. For this purpose, the wave vector k in (7.4) is first specified using the ring boundary condition, as sketched in Figure 7.3. In this boundary condition, the electron leaves the last cell in the ring to reenter into the first one and the number of unit cells N is taken to be very large. This kind of boundary condition is often used for considering the motion of the electron in the bulk crystal, while neglecting the edge effects. From this ring boundary condition, one can write

$$e^{ikdN} = 1 \equiv e^{i2\pi n}, \tag{7.9}$$

where n denotes a positive integer. The wave vector is thus determined by

$$k_n = \frac{2\pi n}{dN} = \frac{2\pi n}{L}, \tag{7.10}$$

where L is the length of the crystal. It is clear from (7.10) that the k-values become quasi continuous in the limit of large N. The momentum associated with the wave vector, that is, $\hbar k_n$ is called the crystal momentum.

In the limit in which the well width of the unit cell is infinite, the modulation function $u(x)$ in (7.4) should be put to unity. In this case, the Bloch wavefunction $\varphi(x)$

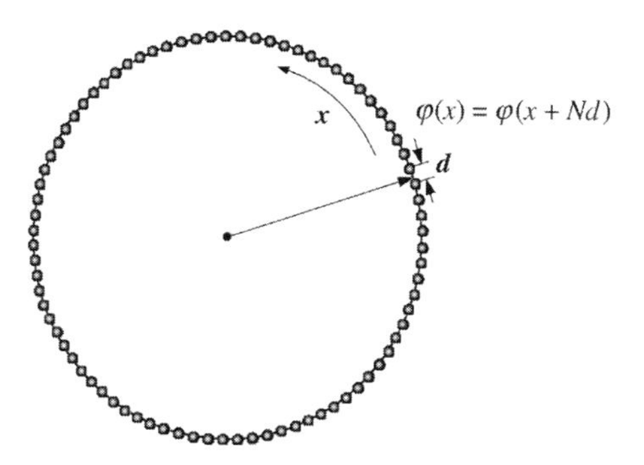

Figure 7.3 The graphical illustration of the ring boundary condition in 1D crystal.

reduces to that of a free particle, as it should. However, in the presence of periodic potential $V(x)$, $\varphi(x)$ should obviously be a modulated plane wave and this modulation is accounted for by $u(x)$. Since $u(x)$ should also be periodic, the probability density of the wavefunction is identical in all unit cells.

With this general picture in mind, consider the K–P potential, as sketched in Figure 7.2. One can write the wavefunction in the usual manner as

$$\varphi = \begin{cases} Ae^{ik_1x} + Be^{-ik_1x}, & 0 \le x \le a, \\ Ce^{-\varkappa x} + De^{\varkappa x}, & a \le x \le d, \end{cases} \tag{7.11}$$

See (6.1), for example. Here, the analysis is focused on and confined to the case where the electron energy is less than the barrier height, that is, $E < V$. Then, k_1 and $\varkappa$ in the well and barrier are given by

$$k_1 = [2mE/\hbar^2]^{1/2}, \tag{7.12a}$$

$$\varkappa = [2m(V-E)/\hbar^2]^{1/2}. \tag{7.12b}$$

Periodic Boundary Condition Now, the periodic boundary conditions for the modulation function $u(x)$ should be first considered, namely, $u(0^+) = u(d^-)$ and $u'(0^+) = u'(d^-)$. The former condition can be expressed in terms of $\varphi(x)$ from (7.4), that is, via the relation $u(x) = \varphi(x)\exp{-ikx}$ as

$$\varphi(0^+) = e^{-ikd}\varphi(d^-), \tag{7.13a}$$

while the latter condition is specified via $u'(x) = \varphi'(x)[\exp-(ikx)]-iku(x)$:

$$\varphi'(0^+) = e^{-ikd}\varphi'(d^-). \tag{7.13b}$$

In (7.13b), the second term in $u'(x)$ is automatically canceled from both sides in view of (7.13a). Hence, inserting (7.11) into these periodic boundary conditions of (7.13), one can write

$$A + B = e^{-ikd}(Ce^{-\varkappa d} + De^{\varkappa d}), \tag{7.14a}$$

$$ik_1(A-B) = -\varkappa e^{-ikd}(C^{-\varkappa d} - De^{\varkappa d}). \tag{7.14b}$$

Also, the usual boundary condition that $\varphi(x)$ and $\varphi'(x)$ be continuous inside the unit cell at $x = a$ is specified with the use of (7.11) by

$$Ae^{ik_1a} + Be^{-ik_1a} = Ce^{-\varkappa a} + De^{\varkappa a}, \tag{7.15a}$$

$$ik_1(Ae^{ik_1a} - Be^{-ik_1a}) = -\varkappa(C^{-\varkappa a} - De^{\varkappa a}). \tag{7.15b}$$

The analysis of the wavefunction of (7.11) is thus reduced to determining the four constants of integration A, B, C, and D in such a manner that the four boundary conditions given in (7.14) and (7.15) are satisfied. For this purpose, one can first express A and B in terms of C and D from (7.14), obtaining

$$A = \frac{1}{2} e^{-ikd} (\alpha e^{-\varkappa d} C + \alpha^* e^{\varkappa d} D), \tag{7.16a}$$

$$B = \frac{1}{2} e^{-ikd} (\alpha^* e^{-\varkappa d} C + \alpha e^{\varkappa d} D), \tag{7.16b}$$

where

$$\alpha \equiv 1 + \frac{i\varkappa}{k_1}. \tag{7.17}$$

One can likewise find A and B in terms of C and D from (7.15) as

$$A = \frac{1}{2} e^{-ik_1 a} (\alpha e^{-\varkappa a} C + \alpha^* e^{\varkappa a} D), \tag{7.18a}$$

$$B = \frac{1}{2} e^{ik_1 a} (\alpha^* e^{-\varkappa a} C + \alpha e^{\varkappa a} D). \tag{7.18b}$$

Next, one can eliminate A and B by equating the right-hand sides of (7.16a), (7.18a), (7.16b), and (7.18b) and regrouping the terms, obtaining the coupled equations for C and D:

$$\begin{pmatrix} a_{11} & a_{12} \\ a_{21} & a_{22} \end{pmatrix} \begin{pmatrix} C \\ D \end{pmatrix} = 0, \tag{7.19}$$

with

$$a_{11} = \alpha(e^{-ikd - \varkappa d} - e^{-ik_1 a - \varkappa a}), \tag{7.20a}$$

$$a_{12} = \alpha^*(e^{-ikd + \varkappa d} - e^{-ik_1 a + \varkappa a}), \tag{7.20b}$$

$$a_{21} = \alpha^*(e^{-ikd - \varkappa d} - e^{ik_1 a - \varkappa a}), \tag{7.20c}$$

$$a_{22} = \alpha(e^{-ikd + \varkappa d} - e^{ik_1 a + \varkappa a}). \tag{7.20d}$$

Since the coupled equation (7.19) is homogeneous, it readily follows from the Kramer's rule that the constants C and D are zero, unless the secular equation is satisfied, namely, the determinant of the coupling matrix is zero:

$$\begin{vmatrix} a_{11} & a_{12} \\ a_{21} & a_{22} \end{vmatrix} = 0. \tag{7.21}$$

It is important to notice at this point that if C and D are zero, so are A and B, as clearly follows from (7.16) or (7.18) and, consequently, the wavefunction under consideration becomes trivial. Therefore, the secular equation (7.21) is a crucial condition to be satisfied to make the wavefunction physical and nontrivial. The secular equation (7.21) can be specified in a routine manner by expanding the determinant. One finds after a lengthy but straightforward algebra

$$\cos k_1 a \cosh \varkappa b - \frac{k_1^2 - \varkappa^2}{2k_1\varkappa} \sin k_1 a \sinh \varkappa b = \cos kd. \tag{7.22}$$

In obtaining (7.22), use has been made of the identity

$$e^{-2ikd} + 1 \equiv 2e^{-ikd} \cos kd.$$

It is important to note that (7.22) represents the implicit dispersion relation between the energy of the electron E and the wave vector k with V, a, b, and d $(= a + b)$ entering as the crystal parameters. Once E is found as an explicit function of k or vice versa from (7.22), both the wavefunction $\varphi(x)$ in (7.11) and the modulation function $u(x)$ in (7.4) can be readily obtained.

7.2
E–k Dispersion and Energy Bands

As pointed out, the transcendental equation (7.22) implicitly correlates E with k and one has to resort to graphical or numerical means to find E as an explicit function of k or vice versa. The $E-k$ dispersion inherent in (7.22) is rich in physical contents and is discussed in this section by analyzing the $E-k$ curves by graphical means. Figure 7.4 plots the left-hand side of (7.22) versus E/V and, as can be clearly seen from the figure, this curve oscillates with diminishing amplitudes with increasing E. Also indicated in the figure are the values of $\cos kd$ appearing on the right-hand side of (7.22). Based on these two plots, the general features of the $E-k$ behavior can be brought out.

Allowed Bands and Forbidden Gaps Since $|\cos kd| \leq 1$, it is clear from Figure 7.4 that only for those E values for which the left-hand side of (7.22) falls within the limited bounds of $\cos kd$, there exists one-to-one correspondence between real k and real E. In

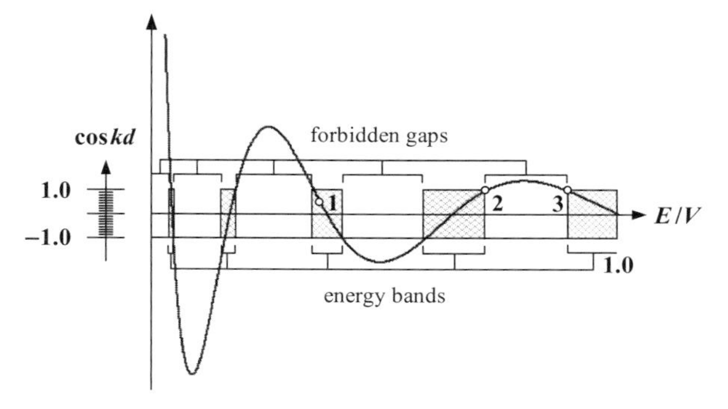

Figure 7.4 The graphical analysis of the dispersion relation. The left-hand side of (7.22) is plotted versus E/V, together with typical values of $\cos kd$. The allowed energy bands and forbidden gaps are demarcated by the condition $\cos kd = \pm 1$.

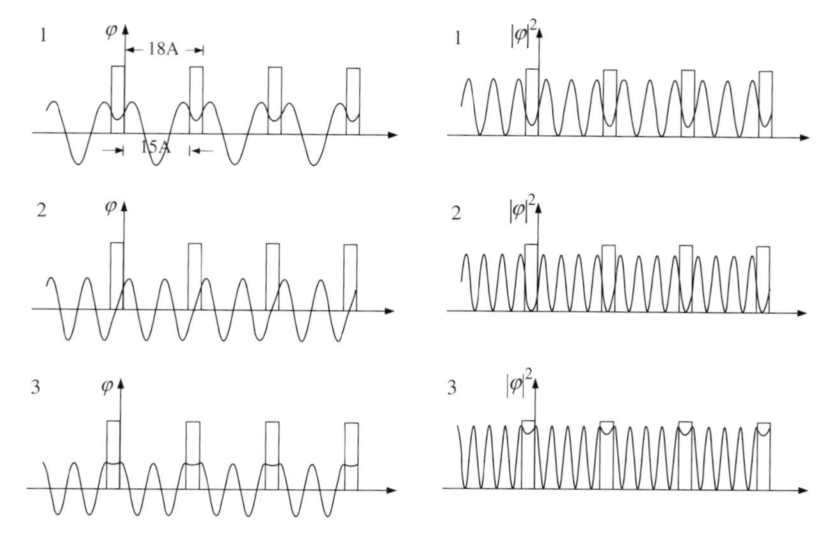

Figure 7.5 Typical Bloch wavefunctions in energy bands and corresponding probability densities for different values of E and k pair: E chosen from within an energy band (1), at the top of given band (2), and at the bottom of next higher lying band (3).

this range of E, the electron can freely propagate in the crystal with a real propagation vector k. The energy range in which this condition is met is called the allowed band, as indicated in the figure. On the other hand, for those E values for which the magnitude of the left-hand side of (7.22) is greater than unity, the corresponding k becomes complex and the free propagation ceases to exist. Such energy ranges are called the forbidden gaps.

As a consequence, the energy spectrum of the electron in 1D crystal consists of a series of allowed bands, separated by forbidden gaps. Moreover, the width of the allowed band is shown broadened with increasing E. By the same token, the width of a given allowed energy band decreases with increasing potential barrier V, that is, with tighter binding of electrons by host atoms.

Wavefunctions in Energy Bands Now that allowed energy bands and forbidden gaps have been shown as inherent property of the dispersion relation, the Bloch wavefunction therein is discussed in more detail. The procedure for finding $\varphi(x)$ is as follows. One first chooses a pair of E and k values satisfying the dispersion relation (7.22) and inserts those values in (7.19). The two coupled equtions then become redundant by the very definition of the secular equation (see (7.19) and (7.21)) and, therefore, one can express C in terms of D or vice versa via either $C = -(a_{12}/a_{11})D$ or $C = -(a_{22}/a_{21})D$. Once C is related to D, the constants A and B can also be expressed in terms of D, using (7.16). In this way, $\varphi(x)$ is specified in terms of D for given E and k. The constant D can then be used to normalize the wavefunction, if required.

Figure 7.5 shows typical wavefunctions thus found for different values of E and k pair, together with corresponding probability densities. Indeed, these wavefunctions in one unit cell are similar in shape to bound-state wavefunctions in quantum wells, as expected. It is interesting to notice that the wavefunctions at the top of a given energy band (point 2, Figure 7.4) and at the bottom of the next higher lying energy band (point 3, Figure 7.4) are antisymmetric or symmetric, centered at the midpoint in the quantum well. It is also interesting that the latter wavefunction has higher probability density in the potential barrier region.

The $E-k$ Dispersion in Allowed Bands It is also clear from (7.22) that a given E in allowed bands is associated with multiple k values, namely, $k + 2\pi n/d$ with n denoting any integer. Conversely, for given k, there exist multiple E values, as can be clearly observed from Figure 7.4. The resulting $E-k$ curves are shown in Figure 7.6, in which E is matched one-to-one with $k + 2\pi n/d$ and is plotted versus $kd + 2\pi n$. Note that E is an even function of kd, as is clear from (7.22). Specifically, a given E on the left-hand side of (7.22) is matched by both kd and $-kd$ on the right-hand side. Also, for comparison, the equivalent free particle dispersion,

$$E = \frac{[\hbar(k + 2\pi n/d)]^2}{2m}, \tag{7.23}$$

has been plotted using the momentum $p = \hbar(k + 2\pi n/d)$. As clearly shown in the figure, the free particle $E-k$ curve roughly follows the crystal $E-k$ curve but the two differ substantially near the band edges.

Since k is determined to within an integer multiple of $2\pi/d$, it suffices to draw all $E-k$ curves obtained in entire allowed bands in the single interval, $-\pi \le kd \le \pi$, called the first Brillouin zone. This is done by collecting all $E-k$ curves from other Brillouin zones and displacing the curves horizontally by $\pm 2\pi n/d$. The resulting reduced zone $E-k$ curves are also shown in Figure 7.6.

It is interesting to notice that at the band edges, the energy as a function of k has a zero slope, that is, $\partial E(k)/\partial k = 0$. This can be seen on a general ground by differentiating both sides of (7.22) with respect to k, obtaining $\partial E(k)/\partial k \propto \sin kd$. At band edges $kd = n\pi$ and therefore $\sin kd = 0$. This carries an interesting

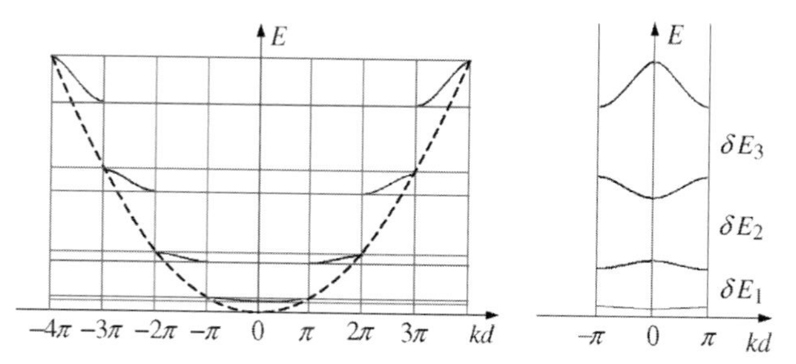

Figure 7.6 The $E-k$ dispersion curves: E versus $k + 2\pi n/d$ (left) and the first four dispersion curves in reduced Brillouin zone (right). Also shown for comparison is the dispersion curve for a free particle, $E = (\hbar^2/2m)\ (k + 2\pi n/d)^2$ (dotted line).

implication, namely, that near the band edges, $E \propto k^2$ and the energy is therefore well approximated by that of a free particle. This in turn suggests that the electron should indeed behave as a free particle near band edges.

Also, at the edge of the forbidden gap, $\cos kd = \pm 1$ so that $kd \equiv (2\pi/\lambda)d = n\pi$ or, equivalently, $2d = n\lambda$ with n denoting an integer. Hence, in this case, the round-trip distance of the unit cell is an integer multiple of the de Broglie wavelength of the electron. This condition is clearly equivalent to the one-dimensional Bragg reflection, representing the constructive interference of reflected waves. The wave then cannot penetrate into the next cell and propagate but becomes evanescent. In this case, the wavefunction degenerates into a standing wave, consisting of both forward and backward components with equal amplitudes. Naturally, there are two ways of forming standing waves, that is, even and odd parity wavefunctions with two different energy eigenvalues, as already pointed out. The resulting splitting of energy levels accounts for the top and bottom of the energy gap. This can be clearly observed from Figure 7.5, in which wavefunctions found at the top of a given energy band and at the bottom of the next higher lying energy band are plotted (2 and 3 in Figures 7.4 and 7.5). The main difference between these two wavefunctions consists in high and low amplitudes in higher energy potential barrier region. The resulting difference in energy, that is, $\langle \hat{H} \rangle$ in the unit cell, accounts for the difference in energies amounting to the energy gap for the given crystal wave vector k.

Quantum States Per Band The total number of the quantum states per band is an important quantity and can be analyzed using the ring boundary condition (7.10). Clearly, the number of k values or, equivalently, the number of wavefunctions in the dk range is given from (7.10) by

$$dn = \frac{L}{2\pi} dk. \tag{7.24}$$

Multiplying (7.24) by 2 to account for the two quantum states for every k due to the electron spin, namely, the spin up and spin down states and integrating (7.24) over the Brillouin zone, one obtains

$$n = 2\frac{L}{2\pi} \int_{-\pi/d}^{\pi/d} dk = \frac{2L}{d} = 2N, \tag{7.25}$$

where N is the total number of unit cells in the crystal. Thus, the total number of the quantum states per band is given by the number of unit cells constituting the crystal.

7.2.1
The Dispersion Curves and Electron Motion

The dispersion curves plotted in Figure 7.6 for K–P potential provide a convenient means by which to discuss the motion of the electron in the crystal. To discuss the motion, the $E-k$ dispersion of a free particle is first considered as a reference, that is,

$$E = \hbar^2 k^2/2m, \quad \hbar k \equiv p. \tag{7.26}$$

As noted, the velocity of the particle is obtained in this case by

$$v = \frac{1}{\hbar}\frac{dE}{dk} = \frac{\hbar k}{m} = \frac{p}{m} \tag{7.27}$$

and can be identified with the group velocity with which the envelope of the wave packet is propagating, namely,

$$v_g = \frac{\partial \omega}{\partial k} = \frac{1}{\hbar}\frac{\partial E}{\partial k}. \tag{7.28}$$

This means that the propagation velocity of the electron in the crystal is to be described by the slope of the E–k curve.

Also, when an external electric field E is applied, the electron with charge $-e$ gains energy δE from the field in time δt as

$$\delta E = -e\mathrm{E} v_g \delta t, \tag{7.29}$$

where $-e\mathrm{E}$ is the force acting on electron and $v_g \delta t$ is the displacement made in time δt. The energy gain can also be expressed using (7.28) as

$$\delta E \equiv \frac{dE}{dk}\delta k = \hbar v_g \delta k. \tag{7.30}$$

Hence, equating the right-hand sides of (7.29) and (7.30) results in the usual equation of motion:

$$\frac{\hbar \delta k}{dt} \equiv \frac{dp}{dt} = -e\mathrm{E}. \tag{7.31}$$

Now the factor $\delta k/dt$ in (7.31) can be related to the acceleration of the electron with the aid of (7.28) by

$$a \equiv \frac{dv_g}{dt} = \frac{1}{\hbar}\frac{d}{dt}\left(\frac{\partial E(k)}{\partial k}\right) = \frac{1}{\hbar}\frac{\partial^2 E(k)}{\partial k^2}\frac{\partial k}{\partial t}. \tag{7.32}$$

Hence, eliminating $\delta k/dt$ from (7.32) and (7.31), one can write

$$\left[\frac{1}{\hbar^2}\frac{\partial^2 E(k)}{\partial k^2}\right]^{-1} a = -e\mathrm{E} \tag{7.33}$$

and can formally identify (7.33) with Newton's equation of motion, relating the force with acceleration via the mass. Thus, the effective mass of the electron in the crystal is to be represented in terms of the operation

$$m_n = \left(\frac{1}{\hbar^2}\frac{\partial^2 E(k)}{\partial k^2}\right)^{-1}. \tag{7.34}$$

7.3
Energy Bands and Resonant Tunneling

The energy bands that naturally follow from the dispersion relation (7.22) can be further elaborated from the standpoint of the resonant tunneling of electrons in the periodic potential. For this purpose, first consider the limiting case, where the barrier height V becomes infinite. Then, the quantities $\varkappa$, $\sinh \varkappa b$, and $\cosh \varkappa b$ appearing in (7.22) all diverge (see (7.12)) and, consequently, the dispersion relation (7.22) reduces simply to

$$\sin k_1 a = 0.$$

This leads with the use of (7.12) to

$$k_1 a = [2mE/\hbar^2]^{1/2} a = n\pi, \quad n = 1, 2, \ldots. \tag{7.35}$$

Therefore, the energy levels of the Bloch wavefunctions are given by

$$E_n = \frac{\hbar^2 \pi^2}{2ma^2} n^2 \tag{7.36}$$

and (7.36) is identical to the expression of the quantized energy levels in infinite square well potential of width a (see (4.9)). This is to be expected, since the electron in this case is strictly confined to one quantum well and is not coupled with other unit cells.

Another interesting limiting case consists of increasing the potential barrier width b to a large value, so that the coupling between the two neighboring cells is minimized. In this limit of large b, $\sinh \varkappa b = \cosh \varkappa b \to \infty$ and the dispersion relation (7.22) reduces to

$$\tan 2\xi = \frac{2\xi\eta}{\xi^2 - \eta^2}, \quad k_1 a \equiv 2\xi, \ \varkappa a \equiv 2\eta, \tag{7.37}$$

and (7.37) can be recast with the use of the trigonometric identity,

$$\tan 2\xi = 2\tan\xi/(1 - \tan^2\xi), \tag{7.38}$$

into a quadratic equation for $\tan\xi$ as

$$\tan^2\xi + \frac{\xi^2 - \eta^2}{\xi\eta} \tan\xi - 1 = 0. \tag{7.39}$$

Hence, one can specify $\tan\xi$ explicitly by formally solving this quadratic equation, obtaining

$$\xi \tan\xi = \eta, \tag{7.40a}$$

$$\xi \cot\xi = -\eta \tag{7.40b}$$

as positive and negative branches of the solution. Clearly, (7.40a) and (7.40b) represent the reduced version of the dispersion relation (7.22) and are identical to the dispersion relations and/or the quantization conditions in quantum well of width a corresponding to even and odd parity eigenfunctions (see (4.45) and (4.48)).

In light of these two limiting cases, the nature of the extended Bloch wavefunction and the energy bands can be understood from a simple dynamic point of view.

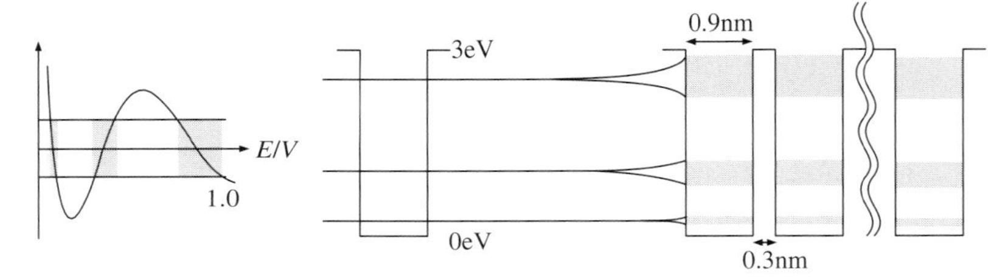

Figure 7.7 A schematic illustration of bound-state energy levels in isolated quantum well being broadened into corresponding bands in 1D crystal. Also shown are the corresponding dispersion curves.

Specifically, it is now apparent that the energy bands are rooted in the same energy quantization conditions as those discrete energy levels in isolated quantum wells. The only difference between the two consists in the fact that those discrete energy levels are broadened into bands obviously due to the overlapping of the eigenfunctions in adjacent cells, as illustrated in Figure 7.7.

Thus, the fact that an electron behaves as a free particle in allowed bands can be interpreted in terms of the resonant tunneling. That is, an electron in allowed bands automatically satisfies the condition of the resonant tunneling by occupying one of the eigenstates of the quantum well. Therefore, the electron can tunnel through the string of potential barriers with 100% probability. In this context, the potential barriers become transparent and the electron moves as if it is a free particle. Inasmuch as the resonant tunneling is rooted in the wave nature of electrons, the allowed bands are deeply rooted in the duality of matter.

7.4 Problems

7.1. The dispersion relation derived in (7.22) can be simplified by considering the limiting case, in which the barrier height V is taken to be infinite while the width b is taken infinitesimally small such that the product $V \times b$ is finite.

(a) Show in this limit that the dispersion relation (7.22) reduces to

$$P\frac{\sin k_1 a}{k_1 a} + \cos k_1 a = \cos ka,$$

where

$$P \equiv \frac{mVba}{\hbar^2}.$$

(b) Taking P as a parameter, analyze the simplified dispersion relation and discuss the behavior of the first few subbands and/or energy bands in terms of the parameters V, a, and b.

7.2. Consider a simple characteristic matrix equation

$$\begin{pmatrix} 1 & 2 \\ 2 & 1 \end{pmatrix}\begin{pmatrix} x_1 \\ x_2 \end{pmatrix} = \lambda\begin{pmatrix} x_1 \\ x_2 \end{pmatrix}$$

or equivalently, the coupled equation,

$$\begin{aligned}(1-\lambda)x_1 + 2x_2 &= 0 \\ 2x_1 + (1-\lambda)x_2 &= 0.\end{aligned} \tag{7.41}$$

(a) Show that the solution of the coupled equation is trivial unless the secular equation is satisfied,

$$\begin{vmatrix} 1-\lambda & 2 \\ 2 & 1-\lambda \end{vmatrix} = 0. \tag{7.42}$$

(b) Show that the two characteristic roots $\lambda_{\pm}$ found from (7.42) yield the infinite number of solutions as long as x_1 and x_2 are related by $x_2 = \pm x_1$.

(c) Show that the relation $x_2 = \pm x_1$ can be found from any one of the two equations in (7.41).

(d) Show that if the normalization condition is imposed, that is, $x_1^2 + x_2^2 = 1$, the solution is given by

$$X_1 = \frac{1}{\sqrt{2}}\begin{pmatrix} 1 \\ 1 \end{pmatrix}; \quad X_2 = \frac{1}{\sqrt{2}}\begin{pmatrix} 1 \\ -1 \end{pmatrix}.$$

7.3.

(a) Starting from (7.14) and (7.15), fill in the algebraic steps and reproduce the coupled 2×2 matrix equation (7.19) and (7.20).

(b) By expanding the secular equation, derive the transcendental dispersion relation (7.22).

(c) Perform the graphical analysis of the dispersion relation using $V = 3$ eV, $a = 1.0$ nm, $b = 0.3$ nm, and $m = 0.2\, m_0$, m_0 being the electron rest mass.

(d) Choose a few pairs of E, k values from the allowed energy bands and construct the corresponding Bloch wavefunctions.

(e) Are there any similarities existing between problems 2 and 3?

7.4. Consider a Kronig–Penny periodic potential in which the unit cell parameters are given by the barrier height $V = 1.5$ eV, width $b = 0.5$ nm, and well width $a = 0.5$ nm. Find the width of the first and second allowed bands.

7.5. The superlattice structures can be represented by periodic potentials consisting of quantum wells and barrier potentials for both electrons and holes, as sketched in Figure 7.8.

(a) For given potential barrier of 0.3 eV, design the well width a such that the first two subbands for electrons are separated by 40 meV. Take effective mass of electron $m_n \simeq 0.07 m_0$, with m_0 denoting the rest mass.

(b) Estimate the widths of subbands in (a).

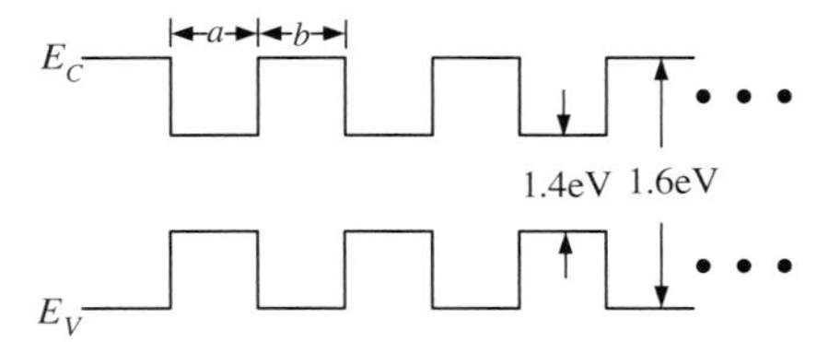

Figure 7.8 Superlattice structure comprised of quantum wells for electrons and holes formed in between two potential barriers.

Suggested Reading

1 Singh, J. (1996) *Quantum Mechanics, Fundamentals & Applications to Technology*, John Wiley & Sons, Inc.
2 Yariv, A. (1982) *An Introduction to Theory and Applications of Quantum Mechanics*, John Wiley & Sons, Inc.
3 Kroemer, H. (1994) *Quantum Mechanics for Engineering, Materials Science, and Applied Physics*, International edn, Prentice Hall.
4 Blakemore, J.S. (1985) *Solid State Physics*, 2nd edn, Cambridge University Press.
5 Liboff, R.L. (2002) *Introductory Quantum Mechanics*, 4th edn, Addison-Wesley Publishing Company, Reading, MA.

8
The Harmonic Oscillator

The harmonic oscillator constitutes a key component in a number of important dynamic systems, and the understanding of its dynamics is therefore essential for studying various disciplines in science and engineering. The electromagnetic field, for example, is to be viewed and understood in terms of the ensemble of harmonic oscillators. In molecules, atomic positions are configured such that the overall interaction potential energies are at the minimum level. Near those minimum points, the potential energy is evidently well approximated by those of HO and thus HO provides the basis for understanding and describing molecular dynamics as well. In addition, the mathematical techniques used for treating HO amply demonstrate the elegant methodology of the mathematical physics developed over centuries. Also, the quantization of the electromagnetic field is formulated in the same context as the operator treatment of HO.

8.1
Energy Eigenequation

Consider a particle of mass m attached to a spring with spring constant k, as shown in Figure 8.1. As discussed in Chapter 1, the Hamiltonian is given by

$$\hat{H}(x) = -\frac{\hbar^2}{2m}\frac{\partial^2}{\partial x^2} + \frac{1}{2}kx^2, \quad k \equiv m\omega^2, \quad \hat{p} = -i\hbar\frac{\partial}{\partial x} \tag{8.1}$$

where ω denotes the characteristic frequency. The analysis of harmonic oscillator (HO) primarily consists of solving the energy eigenequation

$$\left(-\frac{\hbar^2}{2m}\frac{\partial^2}{\partial x^2} + \frac{1}{2}kx^2\right)u = Eu. \tag{8.2}$$

To solve the differential equation (8.2), it is convenient to first introduce the dimensionless displacement ξ and the energy parameter λ as

$$\xi = \alpha x, \quad \alpha \equiv \left(\frac{mk}{\hbar^2}\right)^{1/4} = \left(\frac{m\omega}{\hbar}\right)^{1/2}, \tag{8.3}$$

Introductory Quantum Mechanics for Semiconductor Nanotechnology. Dae Mann Kim

ISBN: 978-3-527-40975-4

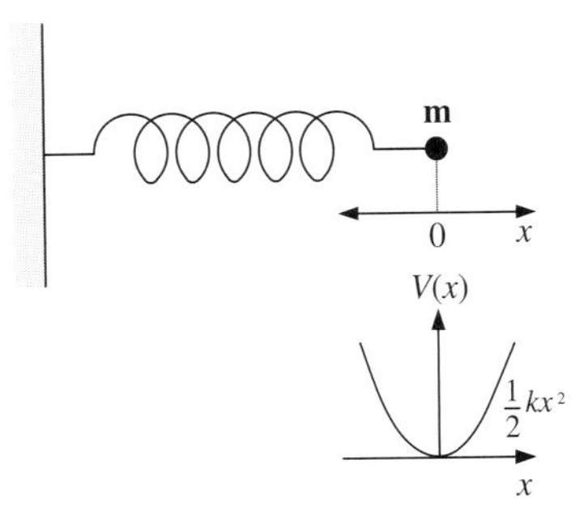

Figure 8.1 The harmonic oscillator: a particle of mass m attached to the spring with constant k, providing the potential energy $kx^2/2$.

$$\lambda \equiv 2E/\hbar\omega, \tag{8.4}$$

and simplify (8.2) by multiplying both sides by $2(m/k)^{1/2}/\hbar = 2/\hbar\omega$, obtaining

$$\frac{d^2}{d\xi^2}u + (\lambda - \xi^2)u = 0. \tag{8.5}$$

In the limit where ξ is large, the constant λ can be neglected compared with ξ and the solution is well approximated by $u(\xi) \approx \exp{-\xi^2/2}$, as can be easily verified by substituting it into (8.5). Thus, look for the solution in the form

$$u(x) = e^{-\xi^2/2}H(\xi), \quad \xi \equiv \alpha x. \tag{8.6}$$

Inserting (8.6) into (8.5) and performing a lengthy but elementary algebra, one obtains the differential equation governing $H(\xi)$:

$$H'' - 2\xi H' + (\lambda - 1)H = 0, \tag{8.7}$$

where the primes denote the differentiation with respect to ξ.

A powerful technique for solving an ordinary differential equation is the method of series solution and the technique is used here for solving (8.7). Thus, take $H(\xi)$ as a polynomial in ξ, that is,

$$H(\xi) = \xi^s \sum_{n=0}^{\infty} a_n \xi^n, \quad a_0 \neq 0. \tag{8.8}$$

Here, a_0 is taken nonzero, so that the index s does not lose its meaning. Differentiating (8.8) with respect to ξ, one can write

$$H'(\xi) = \sum_{n=0}^{\infty} (s+n)a_n \xi^{s+n-1}, \tag{8.9}$$

$$\begin{aligned} H''(\xi) &= \sum_{n=0}^{\infty} (s+n)(s+n-1)a_n \xi^{s+n-2} \\ &= s(s-1)a_0\xi^{s-2} + (s+1)sa_1\xi^{s-1} + \sum_{n=0}^{\infty}(s+n+2)(s+n+1)a_{n+2}\xi^{s+n}. \end{aligned} \tag{8.10}$$

In (8.10), the first two terms have been taken out of the summation and the remaining terms have been summed over with new dummy variable $n-2 \rightarrow n$.

When (8.8)–(8.10) are inserted into (8.7) and the polynomial terms are regrouped, the differential equation for $H(\xi)$ is recast into a form

$$s(s-1)a_0\xi^{s-2} + (s+1)sa_1\xi^{s-2} + \sum_{n=0}^{\infty}\{[(s+n+2)(s+n+1)a_{n+2}] \\ -[2(s-n)+1+\lambda]a_n\}\xi^{s+n} = 0. \tag{8.11}$$

Thus, solving the differential equation (8.7) has been reduced to satisfying (8.11) for arbitrary powers in and values of ξ. This can be done by imposing the conditions

$$s(s-1)a_0 = 0, \quad (s+1)sa_1 = 0, \tag{8.12}$$

$$(s+n+2)(s+n+1)a_{n+2} = [2(s+n)+1-\lambda]a_n. \tag{8.13}$$

Here, the two conditions in (8.12) are called the indicial equations, while (8.13) provides the recurrence relation connecting higher order coefficients to lower order ones in a recursive manner.

Since $a_0 \neq 0$, the first indicial equation requires that $s = 0, 1$. For $s = 0$, both indicial equations are satisfied and $H(\xi)$ can therefore be expressed in terms of two polynomials of infinite order with a_0 and a_1 specifying all higher order coefficients via the recurrence relation (8.13):

$$H(\xi) = a_0\left(1 + \frac{a_2}{a_0}\xi^2 + \frac{a_4}{a_2}\frac{a_2}{a_0}\xi^4 + \cdots\right) + a_1\xi\left(1 + \frac{a_3}{a_1}\xi^2 + \cdots\right). \tag{8.14}$$

With the substitution of (8.14) into (8.6), the solution for $u(\xi)$ is completed.

8.1.1
Eigenfunction and Energy Quantization

Although $u(\xi)$ has been found, it is necessary to examine the asymptotic behavior of $H(\xi)$ for large ξ to ensure that $u(\xi)$ is physically well behaving. To examine this behavior, first consider the Taylor expansion of the exponential function, namely,

$$\exp \xi^2 = \sum_{n=0}^{\infty}\frac{\xi^{2n}}{n!}. \tag{8.15}$$

In this series, the ratio of two successive expansion coefficients is given by $a_{n+1}/a_n = n!/(n+1)! = 1/n$ for large n. The corresponding ratio for the two polynomials of $H(\xi)$, as determined from the recurrence relation (8.13), is also the same, that is, $a_{n+2}/a_n = 1/n$. This indicates that $H(\xi)$ diverges as $H(\xi) \approx \exp\xi^2$ and so does $u(\xi)$ since $u(\xi) \approx H(\xi)\exp-\xi^2/2 \approx \exp\xi^2/2$. This is physically unacceptable, since the probability density of the oscillator should obviously vanish for $\xi \propto x \rightarrow \infty$ rather than diverge.

Therefore, the solution of $H(\xi)$ has to be tailored to render $u(\xi)$ physically well behaving. This can be done by terminating the a_0- series in (8.14) at a finite order while eliminating the second polynomial series by putting $a_1 = 0$. The termination condition is determined from the recurrence relation (8.13) by

$$0 \equiv a_{n+2} = \frac{2n+1-\lambda}{(n+1)(n+2)} a_n \tag{8.16}$$

and (8.16) in turn requires that the parameter, λ in (8.4) satisfy the condition

$$\lambda_n = 2n+1; \quad n = 0, 2, 4, \ldots \tag{8.17}$$

The polynomial $H(\xi)$ then consists of even powers of ξ, which makes $u(\xi)$ an even function of ξ (see (8.6)).

For $s = 1$, the physically acceptable energy eigenfunction $u(\xi)$ can likewise be obtained by again putting $a_1 = 0$ and terminating the a_0-series, namely,

$$0 \equiv a_{n+2} = \frac{2n+3-\lambda}{(n+2)(n+3)} a_n, \tag{8.18a}$$

$$\lambda_n = 2n+3, \quad n = 0, 2, 4, \ldots. \tag{8.18b}$$

In this case, $u(\xi)$ is odd in ξ (see (8.6) and (8.8)).

The two termination conditions (8.17) and (8.18b) can be combined into one as

$$\lambda_n = 2n+1, \quad n = 0, 1, 2, 3, 4, \ldots \tag{8.19}$$

and (8.19) naturally leads, when combined with (8.4), to the energy quantization, namely,

$$E_n = \hbar\omega\left(n + \frac{1}{2}\right), \quad n = 0, 1, 2, 3 \ldots. \tag{8.20}$$

These quantized energy levels of the harmonic oscillator are in distinct contrast with the classical description, in which the energy is an analog quantity and can vary in continuous fashion by any infinitesimal amount, as discussed in Chapter 1. Also, it is both interesting and important to note that the energy quantization is the natural consequence arising from the obvious requirement of the wavefunction physically well behaving. Also, the quantized energy levels are equally spaced and are separated by the characteristic frequency of the oscillator, namely, $\hbar\omega$, as shown in Figure 8.2. The integer n appearing in the quantized energy level (8.20) is called the quantum number.

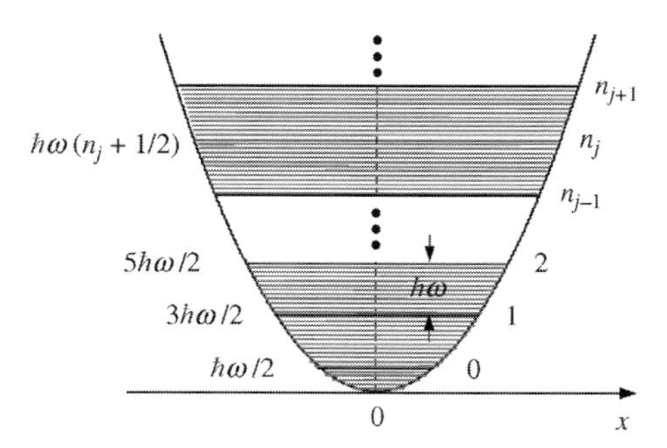

Figure 8.2 The oscillator energy spectrum. The discrete energy levels are separated by the quantum of energy $\hbar\omega$ in contrast with classical oscillator oscillating with any amount of energy E.

8.1.2
The Ground-State and Zero-Point Energy

The ground-state energy representing the lowest energy level with the quantum number $n = 0$, that is,

$$E_0 = \hbar\omega/2, \tag{8.21}$$

is not zero but has a finite value, called the zero-point energy. This is again in marked contrast with the classical oscillator whose total energy at rest in the equilibrium position is zero. This finite zero-point energy is again rooted in the duality of matter, in particular, the uncertainty relation.

In fact, the ground state is characterized by the minimum uncertainty relation, namely,

$$\Delta x \Delta p = \hbar/2. \tag{8.22}$$

This fact can be seen by taking the total energy to consist of the potential and kinetic energies arising from the uncertainties in position and momentum, namely, $\Delta x,\ \Delta p$:

$$E = \frac{1}{2}k\Delta x^2 + \frac{\Delta p^2}{2m}. \tag{8.23}$$

Combining (8.22) and (8.23), E can be expressed as a function of, say, Δp only, namely,

$$E = \frac{1}{2}\left[\frac{k\hbar^2}{4\Delta p^2} + \frac{\Delta p^2}{m}\right]. \tag{8.24}$$

The minimum value of E as a function of Δp is therefore obtained by

$$0 = \frac{\partial E}{\partial \Delta p^2} = -\frac{k\hbar^2}{4\Delta p^4} + \frac{1}{m}. \tag{8.25}$$

Hence, the minimum level of energy occurs for

$$\Delta p^2 = \left(\frac{k\hbar^2 m}{4}\right)^{1/2} = \frac{\hbar\omega m}{2}, \quad k \equiv m\omega^2. \tag{8.26}$$

Inserting (8.26) in (8.24), one finds $E_{\min} = \hbar\omega/2$, in agreement with the zero-point energy.

8.2
The Properties of Eigenfunctions

8.2.1
Hermite Polynomials and Recurrence Relations

The polynomial $H_n(\xi)$ appearing in the expression of $u_n(\xi)$ is called Hermite polynomial and has been investigated extensively as a special function in mathematical physics. Its properties are briefly discussed in this section. When the energy quantization condition (8.19) is inserted into (8.7), the differential equation reads as

$$H''_n - 2\xi H'_n + 2nH_n = 0, \tag{8.27}$$

with n denoting a positive integer. This differential equation is known as Hermite differential equation and the finite-order polynomial solutions of this equation are called Hermite polynomials.

The properties of $H_n(\xi)$ can be conveniently discussed with the use of its generating function given by

$$G(\xi, s) \equiv e^{\xi^2-(s-\xi)^2} = e^{-s^2+2s\xi} \equiv \sum_{n=0}^{\infty} \frac{H_n(\xi)s^n}{n!}. \tag{8.28}$$

First, the recurrence relations existing among Hermite polynomials can be found by differentiating both sides of $G(\xi, s)$ with respect to ξ, obtaining

$$\frac{\partial}{\partial \xi} G(\xi, s) = 2se^{-s^2+2s\xi} = 2\sum_{n=0}^{\infty} \frac{H_n(\xi)s^{n+1}}{n!} \equiv \sum_{n=0}^{\infty} \frac{H'_n(\xi)s^n}{n!}. \tag{8.29}$$

In obtaining (8.29), (8.28) has again been used after the differentiation of the left-hand side of (8.28). Thus, singling out the coefficients of the equal powers of s from both sides of the equation, one finds

$$H'_n = 2nH_{n-1}. \tag{8.30}$$

By the same token, the differentiation of $G(\xi, s)$ with respect to s yields

$$\begin{aligned}\frac{\partial}{\partial s} G(\xi, s) &= (-2s+2\xi)e^{-s^2+2s\xi} = (-2s+2\xi)\sum_{n=0}^{\infty} \frac{H_n s^n}{n!} \\ &\equiv \sum_{n=1}^{\infty} \frac{H_n s^{n-1}}{(n-1)!} = \sum_{n=0}^{\infty} \frac{H_{n+1} s^n}{n!}.\end{aligned} \tag{8.31}$$

In (8.31), a new dummy variable has been used, namely, $n-1 \rightarrow n$ after differentiating the right-hand side of (8.28) with respect to s. Thus again, singling out the coefficients of the equal powers of s from both sides results in

$$\xi H_n = \frac{1}{2} H_{n+1} + nH_{n-1}. \tag{8.32}$$

The two recurrence relations (8.30) and (8.32) are rather useful for performing various calculations involving $u_n(\xi)$.

In addition, from the generating function (8.28), one can single out $H_n(\xi)$ by performing differentiation n times with respect to s and letting $s \rightarrow 0$. In this case, those terms with power in s less than n drop out in the process of differentiation, while those terms with power in s larger than n are made zero in the process of letting $s \rightarrow 0$. Thus, one can write

$$
\begin{aligned}
H_n(\xi) &\equiv \frac{\partial^n}{\partial s^n} e^{\xi^2-(s-\xi)^2}|_{s=0} \\
&= e^{\xi^2}(-)^n \frac{\partial^n}{\partial \xi^n} e^{-(s-\xi)^2}|_{s=0} \\
&= (-)^n e^{\xi^2} \frac{\partial^n}{\partial \xi^n} e^{-\xi^2}.
\end{aligned}
\tag{8.33}
$$

This operational representation of $H_n(\xi)$ is known as Rodrigue's formula and serves as an efficient means for generating $H_n(\xi)$. For example, one can write by performing mere differentiations entailed in (8.33):

$$
H_0 = 1, \quad H_1 = 2\xi, \quad H_2 = 4\xi^2 - 2, \ldots . \tag{8.34}
$$

Here, the even and odd order polynomials are obtained alternatively by even and odd number of differentiations with respect to ξ.

8.2.2 The Orthogonality of Energy Eigenfunctions

The nth energy eigenstate of the oscillator is thus given by

$$
u_n(\xi) = N_n e^{-\xi^2/2} H_n(\xi), \tag{8.35}
$$

with N_n denoting the normalization constant. Clearly, the parity of $u_n(\xi)$ is determined by the quantum number n, which also characterizes the energy eigenvalues and eigenfunctions. To find N_n and also to show at the same time the orthogonality of the eigenfunctions, consider the integral involving the product of two generating functions:

$$
\int_{-\infty}^{\infty} d\xi e^{-s^2+2s\xi} e^{-t^2+2t\xi} e^{-\xi^2} = \sum_n \frac{s^n}{n!} \sum_m \frac{t^m}{m!} \int_{-\infty}^{\infty} d\xi H_n H_m e^{-\xi^2}. \tag{8.36}
$$

Now, one can readily perform the integral on the left-hand side with the use of (1.24), obtaining

$$
\int_{-\infty}^{\infty} d\xi e^{-s^2+2s\xi} e^{-t^2+2t\xi} e^{-\xi^2} = \sqrt{\pi} e^{2ts} = \sqrt{\pi} \sum_n \frac{2^n t^n s^n}{n!}. \tag{8.37}
$$

In (8.37), the exponential function $\exp(2ts)$ has been Taylor expanded. Thus, the double sum on the right-hand side of (8.36) has been reduced to the single sum in (8.37), which clearly indicates that only the diagonal terms in (8.36) are nonzero, namely,

$$
\int_{-\infty}^{\infty} d\xi H_n H_m e^{-\xi^2} = \sqrt{\pi} 2^n n! \delta_{nm}, \tag{8.38}
$$

with δ_{nm} denoting the Kronecker delta function, namely, $\delta_{nm} = 1$ for $n = m$ and zero otherwise. Thus, the orthogonality of the energy eigenfunctions, $\{u_n(\xi)\}$, has been proven in (8.38).

The normalization constant can also be readily found from (8.35) and (8.38):

$$1 \equiv N_n^2 \int_{-\infty}^{\infty} dx e^{-\xi^2} H_n^2 = \frac{N_n^2}{\alpha} \int_{-\infty}^{\infty} d\xi e^{-\xi^2} H_n^2 = \frac{N_n^2}{\alpha} \sqrt{\pi} 2^n n! \tag{8.39}$$

so that one can write the normalized eigenfunction as

$$u_n(x) = \left(\frac{\alpha}{\sqrt{\pi} 2^n n!}\right)^{1/2} e^{-\xi^2/2} H_n(\xi), \quad \xi \equiv \alpha x, \ \alpha = [m\omega/\hbar]^{1/2}. \tag{8.40}$$

8.2.3
Probability Densities and Correspondence Principle

A few examples of the energy eigenfunctions are plotted in Figure 8.3, together with corresponding probability densities. There are a few interesting features to be noticed in this figure. First, there is a small but finite probability of finding the oscillator in classically forbidden region and the degree of penetration increases with increasing n, that is, with increasing energy. Also, the peak of the probability densities shifts from the center at $x \approx 0$ for small n to the edges at $x \approx x_0$ for large n. This behavior can be compared with the classical analog of the probability density.

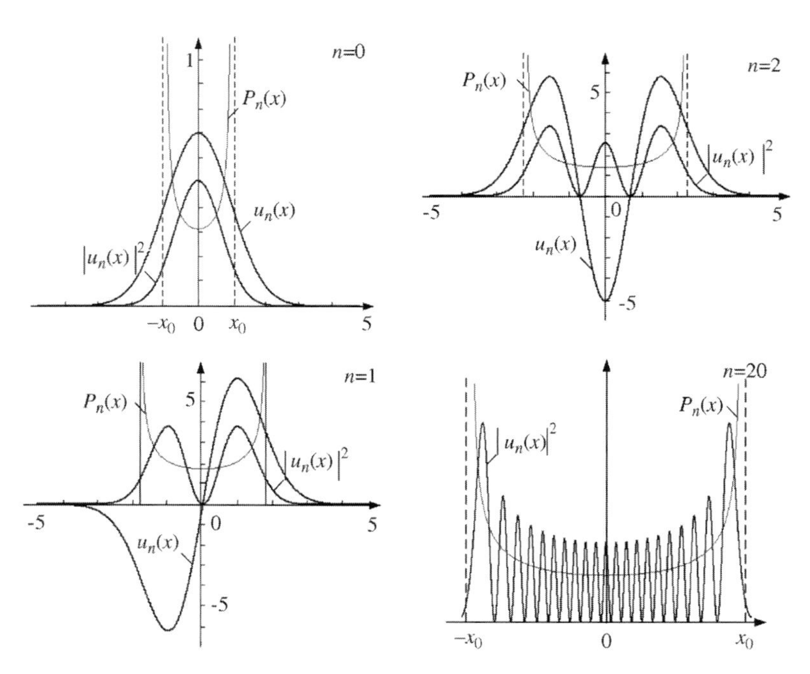

Figure 8.3 Typical harmonic oscillator eigenfunctions and corresponding probability densities. Also shown are the classical turning points and/or the amplitude of oscillation x_0 for each energy level, namely, $kx_0^2/2 = E$, and the classical analog of the probability density $P_n(x)$.

Classically, the probability of finding the oscillator between x and $x + dx$ is given by the dwell time of the oscillator at x during one period of oscillation, namely,

$$P = \frac{dt(x)}{T} = 2\frac{dx/\dot{x}}{T}, \quad T \equiv 2\pi/\omega, \tag{8.41}$$

where the factor 2 is introduced to account for the fact that the oscillator passes through dx element twice during one period of oscillation. Now, that the harmonic oscillator is a conservative system, the total energy is generally given by

$$\frac{1}{2}kx_0^2 = \frac{1}{2}m\dot{x}^2 + \frac{1}{2}kx^2, \tag{8.42}$$

where x_0 is the amplitude of oscillation. Thus, one can find $\dot{x}$ as a function of x from (8.42) as

$$\dot{x} = \left[k(x_0^2 - x^2)/m\right]^{1/2} = \omega(x_0^2 - x^2)^{1/2}, \quad \omega^2 \equiv k/m. \tag{8.43}$$

Inserting (8.43) into (8.41), one finds the classical probability of finding the oscillator at x as

$$P = \frac{1}{\pi}\frac{dx}{(x_0^2 - x^2)^{1/2}}. \tag{8.44}$$

It therefore follows from (8.44) that the probability attains the minimum value near the origin, where the oscillator moves with the largest speed (see (8.43)), while it attains the maximum value at $x = x_0$ where the oscillator is momentarily at rest before reversing its motion. This classical probability is also plotted in Figure 8.3 for comparison. As is clear from this figure, the classical probability profile is in general agreement with the locus of the subpeaks of the probability density, $|u_n(x)|^2$, for large n. This agreement is generally referred to as the correspondence principle. However, for small n, there exists a marked difference between the two probabilities.

8.2.4 The Uncertainty Relation

The uncertainties in x and p associated with the energy eigenstate $u_n(x)$ can be evaluated by using the recurrence relations (8.30) and (8.32) and in terms of the associated variances. Evidently, the average value of x is zero in view of the definite parity possessed by $u(x)$, regardless of whether it is even or odd, that is,

$$\langle x \rangle = \langle u_n | x | u_n \rangle = 0, \tag{8.45a}$$

while one can write

$$\begin{aligned} \langle x^2 \rangle \equiv \langle u_n | x^2 | u_n \rangle &= \int_{-\infty}^{\infty} dx u_n^*(x) x^2 u_n(x) \\ &= \frac{N_n^2}{\alpha^3} \int_{-\infty}^{\infty} d\xi e^{-\xi^2/2} H_n(\xi) \xi^2 e^{-\xi^2/2} H_n(\xi), \quad \xi = \alpha x. \end{aligned} \tag{8.45b}$$

Now, with the use of the recurrence relation (8.32), one can write

$$[\xi H_n(\xi)]^2 = \left[\frac{1}{2}H_{n+1}(\xi) + nH_{n-1}(\xi)\right]^2$$

and insert it in (8.45b) and perform the integration with the use of orthonormality of the eigenfunctions, (8.38), obtaining

$$\langle x^2 \rangle = \frac{\hbar}{m\omega}\left(n + \frac{1}{2}\right). \tag{8.45c}$$

Hence, one can write

$$\Delta x^2 \equiv \langle (x - \langle x \rangle)^2 \rangle = \langle x^2 \rangle - \langle x \rangle^2 = \frac{\hbar}{m\omega}\left(n + \frac{1}{2}\right). \tag{8.46}$$

One can likewise find again by using the recurrence relations and the orthonormality of eigenfunctions,

$$\langle p_x \rangle = \langle u_n | -i\hbar \frac{\partial}{\partial x} | u_n \rangle = 0, \tag{8.47a}$$

while

$$\begin{aligned} \langle p^2 \rangle &= N_n^2 \alpha (i\hbar)^2 \int_{-\infty}^{\infty} d\xi e^{-\xi^2/2} H_n(\xi) \left(\frac{\partial}{\partial \xi}\right)^2 \left[e^{-\xi^2/2} H_n(\xi)\right] \\ &= m\omega\hbar\left(n + \frac{1}{2}\right). \end{aligned} \tag{8.47b}$$

Hence, there results

$$\Delta p^2 \equiv \langle (p - \langle p \rangle)^2 \rangle = \langle p^2 \rangle - \langle p \rangle^2 = m\omega\hbar\left(n + \frac{1}{2}\right). \tag{8.48}$$

Hence, the uncertainty relation for the position and momentum in the nth eigenstate $u_n(x)$ is given by

$$\Delta x \Delta p = \hbar\left(n + \frac{1}{2}\right). \tag{8.49}$$

The ground state in which $n = 0$ therefore is shown to possess the minimum uncertainty value of $\hbar/2$, in agreement with the previous analysis.

8.2.5
Useful Matrix Elements

As pointed out, the harmonic oscillator constitutes a key element in a number of important dynamic systems and the matrices connecting its eigenfunctions are extensively utilized. The integrals required for evaluating these matrix elements can again be conveniently carried out with the use of recurrence relations (8.30) and (8.32). For instance, the matrix element

$$\langle u_n|x|u_m\rangle = \frac{N_n N_m}{\alpha^2}\int_{-\infty}^{\infty} d\xi e^{-\xi^2/2} H_n(\xi)\xi e^{-\xi^2/2} H_m(\xi) \tag{8.50}$$

is often used, for describing the coupling between light and matter. One can readily perform the integration using (8.32) and (8.38), obtaining

$$\langle u_l|x|u_{l'}\rangle = \begin{cases} (l+1)^{1/2}/(2m\omega/\hbar)^{1/2}, & l' = l+1, \\ l^{1/2}/(2m\omega/\hbar)^{1/2}, & l' = l-1, \\ 0, & \text{otherwise.} \end{cases} \tag{8.51}$$

One can similarly find

$$\langle u_l|\frac{\partial}{\partial x}|u_{l'}\rangle = \begin{cases} (m\omega/2\hbar)^{1/2}(l+1)^{1/2}, & l' = l+1, \\ -(m\omega/2\hbar)^{1/2} l^{1/2}, & l' = l-1, \\ 0, & \text{otherwise.} \end{cases} \tag{8.52}$$

8.2.6
Oscillator in a Linearly Superposed State

When the harmonic oscillator is in a mixed state, consisting of ground and first excited states with equal probability, for example, the oscillator is described by the wavefunction

$$\psi(x,t) = \frac{1}{\sqrt{2}}(e^{-i\omega t/2}u_0 + e^{-i3\omega t/2}u_1). \tag{8.53}$$

Here the oscillatory time factor $\exp - iEt/\hbar$ for each eigenstate has been incorporated (see (3.11)). The corresponding probability density

$$\psi^*\psi = \frac{1}{2}(u_0^2 + u_1^2 + 2u_0u_1 \cos\omega t) \tag{8.54}$$

then consists of the time-independent background u_0^2, u_1^2, superposed by an oscillating term, as shown in Figure 8.4. In this linearly superposed state of eigenfunctions, the probability density exhibits the oscillatory behavior in time.

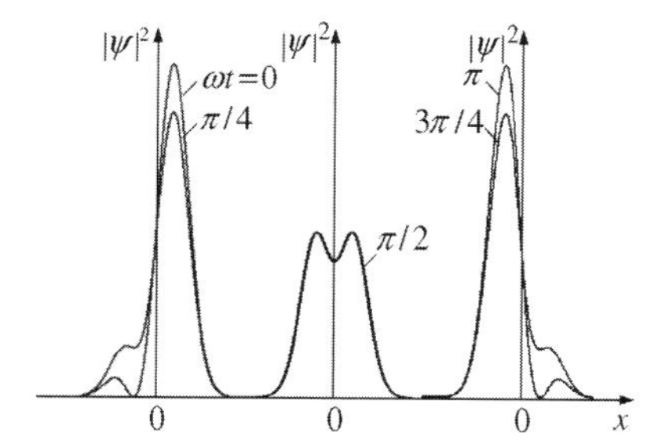

Figure 8.4 The probability densities, at different times, of a superposed state, consisting of ground and first excited states with equal probability. The center of the probability density oscillates in time.

This can be made more explicit by finding the expectation values of x and p:

$$\langle x\rangle = \frac{1}{2}\left\langle e^{-i\omega t/2}u_0 + e^{-i3\omega t/2}u_1|x|e^{-i\omega t/2}u_0 + e^{-i3\omega t/2}u_1\right\rangle$$
$$= x_{12}\cos\omega t, \quad x_{12} \equiv \langle u_0|x|u_1\rangle = 1/(2m\omega/\hbar)^{1/2}, \tag{8.55}$$

$$\langle p_x\rangle = \frac{1}{2}\left\langle e^{-i\omega t/2}u_0 + e^{-i3\omega t/2}u_1|-i\hbar\frac{\partial}{\partial x}|e^{-i\omega t/2}u_0 + e^{-i3\omega t/2}u_1\right\rangle$$
$$= -p_{12}\sin\omega t, \quad p_{12} \equiv \hbar\langle u_0|\frac{\partial}{\partial x}|u_1\rangle = (m\omega\hbar/2)^{1/2}. \tag{8.56}$$

In deriving (8.55) and (8.56), the orthogonality of the eigenfunctions has been used and the matrix elements, derived in (8.51) and (8.52), have also been used for defining the amplitudes of oscillation of the position and momentum. Clearly, these average quantities oscillate in time in a manner similar to the classical oscillator.

8.3 The Operator Treatment

The Hamiltonian of the harmonic oscillator can also be expressed in the operator format and the operator treatment of the oscillator finds extensive applications, for example, in the quantum treatment of electromagnetic fields. Thus, consider the operators defined as

$$a = \frac{1}{\sqrt{2}}\left[\alpha x + i\frac{1}{\hbar\alpha}p\right] = \frac{1}{\sqrt{2}}\left(\xi + \frac{\partial}{\partial\xi}\right), \tag{8.57}$$

$$a^+ = \frac{1}{\sqrt{2}}\left[\alpha x - i\frac{1}{\hbar\alpha}p\right] = \frac{1}{\sqrt{2}}\left(\xi - \frac{\partial}{\partial\xi}\right), \tag{8.58}$$

where ξ is the dimensionless displacement introduced in (8.3) and a and a^+ thus introduced are called the lowering and raising operators, respectively. One can find the commutation relation of this pair of operators from that of x and p:

$$\begin{aligned}[a, a^+] &= \left[\frac{\alpha}{\sqrt{2}}x + i\frac{1}{\sqrt{2}\hbar\alpha}p, \frac{\alpha}{\sqrt{2}}x - i\frac{1}{\sqrt{2}\hbar\alpha}p\right] \\ &= \frac{1}{2}\left(\frac{-i}{\hbar}\right)[x, p] + \frac{1}{2}\left(\frac{i}{\hbar}\right)[p, x] = 1,\end{aligned} \tag{8.59}$$

where use has been made of $[x, p] = i\hbar$ and $[x, x] = [p, p] = 0$.

One can invert (8.57) and (8.58) and express x, p in terms of a, a^+, obtaining

$$x = \frac{1}{\sqrt{2}\alpha}(a + a^+), \tag{8.60}$$

$$p = \frac{i\hbar\alpha}{\sqrt{2}}(a^{+} - a). \tag{8.61}$$

Hence, the Hamiltonian of the oscillator can also be expressed in terms of a and a^{+} as

$$\hat{H} = \frac{p^2}{2m} + \frac{1}{2}kx^2 = \hbar\omega\left(a^{+}a + \frac{1}{2}\right). \tag{8.62}$$

In (8.62), use has been made of $k/\alpha^2 = (\hbar\alpha)^2/m = \hbar\omega$ (see (8.3)), together with the commutation relation of (8.59) for converting aa^{+} in terms of $a^{+}a$, namely, $aa^{+} = a^{+}a + 1$. The operator $a^{+}a$ thus introduced in (8.62) is called the number operator. It is clear from (8.62) that $\hat{H}$ commutes with $a^{+}a$ and consequently the number operator can share the eigenfunction of $\hat{H}$, namely $u_n(x)$.

Raising, Lowering, and Number Operators The operators a, a^{+}, and $a^{+}a$, when operating on $u_n(x)$, lead to interesting results. To show this, first consider a operating on $u_n(x)$:

$$au_n = \frac{1}{\sqrt{2}}N_n\left(\xi + \frac{\partial}{\partial\xi}\right)H_n e^{-\xi^2/2} = \frac{1}{\sqrt{2}}N_n H'_n e^{-\xi^2/2} = \sqrt{n}u_{n-1}. \tag{8.63}$$

In (8.63), (8.40) was used for $u_n(x)$ and (8.30) was also used for converting H'_n to H_{n-1}. Thus, the net effect of a operating on $u_n(x)$ is to convert $u_n(x)$ to $u_{n-1}(x)$, that is, to lower the energy eigenstate from n to $n-1$, hence the name the lowering operator.

By the same token, a^{+} acting on $u_n(x)$ yields with additional use of the recurrence relation (8.32)

$$a^{+}u_n = \frac{1}{\sqrt{2}}N_n\left(\xi - \frac{\partial}{\partial\xi}\right)H_n e^{-\xi^2/2} = \frac{1}{\sqrt{2}}N_n(2\xi H_n - H'_n)e^{-\xi^2/2} = \sqrt{n+1}u_{n+1}. \tag{8.64}$$

Hence, a^{+} is shown to transform the eigenstate $u_n(x)$ to $u_{n+1}(x)$ and raise the energy level by $\hbar\omega$, hence the name the raising operator.

One can also find the net effect of operating $a^{+}a$ on $u_n(x)$ by using (8.63) and (8.64) in succession, obtaining

$$a^{+}au_n = \sqrt{n}a^{+}u_{n-1} = \sqrt{n}\sqrt{n-1+1}u_{n-1+1} = nu_n. \tag{8.65}$$

It has therefore become clear that $u_n(x)$ is also the eigenfunction of the number operator $a^{+}a$ with n providing the eigenvalue.

Creation, Annihilation Operators, and Phonons In view of the roles of the operators a, a^{+}, and $a^{+}a$ acting on $u_n(x)$, the eigenenergy associated with u_n can be interpreted from an interesting point of view. Specifically, it is now possible to introduce the basic quantum of energy, called the phonon, and view the eigenenergy to consist of n number of phonons present therein.

In this context, the operator, a lowering u_n to u_{n-1} can be viewed as destroying one phonon and is often called the annihilation operator. By the same token, the operator

a^+ raising u_n to u_{n+1} can be viewed as creating one phonon and is also called the creation operator.

With the use of these creation and annihilation operators, it is also possible to generate the energy eigenfunctions, based purely on the differentiation rather than solving the energy eigenequation. The starting point of this line of approach to find $u_n(x)$ is the basic fact that when a acts on the ground state, it pushes the state out of the Hilbert space of the eigenfunction, namely,

$$au_0 = \sqrt{0}u_{0-1} = 0. \tag{8.66}$$

Now (8.66) can be expressed using (8.57) as

$$\left(\xi + \frac{\partial}{\partial \xi}\right) u_0 = 0 \tag{8.67}$$

and this differential equation is easily solved to yield

$$u_0 \propto e^{-\xi^2/2} = N_0 e^{-\xi^2/2}, \tag{8.68}$$

where the constant of integration, N_0, can be used to satisfy the normalization condition, namely,

$$1 = N_0^2 \int_{-\infty}^{\infty} dx u_0^2 = \frac{N_0^2}{\alpha} \int_{-\infty}^{\infty} d\xi e^{-\xi^2} = \frac{N_0^2}{\alpha} \sqrt{\pi}. \tag{8.69}$$

Thus, the normalized ground-state eigenfunction reads as

$$u_0(\xi) = \left(\frac{\alpha}{\sqrt{\pi}}\right)^{1/2} e^{-\xi^2/2} \tag{8.70}$$

in agreement with (8.40).

Once u_0 is found, the excited states can be systematically generated by creating phonons one by one in succession by applying the creation operators repeatedly. For instance, the first excited state is simply obtained by creating one phonon in the ground state by using (8.64), namely,

$$\begin{aligned} u_1 \equiv \frac{1}{\sqrt{1}} a^+ u_0 &= \frac{1}{\sqrt{1}} \frac{1}{\sqrt{2}} \left(\xi - \frac{\partial}{\partial \xi}\right) u_0(\xi) \\ &= \left(\frac{\alpha}{\sqrt{\pi} 1! 2}\right)^{1/2} e^{-\xi^2/2} H_1(\xi). \end{aligned} \tag{8.71}$$

In (8.71), the recurrence relations (8.30) and (8.32) have been used together with (8.40) for $u_n(x)$. One can generate the nth eigenstate by applying the creation operator n times, namely,

$$u_n(\xi) = \frac{1}{\sqrt{n!}} (a^+)^n u_0(\xi), \quad a^+ = \frac{1}{\sqrt{2}} \left(\xi - \frac{\partial}{\partial \xi}\right), \tag{8.72}$$

and find the energy eigenfunction of the harmonic oscillator simply by differentiation instead of integration.

8.4 Problems

8.1. (a) Starting with the energy eigenequation of the harmonic oscillator (8.2), fill in the algebraic steps and reduce the equation to

$$u'' + (\lambda - \xi^2)u = 0, \quad \xi = \alpha x, \ \alpha \equiv \left(\frac{mk}{\hbar^2}\right)^{1/4} = \left(\frac{m\omega}{\hbar}\right)^{1/2}, \ \lambda \equiv 2E/\hbar\omega. \tag{8.73}$$

(b) By putting $u(x) = \exp-(\xi^2/2)H(\xi)$, show that (8.73) is transformed into the differential equation for H as

$$H'' - 2\xi H' + (\lambda - 1)H = 0. \tag{8.74}$$

8.2. (a) Using the recurrence relations (8.30) and (8.32), evaluate the matrix elements,

$$\langle u_n|x^2|u_{n'}\rangle; \quad \langle u_n|\hat{p}_x^2|u_{n'}\rangle.$$

(b) Using the results obtained in (a), evaluate the variances

$$\Delta x^2 = \langle u_n|(x - \langle x\rangle)^2|u_n\rangle; \quad \Delta p_x^2 = \langle u_n|(p_x - \langle p_x\rangle)^2|u_n\rangle$$

and show that the uncertainty relation operative in u_n is given by

$$\Delta x \Delta p_x = \hbar\left(n + \frac{1}{2}\right).$$

(c) Following steps given in the text, derive (8.51) and (8.52).

8.3. (a) For a classical oscillator with mass m and spring constant k and oscillating with an amplitude x_0, find the kinetic and potential energies averaged over one period of oscillation and compare the results with the total energy.

(b) For a quantum oscillator, find the average kinetic and potential energies in nth eigenstate and compare the average values with the total energy of the eigenstate.

(c) Discuss the similarities or differences between the two descriptions.

8.4. Consider a three-dimensional harmonic oscillator with the Hamiltonian

$$\hat{H} = -\frac{\hbar^2}{2m}\nabla^2 + \frac{1}{2}k_x x^2 + \frac{1}{2}k_y y^2 + \frac{1}{2}k_z z^2.$$

(a) Write down the energy eigenequation and using the separation of variable technique, show that the 3D eigenequation is reduced to three independent 1D eigenequations.

(b) Find the eigenfunctions and the eigenvalues of the 3D oscillator.

(c) For the case $k_x = k_y = k_z = k$, discuss the energy spectrum. Find the degeneracy of the first, second, and third excited states by finding the number of quantum states having the same energy.

8.5. The HI molecule can be approximately modeled as composed of heavier iodine atom fixed and lighter hydrogen atom undergoing a vibration centered around the equilibrium harmonic potential. If the effective force constant is given by $k = 313.8$ N/m, determine the energy spacing and the frequency of photon required to induce the transitions between two neighboring states.

8.6. Consider again two atoms with masses m_1 and m_2 and coupled via an effective spring constant k.

(a) Show that the Hamiltonian can be represented by

$$H = \frac{1}{2}M\dot{X}^2 + \frac{1}{2}\mu\dot{x}^2 + \frac{1}{2}kx^2,$$

with M and μ denoting the total and reduced mass of the two atoms (see Problem 1.3).

(b) Write down the energy eigenequation and using the separation of variable technique, find the eigenfunction and eigenenergy in terms of the motions of the center of mass and internal vibration.

8.7. The vibrational spectra of molecules can be observed by means of the infrared spectroscopy. The carbon monoxide (CO) molecule is to be viewed as the two atoms coupled via a spring with an effective constant k. The energy spacing between two neighboring vibration states is observed to be given by the wavenumber $1/\lambda = 2170\,\text{cm}^{-1}$.

(a) Taking the masses of C and O to be 12 and 16 atomic units, determine k that is a measure of the bond stiffness.

(b) Find the zero-point energy.

8.8. Representing x and p in terms of raising and lowering operators a^+ and a (see (8.60) and (8.61)) and using the commutation relation of a^+, a (see (8.59)), derive the oscillator Hamiltonian given in (8.62).

8.9. (a) Consider a function $F(x) = e^{xa^+a}ae^{-xa^+a}$ so that $F(0) = a$.
Show that

$$\begin{aligned}\frac{dF(x)}{dx} &= e^{xa^+a}[a^+a, a]e^{-xa^+a}\\ &= -e^{xa^+a}ae^{-xa^+a} = -F(x), \quad [a^+a, a] = a.\end{aligned}$$

Solving the differential equation for $F(x)$ shown that

$$F(x) = F(0)e^{-x}.$$

That is,

$$e^{xa^+a}ae^{-xa^+a} = ae^{-x}.$$

(b) Show using a similar method

$$e^{xa^+a}a^+e^{-xa^+a} = a^+e^x.$$

8.10. (a) Using (8.72), generate oscillator eigenfunctions u_1, u_2, u_3, and u_4.
(b) Prove by induction (8.72).

Suggested Reading

1 Yariv, A. (1982) *An Introduction to Theory and Applications of Quantum Mechanics*, John Wiley & Sons, Inc.
2 Singh, J. (1996) *Quantum Mechanics, Fundamentals & Applications to Technology*, John Wiley & Sons, Inc.
3 Gasiorowicz, S. (2003) *Quantum Physics*, 3rd edn, John Wiley & Sons, Inc.

9
Angular Momentum

The angular momentum has played an essential role in the formulation of quantum mechanics. This is evidenced by the fact that the epoch-making theory of hydrogen atom by Bohr starts with quantizing the angular momentum as a basic hypothesis. Understandably, the angular momentum also plays a key role in the wave mechanical treatment of the H-atom. In the wave mechanical treatment, however, the quantization of the angular momentum is not an assumption but a natural consequence of the angular momentum eigenfunctions physically well behaving. More important, the eigenfunctions of the angular momentum provide a powerful basis for describing the atomic and molecular structures and the bonding of molecules. It is thus important to go through a thorough analysis of the angular momentum operators and eigenfunctions.

9.1
Angular Momentum Operators

The angular momentum vector $\hat{\boldsymbol{l}}$ of a particle moving with the linear momentum $\boldsymbol{p} = m\boldsymbol{v}$ at a position $\boldsymbol{r}$ is defined in terms of the vector product of $\boldsymbol{r}$ and $\boldsymbol{p}$, as shown in Figure 9.1, namely,

$$\hat{\boldsymbol{l}} = \boldsymbol{r} \times \boldsymbol{p}. \tag{9.1}$$

In Cartesian coordinate frame, $\hat{\boldsymbol{l}}$ decomposes into the three components as

$$\hat{l}_x = yp_z - zp_y = -i\hbar\left(y\frac{\partial}{\partial z} - z\frac{\partial}{\partial y}\right), \tag{9.2a}$$

$$\hat{l}_y = zp_x - xp_z = -i\hbar\left(z\frac{\partial}{\partial x} - x\frac{\partial}{\partial z}\right), \tag{9.2b}$$

$$\hat{l}_z = xp_y - yp_x = -i\hbar\left(x\frac{\partial}{\partial y} - y\frac{\partial}{\partial x}\right), \tag{9.2c}$$

Introductory Quantum Mechanics for Semiconductor Nanotechnology. Dae Mann Kim

ISBN: 978-3-527-40975-4

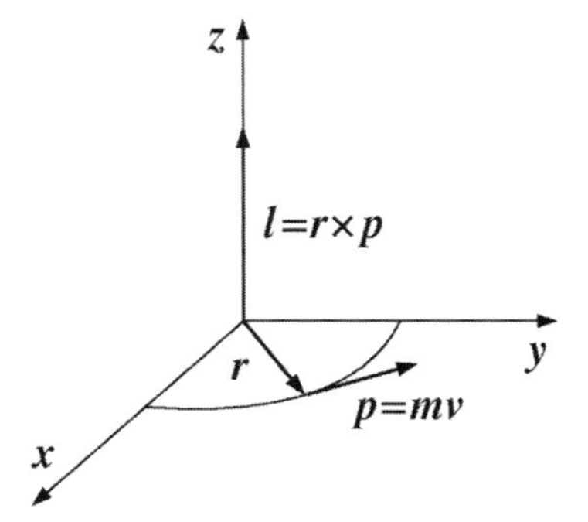

Figure 9.1 The angular momentum as a vector product of $\boldsymbol{r}$ and $\boldsymbol{p}$.

where the cyclic properties of the unit vectors $\hat{x}$, $\hat{y}$, and $\hat{z}$, namely, $\hat{x} \times \hat{y} = \hat{z}$, $\hat{y} \times \hat{z} = \hat{x}$, and $\hat{z} \times \hat{x} = \hat{y}$ have been used together with the operator representation of the linear momentum.

In the spherical coordinate frame (Figure 9.2) in which x, y, and z are related to r,θ,φ as

$$x = r \sin\theta \cos\varphi, \tag{9.3a}$$

$$y = r \sin\theta \sin\varphi, \tag{9.3b}$$

$$z = r \cos\theta, \tag{9.3c}$$

so that

$$dx = \sin\theta \cos\varphi \, dr + r\cos\theta \cos\varphi \, d\theta - r \sin\theta \sin\varphi \, d\varphi, \tag{9.4a}$$

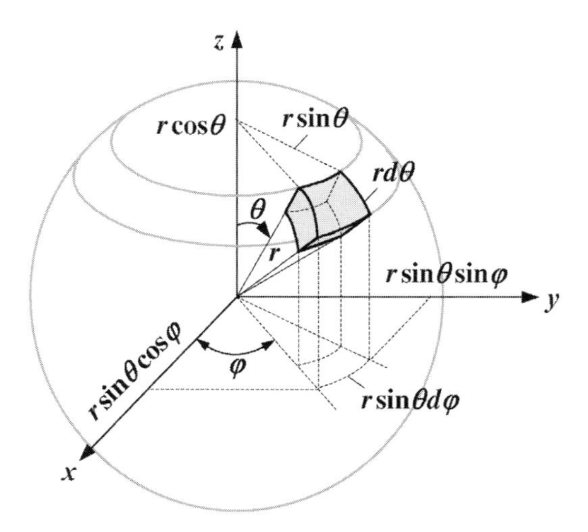

Figure 9.2 The spherical coordinate frame, with variables $\boldsymbol{r}$, θ, φ.

$$dy = \sin\theta \sin\varphi \, dr + r\cos\theta \sin\varphi \, d\theta + r\sin\theta \cos\varphi \, d\varphi, \tag{9.4b}$$

$$dz = \cos\theta \, dr - r\sin\theta \, d\theta. \tag{9.4c}$$

Inverting (9.4), one finds

$$dr = \sin\theta \cos\varphi \, dx + \sin\theta \sin\varphi \, dy + \cos\theta \, dz, \tag{9.5a}$$

$$d\theta = \frac{1}{r}(\cos\theta \cos\varphi \, dx + \cos\theta \sin\varphi \, dy - \sin\theta \, dz), \tag{9.5b}$$

$$d\varphi = \frac{1}{r\sin\theta}(-\sin\varphi \, dx + \cos\varphi \, dy). \tag{9.5c}$$

Hence, the derivatives with respect to x, y, and z are transcribed into r, θ, and φ using (9.5) as

$$\begin{aligned} \frac{\partial}{\partial x} &= \frac{\partial}{\partial r}\left(\frac{\partial r}{\partial x}\right) + \frac{\partial}{\partial \theta}\left(\frac{\partial \theta}{\partial x}\right) + \frac{\partial}{\partial \varphi}\left(\frac{\partial \varphi}{\partial x}\right) \\ &= \sin\theta \cos\varphi \frac{\partial}{\partial r} + \frac{1}{r}\cos\theta \cos\varphi \frac{\partial}{\partial \theta} - \frac{\sin\varphi}{r\sin\theta}\frac{\partial}{\partial \varphi}, \end{aligned} \tag{9.6a}$$

$$\frac{\partial}{\partial y} = \sin\theta \sin\varphi \frac{\partial}{\partial r} + \frac{1}{r}\cos\theta \sin\varphi \frac{\partial}{\partial \theta} + \frac{\cos\varphi}{r\sin\theta}\frac{\partial}{\partial \varphi}, \tag{9.6b}$$

$$\frac{\partial}{\partial z} = \cos\theta \frac{\partial}{\partial r} - \frac{\sin\theta}{r}\frac{\partial}{\partial \theta}. \tag{9.6c}$$

Thus, the z-component of the angular momentum $\hat{l}_z$, for example, reads with the use of (9.2) and (9.6) as

$$\hat{l}_z = -i\hbar r\left(\sin\theta \cos\varphi \frac{\partial}{\partial y} - \sin\theta \sin\varphi \frac{\partial}{\partial x}\right) = -i\hbar \frac{\partial}{\partial \varphi}. \tag{9.7}$$

Similarly, one can write

$$\hat{l}_x = i\hbar\left(\sin\varphi \frac{\partial}{\partial \theta} + \cot\theta \cos\varphi \frac{\partial}{\partial \varphi}\right), \tag{9.8}$$

$$\hat{l}_y = i\hbar\left(-\cos\varphi \frac{\partial}{\partial \theta} + \cot\theta \sin\varphi \frac{\partial}{\partial \varphi}\right). \tag{9.9}$$

Using (9.7)–(9.9), one can also write after a lengthy but straightforward algebra

$$\hat{l}^2 = \hat{l}_x^2 + \hat{l}_y^2 + \hat{l}_z^2 = -\hbar^2\left[\frac{1}{\sin\theta}\frac{\partial}{\partial \theta}\left(\sin\theta \frac{\partial}{\partial \theta}\right) + \frac{1}{\sin^2\theta}\frac{\partial^2}{\partial \varphi^2}\right]. \tag{9.10}$$

Next, the commutation relation of $\hat{l}_x$ and $\hat{l}_y$ can be obtained from the commutation relations of x, y, z and p_x, p_y, p_z (see (3.59)):

$$\begin{aligned}[\hat{l}_x, \hat{l}_y] &\equiv [(yp_z - zp_y), (zp_x - xp_z)] = [yp_z, zp_x] + [zp_y, xp_z] \\ &= yp_x[p_z, z] + p_y x[z, p_z] = i\hbar(xp_y - yp_x) \equiv i\hbar\hat{l}_z\end{aligned} \quad (9.11)$$

Similarly, one also finds

$$[\hat{l}_y, \hat{l}_z] = i\hbar\hat{l}_x, \quad (9.12)$$

$$[\hat{l}_z, \hat{l}_x] = i\hbar\hat{l}_y. \quad (9.13)$$

It is clear from (9.7) through (9.10) that

$$[\hat{l}^2, \hat{l}_z] = [\hat{l}^2, \hat{l}_x] = [\hat{l}^2, \hat{l}_y] = 0. \quad (9.14)$$

9.2 Eigenfunctions of $\hat{l}_z$ and $\hat{l}^2$ and Spherical Harmonics

Eigenfunction of $\hat{l}_z$ Now that the angular momentum operators are explicitly specified, the analysis of the eigenfunctions of these operators is in order. Thus, consider the eigenequation of $\hat{l}_z$, which is given from (9.7) by

$$-i\hbar\frac{\partial}{\partial\varphi}u(\varphi) = l_z u(\varphi), \quad (9.15)$$

where l_z denotes the eigenvalue. The eigenequation can be easily solved to yield

$$u(\varphi) \propto e^{i(l_z/\hbar)\varphi} = \frac{1}{\sqrt{2\pi}}e^{i(l_z/\hbar)\varphi}, \quad (9.16)$$

where the constant of integration was used to normalize the eigenfunction over the interval of the azimuthal angle, φ, from 0 to 2π.

Obviously, the eigenfunction $u(\varphi)$ and its probability density should be single valued and one therefore has to impose the boundary condition, namely,

$$u(\varphi) = u(\varphi + 2\pi). \quad (9.17)$$

When this condition is applied to (9.16), the eigenvalue l_z should satisfy the condition $(l_z/\hbar)2\pi = 2\pi m$ with m denoting any integer. Hence, l_z varies digitally in unit of $\hbar$, that is,

$$l_z = m\hbar, \quad m = 0, \pm1, \pm2, \ldots, \quad (9.18)$$

and one can write

$$u_m(\varphi) = \frac{1}{\sqrt{2\pi}}e^{im\varphi}, \quad m = 0, \pm1, \pm2, \ldots. \quad (9.19)$$

Eigenfunctions of $\hat{l}^2$ and Spherical Harmonics Next, consider the eigenequation of $\hat{l}^2$, which reads from (9.10) as

$$-\hbar^2\left[\frac{1}{\sin\theta}\frac{\partial}{\partial\theta}\left(\sin\theta\frac{\partial}{\partial\theta}\right)+\frac{1}{\sin^2\theta}\frac{\partial^2}{\partial\varphi^2}\right]Y_\beta(\theta,\varphi)=\beta\hbar^2 Y_\beta(\theta,\varphi), \tag{9.20}$$

where $Y_\beta(\theta,\varphi)$ and $\beta\hbar^2$ are the eigenfunction and eigenvalue, respectively. Now, since $\hat{l}^2$ and $\hat{l}_z$ commute (see (9.14)), these two operators can share a common eigenfunction and one can therefore look for the solution in the form

$$Y_{\beta m}(\theta,\varphi)\propto u(\varphi)=u(\varphi)P_{\beta m}(\theta), \tag{9.21}$$

with the function $P_{\beta m}(\theta)$ playing the role of the proportionality constant for the angular variable φ. Inserting (9.21) into (9.20) results in

$$\left[\frac{1}{\sin\theta}\frac{\partial}{\partial\theta}\left(\sin\theta\frac{\partial}{\partial\theta}\right)-\frac{m^2}{\sin^2\theta}\right]P_{\beta m}(\theta)=-\beta P_{\beta m}(\theta). \tag{9.22}$$

To proceed further, it is convenient to recast (9.22) in terms of a new variable $w=\cos\theta$, obtaining

$$\frac{d}{dw}(1-w^2)\frac{dP_{\beta m}}{dw}+\left[\beta-\frac{m^2}{1-w^2}\right]P_{\beta m}=0. \tag{9.23}$$

The differential equation (9.23) has singularities at $w=\pm 1$ and it is thus necessary to examine the behavior of the eigenfunction near those singular points. Near $w\approx 1$, for example, one can put $w+1\approx 2$ and (9.23) simplifies as

$$P''_{\beta m}-\frac{1}{1-w}P'_{\beta m}-\frac{m^2}{4(1-w)^2}P_{\beta m}=0, \tag{9.24}$$

where the primes denote the differentiation with respect to w and the constant term β has been neglected, compared with the diverging term. One may try a series solution in powers of $1-w$, that is,

$$P_{\beta m}=(1-w)^\alpha\sum_{n=0}^{\infty}a_n(1-w)^n,\quad a_0\neq 0. \tag{9.25}$$

Inserting (9.25) into (9.24) and singling out the lowest order term in $1-w$, namely, $(1-w)^{\alpha-2}$, one finds

$$\left[\alpha(\alpha-1)+\alpha-\frac{m^2}{4}\right]a_0=0,$$

so that

$$\alpha=\pm\frac{m}{2}.$$

Obviously, the negative branch of α makes the series solution (9.25) diverge and should therefore be discarded. Hence, one may write

$$P(w \to 1) = (1-w)^{|m|/2}[a_0 + a_1(1-w) + \cdots]. \tag{9.26}$$

One can carry out a similar analysis near the singular point at $w = -1$, obtaining

$$P(w \to -1) = (1+w)^{|m|/2}[a_0 + a_1(1+w) + \cdots]. \tag{9.27}$$

Next, using (9.26) and (9.27), one may look for the solution of the equation (9.23) in the form

$$P_{\beta m} = [(1-w)(1+w)]^{|m|/2} Z_{\beta m}(w) \tag{9.28}$$

and insert (9.28) into (9.23), obtaining after a lengthy but straightforward algebra

$$(1-w^2)Z''_{\beta m} - 2(|m|+1)wZ'_{\beta m} + [\beta - |m|(|m|+1)]Z_{\beta m} = 0. \tag{9.29}$$

The differential equation (9.29) can now be solved in routine manner in powers of w by putting

$$Z_{\beta m} = \sum_{k=0}^{\infty} a_k w^k, \tag{9.30}$$

in which case one can write

$$Z'_{\beta m} = \sum_{k=0}^{\infty} a_k k w^{k-1}, \tag{9.31}$$

$$Z''_{\beta m} = \sum_{k=2}^{\infty} a_k k(k-1)w^{k-2} = \sum_{k=0}^{\infty} a_{k+2}(k+2)(k+1)w^k. \tag{9.32}$$

In (9.32), a new dummy variable has been used, namely, $k \to k+2$.

Inserting (9.30)–(9.32) into (9.29) and again performing a lengthy but straightforward algebra, one finds

$$\sum_{k=0}^{\infty} \{a_k[-k(k-1) - 2k(|m|+1) + \beta - |m|(|m|+1)] + a_{k+2}(k+2))(k+1)\}w^k = 0. \tag{9.33}$$

Thus, solving the second-order differential equation, (9.29) has been reduced to satisfying (9.33) for arbitrary powers and values of w. Evidently, this can be done by means of the recurrence relation

$$a_{k+2} = \frac{(k+|m|)(k+|m|+1) - \beta}{(k+2)(k+1)} a_k \tag{9.34}$$

and the solution can generally be expressed in terms of two polynomials, each characterized by a_0 and a_1, respectively. Here, a_0 and a_1 are the two constants of integration and further specify the coefficients of higher order terms in w via the recurrence relation (9.34). Thus, one can write

$$Z_{\beta m} = a_0\left(1+\frac{a_2}{a_0}w^2+\cdots\right)+a_1 w\left(1+\frac{a_3}{a_1}w^2+\cdots\right). \tag{9.35}$$

The two infinite series thus obtained in (9.35) diverge at $w = \pm 1$ and it is again necessary to make the solution physically acceptable. This can be done by terminating the series with a_0 constant and putting $a_1 = 0$ or vice versa. In this way, the solution obtained in (9.35) is reduced to finite-order polynomial in w, either even or odd. The termination condition is given from the recurrence relation (9.34) by

$$\beta = (k+|m|)(k+|m|+1) \tag{9.36}$$

or

$$\beta = l(l+1), \quad l \equiv k+|m|. \tag{9.37}$$

Since $k \geq 0$, $l \geq |m|$ and for the given l, m ranges from $-l$ to l in steps of 1, that is,

$$m = -l, -l+1, \cdots -1, 0, 1, 2, \cdots l-1, l. \tag{9.38}$$

When the termination condition (9.37) is inserted into (9.23), the resulting differential equation

$$\frac{d}{dw}(1-w^2)\frac{dP_l^m}{dw}+\left[l(l+1)-\frac{m^2}{1-w^2}\right]P_l^m = 0 \tag{9.39}$$

is known as the associated Legendre differential equation. For the special case in which $m = 0$, the equation reduces to the (ordinary) Legendre differential equation. Also, the finite-order polynomials $P_l^0 (\equiv P_l)$ and P_l^m satisfying the differential equation are called Legendre and associated Legendre polynomials, respectively.

These polynomials constitute another well-known special function in mathematical physics and can again be generated by using the Rodrigue's formula:

$$P_l(w) = \frac{1}{2^l l!}\left(\frac{d}{dw}\right)^l (w^2-1)^l, \quad w = \cos\theta \tag{9.40}$$

Once P_l is known, P_l^m with $m > 0$ is obtained by performing the operation

$$P_l^m(w) = (-)^m(1-w^2)^{m/2}\frac{d^m}{dw^m}P_l(w), \quad P_l^{-m}(w) = (-)^m P_l^m(w) \tag{9.41}$$

These polynomials are also orthogonal, namely,

$$\int_{-1}^{1} dw P_l^m(w) P_{l'}^m = \frac{2}{2l+1}\frac{(l+|m|)!}{(l-|m|)!}\delta_{ll'}, \tag{9.42}$$

and one can therefore write the normalized eigenfunction of both $\hat{l}_z$ and $\hat{l}^2$ as

$$Y_l^m(\theta,\varphi) = (-)^m\left[\frac{2l+1}{4\pi}\frac{(l-|m|)!}{(l+|m|)!}\right]^{1/2} P_l^m(\theta)e^{im\varphi}, \tag{9.43a}$$

with

$$Y_l^{-m}(\theta, \varphi) = (-)^m Y_l^m(\theta, \varphi)^*. \tag{9.43b}$$

The functions specified in (9.43) are known as the spherical harmonics and constitute one of the most important and extensively utilized eigenfunctions.

The Spatial Quantization Using these spherical harmonics, the angular momentum eigenequations (9.15) and (9.20) are compactly denoted as

$$\hat{l}_z|lm\rangle = \hbar m|lm\rangle, \quad |lm\rangle \equiv Y_l^m \equiv Y_{lm}, \tag{9.44}$$

$$\hat{l}^2|lm\rangle = \hbar^2 l(l+1)|lm\rangle. \tag{9.45}$$

The absolute value of the angular momentum is thus specified in units of $\hbar$ and by a discrete set of numbers $\sqrt{l(l+1)}$, $l = 0, 1, 2, \ldots$. Also, for the given l, the z-component of the angular momentum varies digitally in unit steps of $\hbar$ over the range from $-l$ to l. This is illustrated graphically in Figure 9.3, in which a given magnitude of the angular momentum precesses around the z-axis in discrete orientations, so that its projection onto the z-axis varies in steps of $\hbar$. This unique feature of the spatial quantization of angular momentum is again in marked contrast with the classical theory where both the magnitude and orientation of the angular momentum can vary in continuous fashion by any infinitesimal amount.

When the angular momentum is in an eigenstate $|lm\rangle$, its magnitude and z-component are known exactly, as evident from (9.44) and (9.45), that is,

$$\left\langle lm|\hat{l}^2|lm\right\rangle = l(l+1)\hbar^2, \tag{9.46}$$

$$\left\langle lm|\hat{l}_z|lm\right\rangle = m\hbar. \tag{9.47}$$

However, it is also clear from (9.8) and (9.9) that

$$\left\langle lm|\hat{l}_x|lm\right\rangle = \left\langle lm|\hat{l}_y|lm\right\rangle = 0. \tag{9.48}$$

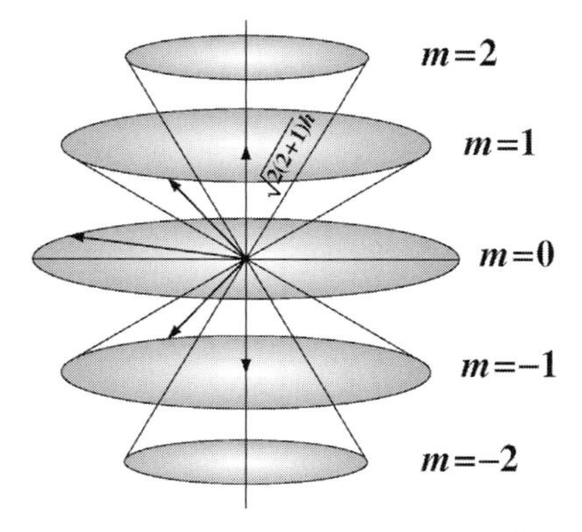

Figure 9.3 The spatial quantization of $\hat{l}_z$ for $l = \hbar\sqrt{2 \cdot 3}$.

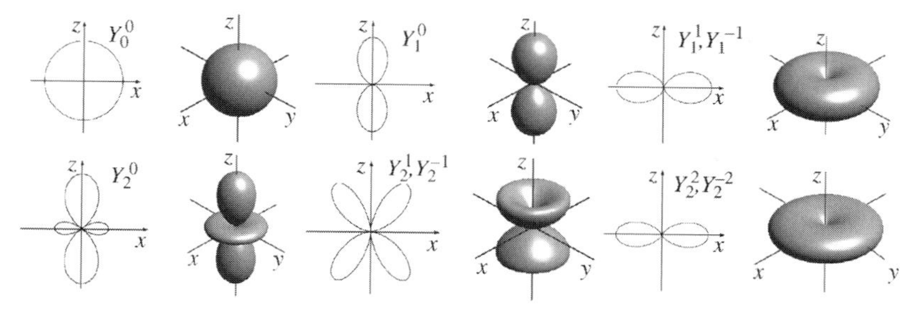

Figure 9.4 The three-dimensional and projected polar plots of spherical harmonics, Y_0^0, Y_1^0, $Y_1^{\pm 1}$, Y_2^0, $Y_2^{\pm 1}$, and $Y_2^{\pm 2}$.

Table 9.1 Spherical harmonics.

l	m	$Y_l^m(\theta,\varphi) \equiv Y_{lm}(\theta,\varphi)$
0	0	$Y_{00} = \dfrac{1}{2\pi^{1/2}}$
1	0	$Y_{10} = \dfrac{1}{2}(3/\pi)^{1/2} \cos\theta$
	± 1	$Y_{1\pm 1} = \mp\dfrac{1}{2}(3/2\pi)^{1/2} \sin\theta e^{\pm i\varphi}$
2	0	$Y_{20} = \dfrac{1}{4}(5/\pi)^{1/2}(3\cos^2\theta - 1)$
	± 1	$Y_{2\pm 1} = \mp\dfrac{1}{2}(15/2\pi)^{1/2}\cos\theta \sin\theta e^{\pm i\varphi}$
	± 2	$Y_{2\pm 2} = \dfrac{1}{4}(15/2\pi)^{1/2} \sin^2\theta e^{\pm 2i\varphi}$

That is, the information of the projected x and y components of the angular momentum has been wiped away. This is rooted in the noncommuting behavior of the three components of the angular momentum.

As noted, the spherical harmonics provide a powerful means of describing the atomic and molecular orbitals, which play a key role in the study of the molecular structures and chemical bonding. Typical polar plots of these functions are presented in Figure 9.4. Also, in Table 9.1 are tabulated some of the typical examples of the spherical harmonics for different l and m values.

9.3 Problems

9.1. The angular momentum of a particle is defined in terms of its position and momentum as

$$\hat{\boldsymbol{l}} = \boldsymbol{r} \times \boldsymbol{p} = (\hat{x}x + \hat{y}y + \hat{z}z) \times ((\hat{x}p_x + \hat{y}p_y + \hat{z}p_z).$$

(a) Using the cyclic properties of the unit vectors $\hat{x} \times \hat{y} = \hat{z}$, $\hat{y} \times \hat{z} = \hat{x}$, and $\hat{z} \times \hat{x} = \hat{y}$, show that

$$\hat{l}_x = yp_z - zp_y; \quad \hat{l}_y = zp_x - xp_z; \quad \hat{l}_z = xp_y - yp_x.$$

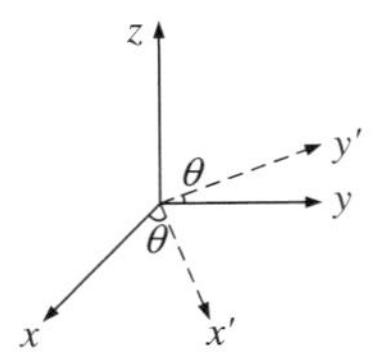

Figure 9.5 A new Cartesian frame resulting from rotation around z-axis by θ.

(b) Using commutation relation existing between position and momentum, such as x and p_x, show that

$$[\hat{l}_x, \hat{l}_y] = i\hbar\hat{l}_z; \quad [\hat{l}_y, \hat{l}_z] = i\hbar\hat{l}_x; \quad [\hat{l}_z, \hat{l}_x] = i\hbar\hat{l}_y.$$

9.2. Using the expressions for $\hat{l}_x$, $\hat{l}_y$, and $\hat{l}_z$ given in (9.7)–(9.9),

(a) Show that

$$\hat{l}^2 = \hat{l}_x^2 + \hat{l}_y^2 + \hat{l}_z^2 = -\hbar^2\left[\frac{1}{\sin\theta}\frac{\partial}{\partial\theta}(\sin\theta\frac{\partial}{\partial\theta}) + \frac{1}{\sin^2\theta}\frac{\partial^2}{\partial\varphi^2}\right].$$

(b) Also show that

$$[\hat{l}^2, \hat{l}_z] = [\hat{l}^2, \hat{l}_x] = [\hat{l}^2, \hat{l}_y] = 0.$$

9.3. Consider the Cartesian coordinate frame shown in Figure 9.5 with unit vectors, $\hat{x}$, $\hat{y}$, and $\hat{z}$, respectively. If the frame is rotated around the z-axis by an angle, say, $\theta = \pi/4$, the corresponding unit vectors can be expressed as

$$\begin{pmatrix} \hat{x}' \\ \hat{y}' \end{pmatrix} = \begin{pmatrix} \cos\theta & \sin\theta \\ -\sin\theta & \cos\theta \end{pmatrix}\begin{pmatrix} \hat{x} \\ \hat{y} \end{pmatrix} = \begin{pmatrix} 1/\sqrt{2} & 1/\sqrt{2} \\ -1/\sqrt{2} & 1/\sqrt{2} \end{pmatrix}\begin{pmatrix} \hat{x} \\ \hat{y} \end{pmatrix},$$

as clearly indicated in Figure 9.5.

(a) Show that the new set of unit vectors $\hat{x}'$, $\hat{y}'$, and $\hat{z}'$ are orthonormal.

(b) Show that Y_1^0, Y_1^1, and Y_1^{-1} are orthonormal.

(c) Show that the set of new eigenfunctions

$$|p_z\rangle \equiv Y_1^0; \quad |p_x\rangle \equiv \frac{1}{\sqrt{2}}(Y_1^1 + Y_1^{-1}); \quad |p_y\rangle \equiv \frac{1}{\sqrt{2}i}(Y_1^1 - Y_1^{-1})$$

are also orthonormal.

(d) Is there any analogy existing between $\hat{x}$, $\hat{y}$, $\hat{z}$ and $\hat{x}'$, $\hat{y}'$, $\hat{z}'$ on the one hand and between Y_1^0, Y_1^1, Y_1^{-1} and $|p_z\rangle$, $|p_x\rangle$, $|p_y\rangle$ on the other?

(e) Find

$$\left\langle p_z|\hat{l}^2|p_z\right\rangle; \quad \left\langle p_z|\hat{l}_z|p_z\right\rangle; \quad \left\langle p_x|\hat{l}^2|p_x\right\rangle; \quad \left\langle p_x|\hat{l}_z|p_x\right\rangle; \quad \left\langle p_y|\hat{l}^2|p_y\right\rangle; \quad \left\langle p_y|\hat{l}_z|p_y\right\rangle$$

and compare the results with those results obtained for Y_1^0, Y_1^1, and Y_1^{-1}.

Suggested Reading

1 Yariv, A. (1982) *An Introduction to Theory and Applications of Quantum Mechanics*, John Wiley & Sons, Inc.
2 Gasiorowicz, S. (2003) *Quantum Physics*, 3rd edn, John Wiley & Sons, Inc.
3 Haken, H., Wolf, H.C., and Brewer, W.D. (2007) *The Physics of Atoms and Quanta: Introduction to Experiments and Theory*, 7th revised and enlarged edn, Springer.
4 Karplus, M. and Porter, R.N. (1970) *Atoms and Molecules: An Introduction for Students of Physical Chemistry*, Addison-Wesley Publishing Company.

10
Hydrogen Atom: The Schrödinger Treatment

The hydrogen atom is the simplest atomic system, but its understanding is essential for comprehending other atoms, molecules, and the materials in general. The versatility of the Schrödinger equation was first demonstrated by the successful description of the H-atom. The information emanating from the wave mechanical treatment of the H-atom is richer than what was provided by Bohr's celebrated theory of the H-atom. Moreover, in the wave mechanical treatment, the quantization of the energy as well as the angular momentum is not based upon *ad hoc* assumptions but on a natural consequence of eigenfunctions physically well behaving. The wavefunctions of the H-atom further provide the basis by which to describe the atomic and molecular orbitals, chemical bonding, and ultimately the properties of material.

In addition, the H-atom offers one of those rare examples that can be treated in a closed analytical form. The mathematical analysis required for treating the H-atom is lengthy but a substantial portion of it has already been covered in connection with the angular momentum in Chapter 9.

10.1
Two-Body Central Force Problem

The hydrogen atom consists of a proton and an electron, bound by an attractive Coulomb force. Although simplest in structure, the H-atom theory provides essential ingredients and the basic foundation for understanding all other atoms and molecules. In addition, the theory introduces the concept of spectroscopy, the novel phenomena of the angular momentum, and the atomic orbitals. Also, the H-atom provides the typical example of the two-body central force problem and offers one of those rare examples that can be treated exactly and analytically.

Thus, consider a system of two particles of masses m_1 and m_2 interacting via a central force. The equations of motion of these two particles are then given by

$$m_1\ddot{\boldsymbol{r}}_1 = \boldsymbol{f}(|\boldsymbol{r}_1 - \boldsymbol{r}_2|), \tag{10.1}$$

$$m_2\ddot{\boldsymbol{r}}_2 = -\boldsymbol{f}(|\boldsymbol{r}_1 - \boldsymbol{r}_2|), \tag{10.2}$$

Introductory Quantum Mechanics for Semiconductor Nanotechnology. Dae Mann Kim

ISBN: 978-3-527-40975-4

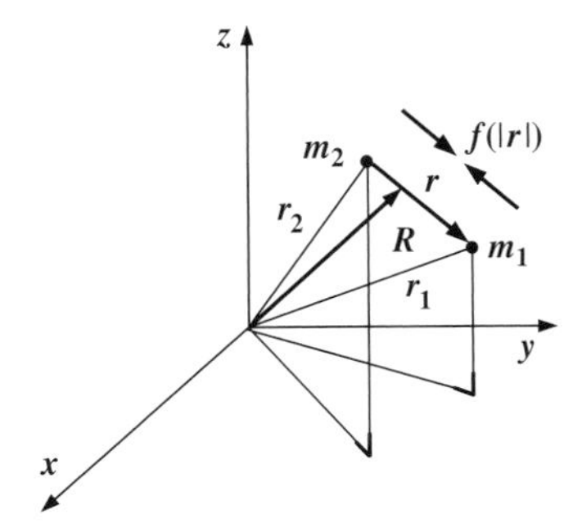

Figure 10.1 A system of two particles bound by a central force: $\boldsymbol{r}_1$ and $\boldsymbol{r}_2$ can be replaced by the center of mass and relative coordinates $\boldsymbol{R}$ and $\boldsymbol{r}$.

where $\boldsymbol{r}_1$ and $\boldsymbol{r}_2$ are the positions of these two particles and the interaction force depends only on the distance between the two and acts on both in opposite directions, as shown in Figure 10.1. Thus, the addition of (10.1) and (10.2) leads to

$$m_1\ddot{\boldsymbol{r}}_1 + m_2\ddot{\boldsymbol{r}}_2 = 0. \tag{10.3}$$

Equation 10.3 can be recast into

$$M\ddot{\boldsymbol{R}} = 0, \tag{10.4}$$

where

$$\boldsymbol{R} = \frac{m_1\boldsymbol{r}_1 + m_2\boldsymbol{r}_2}{M}, \qquad M \equiv m_1 + m_2, \tag{10.5}$$

is the center of mass coordinate with M denoting the total mass (see Figure 10.1).

Dividing (10.1) and (10.2) by the respective masses and subtracting the latter equation from the former results in

$$\ddot{\boldsymbol{r}}_1 - \ddot{\boldsymbol{r}}_2 = \left(\frac{1}{m_1} + \frac{1}{m_2}\right)\boldsymbol{f}(|\boldsymbol{r}_1 - \boldsymbol{r}_2|). \tag{10.6}$$

Or

$$\mu\ddot{\boldsymbol{r}} = \boldsymbol{f}(r), \tag{10.7}$$

where

$$\boldsymbol{r} = \boldsymbol{r}_1 - \boldsymbol{r}_2 \tag{10.8}$$

is the relative displacement of the particle 1 with respect to particle 2 and μ defined as

$$\frac{1}{\mu} \equiv \frac{1}{m_1} + \frac{1}{m_2} \tag{10.9}$$

is called the reduced mass. Thus, the motion of two particles bound by a central force naturally partitions into the motion of the center of mass, moving as a free particle with total mass M (see (10.4)) and the relative motion of a fictitious particle with the reduced mass μ moving with respect to a fixed force center (see (10.7)). In the limit where $m_1 \ll m_2$, $\mu \approx m_1$ and the relative motion is reduced essentially to the motion of lighter particle with respect to the heavier one.

It is intuitively clear and can also be shown rigorously that the kinetic energy of this two particle system can be expressed as the sum of the kinetic energies of the center of mass and the relative motion:

$$K = \frac{p_1^2}{2m_1} + \frac{p_2^2}{2m_2} = \frac{P^2}{2M} + \frac{p^2}{2\mu}, \qquad \boldsymbol{P} = M\boldsymbol{R},\ \boldsymbol{p} = \mu\boldsymbol{r}. \tag{10.10}$$

Hence, the Hamiltonian of the system reads as

$$\hat{H} = -\frac{\hbar^2}{2M}\nabla_R^2 - \frac{\hbar^2}{2\mu}\nabla^2 + V(r), \qquad r \equiv |\boldsymbol{r}_1 - \boldsymbol{r}_2|, \tag{10.11}$$

where

$$\nabla_R^2 = \frac{\partial^2}{\partial X^2} + \frac{\partial^2}{\partial^2 Y} + \frac{\partial^2}{\partial Z^2}; \quad \nabla^2 = \frac{\partial^2}{\partial x^2} + \frac{\partial^2}{\partial y^2} + \frac{\partial^2}{\partial z^2}$$

are the Laplacian operators involving $\boldsymbol{R}$ and $\boldsymbol{r}$ coordinates, respectively.

The wavefunction describing the two particle system can therefore be partitioned into $\boldsymbol{R}$ and $\boldsymbol{r}$:

$$\begin{aligned} \psi(\boldsymbol{R}, \boldsymbol{r}, t) &= \varphi_{\mathrm{CM}}(\boldsymbol{R}, t)\varphi(\boldsymbol{r}, t) \\ &= e^{-i(E_{\mathrm{CM}}/\hbar)t} u_{\mathrm{CM}}(\boldsymbol{R}) e^{-i(E/\hbar)t} u(\boldsymbol{r}). \end{aligned} \tag{10.12}$$

Inserting (10.12) into the Schrödinger equation,

$$i\hbar\frac{\partial}{\partial t}\psi(\boldsymbol{R}, \boldsymbol{r}, t) = \left[-\frac{\hbar^2}{2M}\nabla_R^2 - \frac{\hbar^2}{2\mu}\nabla^2 + V(r)\right]\psi(\boldsymbol{R}, \boldsymbol{r}, t), \tag{10.13}$$

and separating the variables $\boldsymbol{R}$ and $\boldsymbol{r}$ in the usual manner, one finds

$$-\frac{\hbar^2}{2M}\nabla_R^2 u_{\mathrm{CM}}(\boldsymbol{R}) = E_{\mathrm{CM}} u(\boldsymbol{R}), \tag{10.14}$$

$$\left[-\frac{\hbar^2}{2\mu}\nabla^2 + V(r)\right] u(\boldsymbol{r}) = E u(\boldsymbol{r}), \tag{10.15}$$

where the total energy consists of the kinetic energy of the center of mass and the energy of the internal motion, namely,

$$E_T = E_{\mathrm{CM}} + E. \tag{10.16}$$

As discussed in Section 4.2, (10.14) is the eigenequation of a free particle moving in three dimension and one can therefore write the wavefunction of the center of mass in the form

$$u_{\mathrm{CM}}(\boldsymbol{R}) \propto e^{i\boldsymbol{K}\cdot\boldsymbol{R}}; \quad \hbar^2 K^2/2M = E_{\mathrm{CM}}. \tag{10.17}$$

This leaves the bulk of the analysis of the H-atom to finding the eigenfunction $u(\mathbf{r})$ of the internal relative motion. Since the mass of the proton is much larger than that of the electron, the reduced mass μ is practically equal to the electron mass and the internal motion essentially consists of the motion of the electron with respect to the proton. However, the reduced mass μ can be substantially different from the electron mass in other hydrogenic atomic systems.

10.2
The Hydrogenic Atom

Internal Motion and Bound State The attractive Coulomb potential binding the electron to the nucleus in the hydrogenic atom is generally given by

$$V(r) = -\frac{Ze_M^2}{r}, \quad e_M^2 \equiv \frac{e^2}{4\pi\varepsilon_0}, \tag{10.18}$$

where Z is the atomic number and the Coulomb constant $1/4\pi\varepsilon_0$ is compacted into the shorthand notation e_M^2. Thus, Hamiltonian for the internal motion is given from (10.15) and (10.18) by

$$\hat{H} = -\frac{\hbar^2}{2\mu}\nabla^2 + V(r). \tag{10.19}$$

As detailed in Chapter 9, the Laplacian ∇^2 involving x, y, and z can be transcribed into the spherical coordinates as

$$\nabla^2 = \frac{1}{r^2}\frac{\partial}{\partial r}r^2\frac{\partial}{\partial r} + \frac{1}{r^2}\left[\frac{1}{\sin\theta}\frac{\partial}{\partial\theta}\sin\theta\frac{\partial}{\partial\theta} + \frac{1}{\sin^2\theta}\frac{\partial^2}{\partial\varphi^2}\right]. \tag{10.20}$$

It is interesting to note in (10.20) that the operators involving angular variables are identical to $-\hat{l}^2/\hbar^2$, when lumped together (see (9.10)). Thus, one can compact (10.20) as

$$\nabla^2 = \frac{1}{r^2}\frac{\partial}{\partial r}r^2\frac{\partial}{\partial r} - \frac{1}{r^2}\frac{1}{\hbar^2}\hat{l}^2 \tag{10.21}$$

and the Hamiltonian is therefore compacted as

$$\hat{H} = -\frac{\hbar^2}{2\mu}\frac{1}{r^2}\frac{\partial}{\partial r}r^2\frac{\partial}{\partial r} + \frac{1}{2\mu r^2}\hat{l}^2 + V(r). \tag{10.22}$$

The energy eigenequation for the internal motion thus reads as

$$\left[-\frac{\hbar^2}{2\mu}\frac{1}{r^2}\frac{\partial}{\partial r}r^2\frac{\partial}{\partial r} + \frac{1}{2\mu r^2}\hat{l}^2 + V(r)\right]u(r,\theta,\varphi) = Eu(r,\theta,\varphi). \tag{10.23}$$

The first step for analyzing this second-order partial differential equation is to note from (10.22) that $\hat{H}$ and $\hat{l}^2$ commute and, therefore, the two operators can share the common eigenfunction, namely, the spherical harmonics:

$$u(\mathbf{r}) \propto Y_l^m(\theta, \varphi) = Y_l^m(\theta, \varphi) R(r). \tag{10.24}$$

Here, the radial part of the wavefunction $R(r)$ is to be viewed as a proportionality constant for angular variables θ and φ. Since $Y_l^m(\theta, \varphi)$ is the eigenfunction of $\hat{l}^2$ (see (9.45)), when (10.24) is inserted into (10.23), $Y_l^m(\theta, \varphi)$ is automatically canceled from both sides and (10.23) simply reduces to the differential equation involving only the radial component of the wavefunction:

$$\left[-\frac{\hbar^2}{2\mu}\frac{1}{r^2}\frac{\partial}{\partial r}r^2\frac{\partial}{\partial r} + V_{\text{eff}}(r)\right]R(r) = ER(r), \tag{10.25}$$

where the effective potential

$$V_{\text{eff}}(r) = V(r) + \frac{1}{2\mu r^2}\hbar^2 l(l+1), \quad V(r) = -\frac{Ze_M^2}{r}, \tag{10.26}$$

is comprised of the attractive Coulomb potential (10.18) and a repulsive term, arising from the rotation-induced centrifugal force. For large r, the attractive term $\propto -1/r$ dominates over the repulsive term $\propto 1/r^2$, while for small r, the latter term dominates over the former. Hence, these two terms of opposite polarity combine to give rise to a potential well, as shown in Figure 10.2. It is in this potential well that the bound states of the atom are formed. For $l = 0$, however, only the attractive Coulomb force binds the electron to the proton.

Radial Wavefunction As shown in Figure 10.2, it is convenient to take the energy level of the atom zero when the electron is separated from the nucleus. Then, for the bound state, one should put

$$E = -|E|. \tag{10.27a}$$

Next, introduce the dimensionless radial variable

$$\varrho = \alpha r, \quad \alpha^2 \equiv \frac{8\mu|E|}{\hbar^2}, \tag{10.27b}$$

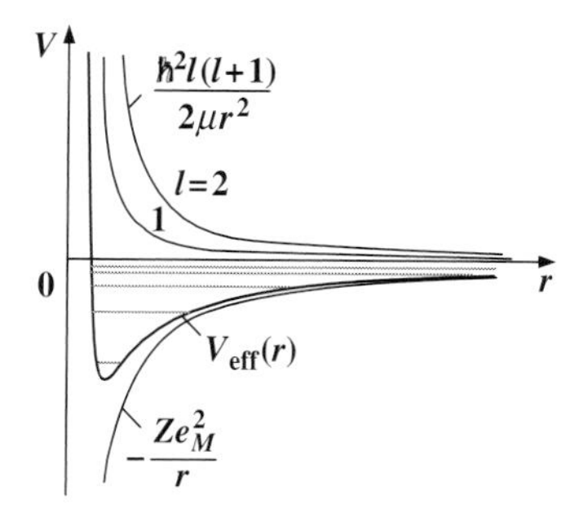

Figure 10.2 The effective potential energy resulting from the attractive Coulomb force and repulsive centrifugal force.

and additional dimensionless parameter

$$\lambda \equiv \frac{Ze_M^2}{\hbar}\left(\frac{\mu}{2|E|}\right)^{1/2}. \tag{10.27c}$$

Then the radial equation (10.25) is expressed in terms of ϱ and λ as

$$\left[\frac{1}{\varrho^2}\frac{d}{d\varrho}\varrho^2\frac{d}{d\varrho}+\frac{\lambda}{\varrho}-\frac{1}{4}-\frac{l(l+1)}{\varrho^2}\right]R_{\lambda l}(\varrho)=0 \tag{10.28}$$

and the radial wavefunction depends parametrically on λ and l.

To analyze (10.28), it is again convenient to first examine its asymptotic behavior. In the limit of large ϱ, (10.28) reduces approximately to

$$R''_{\lambda l}-\frac{1}{4}R_{\lambda l}=0. \tag{10.29}$$

One can then find two possible solutions of (10.29), namely,

$$R_{\lambda l}=e^{\pm\varrho/2}.$$

Obviously, the positive branch of the solution has to be discarded to prevent the solution from diverging at large ϱ. With this fact in mind, one may thus look for the solution of (10.28) in the form

$$R_{\lambda l}(\varrho)=\varrho^s L(\varrho)e^{-\varrho/2}, \tag{10.30}$$

where $L(\varrho)$ is again taken to be a polynomial in ϱ. Inserting (10.30) into (10.28) and carrying out a lengthy but straightforward algebra, one finds

$$\varrho^2 L''+\varrho[2(s+1)-\varrho]L'+[\varrho(\lambda-1-s)+s(s+1)-l(l+1)]L=0. \tag{10.31}$$

This differential equation can obviously be simplified by choosing the index s such that

$$s(s+1)-l(l+1)=0,$$

that is,

$$s=l,\,-(l+1).$$

But to prevent $R_{\lambda l}(\varrho)$ from diverging at $\varrho\approx 0$, s should be confined to $s=l$. With this choice, (10.31) reduces to

$$\varrho L''+[2(l+1)-\varrho]L'+(\lambda-l-1)L=0 \tag{10.32}$$

and is ready for the usual series solution. Thus put

$$L=\sum_{m=0}^{\infty}a_m\varrho^m, \tag{10.33a}$$

so that

$$L' = \sum_{m=0}^{\infty} a_m m \varrho^{m-1} = \sum_{m=0}^{\infty} a_{m+1}(m+1)\varrho^m, \tag{10.33b}$$

$$L'' = \sum_{m=0}^{\infty} a_m m(m-1)\varrho^{m-2} = \sum_{m=0}^{\infty} a_{m+1}(m+1)m\varrho^{m-1}. \tag{10.33c}$$

In (10.33b) and (10.33c), new dummy variables have been used, that is, $m \to m+1$. Substituting (10.33) into (10.32) and regrouping the terms results in

$$\sum_{m=0}^{\infty}\{[(m+1)m+2(l+1)(m+1)]a_{m+1}-[m-(\lambda-l-1)]a_m\}\varrho^m = 0. \tag{10.34}$$

Hence, $L(\varrho)$ can be completely specified in terms of a_0 and the recurrence relation

$$a_{m+1} = a_m \frac{m+l+1-\lambda}{(m+1)(m+2l+2)}. \tag{10.35}$$

However, the resulting infinite series diverges as $L(\varrho) \approx \exp\varrho$, and $R_{\lambda l}(\varrho)$ will diverge unless the series is terminated at a finite order (see (10.30)). The required termination condition is given from (10.35) by

$$\lambda = m+l+1 \equiv n, \tag{10.36}$$

where n is a positive integer. Since the lower limit of the dummy variable m is 0,

$$n \geq l+1. \tag{10.37}$$

When the termination condition (10.36) is inserted into the differential equation (10.32), it reduces to the Laguerre differential equation, namely,

$$\varrho L''_{nl} + [2(l+1)-\varrho]L'_{nl} + (n-l-1)L_{nl} = 0. \tag{10.38}$$

The finite-order series solution L_{nl} is called the associated Laguerre polynomial and is specified in terms of two parameters n and l. In fact, the standard form of the Laguerre equation reads as

$$\varrho L_q^{p''} + [p+1-\varrho]L_q^{p'} + (q-p)L_q^p = 0. \tag{10.39}$$

Hence, comparing (10.38) and (10.39), one can make the identifications $p = 2l+1$ and $q = n+l$. And the polynomial $L_{nl}(\varrho)$ in (10.38) can be expressed in the standard notation as

$$L_{nl}(\varrho) = L_{n+l}^{2l+1}(\varrho). \tag{10.40}$$

The radial wavefunction (10.30) can therefore be expressed from (10.30) as

$$R_{nl} = N_{nl}e^{-\varrho/2}\varrho^l L_{n+l}^{2l+1}(\varrho), \tag{10.41}$$

with N_{nl} denoting the normalization constant.

Energy Quantization Now that the mathematical analysis of the energy eigenfunction of the internal motion has been completed, the physical discussion of the wavefunction, in particular, the discussion of the information contained therein is in order. First, the constraint (10.36) imposed on the parameter λ to terminate the series so that the wavefunction becomes physically acceptable yields, when combined with its definition in (10.27c),

$$\frac{Ze_M^2}{\hbar}\left(\frac{\mu}{2|E_n|}\right)^{1/2} = n. \tag{10.42}$$

Therefore, the energy quantization naturally follows from (10.42) as

$$E_n \equiv -|E_n| = -\frac{\mu Z^2 e_M^4}{2\hbar^2}\frac{1}{n^2} = -\frac{Z^2 e_M^2}{2a_0}\frac{1}{n^2}, \tag{10.43}$$

where

$$a_0 \equiv \frac{\hbar^2}{\mu e_M^2} = \frac{\hbar^2}{e_M^2 m_e}\left(1+\frac{m_e}{m_N}\right) = a_B\left(1+\frac{m_e}{m_N}\right) \tag{10.44}$$

is essentially the Bohr radius $a_B(=0.053\,\mathrm{nm})$, however with the reduced mass replacing the electron mass m_e. Because the mass of the nucleus is much greater than m_e, one can put $\mu \approx m_e$, since $1/\mu = 1/m_e + 1/m_N \approx 1/m_e$ (see (10.9)), but μ could be substantially different from m_e in some of the hydrogenic atoms, especially with the nuclear mass much lighter than the proton. Clearly, the quantized energy level derived in (10.43) is in exact agreement with the corresponding expression resulting from Bohr's H-atom theory, discussed in Chapter 2 (see (2.31) and (2.34)).

Energy Eigenfunction With the energy eigenvalue explicitly specified in (10.43), the parameter α introduced in (10.27b) now reads as

$$\alpha_n^2 = \frac{8\mu}{\hbar^2}|E_n| = \frac{8\mu}{\hbar^2}\frac{Z^2 e_M^2}{2a_0}\frac{1}{n^2} = \left(\frac{2Z}{a_0 n}\right)^2. \tag{10.45}$$

Thus, the parameter α_n naturally scales the radial atomic dimensions in terms of the Bohr radius a_0, the atomic number Z, and the energy level of the bound state n. This can be shown more explicitly by expressing the radial variable ϱ using (10.27b) and (10.45), namely,

$$\varrho \equiv \alpha_n r = \left(\frac{2Z}{a_0 n}\right) r. \tag{10.46}$$

Thus, the energy eigenfunction resulting from (10.24), (10.41), and (10.46), namely,

$$u_{nlm}(r,\theta,\varphi) = N_{nl} R_{nl}(r) Y_l^m(\theta,\varphi),$$

can be normalized as

$$1 = N_{lm}^2 \int_0^\infty r^2 dr \int_0^{2\pi} d\varphi \int_0^\pi d\theta \sin\theta \left|Y_l^m\right|^2 [R_{nl}]^2$$
$$= N_{lm}^2 \frac{1}{\alpha_n^3} \int_0^\infty d\varrho \varrho^2 [R(\varrho)]^2. \tag{10.47}$$

In (10.47), the angular integration automatically yields unity, as discussed in Chapter 9. When (10.47) is combined with the well-known integral

$$\int_0^\infty d\varrho \varrho^2 e^{-\varrho} \varrho^{2l} [L_{n+l}^{2l+1}(\varrho)]^2 = \frac{2n[(n+l)!]^3}{(n-l-1)!}, \tag{10.48}$$

the constant N_{lm} is readily determined.

The normalized radial wavefunction (10.41) therefore reads as

$$R_{nl}(r) = \left[\left(\frac{2Z}{na_0}\right)^3 \frac{(n-l-1)!}{2n[(n+l)!]^3}\right]^{1/2} e^{-\varrho/2} \varrho^l L_{n+l}^{2l+1}(\varrho), \quad \varrho \equiv \left(\frac{2Z}{a_0 n}\right) r \tag{10.49}$$

and the normalized energy eigenfunction is completely specified in terms of the normalized radial wavefunction (10.49) and the spherical harmonics:

$$u_{nlm} = R_{nl}(r) Y_l^m(\theta, \varphi). \tag{10.50}$$

The typical examples R_{nl} and u_{nlm} functions are listed in Tables 10.1 and 10.2.

Table 10.1 Hydrogenic radial wavefunctions.

$a_0 \equiv \frac{\hbar^2}{\mu e_M^2} = \frac{\hbar^2}{e_M^2 m_e}\left(1 + \frac{m_e}{m_N}\right) = a_B\left(1 + \frac{m_e}{m_N}\right)$

$\varrho \equiv \left(\frac{2Z}{a_0 n}\right) r$

n	**l**	**Orbital**	$R_{nl}(r)$
1	0	1s	$(Z/a_0)^{3/2} 2e^{-\varrho/2}$
2	0	2s	$(Z/a_0)^{3/2}(1/8)^{1/2}(2-\varrho)e^{-\varrho/2}$
2	1	2p	$(Z/a_0)^{3/2}(1/24)^{1/2}\varrho e^{-\varrho/2}$
3	0	3s	$(Z/a_0)^{3/2}(1/243)^{1/2}(6-6\varrho+\varrho^2)e^{-\varrho/2}$
3	1	3p	$(Z/a_0)^{3/2}(1/486)^{1/2}(4-\varrho)\varrho e^{-\varrho/2}$
3	2	3d	$(Z/a_0)^{3/2}(1/2430)^{1/2}\varrho^2 e^{-\varrho/2}$

a) a_B: Bohr radius (0.053 nm); m_e: electron mass; m_N: mass of the nucleus.

Table 10.2 Hydrogenic energy eigenfunctions.

$a_0 \equiv \frac{\hbar^2}{\mu e_M^2} = \frac{\hbar^2}{e_M^2 m_e}(1+\frac{m_e}{m_N}) = a_B(1+\frac{m_e}{m_N})$

$\varrho \equiv (\frac{2Z}{a_0 n})r$

Orbital	u_{nlm}
1s	$u_{100} = \frac{(Z/a_0)^{3/2}}{\pi^{1/2}} e^{-Zr/a_0}$
2s	$u_{200} = \frac{(Z/a_0)^{3/2}}{(32\pi)^{1/2}} (2-\frac{Zr}{a_0}) e^{-Zr/2a_0}$
2p	$u_{210} = \frac{(Z/a_0)^{3/2}}{(32\pi)^{1/2}} \frac{Zr}{a_0} e^{-Zr/2a_0} \cos\theta$
2p	$u_{21\pm1} = \frac{(Z/a_0)^{3/2}}{(64\pi)^{1/2}} \frac{Zr}{a_0} e^{-Zr/2a_0} \sin\theta e^{\pm i\varphi}$
3s	$u_{300} = \frac{(Z/a_0)^{3/2}}{81(3\pi)^{1/2}} (27-18\frac{Zr}{a_0}+2\frac{Z^2r^2}{a_0^2}) e^{-Zr/3a_0}$
3p	$u_{310} = \frac{2^{1/2}(Z/a_0)^{3/2}}{81(\pi)^{1/2}} (6-\frac{Zr}{a_0}) \frac{Zr}{a_0} e^{-Zr/3a_0} \cos\theta$
3p	$u_{31\pm1} = \frac{(Z/a_0)^{3/2}}{81(\pi)^{1/2}} (6-\frac{Zr}{a_0}) \frac{Zr}{a_0} e^{-Zr/3a_0} \sin\theta e^{\pm i\varphi}$
3d	$u_{320} = \frac{(Z/a_0)^{3/2}}{81(6\pi)^{1/2}} \frac{Z^2r^2}{a_0^2} e^{-Zr/3a_0} (3\cos^2\theta-1)$
3d	$u_{32\pm1} = \frac{(Z/a_0)^{3/2}}{81(\pi)^{1/2}} \frac{Z^2r^2}{a_0^2} e^{-Zr/3a_0} \sin\theta\cos\theta e^{\pm i\varphi}$
3d	$u_{31\pm2} = \frac{(Z/a_0)^{3/2}}{162(\pi)^{1/2}} \frac{Z^2r^2}{a_0^2} e^{-Zr/3a_0} \sin^2\theta e^{\pm 2i\varphi}$

a) a_B: Bohr radius (0.053 nm); m_e: electron mass; m_N: mass of the nucleus.

10.3 The Atomic Orbital

Hierarchy of Quantum Numbers The one-electron wavefunction (10.50) is called the atomic orbital and it carries a wealth of information. Clearly, u_{nlm} is characterized by three quantum numbers: the principal quantum number n ranging $1, 2, 3, \ldots$ (see (10.36); for the given n, the angular momentum quantum number l ranging $0, 1, 2, \ldots, n-1$ (see (10.37)); and for the given l, the magnetic quantum number m ranging $-l, -l+1, \ldots, 0, \ldots, l-1, l$ (see (9.38)). The quantum states associated with $l = 0, 1, 2, 3, \ldots$ are designated as s, p, d, f states in spectroscopy.

Degeneracy As pointed out, the energy of the quantum state u_{nlm} has been quantized as the natural consequence of the wavefunction physically well behaving. Also, the energy eigenvalues have been shown to depend only on the principal quantum number n, as is clear from (10.43). This points to the multiple quantum states u_{nlm} having the same energy at the given n but having different l and m. The number of such degenerate states for the given n can be found by summing over l from 0 to $n-1$, with each l accompanied by $2l+1$ possible values of m. Also, for the

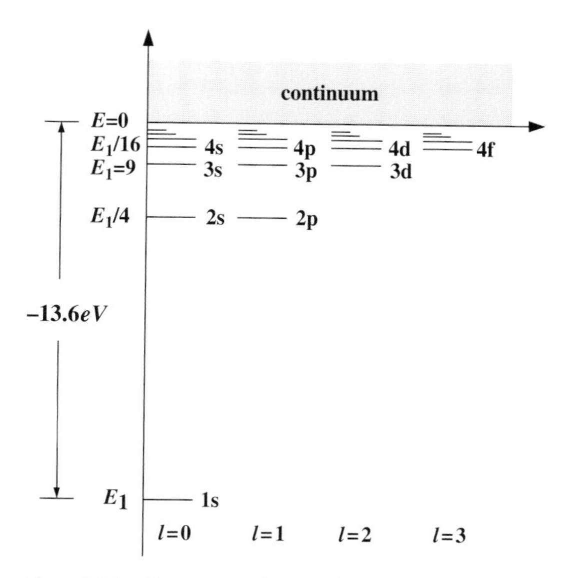

Figure 10.3 The quantized energy level diagram of hydrogen atom and the spectroscopic notations.

given l and m, there are two possible spin states, spin up and spin down. Thus, one can find the total number of degenerate states for the given n by

$$g_n = 2\sum_{l=0}^{n-1}(2l+1) = 2\left\{\left[2\frac{n(n-1)}{2}\right] + n)\right\} = 2n^2. \tag{10.51}$$

These energy levels are sketched in Figure 10.3, together with the corresponding quantum numbers and spectroscopic notations. The eigenstate u_{100} with the lowest energy level is called the ground state. Also the degeneracy is lifted in part by spin-orbit coupling, relativistic refinement, etc.

Reduced Radial Probability Density As discussed, the radial wavefunction R_{nl} is naturally scaled in terms of both the atomic radius a_0 and the principal quantum number n (see (10.46)). This approximately corresponds to Bohr's electron orbits, arising from the angular momentum quantization. The probability density $u^*_{nlm}u_{nlm}$ contains three variables r ,θ, and φ and can thus be viewed as a joint probability density of finding the electron in the volume element between r and $r + dr$, θ and $\theta + d\theta$, φ and $\varphi + d\varphi$, respectively.

Now, given a joint probability function $f(x, y)$, for example, the probability of finding the system between x and $x + dx$ regardless of y variable is obtained by the operation

$$f(x) = \int_{-\infty}^{\infty} dy f(x, y). \tag{10.52}$$

By the same token, the reduced probability density of finding the electron between r and $r + dr$ regardless of the angular variables is obtained by performing the integration

$$P(r)dr = \int_0^{2\pi} d\varphi \int_0^{\pi} \sin\theta d\theta r^2 dr u_{nlm}^* u_{nlm} = \left|R_{nl}^2\right| r^2 dr. \tag{10.53}$$

Figure 10.4 shows the radial wavefunction R_{nl} and its probability density $R_{nl}^* R_{nl}$ versus r for different n and l. For the s states ($l = 0$), these two curves do not vanish at $r = 0$ but attain the maximum values at $r = 0$. This is due in part to the electron being tightly bound to the nucleus via the attractive Coulomb force alone, without the rotation-induced repulsive centrifugal force. Nevertheless, this does not mean that the electron collapses into the nucleus. When the wavefunction is sharply localized near $r \approx 0$, its slope representing the momentum increases in which case the uncertainty principle dictates that the electron cannot be contained in the volume provided by the nucleus. Rather a bound state is formed near $r \approx 0$ at which the total energy is still below the zero level (see Figure 10.2). Also, for the given n, the peak of $R_{nl}^* R_{nl}$ tends to be pushed out from the nucleus with increasing l, as the centrifugal force is increased.

Also plotted in Figure 10.4, for comparison, are the reduced radial probability densities $P(r) = R_{nl}^* R_{nl} r^2$ versus r. Indeed, $P(r)$ vanishes at $r = 0$, as it should for all eigenstates, including the s-states. This is due to the factor r^2 that naturally enters into the definition of the reduced radial probability density (see (10.53)). These radial profiles of the atomic orbitals clearly exhibit the gross picture of the electronic clouds around the nucleus, with the peak values roughly reflecting the Bohr's electron orbits around the nucleus. As expected, the value of r at which $P(r)$ is peaked increases with increasing n and l.

The Boundary Surface The distributed nature of electron charge cloud around the nucleus is also well described by the probability density as represented by the darkness of shading, as illustrated for the ground state in Figure 10.5. Also, the effective size of the atom can be conveniently quantified by the boundary surface representation. In this representation, the surface is defined as the constant probability surface within which there is, say, 90% probability of finding the electron. This is also shown in Figure 10.5 for the ground state. Naturally, the radius R of the boundary surface is found in terms of the reduced radial probability density in (10.53) as

$$0.9 = \int_0^R dr P(r). \tag{10.54}$$

The s- and p-Orbitals The s-orbitals are spherically symmetric but other orbitals for $l \neq 0$ are distinctly nonsymmetric and in fact are orientation sensitive. For example, the p-orbitals with $l = 1$ for the given n are associated with three possible m values, namely, $m = 0, \ \pm 1$. The corresponding wavefunctions are given by

$$|p_z\rangle \equiv R_{n1} Y_1^0 = (3/4\pi)^{1/2} R_{n1} \cos\theta, \tag{10.55a}$$

$$|p_+\rangle \equiv R_{n1} Y_1^1 = -(3/8\pi)^{1/2} R_{n1} \sin\theta e^{i\varphi}, \tag{10.55b}$$

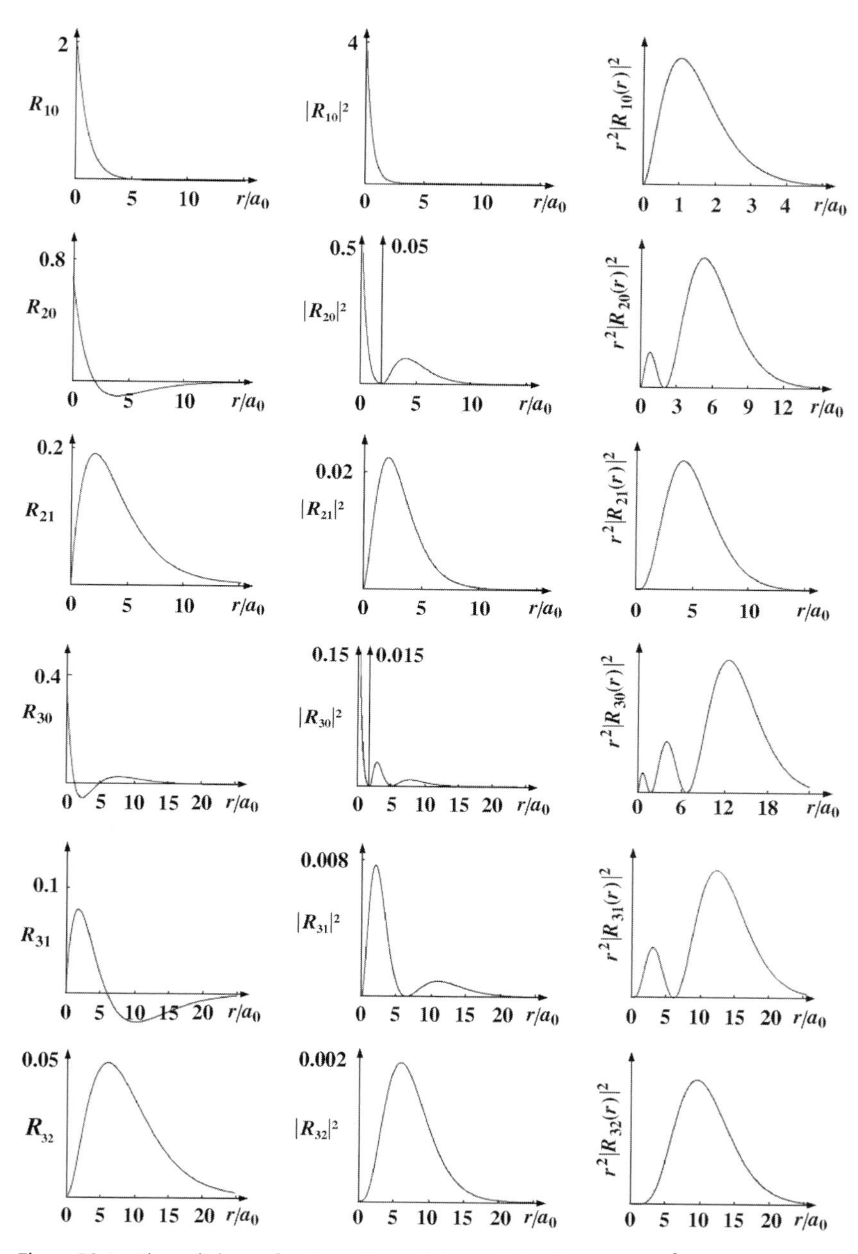

Figure 10.4 The radial wavefunctions R_{nl}, radial probability densities $|R_{nl}|^2$, and the reduced probability densities $|R_{nl}|^2 r^2$ for different n and l.

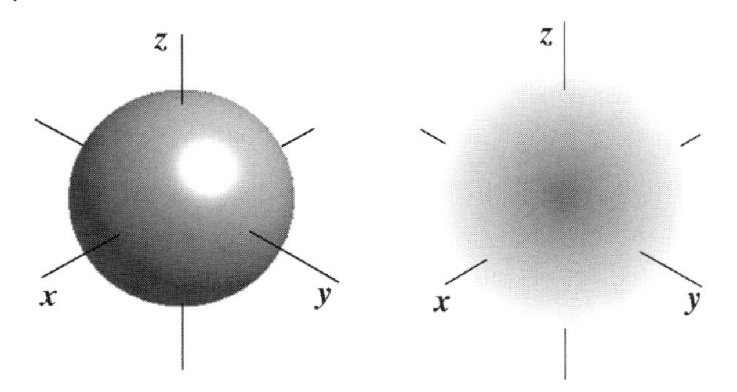

Figure 10.5 The ground-state electron cloud as represented by the constant probability surface and also by darkness of shading.

$$|p_-\rangle \equiv R_{n1} Y_1^{-1} = (3/8\pi)^{1/2} R_{n1} \sin\theta e^{-i\varphi}. \tag{10.55c}$$

These p-orbitals are rather useful for examining atomic and molecular structures and are shown in Figure 10.6. The two complex eigenfunctions (10.55b) and (10.55c) can be combined into a new set of real orthonormal eigenfunctions as

$$|p_x\rangle \equiv \frac{1}{\sqrt{2}}(|p_-\rangle - |p_+\rangle) = (3/4\pi)^{1/2} R_{n1} \sin\theta \cos\varphi, \tag{10.56a}$$

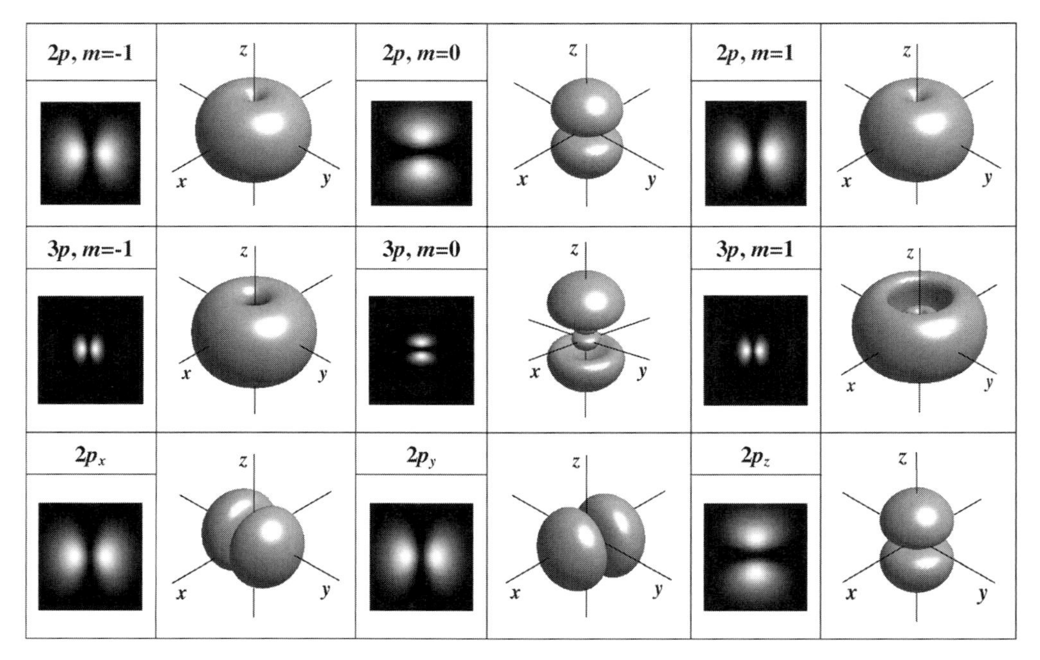

Figure 10.6 The boundary surfaces of p-orbitals for $n = 2, 3$ and of p_x,p_y,p_z states with corresponding side projections.

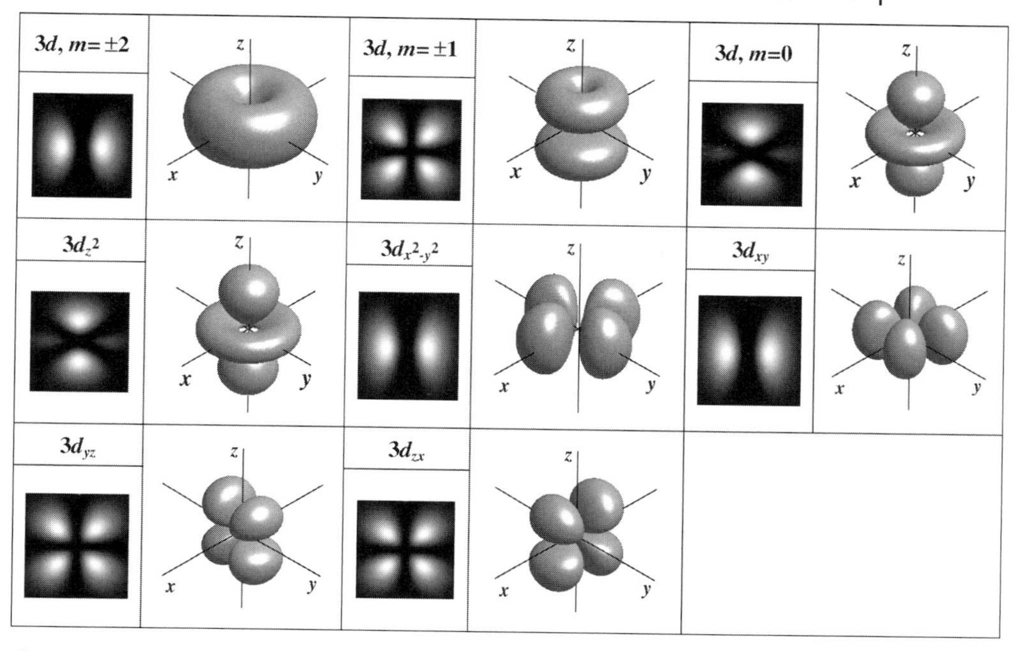

Figure 10.7 The boundary surfaces of d-orbitals and of five real d-orbitals with corresponding side projections.

$$|p_y\rangle \equiv \frac{i}{\sqrt{2}}(|p_-\rangle + |p_+\rangle) = (3/4\pi)^{1/2} R_{n1} \sin\theta \sin\varphi \tag{10.56b}$$

(see Table 9.1). These new p-orbitals when combined with the p_z-orbital in (10.55a) are again extensively utilized for analyzing the atoms and molecules and are also shown in Figure 10.6. The transformation of three p-orbitals (10.55a)–(10.55c) into a new set consisting of (10.56a), (10.56b), and (10.55a) is similar in context to transforming one set of basis vectors into another via rotation.

Also, there are five d-orbitals ($l = 2$) for $n > 3$, corresponding to $m = 0, \ \pm 1, \ \pm 2$. These orbitals can again be combined to give a new set of real orthonormal eigenfunctions and these wavefunctions are shown in Figure 10.7 and tabulated in Table 10.3.

10.4
Virial Theorem and Doppler Shift

Virial Theorem The expectation value of the potential energy in the ground state is readily obtained using u_{100} given in Table 10.2, namely,

$$\begin{aligned}\langle 100|V|100\rangle &= (Z^3/\pi a_0^3)\int_0^{2\pi} d\varphi \int_{-1}^{1} d\mu \int_0^{\infty} r^2 dr e^{-2Zr/a_0}\left(-\frac{Ze_M^2}{r}\right), \quad \mu = \cos\theta, \\ &= -\frac{Z^2 e_M^2}{a_0} = 2E_1,\end{aligned} \tag{10.57a}$$

Table 10.3 The five real d-orbitals.

Orbital	Wavefunctions
$d_{z^2} = d_0$	$(5/16\pi)^{1/2} R_{n2}(r)(3\cos^2\theta - 1)$ $= (5/16\pi)^{1/2} R_{n2}(r)[(3z^2 - r^2)/r^2]$
$d_{x^2-y^2} = \frac{1}{\sqrt{2}}(d_{+2} + d_{-2})$	$(15/16\pi)^{1/2} R_{n2}(r)[(x^2 - y^2)/r^2]$
$d_{xy} = \frac{1}{i\sqrt{2}}(d_{+2} - d_{-2})$	$(15/4\pi)^{1/2} R_{n2}(r)(xy/r^2)$
$d_{yz} = \frac{1}{i\sqrt{2}}(d_{+1} + d_{-1})$	$-(15/4\pi)^{1/2} R_{n2}(r)(yz/r^2)$
$d_{zx} = \frac{1}{\sqrt{2}}(d_{+1} - d_{-1})$	$-(15/4\pi)^{1/2} R_{n2}(r)(zx/r^2)$
$\cos 2\varphi = \cos^2\varphi - \sin^2\varphi$	
$\sin 2\varphi = 2\sin\varphi\cos\varphi$	
$z = r\sin\theta;\quad x = r\sin\theta\cos\varphi;\quad y = r\sin\theta\sin\varphi$	

while the average kinetic energy therein is obtained from (10.19) and (10.21) as

$$\langle 100|K|100\rangle = -\frac{\hbar^2}{2\mu}\frac{Z^3}{\pi a_0^3}\int_0^{2\pi} d\varphi \int_{-1}^{1} d\mu \int_0^{\infty} r^2 dr e^{-Zr/a_0}\left(\frac{1}{r^2}\frac{d}{dr}r^2\frac{d}{dr}\right)e^{-Zr/a_0}$$
$$= \frac{Z^2 e_M^2}{2a_0} = -E_1. \tag{10.57b}$$

Hence, one can write

$$\langle V\rangle_{100} = -2\langle K\rangle_{100} = 2E_1. \tag{10.58}$$

In evaluating (10.57b), use has been made of (10.21) for the kinetic energy, $-(\hbar^2/2\mu)\nabla^2$ with $l = 0$. The relationship existing between $\langle K\rangle$ and $\langle V\rangle$ in (10.58) is called the virial theorem. In fact, the theorem is valid for any two-body central force problem and the same relationship can be shown to hold true for any eigenstate u_{nlm} from the general property of Laguerre polynomial, that is,

$$\langle V\rangle_{nlm} = 2E_n = -\frac{Z^2 e_M^2}{a_0}\frac{1}{n^2}, \tag{10.59a}$$

$$\langle K\rangle_{nlm} = -E_n = \frac{Z^2 e_M^2}{2a_0}\frac{1}{n^2}. \tag{10.59b}$$

Doppler Shift When the atom is in an excited state, it emits radiation when the electron makes the transition from an upper initial state n to a lower final state n' to conserve the energy. When the atom moves while emitting the radiation, the atomic motion shifts the frequency of the emitted radiation. This effect is known as the Doppler shift. The schematics for such emission process are shown in Figure 10.8.

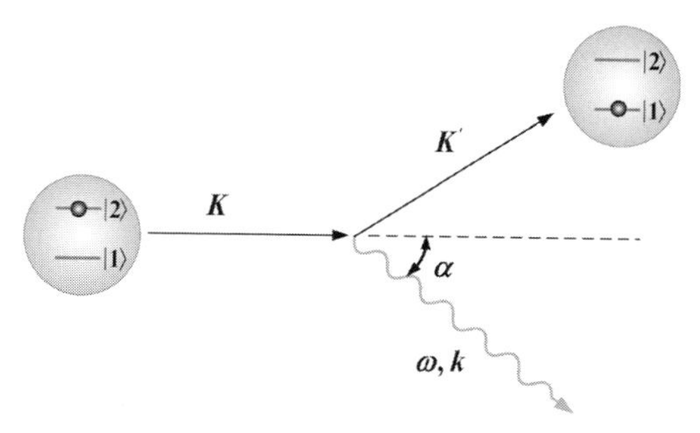

Figure 10.8 The Doppler shift of radiation emitted by a moving atom. The electron in an upper state, $|2\rangle$, makes a transition to a lower state, $|1\rangle$, emitting radiation to conserve energy and momentum.

The Doppler shift can be described by using the total wavefunction of the atom, consisting of wavefunctions of the center of mass and the internal motion (see (10.12)). During the emission of radiation, both the energy and momentum are conserved:

$$\frac{\hbar^2 K^2}{2M} + E_n = \frac{\hbar^2 K'^2}{2M} + E_{n'} + \hbar\omega, \tag{10.60}$$

$$\hbar \boldsymbol{K} = \hbar \boldsymbol{K}' + \hbar \boldsymbol{k}. \tag{10.61}$$

Here, $\boldsymbol{K}$ and $\boldsymbol{K}'$ are the wave vectors of the center of mass before and after the emission, E_n and $E_{n'}$ are the internal atomic energies before and after the emission, and $\boldsymbol{k}$ and ω the wave vector and frequency of the emitted photon, respectively (Figure 10.8).

The energy of the photon $\hbar\omega$ is then given from (10.60) by

$$\omega = \omega_0 + \frac{\hbar}{2M}(K^2 - K'^2), \quad \hbar\omega_0 = E_n - E_{n'}, \tag{10.62}$$

and consists of the atomic transition frequency ω_0 plus the difference between the kinetic energies of the center of mass before and after the emission. This difference is found from (10.61) with the operation

$$(\boldsymbol{K} - \boldsymbol{k}) \cdot (\boldsymbol{K} - \boldsymbol{k}) = \boldsymbol{K}' \cdot \boldsymbol{K}'$$

or by regrouping the terms

$$K^2 - K'^2 = 2Kk\cos\alpha - k^2, \quad K \cdot k = Kk\cos\alpha \tag{10.63}$$

Hence, inserting (10.63) into (10.62) and identifying the momentum of the photon and the velocity of the atom, namely,

$$\hbar k = p = \hbar\omega/c,$$

$$\hbar K/M = v,$$

one finds

$$\begin{aligned}\omega &= \omega_0 + \frac{\hbar}{2M}(2Kk\cos\alpha - k^2) \\ &= \omega_0 + \omega\left(\frac{v}{c}\cos\alpha - \frac{\hbar\omega}{2Mc^2}\right).\end{aligned} \tag{10.64}$$

The emitted frequency can therefore be found exactly in terms of the atomic transition frequency and other parameters by solving this quadratic equation for ω. However, in view of the fact that $|\omega - \omega_0| \ll \omega, \omega_0$, it suffices to replace ω appearing on the right-hand side by ω_0 and find the difference in frequency as

$$\omega - \omega_0 \simeq \omega_0\left(\frac{v}{c}\cos\alpha - \frac{\hbar\omega_0}{2Mc^2}\right). \tag{10.65}$$

Here, the first term proportional to v is called the Doppler shift of the first kind, while the second term is known as the Doppler shift of the second kind. The first term can either increase or decrease the emitted frequency, depending on whether the atom is moving toward or away from the detector. The second term accounts for the minute shift of the frequency due to the recoil suffered by the atom.

Hence, the radiation emanating from an ensemble of atoms at thermodynamic equilibrium at temperature T and detected on the x–y plane should be of the Gaussian spectral profile, namely,

$$|E(\omega)|^2 \propto \exp - \frac{(\omega - \omega_0)^2}{2\Delta\omega^2}, \tag{10.66a}$$

where the variance resulting from the random atomic motion in the z-direction is to be obtained using (10.65) and (1.23) as

$$\begin{aligned}(\Delta\omega)^2 &= \langle(\omega - \omega_0)^2\rangle \\ &\simeq \left(\frac{M}{2\pi k_B T}\right)^{3/2} \int_{-\infty}^{\infty} e^{-\beta v_x^2} dv_x \int_{-\infty}^{\infty} dv_y e^{-\beta v_y^2} \int_{-\infty}^{\infty} dv_z e^{-\beta v_z^2}\left(\omega_0 \frac{v_z}{c}\right)^2 \\ &= \omega_0^2\left(\frac{k_B T}{Mc^2}\right), \quad \beta = \frac{M}{2k_B T}.\end{aligned} \tag{10.66b}$$

In (10.66b), the atomic recoil term in (10.65) is neglected and M is the mass of the atom. This kind of broadening of the emitted frequency is called the inhomogeneous line broadening.

10.5 Problems

10.1. Consider a system of particles with masses m_1 and m_2 and bound together by a central force.

(a) Show that the total kinetic energy of the system can be expressed in terms of kinetic energies of center of mass and relative motion, namely,

$$\frac{p_1^2}{2m_1} + \frac{p_2^2}{2m_2} = \frac{P^2}{2M} + \frac{p^2}{2\mu},$$

where

$$\boldsymbol{p}_1 = m_1\dot{\boldsymbol{r}}_1; \quad \boldsymbol{p}_2 = m_1\dot{\boldsymbol{r}}_2,$$

$$\boldsymbol{R} \equiv \frac{m_1\boldsymbol{r}_1 + m_2\boldsymbol{r}_2}{M}, \quad M \equiv m_1 + m_2; \quad \boldsymbol{P} = M\dot{\boldsymbol{R}},$$

$$\mu^{-1} \equiv m_1^{-1} + m_2^{-1}, \quad \boldsymbol{p} = \mu\dot{\boldsymbol{r}}.$$

(b) Find the reduced mass μ in two limiting cases, where $m_2 \gg m_1$ and $m_2 = m_1$. For algebraic simplicity, you may consider the 1D motion in the x-direction and generalize the result.

10.2. Show that the Laplacian operator ∇^2 can be transformed from x, y, and z to r, θ, and φ coordinates as

$$\frac{\partial^2}{\partial x^2} + \frac{\partial^2}{\partial y^2} + \frac{\partial^2}{\partial z^2} = \frac{1}{r^2}\frac{\partial}{\partial r}r^2\frac{\partial}{\partial r} + \frac{1}{r^2}\left[\frac{1}{\sin\theta}\frac{\partial}{\partial\theta}\sin\theta\frac{\partial}{\partial\theta} + \frac{1}{\sin^2\theta}\frac{\partial^2}{\partial\varphi^2}\right]$$

with $x = r\sin\theta\cos\varphi$, $y = r\sin\theta\sin\varphi$, and $z = r\cos\theta$.
Hint: Make use of (9.3)–(9.6).

10.3. Find the effective Bohr radius a_0 in (10.44) for the following cases:

(a) Singly ionized helium He^+ (a neutral helium consists of two protons and two neutrons at the nucleus and two electrons revolving around it).

(b) Positronium, consisting of a positron and an electron bound together by an attractive Coulomb force (the positron has the same mass as electron but a positive charge $+e$).

(c) Two neutron system bound by the gravitational force.

10.4. (a) Calculate the radii of the boundary surfaces of 1s, 2s, and 3s states in hydrogen atom.

(b) Find the mean radius $\langle r\rangle$, the mean square radius $\langle r^2\rangle$, and the variance $\langle(r-\langle r\rangle)^2\rangle$ in 1s, 2s, and 3s states in hydrogen atom.

10.5. Given the threefold degeneracy in 2p state, that is, $n = 2$ and $l = 1$, the eigenfunctions can be regrouped into

$$|x\rangle = N(|u_{211}\rangle + |u_{21-1}\rangle),$$

$$|y\rangle = N(|u_{211}\rangle - |u_{21-1}\rangle),$$

$$|z\rangle = |u_{210}\rangle.$$

(a) Determine the normalization constant N.
(b) Show that these new set of eigenfunctions are orthonormal.

10.6. Using the ground-state wavefunction of the hydrogen atom,
(a) Show that the average kinetic and potential energies are related by

$$\left\langle \frac{p^2}{2\mu} \right\rangle_{100} = -\frac{1}{2}\langle V \rangle_{100}, \quad V = \frac{e_M^2}{r}.$$

(b) Find the average kinetic and potential energies in 2s and 2p states in H-atom.

10.7. (a) Find the probability that the electron in the ground state of hydrogen atom is located at a distance from the nucleus greater than a_B, a_B being the Bohr radius.
(b) Find the probability that the electron in 2s and 2p states of hydrogen atom is located beyond $2a_B$ from the nucleus.

10.8. (a) Determine the longest and shortest wavelengths and wavenumbers in Lyman, Bahmer, Paschen, and Bracket series in H-atom. (The wave-number is defined as $1/\lambda$.)
(b) Repeat the same calculation for the positronium and singly ionized helium atom.

10.9. An electron makes a transition from $n = 2$ to $n = 1$ states by emitting a radiation.
(a) Find the frequency of radiation emitted from hydrogen and deuterium atoms, respectively.
(b) Find the recoil of these two atoms due to the emission of radiations.
(c) If these atoms are to be optically excited from $n = 1$ to $n = 3$ states, what frequencies will be required?

10.10. The phosphorus atom, when incorporated into silicon as a donor atom, is to be modeled as a hydrogen-like atom, consisting of an outermost electron in $n = 3$ state and bound by a proton in the nucleus. The dielectric constant of the medium in between the electron and proton is $\varepsilon_r = 11.9$ and the effective mass of the electron is $m_n \simeq 1.1\, m_0$, with m_0 denoting the rest mass.
(a) Calculate the ionization energy required to knock out the electron from the ground state.
(b) Find Bohr radius of the atom.
(c) Find de Broglie wavelength of the electron in the ground state.

Suggested Reading

1 Yariv, A. (1982) *An Introduction to Theory and Applications of Quantum Mechanics*, John Wiley & Sons, Inc.
2 Haken, H. and Wolf, H.C. (2004) *Molecular Physics and Elements of Quantum Chemistry: Introduction to Experiments and Theory*, 2nd edn, Springer.
3 Haken, H., Wolf, H.C., and Brewer, W.D. (2007) *The Physics of Atoms and Quanta: Introduction to Experiments and Theory*, 7th revised and enlarged edn, Springer.
4 Gasiorowicz, S. (2003) *Quantum Physics*, 3rd edn, John Wiley & Sons, Inc.
5 Karplus, M. and Porter, R.N. (1970) *Atoms and Molecules: An Introduction for Students of Physical Chemistry*, Addison-Wesley Publishing Company.

11
System of Identical Particles and Many-Electron Atoms

The basic difference between the classical and quantum theories is also manifested in the description of identical particles. Here, the Pauli exclusion principle plays a key role and the symmetrized or antisymmetrized wavefunctions underpin the difference between the two descriptions. Furthermore, the exclusion principle provides the building principles of atoms and molecules, thereby influencing profoundly the physical and chemical properties of the matter in general. The properties of the electron spin are discussed and the helium atom is considered as the prototype example of two spin 1/2 system. Also, the multielectron atoms are briefly discussed, based on the theory of the hydrogen atom and Pauli exclusion principle.

11.1
Two-Electron System

Symmetrized and Antisymmetrized Wavefunctions Consider a system of two electrons bound to a common force center. Classically, it is distinctly possible to tag identical particles such as electrons and follow its evolution in space and time, since the position and momentum of these particles can be specified precisely in classical theories. In quantum mechanics, however, it is impossible to distinguish configurations of identical particles by exchange of two or more particles. This is rooted in the fact that an electron bound to an atom is in essence a charge cloud, which is spread all over the volume. It is therefore impossible to disentangle the overlapped charge cloud for identifying each electron, as illustrated in Figure 11.1.

Thus, consider a system of two noninteracting electrons. The Hamiltonian of the system is given by

$$\hat{H}(1,2) = \hat{H}(1) + \hat{H}(2), \tag{11.1}$$

where each electron (j) is taken to be in eigenstates u_n of $\hat{H}(j)$, that is,

$$\hat{H}(j)u_n(j) = E_n u_n(j), \quad j = 1, 2; \quad n = \alpha, \beta. \tag{11.2}$$

Introductory Quantum Mechanics for Semiconductor Nanotechnology. Dae Mann Kim

ISBN: 978-3-527-40975-4

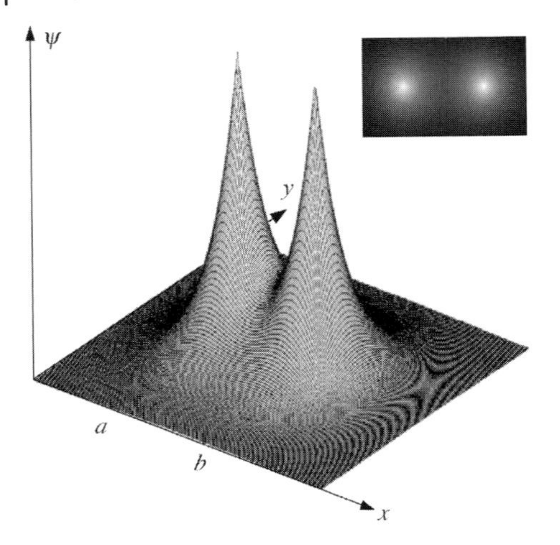

Figure 11.1 The graphical representation of electron charge cloud in the ground state of hydrogen molecule, for example. The charge cloud arises from two electrons attached to two protons and well exhibits the overlapping of the probability densities and the indistinguishable nature of two electrons. The top view of the charge clouds is also shown.

Then the two-electron wavefunction

$$\varphi(1,2) = u_\alpha(1)u_\beta(2) \tag{11.3}$$

satisfies the eigenequation

$$\hat{H}(1,2)\varphi(1,2) = (E_\alpha + E_\beta)\varphi(1,2). \tag{11.4}$$

By the same token, the wavefunction

$$\varphi(2,1) = u_\alpha(2)u_\beta(1) \tag{11.5}$$

also qualifies as the eigenfunction with the same eigenvalue. Moreover, the linear combination of these two eigenfunctions, namely,

$$\varphi_\gamma = \frac{1}{\sqrt{2}}[\varphi(1,2) \pm \varphi(2,1)], \quad \gamma = s, a, \tag{11.6}$$

should also qualify as eigenfunctions. Here, the subscripts s and a denote the symmetric and antisymmetric combinations, respectively.

A question then naturally arises as to which eigenfunction would be appropriate for describing the two noninteracting electron system. To examine the question, consider the permutation operator

$$\hat{P}_{12}f(1,2) \equiv f(2,1). \tag{11.7}$$

The Hamiltonian in (11.1) is symmetric, that is, $\hat{H}(1,2) = \hat{H}(2,1)$, so that it commutes with $\hat{P}_{12}$ and should therefore share the common eigenfunction with $\hat{P}_{12}$. Now the eigenfunction of $\hat{P}_{12}$ satisfies the eigenequation

$$\hat{P}_{12}F(1,2) = \lambda F(1,2), \tag{11.8}$$

with λ denoting the eigenvalue. Applying $\hat{P}_{12}$ one more time to the eigenequation (11.8), one can write

$$\hat{P}_{12}^2 F(1,2) = \lambda \hat{P}_{12} F(1,2) = \lambda^2 F(1,2), \tag{11.9a}$$

while from the definition of permutation,

$$\hat{P}_{12}^2 F(1,2) = \hat{P}_{12} F(2,1) = F(1,2). \tag{11.9b}$$

Hence, it is clear from (11.9a) and (11.9b) that $\lambda^2 = 1$, that is, $\lambda = \pm 1$. It is therefore clear that only the symmetrized and antisymmetrized wavefunctions of (11.6) can be used as the common eigenfunction of $\hat{H}(1,2)$ and $\hat{P}_{12}$.

Fermions and Bosons Particles in nature are generally categorized into two groups. Those with half-odd integer spins, $\hbar/2,\ 3\hbar/2,\ 5\hbar/2, \ldots$, are called fermions and described by the antisymmetric wavefunctions. Electrons, holes, protons, and neutrons, for example, belong to this group of particles. A striking basic feature of these fermions is that no two fermions can share the same set of quantum numbers. This is the celebrated Pauli exclusion principle and constitutes the key building principle for atoms and molecules. On the other hand, those particle having integer spins, $\hbar, 2\hbar, 3\hbar, \ldots$, are called bosons and described by the symmetric wavefunctions. For instance, photons, deuterons, α particles belong to this group of particles.

Given a system of N noninteracting fermions, the wavefunction is conveniently represented by the $N \times N$ Slater determinant:

$$\varphi_a(1,2,\ldots,N) = \frac{1}{\sqrt{N!}} \begin{vmatrix} u_1(1) & u_1(2) & \ldots & u_1(N) \\ \cdot & & & \\ \cdot & & & \\ u_N(1) & u_N(2) & \ldots & u_N(N) \end{vmatrix} \tag{11.10}$$

In this representation, if two quantum numbers are the same, that is, if two rows are identical, the determinant vanishes by definition and so does the wavefunction. This is consistent with the exclusion principle. The same determinant representation can also be used for the system of N noninteracting bosons, with the proviso, however, that the minus signs appearing in the expansion of the determinant are replaced by plus signs.

11.2
Two Spin 1/2 System

As has been noted several times, an electron inherently possesses two spin states, that is, the spin up and spin down states. The spin up state is denoted by

$$\hat{s}_z|\alpha\rangle = \frac{1}{2}\hbar|\alpha\rangle, \tag{11.11}$$

while the spin down state is represented by

$$\hat{s}_z|\beta\rangle = -\frac{1}{2}\hbar|\beta\rangle. \tag{11.12}$$

Here, $\hat{s}_z$ operator formally corresponds to the z-component of the angular momentum operator $\hat{l}_z$. In the same context, one can likewise introduce $\hat{s}^2$ corresponding to $\hat{l}^2$, namely,

$$\hat{s}^2|\gamma\rangle = \frac{1}{2}\left(\frac{1}{2}+1\right)\hbar^2|\gamma\rangle = \frac{3}{4}\hbar^2|\gamma\rangle, \quad \gamma = \alpha, \beta \tag{11.13}$$

(see (9.44) and (9.45)). Figure 11.2 shows the spatial quantization of the spin angular momentum.

In addition, one can introduce the spin–flip operators s_+ and s_-, which flip the spin down state to the spin up state and vice versa:

$$\hat{s}_+|\beta\rangle = |\alpha\rangle, \tag{11.14}$$

$$\hat{s}_-|\alpha\rangle = |\beta\rangle. \tag{11.15}$$

Naturally, these operators s_+ and s_- when acting on the spin up and spin down states, respectively, push the states out of the spin space under consideration, that is,

$$\hat{s}_+|\alpha\rangle = 0; \quad \hat{s}_-|\beta\rangle = 0. \tag{11.16}$$

In further analogy with angular momentum eigenfunctions, the spin up and spin down states are orthonormal:

$$\langle\alpha|\alpha\rangle = \langle\beta|\beta\rangle = 1, \tag{11.17a}$$

$$\langle\alpha|\beta\rangle = \langle\beta|\alpha\rangle = 0. \tag{11.17b}$$

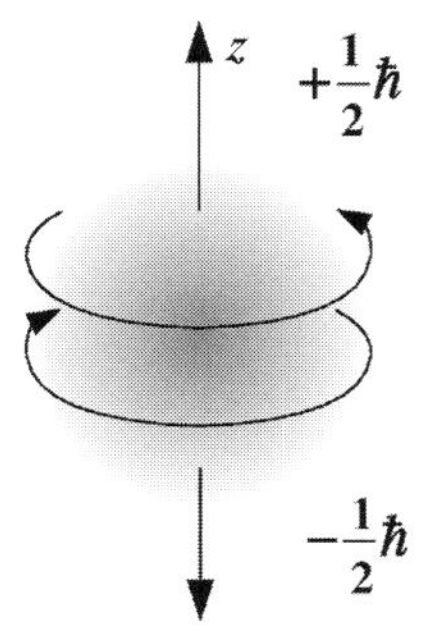

Figure 11.2 The electron spin: spin up and spin down states.

Pauli Spin Matrices The basic properties of the electron spin states are conveniently described by the 2×2 Pauli spin matrices. In this matrix representation, the spin up and spin down states are represented by column matrices:

$$|\alpha\rangle = \begin{pmatrix} 1 \\ 0 \end{pmatrix}; \quad |\beta\rangle = \begin{pmatrix} 0 \\ 1 \end{pmatrix}, \tag{11.18}$$

in which case the orthonormality of the spin states are automatically satisfied, that is,

$$\langle\alpha|\alpha\rangle = (1 \quad 0)\begin{pmatrix} 1 \\ 0 \end{pmatrix} = 1, \tag{11.19a}$$

$$\langle\beta|\beta\rangle = (0 \quad 1)\begin{pmatrix} 0 \\ 1 \end{pmatrix} = 1, \tag{11.19b}$$

$$\langle\alpha|\beta\rangle = (1 \quad 0)\begin{pmatrix} 0 \\ 1 \end{pmatrix} = 0. \tag{11.19c}$$

Next, the spin operators are represented by the Pauli matrices given by

$$\hat{s} \equiv \frac{\hbar}{2}\sigma, \tag{11.20}$$

where

$$\sigma_x = \begin{pmatrix} 0 & 1 \\ 1 & 0 \end{pmatrix}; \quad \sigma_y = \begin{pmatrix} 0 & -i \\ i & 0 \end{pmatrix}; \quad \sigma_z = \begin{pmatrix} 1 & 0 \\ 0 & -1 \end{pmatrix}. \tag{11.21}$$

Then a straightforward matrix multiplication yields

$$\sigma_x^2 = \begin{pmatrix} 0 & 1 \\ 1 & 0 \end{pmatrix}\begin{pmatrix} 0 & 1 \\ 1 & 0 \end{pmatrix} = \begin{pmatrix} 1 & 0 \\ 0 & 1 \end{pmatrix}. \tag{11.22a}$$

Likewise, it readily follows from (11.21) that

$$\sigma_y^2 = \sigma_z^2 = \begin{pmatrix} 1 & 0 \\ 0 & 1 \end{pmatrix}, \tag{11.22b}$$

so that

$$\sigma^2 = \sigma_x^2 + \sigma_y^2 + \sigma_z^2 = \begin{pmatrix} 3 & 0 \\ 0 & 3 \end{pmatrix}. \tag{11.23}$$

Moreover, when these matrices act upon the spin vectors (11.18), they reproduce the basic features of the spin states, namely (11.11)–(11.13):

$$s_z|\alpha\rangle = \frac{\hbar}{2}\begin{pmatrix} 1 & 0 \\ 0 & -1 \end{pmatrix}\begin{pmatrix} 1 \\ 0 \end{pmatrix} = \frac{\hbar}{2}\begin{pmatrix} 1 \\ 0 \end{pmatrix} = \frac{\hbar}{2}|\alpha\rangle, \tag{11.24}$$

$$s_z|\beta\rangle = \frac{\hbar}{2}\begin{pmatrix} 1 & 0 \\ 0 & -1 \end{pmatrix}\begin{pmatrix} 0 \\ 1 \end{pmatrix} = -\frac{\hbar}{2}\begin{pmatrix} 0 \\ 1 \end{pmatrix} = -\frac{\hbar}{2}|\beta\rangle, \tag{11.25}$$

and

$$s^2\begin{pmatrix}1\\0\end{pmatrix}=\frac{\hbar^2}{4}\begin{pmatrix}3&0\\0&3\end{pmatrix}\begin{pmatrix}1\\0\end{pmatrix}=\frac{3\hbar^2}{4}\begin{pmatrix}1\\0\end{pmatrix}, \tag{11.26a}$$

$$s^2\begin{pmatrix}0\\1\end{pmatrix}=\frac{\hbar^2}{4}\begin{pmatrix}3&0\\0&3\end{pmatrix}\begin{pmatrix}0\\1\end{pmatrix}=\frac{3\hbar^2}{4}\begin{pmatrix}0\\1\end{pmatrix}. \tag{11.26b}$$

In addition, one can also define the spin–flip operators as

$$\sigma_+=\frac{1}{2}(\sigma_x+i\sigma_y)=\frac{1}{2}\left[\begin{pmatrix}0&1\\1&0\end{pmatrix}+i\begin{pmatrix}0&-i\\i&0\end{pmatrix}\right]=\begin{pmatrix}0&1\\0&0\end{pmatrix}, \tag{11.27}$$

$$\sigma_-=\frac{1}{2}(\sigma_x-i\sigma_y)=\frac{1}{2}\left[\begin{pmatrix}0&1\\1&0\end{pmatrix}-i\begin{pmatrix}0&-i\\i&0\end{pmatrix}\right]=\begin{pmatrix}0&0\\1&0\end{pmatrix}. \tag{11.28}$$

These operators flip the spin states from the spin up to spin down states or vice versa, namely,

$$\sigma_+|\beta\rangle=\begin{pmatrix}0&1\\0&0\end{pmatrix}\begin{pmatrix}0\\1\end{pmatrix}=\begin{pmatrix}1\\0\end{pmatrix}=|\alpha\rangle, \tag{11.29}$$

$$\sigma_-|\alpha\rangle=\begin{pmatrix}0&0\\1&0\end{pmatrix}\begin{pmatrix}1\\0\end{pmatrix}=\begin{pmatrix}0\\1\end{pmatrix}=|\beta\rangle, \tag{11.30}$$

while σ_+ and σ_- when operating on the spin up and spin down states, respectively, push those states out of the spin space, that is,

$$\sigma_+|\alpha\rangle=\begin{pmatrix}0&1\\0&0\end{pmatrix}\begin{pmatrix}1\\0\end{pmatrix}=\begin{pmatrix}0\\0\end{pmatrix}=0, \tag{11.31a}$$

$$\sigma_-|\beta\rangle=\begin{pmatrix}0&0\\1&0\end{pmatrix}\begin{pmatrix}0\\1\end{pmatrix}=\begin{pmatrix}0\\0\end{pmatrix}=0. \tag{11.31b}$$

Thus, the Pauli spin matrices provide the convenient tool for describing the electron spin. The system of two identical particles or the two spin 1/2 system finds a number of useful applications, as will become clear in Section 11.3.

11.3 Helium Atom

The helium atom consists of two protons (plus two neutrons) in the nucleus and two electrons revolving around the nucleus. Thus, the atom constitutes a three-body problem, as shown in Figure 11.3 and as such an exact analytical treatment of the helium atom is not available. Thus, an approximate theory is presented. The

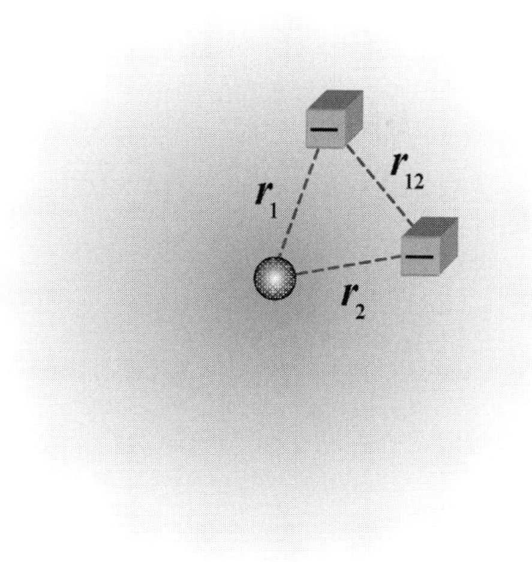

Figure 11.3 The schematic representation of helium atom, consisting of two protons (and two neutrons) at the nucleus ($Z = 2$) and two electrons revolving around the nucleus. Also, shown in the figure are the volume elements of two-electron charge clouds.

Hamiltonian of the atom consists of three terms:

$$\hat{H} = \hat{H}_1 + \hat{H}_2 + \hat{H}_{12}, \tag{11.32a}$$

where

$$\hat{H}_j = -\frac{\hbar^2}{2\mu}\nabla_j^2 - \frac{Ze_M^2}{r_j}, \quad e_M^2 = \frac{e^2}{4\pi\varepsilon_0}, j = 1, 2, \tag{11.32b}$$

is the Hamiltonian of an hydrogenic atom formed by each electron with the common nucleus of the atomic number Z and

$$\hat{H}_{12} = \frac{e_M^2}{r_{12}}, \quad e_M^2 = \frac{e^2}{4\pi\varepsilon_0}, \tag{11.32c}$$

is the repulsive Coulomb interaction term between the two electrons.

Singlet and Triplet States The two electrons in the helium atom constitute a prototype two-fermion system and should therefore be described by antisymmetrized wavefunctions, as pointed out in Section 11.2. In constructing such wavefunctions, it is necessary to first consider the spin states. In so doing, the spin states $|\alpha(j)\rangle$ and $|\beta(j)\rangle$ with $j = 1, 2$ are compactly denoted by $\alpha(j)$ and $\beta(j)$. Then, as discussed in Section 11.1, the spin functions $\alpha(1)\alpha(2)$, $\alpha(1)\beta(2)$, $\beta(1)\alpha(2)$ and $\beta(1)\beta(2)$ all qualify as the eigenfunction of the two spin system. Moreover, these four eigenfunc-

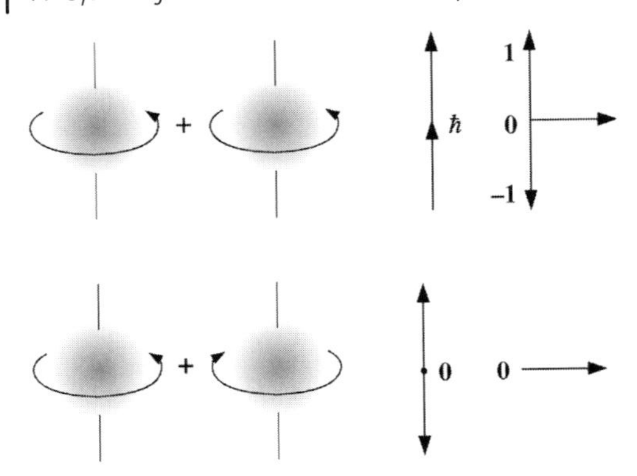

Figure 11.4 The two-electron system comprised of triplet and singlet states. In the former, $m = -1, 0, 1$, while in the latter $m = 0$.

tions can be grouped into symmetric and antisymmetric combinations:

$$\chi_s = \begin{cases} \alpha(1)\alpha(2) \\ \frac{1}{\sqrt{2}}[\alpha(1)\beta(2) + \beta(1)\alpha(2)] \\ \beta(1)\beta(2) \end{cases}, \tag{11.33}$$

$$\chi_a = \frac{1}{\sqrt{2}}[\alpha(1)\beta(2) - \beta(1)\alpha(2)], \tag{11.34}$$

in which case it follows from (11.11) and (11.12) that

$$\sigma_z\chi_s \equiv (s_{1z} + s_{1z})\chi_s = m\hbar\chi_s, \quad m = 1, 0, -1, \tag{11.35a}$$

$$\sigma_z\chi_a \equiv (s_{1z} + s_{1z})\chi_a = m\hbar\chi_a, \quad m = 0. \tag{11.35b}$$

Thus, the symmetrized function χ_s has three possible projections onto the z-axis and is called the triplet state, as shown in Figure 11.4. The antisymmetrized function χ_a, on the other hand, is characterized by $m = 0$ and is called the singlet state.

11.3.1 The Ground State of He

The wavefunction of the ground state can be approximated by

$$\varphi_0(1, 2) = u_{100}(1)u_{100}(2)\chi_a, \quad j \equiv \boldsymbol{r}_j \tag{11.36}$$

in which the two electrons are in the ground state of the hydrogenic atom. The spatial part of the wavefunction $u_{100}(1)u_{100}(2)$, that is, $u_{1s}(1)u_{1s}(2)$ is symmetric and therefore the required antisymmetrization of the two-electron wavefunction should rely upon the singlet spin state χ_a. Also, it is important to point out at the outset that the atomic number Z appearing in (11.32b) is not necessarily equal to 2, representing the two protons in the nucleus. This is because one of these two electrons does not "see" the nuclear charge of $+2e$ owing to the shielding caused by the presence of the other electron. Thus, Z should be viewed as a parameter ranging from 1 to 2, namely, $1 \leq Z \leq 2$, and Z appearing in the wavefunction (11.36) should be used in a similar context.

The energy of the ground state is given to the first-order approximation by

$$E_0^{(1)} = \langle u_{100}(\boldsymbol{r}_1)u_{100}(\boldsymbol{r}_2)\chi_a|\hat{H}_1 + \hat{H}_2 + \hat{H}_{12}|u_{100}(\boldsymbol{r}_1)u_{100}(\boldsymbol{r}_2)\chi_a\rangle = E_0 + \Delta E_0, \tag{11.37}$$

where

$$\begin{aligned} E_0 &= \left\langle u_{100}(\boldsymbol{r}_1)u_{100}(\boldsymbol{r}_2)|\hat{H}_1 + \hat{H}_2|u_{100}(\boldsymbol{r}_1)u_{100}(\boldsymbol{r}_2)\right\rangle \\ &= \sum_{j=1}^{2}\left\langle u_{100}(j)|\hat{H}_j|u_{100}(j)\right\rangle = 2\times\left(-\frac{Z^2e_M^2}{2a_0}\right) \end{aligned} \tag{11.38}$$

is the energy contributed by $\hat{H}_1$ and $\hat{H}_2$ (see (10.43)) and ΔE represents the energy associated with the repulsive Coulomb interaction potential between the two electrons in (11.32), that is,

$$\begin{aligned} \Delta E_0 &= \left\langle u_{100}(1)u_{100}(2)\chi_a|\frac{e_M^2}{r_{12}}|u_{100}(1)u_{100}(2)\chi_a\right\rangle \\ &= e_M^2\left(\frac{Z^3}{\pi a_0^3}\right)^2\int_0^{2\pi}d\varphi_1\int_{-1}^{1}d\mu_1\int_0^{\infty}dr_1r_1^2\int_0^{2\pi}d\varphi_2\int_{-1}^{1}d\mu_2\int_0^{\infty}dr_2r_2^2e^{-2Zr_1/a_0}e^{-2Zr_2/a_0}\frac{1}{r_{12}}, \end{aligned} \tag{11.39}$$

where $\mu_j = \cos\theta_j$, $j = 1, 2$ (see Table 10.2).

The evaluation of this electron–electron interaction term is facilitated by noticing the fact that the inverse relative distance is the generating function of Legendre polynomials:

$$\frac{1}{r_{12}} \equiv \frac{1}{|\boldsymbol{r}_1 - \boldsymbol{r}_2|} = \frac{1}{r_>}\left[1 + \mu\frac{r_<}{r_>} + \frac{1}{2}(3\mu^2 - 1)\left(\frac{r_<}{r_>}\right)^2 + \cdots\right], \tag{11.40}$$

where $\mu \equiv \cos\alpha$ and α is the angle between $\boldsymbol{r}_1$ and $\boldsymbol{r}_2$, and $r_>$ and $r_<$ denote the greater and lesser of r_1 and r_2. The expansion coefficient of the nth power, $(r_</r_>)^n$, in (11.40) is by definition the nth order Legendre polynomial of argument μ. When (11.40) is inserted into (11.39) and the integration is carried out with respect to $\boldsymbol{r}_2$, one can take $\boldsymbol{r}_1$ parallel to the z-direction without any loss of generality. In this case, $\mu \equiv \mu_2 = \cos\theta_2$ and the angular integration can be readily carried out with the use of the orthogonality of the Legendre polynomials (see (9.42)), namely,

$$\int_{-1}^{1} d\mu_2 P_l(\mu_2) P_0(\mu_2) = \frac{2}{2l+1}\delta_{l0}, \quad P_0(\mu_2) = 1.$$

It is therefore clear that the angle-independent first term in (11.40) contributes, while the remaining angle-dependent terms simply vanish upon performing the angular integration. Thus, one can write after performing the angular integrations,

$$\begin{aligned}\Delta E_0 &= \frac{e_M^2}{\pi^2}\left(\frac{Z}{a_0}\right)^6 (4\pi)^2 \int_0^\infty r_1^2 dr_1 e^{-2Zr_1/a_0}\left[\frac{1}{r_1}\int_0^{r_1} r_2^2 dr_2 e^{-2Zr_2/a_0} + \int_{r_1}^\infty r_2 dr_2 e^{-2Zr_2/a_0}\right] \\ &= \frac{5Ze_M^2}{8a_0}.\end{aligned} \tag{11.41}$$

Here, the radial r_2-integration has been divided into two regions, that is, $r_2 \le r_1$ in which $r_1 = r_>$ and $r_2 \ge r_1$ in which $r_2 = r_>$.

Ionization Energy The ground-state energy (11.37) is thus given by

$$E_0^{(1)} = 2 \times \left(-\frac{Z^2 e_M^2}{2a_0}\right) + \frac{5Ze_M^2}{8a_0} = -\frac{Ze_M^2}{a_0}\left(Z - \frac{5}{8}\right) \tag{11.42}$$

and the role of the electron–electron interaction term has become explicitly clear from (11.42). Without this repulsive interaction term, the ground state energy of helium atom would simply consist of two ground-state energies of the hydrogenic atom with $Z = 2$. In this case, the first ionization energy $\text{He} \to \text{He}^+ + e$ and the second ionization energy $\text{He}^+ \to \text{He}^{++} + e$ should be the same theoretically and are given by

$$IP_1 = IP_2 \equiv \left| -\frac{e_M^2}{2a_0} Z^2 \right| = 54.4\ \text{eV}, \quad Z = 2. \tag{11.43}$$

However, the measured values of the first and second ionization energies are $IP_1 = 24.6$ eV and $IP_2 = 54.4$ eV, respectively. It is therefore clear that there is indeed a good agreement between theory and experiment for IP_2. This is expected, since when one electron is left alone after the first ionization, the ionized helium atom is reduced to the hydrogenic atom with $Z = 2$ and the ground-state energy should then be accurately described by the well-known theory of the H-atom.

However, the measured value of the first ionization energy IP_1 is much smaller than IP_2 in clear disagreement with this simple model of the ground state, namely, (11.43). The experimental data have therefore to be compared with the result in (11.42), in which the electron–electron interaction has been taken into account, albeit approximately. Now, the value of IP_1 is by definition the difference

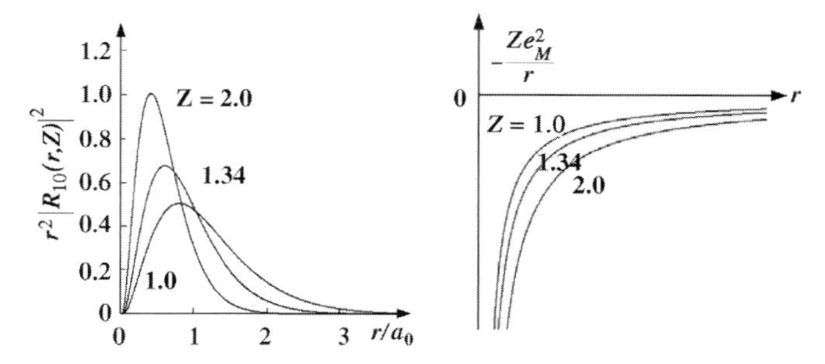

Figure 11.5 The reduced probability density $|R_{nl}|^2 r^2$ and attractive Coulomb potential of electron versus the radial distance for different effective nuclear charges Z.

between the ground-state energies of He^+ and He as found in (11.42), namely,

$$IP_1 \equiv -\frac{Z^2 e_M^2}{2a_0} - \left[-\frac{Ze_M^2}{a_0}\left(Z-\frac{5}{8}\right)\right] = \frac{Ze_M^2}{2a_0}\left(Z-\frac{5}{4}\right). \tag{11.44}$$

Thus, when the helium atomic number $Z = 2$ is substituted in (11.44), the theoretical value of IP_1 amounts to about 20.43 eV and this result better compares with the experimental data. The reason for this improved agreement between theory and experiment consists in replacing Z^2 in (11.43) by $Z(Z-5/4)$ in (11.44), that is to say, the effective reduction of nuclear charge due to screening. This screening effect has been brought about by taking into consideration the electron–electron interaction.

Naturally, the screening should also significantly modify the electron wavefunction and this is illustrated in Figure 11.5 in which the reduced radial probability distribution is plotted versus r with Z used as a varying parameter. Indeed, the average distance of the electron from the nucleus is significantly increased, while the attractive Coulomb potential decreases with decreasing Z. These trends obviously point to the reduced binding energy caused by the screening of the nuclear charge, and smaller IP_1.

11.3.2 The First Excited State of He

The first excited state of the helium atom can be analyzed in a similar manner. In this case, the energy eigenfunction consists of one electron in the ground state, $u_{100}(\mathbf{r}_j) = u_{1s}(j)$, while the other electron in the first excited state, for instance, $u_{200}(\mathbf{r}_{j'}) = u_{2s}(j')$. These wavefunctions can be symmetrized or antisymmetrized and the total two-electron wavefunctions, including the spin, are therefore given by

$$\varphi_s = \frac{1}{\sqrt{2}}[u_{1s}(1)u_{2s}(2) + u_{1s}(2)u_{2s}(1)]\frac{1}{\sqrt{2}}[\alpha(1)\beta(2)-\alpha(2)\beta(1)] \tag{11.45}$$

and

$$\varphi_{a1} = \frac{1}{\sqrt{2}}[u_{1s}(1)u_{2s}(2) - u_{1s}(2)u_{2s}(1)]\alpha(1)\alpha(2), 1 \quad (11.46a)$$

$$\varphi_{a2} = \frac{1}{\sqrt{2}}[u_{1s}(1)u_{2s}(2) - u_{1s}(2)u_{2s}(1)]\frac{1}{\sqrt{2}}[\alpha(1)\beta(2) + \alpha(2)\beta(1)], \quad (11.46b)$$

$$\varphi_{a3} = \frac{1}{\sqrt{2}}[u_{1s}(1)u_{2s}(2) - u_{1s}(2)u_{2s}(1)]\beta(1)\beta(2). \quad (11.46c)$$

It is important to notice from (11.45) and (11.46) that the symmetrized energy eigenfunction is combined with the singlet spin state, while the antisymmetrized energy eigenfunction is combined with the triplet spin states to make the total wavefunction antisymmetric.

Overlap and Exchange Integrals For the singlet state, the energy of the first excited state is approximately given by

$$E_{0s}^{(1)} = \langle \varphi_s | \hat{H}_1 + \hat{H}_2 + \hat{H}_{12} | \varphi_s \rangle = E_0 + \Delta E_s, \quad (11.47)$$

where the spin functions automatically yields unity due to its orthonormality and there is no spin dependency in the Hamiltonian under consideration. Thus,

$$\begin{aligned} E_0 &= \left\langle \varphi_s | \hat{H}_1 + \hat{H}_2 | \varphi_s \right\rangle \\ &= E_{1s} + E_{2s} = -\frac{Z^2 e_M^2}{2a_0} - \frac{Z^2 e_M^2}{2a_0}\frac{1}{4} \end{aligned} \quad (11.48)$$

is the energy arising from the Hamiltonian term $\hat{H}_1 + \hat{H}_2$ and corresponding to 1s and 2s states. And the modification in energy ΔE due to $\hat{H}_{12}$ consists of two integrals:

$$\begin{aligned} \Delta E_s &= \frac{1}{2}\left\langle u_{1s}(1)u_{2s}(2) + u_{1s}(2)u_{2s}(1) | \hat{H}_{12} | u_{1s}(1)u_{2s}(2) + u_{1s}(2)u_{2s}(1) \right\rangle \\ &= J + K, \end{aligned} \quad (11.49)$$

with

$$J \equiv \left\langle u_{1s}(1)u_{2s}(2) | \hat{H}_{12} | u_{1s}(1)u_{2s}(2) \right\rangle = \iint d\boldsymbol{r}_1 d\boldsymbol{r}_2 u_{1s}^2(1)\frac{e_M^2}{r_{12}}u_{2s}^2(2) > 0, \quad (11.50)$$

$$K \equiv \left\langle u_{1s}(1)u_{2s}(2) | \hat{H}_{12} | u_{1s}(2)u_{2s}(1) \right\rangle = \iint d\boldsymbol{r}_1 d\boldsymbol{r}_2 u_{1s}(1)u_{2s}(2)\frac{e_M^2}{r_{12}}u_{1s}(2)u_{2s}(1) > 0. \quad (11.51)$$

It is important to notice in (11.49) that the four integrals involved therein consist in effect of only two integrals, namely, J and K integrals, as can be readily seen by interchanging the variables of integration, $\boldsymbol{r}_1$ and $\boldsymbol{r}_2$. Obviously, the J integral represents the repulsive Coulomb interaction between the two electrons in 1s and 2s states, respectively, and is called the overlap integral. The ***K*** integral, on the other hand, does not carry the usual electron probability densities. Rather the integrand therein simply reflects the exchange of two electrons between the states 1s and 2s, which is required for symmetrizing or antisymmetrizing the wavefunctions (see (11.46)). Thus, the ***K*** integral is genuinely quantum mechanical in nature, rooted in the exclusion principle and is called the exchange integral. These two integrals can be evaluated in a manner similar to the calculation that has been carried out for the ground-state energy. One may thus write

$$E_{0s}^{(1)} = E_0 + J + K. \tag{11.52}$$

For the case of triplet state, one can carry out similar calculations, obtaining

$$E_{0t}^{(1)} = \langle \varphi_a | \hat{H}_1 + \hat{H}_2 + \hat{H}_{12} | \varphi_a \rangle = E_0 + \Delta E_a, \tag{11.53}$$

where

$$\begin{aligned} \Delta E_a &= \frac{1}{2}\Big\langle u_{1s}(1)u_{2s}(2) - u_{1s}(2)u_{2s}(1) | \hat{H}_{12} | u_{1s}(1)u_{2s}(2) - u_{1s}(2)u_{2s}(1)\Big\rangle \\ &= J - K, \end{aligned} \tag{11.54}$$

so that

$$E_{0t}^{(1)} = E_0 + J - K. \tag{11.55}$$

It is therefore clear from (11.52) and (11.55) that the first excited-state energy level for the triplet state is lower than that of the singlet state by an amount $2K$. This can be traced to the fact that in the singlet state, the symmetric combination of the spatial wavefunctions renders the probability density high when the two electrons are close to each other and $\boldsymbol{r}_1 \approx \boldsymbol{r}_2$, as illustrated in Figure 11.6. However, the same probability density becomes minimum for the case of triplet state, in which u_{1s} and u_{2s} are combined in an antisymmetric fashion. Consequently, the repulsive Coulomb interaction between the two electrons is larger in the singlet state, compared with the triplet state, that is,

$$\Big\langle \varphi_a | \frac{e_M^2}{r_{12}} | \varphi_a \Big\rangle < \Big\langle \varphi_s | \frac{e_M^2}{r_{12}} | \varphi_s \Big\rangle, \tag{11.56}$$

and this accounts for the higher excited-state energy. In this way, the spin states significantly affect the energy level, although the spin wavefunctions never enter explicitly in the calculations of the energy level. The energy levels of both ground and first excited states are also sketched in Figure 11.6.

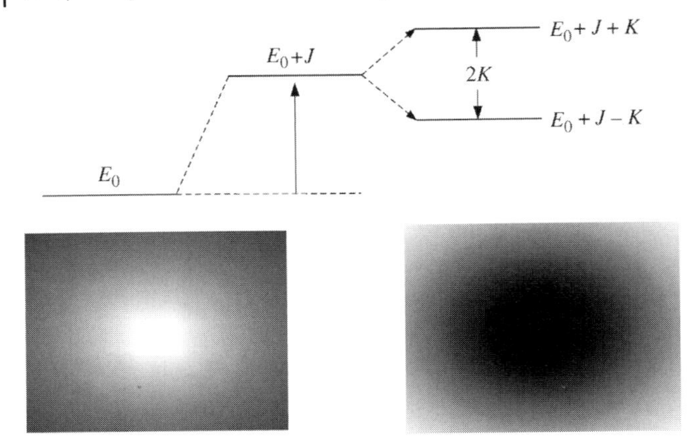

Figure 11.6 The first excited-state energy levels in helium atom, corresponding to symmetrized energy eigenfunction coupled with singlet spin state (top) and antisymmetrized eigenfunction with triplet state (bottom). Also shown are electron charge clouds, corresponding to symmetrized (left) and antisymmetrized (right) wavefunctions. The probability density is commensurate with the degree of brightness.

11.4
The Periodic Table and Structures of Atoms

It is appropriate at this point to briefly discuss the periodic table and atomic structures, based upon the simple theory of hydrogen atom and the Pauli exclusion principle. As clearly seen from the discussion of the helium atom, the theory of the hydrogen atom, in particular its wavefunctions, provide a powerful and convenient base for analyzing other atoms and molecules. The atomic structures are in general well represented by the neutral ground state, which plays an important role in chemical reactions as well as determining other properties of the atom.

The Electron Configuration Naturally, a convenient platform for discussing atomic structures is the periodic table. As well known, the gross features of the periodic table are (i) it consists of rows, called periods, which are comprised of 2, 8, 8, 18, 32, 32 elements from the top row, and (ii) those elements in the same column or group exhibit similar properties, including the first ionization potentials and atomic spectra, for instance. The quantum states of electrons in multielectron atoms are labeled via the two quantum numbers n and l. At the given principal quantum number n, the exclusion principle dictates that the maximum number of electrons in one subshell energy corresponding to the angular momentum quantum number l is $2(2l+1)$, as detailed in Chapter 10.

As discussed, the energy levels of these quantum states are mainly determined by n and are thus grouped by n and the energy levels are further divided into sublevels, depending on l. The subshell energies increase with increasing l, or equivalently with increasing centrifugal force. Naturally, the electrons in a many-electron atom fill the quantum states, starting from the lowest energy level as determined by n. Also, for the

Table 11.1 The first two periods in Periodic Table, showing the ground-state configurations and the first ionization potentials.

H $1s^1$ $IP_1$13.595							He $1s^2$ 24.580
Li $2s^1$ 5.390	Be $2s^2$ 9.320	B $2s^22p^1$ 8.296	C $2s^22p^2$ 11.264	N $2s^22p^3$ 14.54	O $2s^22p^4$ 13.614	F $2s^22p^5$ 17.42	Ne $2s^22p^6$ 21.559

given n, the electrons fill the lowest subshell and once it is filled up, the remaining electrons start filling the next higher lying subshell. The process goes on until all electrons in the atom are accommodated into the quantum states therein.

It is important to point out that the electronic and chemical properties of atoms are mainly determined by those electrons occupying the highest energy level or equivalently residing in the outermost atomic orbitals. These electrons are called the valence electrons. It is also important to notice that the separation in energy of the highest occupied energy level and the lowest unoccupied energy level is an important parameter affecting the property of the atom.

The specification of electrons in an atom with the use of two quantum numbers n and l is called the electron configuration. The electron configuration enables a systematic specification of atoms in the periodic table. This can be illustrated by considering a portion of the table involving only the first two rows (Table 11.1). The single electron in H-atom is naturally specified by $n = 1, l = 0$, and the configuration $1s^1$. The two electrons in the helium atom are then represented by $1s^2$, which also indicates the subshell for $l = 0$ being completely filled up.

The ground states of the atoms comprising the second row starting from hydrogenic Li and ending with closed shell Ne are specified by the two subshells corresponding to $l = 0$, 1 for $n = 2$. The electronic configurations are Li ([He]2s), Be ([He]$2s^2$), B ([He]$2s^22p$), C ([He]$2s^22p^2$), N ([He]$2s^22p^3$), O ([He]$2s^22p^4$), F ([He]$2s^22p^5$), and Ne ([He]$2s^22p^6$). It is noted here that the electron configuration $1s^2$ is often replaced by [He]. The third period starting with hydrogenic Na and ending with Ar can likewise be denoted with Ne serving as the main core.

First Ionization Potential and Electron Affinity The ionization potential, in particular the first ionization potential, is an important parameter of the atom for determining, for example, the nature of the chemical bond. For the case of the hydrogen atom, the ionization potential is simply the energy required to knock out its only electron from the ground state, namely, $IP = e_M^2/2a_0$ (see (10.43)). For the case of the helium atom with two electrons, however, there are two ionization potentials, namely, the first and second ionization potentials. As already discussed, the former was shown to be the energy required to knock out one of those two electrons in the ground state, namely,

$IP_1 = Z_{\text{eff}} e_M^2/2a_0$ with $Z_{\text{eff}} = 2(2-5/4)$ representing the screened charge of the nucleus.

Given an atom A, the first ionization potential IP_1 generally represents the energy required for the process,

$$A \rightarrow A^+ + e^- : IP_1.$$

The inverse process of an atom capturing an electron,

$$A + e^- = A^- : EA,$$

is associated with the electron affinity EA. The electron affinity is the energy released by a free electron at rest when captured by a neutral atom, ending up in a bound-energy state.

The measured values of the first ionization potentials IP_1 are also presented in Table 11.1. As is clear from the data, the first IP increases consistently across a given period but it drops sharply between the periods, that is, as the next period begins. For example, IP_1 of 5.14 eV for Na atom is much smaller than IP_1 of 21.56 eV for Ne, although Na atom has one more proton in the nucleus to pull in the electrons than Ne atom. This behavior of IP_1 data can be semiquantitatively discussed, based upon the ionization energy of the hydrogenic atom (see (10.43)) and the screening of the nuclear charge,

$$IP_1 \simeq Z_{\text{eff}}^2 \frac{e_M^2}{2a_0} \frac{1}{n^2}. \tag{11.57}$$

Thus, for Li atom, for example, the valence electron is in 2s state and the measured IP_1 of 5.39 eV indicates the efficient screening of the nuclear charge of three protons by two inner lying 1s electrons. With increasing atomic number Z and increasing number of protons in the same period, all electrons added reside in the same subshell. Consequently, the screening done on the part of these added electrons is inefficient for one of those valence electrons involved in the first ionization. Thus, IP_1 increases steadily until the closed shell atom of Ne is reached. With the beginning of the new period, starting with hydrogenic Na atom, however, the valence electron is now in 3s state while the rest of the electrons fill up the inner lying subshells, screening efficiently the nuclear charge. As a result, IP_1 drops sharply and becomes comparable with that of Li.

11.5 Problems

11.1. The ground-state wavefunction of the Helium atom is given by

$$\varphi_0(1,2) = u_{100}(1)u_{100}(2)\chi_a, \quad j \equiv \boldsymbol{r}_j$$

with χ_a denoting the singlet spin state.

(a) Show that $\varphi_0(1,2)$ is normalized.

(b) Evaluate the expectation values of the total spin operators

$$\left\langle \varphi_0(1,2)|\hat{S}^2|\varphi_0(1,2)\right\rangle \quad \text{and} \quad \langle \varphi_0(1,2)|\hat{S}_z|\varphi_0(1,2)\rangle,$$

where $\hat{\boldsymbol{S}} = \hat{\boldsymbol{s}}_1 + \hat{\boldsymbol{s}}_2$.

11.2. The first excited singlet and triplet states of the helium atom are given in (11.45) and (11.46) in the text.

(a) Show that these wavefunctions are orthonormal.

(b) Find the expectation values of the total spin operators $\hat{S}^2$ and $\hat{S}_z$ for each wavefunction.

11.3. Consider the three-electron system in which the single-electron eigenstates are given by $u_{100}(1)|\alpha\rangle$, $u_{200}(1)|\beta\rangle$, and $u_{200}(1)|\alpha\rangle$, respectively. Using the Slater determinant, construct the antisymmetrized wavefunction of this three-electron system and find the corresponding energy eigenvalue.

11.4. The lithium atom consists of three protons in the nucleus ($Z = 3$) and three electrons revolving around it.

(a) Write down the Hamiltonian of the lithium atom and show that

$$\varphi(\boldsymbol{r}_1, \boldsymbol{r}_2, \boldsymbol{r}_3) = u_{n1}(\boldsymbol{r}_1)u_{n2}(\boldsymbol{r}_2)u_{n3}(\boldsymbol{r}_3)$$

is the energy eigenfunction with energy $E = E_{n1} + E_{n2} + E_{n3}$, provided the electron–electron interaction terms are neglected. Here, the u functions are the wavefunctions of hydrogenic atom.

(b) The ground-state electron configuration is $1s^2 2s^1$, that is, two electrons in 100 state with spin up and spin down, respectively and the third one in 200 state with spin up or spin down. Write down the wavefunction using the Slater determinant.

(c) Find the energy and the total spin in the ground state of the lithium atom.

11.5. The sodium atom has the atomic number $Z = 11$, that is, it has 11 protons in the nucleus and 11 electrons revolving around the nucleus.

(a) Assign each electron the quantum numbers including the spin.

(b) The observed ionization energy and orbital radius of the atom are 5.14 eV and 0.17 nm, respectively. Explain the data in terms of the screening of the nuclear charge.

11.6. Consider a system of N electrons in a solid. These electrons are often modeled as free particles in a cubic box of volume L^3.

(a) According to Pauli exclusion principle, electrons start filling from the lowest energy level, including the spin states. Find the wavefunction and energy level of the highest electron state at $T = 0$.

(b) Calculate the total amount of energy of this N electron system.

(c) Calculate the energies in (a) and (b) for the specific case of $N = 10^{22}$ and $V = 1\ \text{cm}^3$.

(d) Repeat (a)–(c) for the system of 2D and 1D electrons for $N = 4.6 \times 10^{14}$, $V = 1\ \text{cm}^2$ and $N = 2.15 \times 10^7$, $V = 1$ cm, respectively.

Suggested Reading

1. Haken, H. and Wolf, H.C. (2004) *Molecular Physics and Elements of Quantum Chemistry: Introduction to Experiments and Theory*, 2nd edn, Springer.
2. Haken, H., Wolf, H.C., and Brewer, W.D. (2007) *The Physics of Atoms and Quanta: Introduction to Experiments and Theory*, 7th revised and enlarged edn, Springer.
3. Karplus, M. and Porter, R.N. (1970) *Atoms and Molecules: An Introduction for Students of Physical Chemistry*, Addison-Wesley Publishing Company.
4. Gasiorowicz, S. (2003) *Quantum Physics*, 3rd edn, John Wiley & Sons, Inc.

12
Molecules and Chemical Bonds

The chemical bonding provides the basic mechanism by which molecules are formed and interact with each other. The successful elucidation of this bonding mechanism constitutes one of the remarkable achievements of the quantum mechanics. The bonding is inherently quantum mechanical in nature, whose understanding is essential for studying the molecular structures and for that matter the nanotechnology in general. There are two kinds of chemical bonds, heteropolar and homopolar. In the former, an electron is transferred from one atom to the other, and the resulting two ions of opposite polarity are bound together. In the latter, two neutral atoms are bound. Some of these chemical bonds are singled out for compact discussion.

12.1
Ionized Hydrogen Molecule

The basic mechanism of the chemical bonding is discussed in this section by choosing for consideration the simplest molecule, namely, the ionized hydrogen molecule H_2^+. The molecule consists of one electron interacting simultaneously with two protons, which constitute the two nuclei in the molecule, as sketched in Figure 12.1. In this structure, the two protons repel each other via repulsive Coulomb force, while the electron and either one of two protons attract each other via attractive Coulomb forces. Thus, the basic problem involved in this structure consists in what prevents the two protons to break away from each other and form instead a stable molecule.

The ionized hydrogen molecule is a typical example of a three-body central force problem (Figure 12.1) and the rigorous analytical treatment of the molecule is not available. Thus, an approximate analysis is in order and in so doing, the two protons are taken fixed in space at this point. The effect of repulsive Coulomb interaction between the two protons will be taken into account at an appropriate stage of the discussion. The Hamiltonian then consists of the kinetic energy of the electron plus its Coulomb interactions with two protons (see Figure 12.1):

$$\hat{H} = -\frac{\hbar^2}{2m}\nabla^2 - e_M^2\left(\frac{1}{r_a} + \frac{1}{r_b}\right), \quad e_M^2 = \frac{e^2}{4\pi\varepsilon_0}. \tag{12.1}$$

Introductory Quantum Mechanics for Semiconductor Nanotechnology. Dae Mann Kim

ISBN: 978-3-527-40975-4

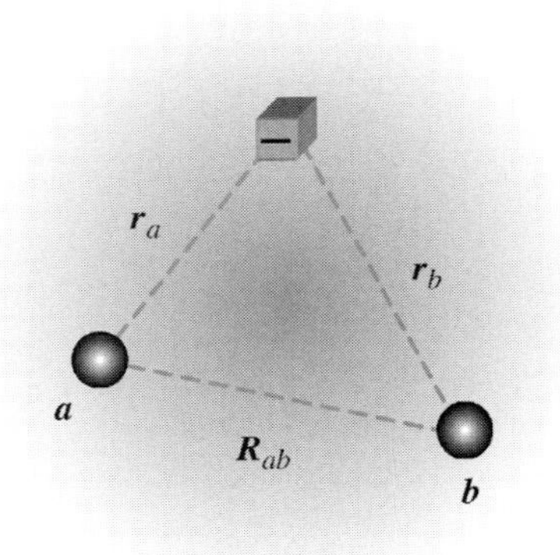

Figure 12.1 The schematic representation of ionized hydrogen molecule, consisting of two protons (*a*, *b*) and the electron charge cloud. The charge volume element simply represents the distributed nature of Coulomb interaction.

And the energy eigenequation then reads as

$$\hat{H}\varphi(\boldsymbol{r}_a, \boldsymbol{r}_b) = E\varphi(\boldsymbol{r}_a, \boldsymbol{r}_b). \tag{12.2}$$

The analysis of this equation is facilitated by partitioning the Hamiltonian into two hydrogenic Hamiltonians, namely,

$$\left(-\frac{\hbar^2}{2m}\nabla^2 - \frac{e_M^2}{r_j}\right)|u_j\rangle = E_0|u_j\rangle, \tag{12.3a}$$

where

$$|u_j\rangle \equiv |u_j(\boldsymbol{r}_j)\rangle, \quad j = a, b, \tag{12.3b}$$

is the eigenfunction of hydrogenic atom with energy E_0. Because there does not exist a closed form solution of (12.2), one may look for the wavefunction of H_2^+ in terms of a linear combination of two hydrogenic wavefunctions, namely,

$$\varphi(\boldsymbol{r}_a, \boldsymbol{r}_b) = c_1|u_a\rangle + c_2|u_b\rangle. \tag{12.4}$$

In this approach, the single electron is taken to form the hydrogenic subsystem with the two nuclei simultaneously by oscillating between the two nuclei or equivalently the electron is taken to be shared by the two nuclei.

Inserting (12.4) into (12.2) and regrouping terms using (2.3) results in

$$c_1\left(\Delta E - \frac{e_M^2}{r_b}\right)|u_a\rangle + c_2\left(\Delta E - \frac{e_M^2}{r_a}\right)|u_b\rangle = 0, \tag{12.5a}$$

with

$$\Delta E \equiv E_0 - E. \tag{12.5b}$$

Thus, solving the equation(12.2) has been reduced to finding the coefficients c_1 and c_2 for both the eigenfunction and the eigenenergy. To proceed further, take the inner product of (12.5a) with respect to $\langle u_a|$, obtaining

$$c_1\Delta E + c_1\left\langle u_a\left| -\frac{e_M^2}{r_b}\right| u_a\right\rangle + c_2\Delta E\left\langle u_a\middle| u_b\right\rangle + c_2\left\langle u_a\left| -\frac{e_M^2}{r_a}\right| u_b\right\rangle = 0, \tag{12.6}$$

where $\langle u_j|u_j\rangle = 1$, $j = a, b$.

One can similarly perform the inner product of (12.5a) with respect to $\langle u_b|$, obtaining

$$c_1\Delta E\left\langle u_b\middle| u_a\right\rangle + c_1\left\langle u_b\left| -\frac{e_M^2}{r_b}\right| u_a\right\rangle + c_2\Delta E + c_2\left\langle u_b\left| -\frac{e_M^2}{r_a}\right| u_b\right\rangle = 0. \tag{12.7}$$

Overlap, Coulomb, and Exchange Integral Now, the integrals appearing in (12.6) and (12.7) divide into three kinds:

$$S = \int d\boldsymbol{r} u_a^*(\boldsymbol{r}_a) u_b(\boldsymbol{r}_b) \equiv \langle u_a|u_b\rangle \equiv \langle u_b|u_a\rangle, \tag{12.8}$$

$$C = \int d\boldsymbol{r} u_a^*\left(-\frac{e_M^2}{r_b}\right) u_a \equiv \left\langle u_a\left| -\frac{e_M^2}{r_b}\right| u_a\right\rangle \equiv \left\langle u_b\left| -\frac{e_M^2}{r_a}\right| u_b\right\rangle < 0, \tag{12.9}$$

$$D = \int d\boldsymbol{r} u_a^*\left(-\frac{e_M^2}{r_a}\right) u_b \equiv \left\langle u_a\left| -\frac{e_M^2}{r_a}\right| u_b\right\rangle \equiv \left\langle u_b\left| -\frac{e_M^2}{r_b}\right| u_a\right\rangle < 0. \tag{12.10}$$

These three integrals carry interesting physical interpretations and significance. S is the overlap integral specifying the measure of overlap between u_a and u_b at the given separation of two nuclei, as depicted in Figure 12.2. C specifies the Coulomb interaction integral between the electron forming hydrogenic atom with one proton

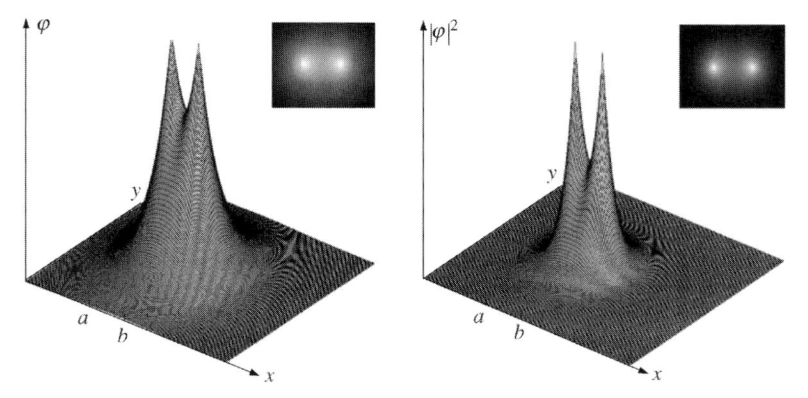

Figure 12.2 The plot of the integrand in the overlap integral. The overlap under consideration is between two wavefunctions centered at protons a and b. The distributed nature of the charge cloud is also apparent from the top view.

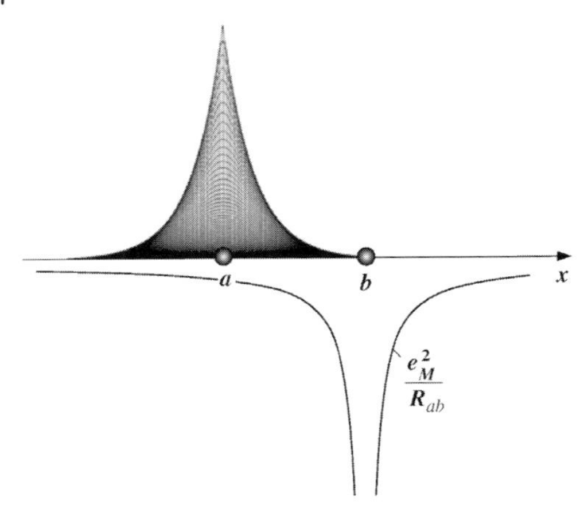

Figure 12.3 The graphical representation of the integrand for the attractive Coulomb interaction between the electron charge cloud associated with the proton at a and the proton at b as a point charge.

and the other proton located nearby as a point charge as shown in Figure 12.3. D denotes the exchange integral representing the interaction between the exchange probability density $u_a^* u_b$ and either one of the two protons. The integral accounts for the attractive Coulomb interaction between the overlapped electron cloud and a proton and is purely quantum mechanical in nature. The exchange integral originates from the two nuclei sharing the same electron or equivalently the electron simultaneously occupying u_a and u_b states, as illustrated in Figure 12.4.

With these shorthand notations for the integrals, the two equations (12.6) and (12.7) are combined into a 2 × 2 matrix equation:

$$\begin{pmatrix} \Delta E + C & \Delta ES + D \\ \Delta ES + D & \Delta E + C \end{pmatrix} \begin{pmatrix} c_1 \\ c_2 \end{pmatrix} = 0. \tag{12.11}$$

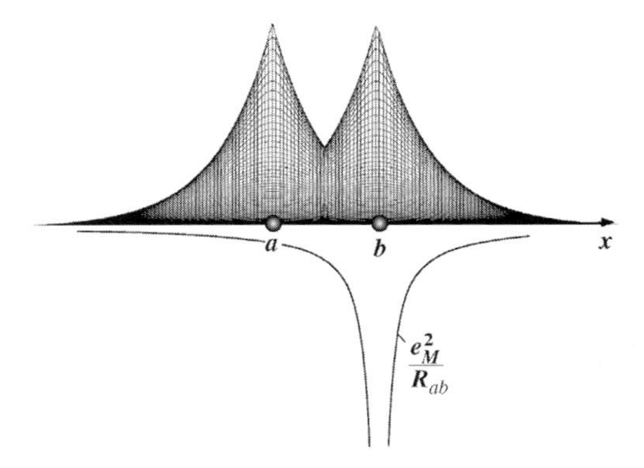

Figure 12.4 The graphical illustration of the integrand in the exchange integral, accounting for the interaction between the exchange probability density and proton a or b as point charge. The integral consists in essence of the Coulomb interaction between the overlapped electron cloud and a proton as a point charge.

It is therefore clear from (12.11) that c_1 and c_2 become trivial unless the secular equation is satisfied, namely, the determinant vanishes:

$$\begin{vmatrix} \Delta E + C & \Delta ES + D \\ \Delta ES + D & \Delta E + C \end{vmatrix} = 0. \tag{12.12}$$

Expanding the determinant (12.12) and solving the resulting quadratic equation for ΔE, one finds

$$\Delta E \equiv E_0 - E = \frac{\pm D - C}{1 \mp S}. \tag{12.13}$$

Inserting ΔE thus found back into the matrix equation (12.11) results in

$$c_1 \pm c_2 = 0,$$

so that

$$c_2 = \mp c_1. \tag{12.14}$$

Hence, the normalized wavefunction of H_2^+ is obtained from (12.4) and (12.14) as

$$\varphi_{\mp}(\boldsymbol{r}_a, \boldsymbol{r}_b) = \frac{1}{\sqrt{2}}[|u_a\rangle \mp |u_b\rangle]. \tag{12.15}$$

The single electron being shared on an equal footing by the two nuclei in H_2^+ has thus been consistently incorporated via the symmetrized and antisymmetrized wavefunctions of the molecule in (12.15). The corresponding eigenenergies are different, however, as is clear from (12.13). The reasons for this difference in energy level can again be traced to the different configurations in the electron charge cloud for symmetrized and antisymmetrized wavefunctions, namely, $\varphi_+(\boldsymbol{r}_a, \boldsymbol{r}_b)$ and $\varphi_-(\boldsymbol{r}_a, \boldsymbol{r}_b)$. This can be clearly seen from Figure 12.5, in which the profiles of $\varphi_+(\boldsymbol{r}_a, \boldsymbol{r}_b)$ and $\varphi_-(\boldsymbol{r}_a, \boldsymbol{r}_b)$ are plotted together with the corresponding probability densities. Indeed, there is a marked difference in the electron probability density, especially in the region between the two nuclei. Note in particular the denser electron charge cloud therein for $\varphi_+(\boldsymbol{r}_a, \boldsymbol{r}_b)$ compared with $\varphi_-(\boldsymbol{r}_a, \boldsymbol{r}_b)$.

Binding Energy of the Molecule The binding energy of H_2^+ can now be analyzed, based on (12.13). For this purpose, it is important to notice that E_0 in (12.13) represents the energy of the electron that is bound to one proton, while the other proton is at infinity. Therefore, the bonding energy of H_2^+ can be defined as the difference between E and E_0, that is,

$$E_b \equiv E - E_0 = \frac{C \mp D}{1 \mp S} + \frac{e_M^2}{R_{ab}} \tag{12.16}$$

where $\mp$ corresponds to $\varphi_-(\boldsymbol{r}_a, \boldsymbol{r}_b)$ and $\varphi_+(\boldsymbol{r}_a, \boldsymbol{r}_b)$, respectively (see (12.13) and (12.15)). It is at this point of the analysis that the repulsive Coulomb interaction between the two protons at finite R_{ab} distance should be brought in. The bonding energy is then clearly shown to depend parametrically on the relative distance between the two nuclei, as it should. More important, R_{ab} critically affects the

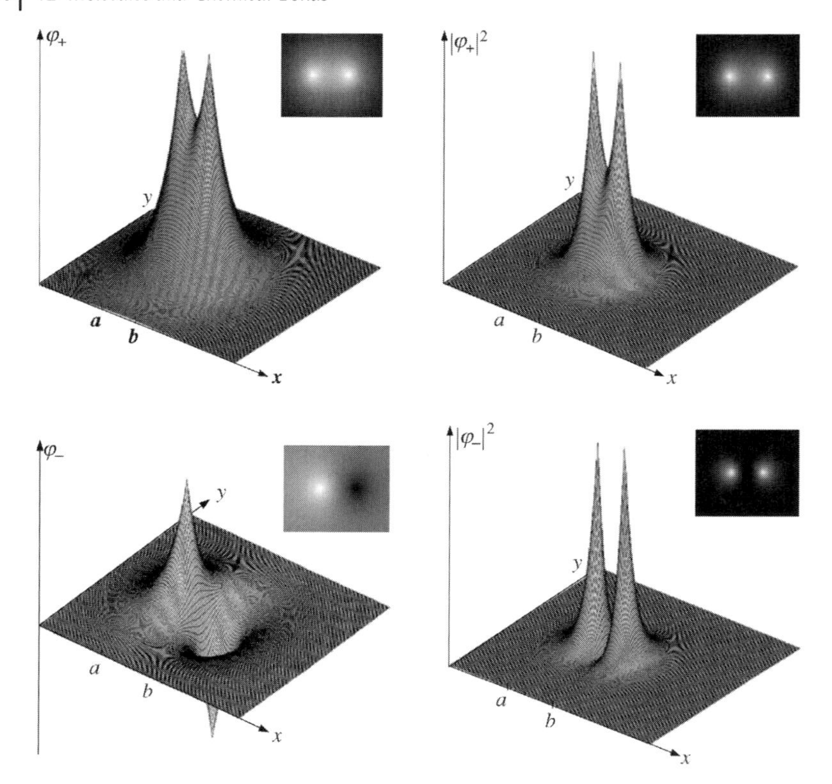

Figure 12.5 The symmetrized (top) and antisymmetrized (bottom) wavefunctions $\varphi_\pm(\boldsymbol{r}_a, \boldsymbol{r}_b)$ and corresponding probability densities. The main difference between the two lies in differing probability densities between the two nuclei. Also shown are the projections from the top of the profiles indicating the overlap by the measure of brightness.

integrals S, C, and D, whose integrand depends sensitively on R_{ab} (see (12.8)–(12.10) and Figure 12.1).

Bonding and Antibonding Since $u_a(\boldsymbol{r}_a)$ and $u_b(\boldsymbol{r}_b)$ are normalized eigenfunctions, the overlap integral is by definition less than unity as long as R_{ab} does not go to zero, that is, $S < 1$. This means that the denominator appearing in (12.16) always remains positive. This leaves the integrals C and D as the critical factors for determining the polarity of E_b. If $E_b < 0$, it definitely suggests that a stable molecule can be formed. If $E_b > 0$, on the other hand, it indicates that there is no molecular formation.

In the limit of $R_{ab} \to 0$, the repulsive Coulomb interaction between the two protons diverges. Concomitantly, C and D reduce simply to integrals, representing the average potential energy of the ground state of hydrogenic atoms (see (12.9) and (12.10)) and therefore converge. Hence, in the limit of small R_{ab}, E_b diverges.

In the limit of large R_{ab}, on the other hand, C simply reduces to the attractive Coulomb interaction between the proton at "b" and the electron attached to the proton "a" or vice versa. Thus, C rapidly approaches the repulsive interaction between

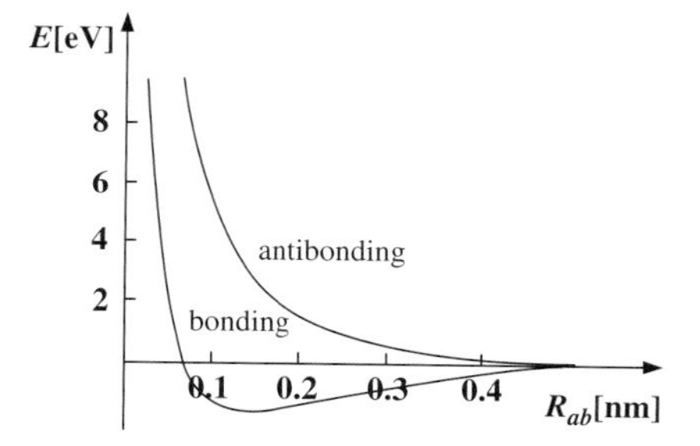

Figure 12.6 The bonding energy of the ionized hydrogen molecule versus the internuclear distance for both triplet (antibonding) and singlet (bonding) states.

two protons same in magnitude but opposite in polarity. This together with the fact that the overlap integral goes to zero, that is, $S \to 0$ for $R_{ab} \to \infty$, makes the terms C and e_M^2/R_{ab} cancel each other out. This leaves the exchange integral D for dictating the polarity of E_b, hence H_2^+ bonding.

Since $D < 0$ by definition, it is clear from (12.16) that the state with the symmetrized wavefunction $\varphi_+(\boldsymbol{r}_a, \boldsymbol{r}_b)$ provides the molecular bonding by rendering $E < E_0$ at a certain range of R_{ab}. Figure 12.6 shows the bonding energy E_b versus R_{ab} for both $\varphi_+(\boldsymbol{r}_a, \boldsymbol{r}_b)$ and $\varphi_-(\boldsymbol{r}_a, \boldsymbol{r}_b)$. Indeed, E_b becomes negative for a range of R_{ab} for the symmetrized state and it is in this range of R_{ab} that the ionized hydrogen molecule is formed. For the case of antisymmetrized wavefunction, E_b is positive for the entire range of R_{ab}, thereby indicating that the molecule cannot be formed in this antisymmetrized state.

The ionized hydrogen molecule being formed in $\varphi_+(\boldsymbol{r}_a, \boldsymbol{r}_b)$ state can be understood, based on the probability density of electrons inherent therein. Specifically, in the symmetrized state, there exists a significant electron charge cloud between the two protons. Thus, the resulting attractive forces between the electron cloud and the two protons more than compensate for the repulsive Coulomb interaction between the two protons. Equivalently, the attractive Coulomb potentials between the electron cloud and the two protons are larger in magnitude than the repulsive Coulomb potential between the two protons, leading to $E_b < 0$.

12.2 Hydrogen Molecule

The analysis of the ionized hydrogen molecule H^+ is next extended for discussing the hydrogen molecule. The H_2 molecule consists of two protons and two electrons, as sketched in Figure 12.7. The Hamiltonian then naturally partitions into two hydrogenic subsystems and additional interaction terms:

$$\hat{H} = \hat{H}_1 + \hat{H}_2 + \hat{V}, \tag{12.17}$$

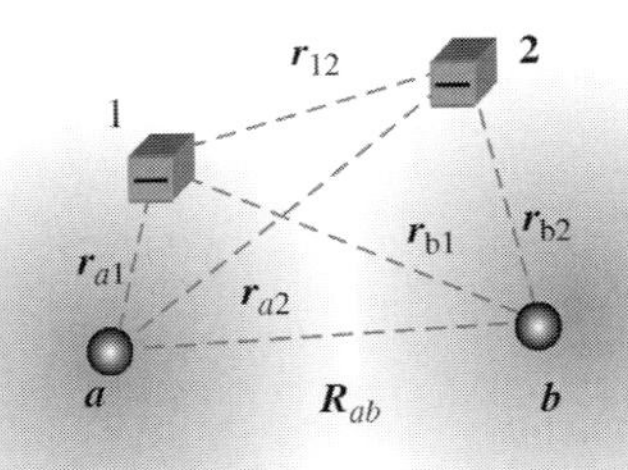

Figure 12.7 A schematic representation of the hydrogen molecule, consisting of two protons and two electrons. Note the distributed nature of Coulomb interaction between protons and electron charge clouds.

where

$$\hat{H}_1 = -\frac{\hbar^2}{2m}\nabla_1^2 - \frac{e_M^2}{r_{a1}}; \quad \hat{H}_2 = -\frac{\hbar^2}{2m}\nabla_2^2 - \frac{e_M^2}{r_{b2}} \tag{12.18a}$$

are two hydrogenic Hamiltonians and the potential

$$\hat{V} = -\frac{e_M^2}{r_{b1}} - \frac{e_M^2}{r_{a2}} + \frac{e_M^2}{R_{ab}} + \frac{e_M^2}{r_{12}}, \quad e_M^2 = \frac{e^2}{4\pi\varepsilon_0}, \tag{12.18b}$$

lumps together the remaining interaction terms between the electron–proton, electron–electron, and proton–proton, respectively (Figure 12.7).

Variational Principle The H_2 molecule constitutes a typical four-body central force problem, whose exact analytical treatment is not available. Thus, one has to resort to an approximate analysis and in this context the variational principle provides a powerful technique for treating the problem of H_2 molecule. The essential features of the variational theory are thus presented first.

Given a dynamical system, the problem essentially consists in solving the energy eigenequation

$$\hat{H}\varphi = E\varphi \tag{12.19}$$

and finding the corresponding energy level:

$$E = \frac{\langle\varphi|\hat{H}|\varphi\rangle}{\langle\varphi|\varphi\rangle} = \frac{\int_{-\infty}^{\infty} dr\varphi^* \hat{H}\varphi}{\int_{-\infty}^{\infty} dr\varphi^*\varphi}. \tag{12.20}$$

If the function φ is the exact solution of (12.19), then E as expressed by (12.20) represents the true energy eigenvalue. On the other hand, if φ is an approximate solution, E resulting from (12.20) should obviously depart from the true value. The variational principle states that the energy values resulting from the approximate wavefunctions are always larger than the true value. Therefore, the degree of accuracy existing among the approximate theories can be compared and assessed, based on the final E values calculated.

Heitler–London Theory With this variational principle as a general guideline, the theory of the H_2 molecule by Heitler and London is presented. In this theory, the two-electron wavefunction of the ground state is taken as

$$|\varphi\rangle = |\varphi_\pm \chi_\mp\rangle, \tag{12.21}$$

where the spatial part of the wavefunction

$$\varphi_\pm = |u_a(1)u_b(2)\rangle \pm |u_b(1)u_a(2)\rangle, \quad j \equiv \mathbf{r}_j \tag{12.22}$$

consists of symmetric and antsymmetric combination of the two ground-state atomic wavefunctions, namely,

$$\hat{H}_{j\alpha}|u_\alpha(j)\rangle = E_0|u_\alpha(j)\rangle, \quad \alpha = a, b, \tag{12.23}$$

with E_0 denoting the ground-state energy of the hydrogenic atom. The spin parts of the wavefunction, χ_-, χ_+, represent the usual singlet and triplet states associated with two-electron spins (see (11.33) and (11.34)). Thus, the total wavefunctions in (12.21) are antisymmetric, as they should.

Next, using the wavefunctions in (12.21) and (12.22), the ground-state energy of the molecule can be calculated. In so doing, the spin functions are deleted with the understanding that φ_+ and φ_- should be combined with the singlet state χ_- and the triplet state χ_+ respectively, as clearly specified in (12.21). Thus, one can write

$$E_\pm = \frac{\langle\varphi_\pm|\hat{H}|\varphi_\pm\rangle}{\langle\varphi_\pm|\varphi_\pm\rangle}. \tag{12.24}$$

In evaluating (12.24), the integrations involved in the denominator can be carried out first in terms of the overlap integral discussed in the previous section:

$$\langle\varphi_\pm|\varphi_\pm\rangle \equiv \langle u_a(1)u_b(2) \pm u_b(1)u_a(2)|u_a(1)u_b(2) \pm u_b(1)u_a(2)\rangle = 2(1 \pm S^2), \tag{12.25a}$$

where

$$S \equiv \langle u_a(1)|u_b(1)\rangle = \langle u_a(2)|u_b(2)\rangle \tag{12.25b}$$

and

$$\langle u_a(i)u_b(j)|u_a(i)u_b(j)\rangle = 1, \quad i, j = 1, 2,\ i \neq j, \tag{12.25c}$$

since u_a and u_b are normalized.

The evaluation of the numerator

$$N = \left\langle \varphi_\pm \left| \hat{H}_1 + \hat{H}_2 + \frac{e_M^2}{R_{ab}} + \frac{(-e_M^2)}{r_{b1}} + \frac{(-e_M^2)}{r_{a2}} + \frac{e_M^2}{r_{12}} \right| \varphi_\pm \right\rangle \tag{12.26}$$

requires 24 integrations in total. Specifically, the six terms appearing in $\hat{H}$ have to be paired with four different combinations of $u_a(i)u_b(j)$ for integration. However, $\hat{H}$ is invariant under $\boldsymbol{r}_1 \leftrightarrow \boldsymbol{r}_2$ and under the interchange of $\boldsymbol{r}_1$ and $\boldsymbol{r}_2$, the four combinations of $u_a(i)u_b(j)$ reduce to two. Thus, the numerator is compacted as

$$N = 2\langle \hat{H} \rangle_1 \pm 2\langle \hat{H} \rangle_2, \tag{12.27}$$

where the first term

$$\begin{aligned} \langle \hat{H} \rangle_1 &\equiv \left\langle u_a(1)u_b(2) \middle| \hat{H}_1 + \hat{H}_2 + \frac{e_M^2}{R_{ab}} + \frac{(-e_M^2)}{r_{b1}} + \frac{(-e_M^2)}{r_{a2}} + \frac{e_M^2}{r_{12}} \middle| u_a(1)u_b(2) \right\rangle \\ &= 2E_0 + \frac{e_M^2}{R_{ab}} + 2C + E_{\mathrm{RI}} \end{aligned} \tag{12.28a}$$

has been specified in terms of the Coulomb interaction integral C and the repulsive Coulomb interaction between two electrons E_{RI}:

$$C = \left\langle u_a(1) \middle| \frac{-e_M^2}{r_{b1}} \middle| u_a(1) \right\rangle = \left\langle u_2(2) \middle| \frac{-e_M^2}{r_{a2}} \middle| u_b(2) \right\rangle, \tag{12.28b}$$

$$E_{RI} = \left\langle u_a(1)u_b(2) \middle| \frac{e_M^2}{r_{12}} \middle| u_a(1)u_b(2) \right\rangle. \tag{12.28c}$$

Similarly, the second term

$$\begin{aligned} \langle \hat{H} \rangle_2 &= \left\langle u_b(1)u_a(2) \middle| \hat{H}_1 + \hat{H}_2 + \frac{e_M^2}{R_{ab}} + \frac{(-e_M^2)}{r_{b1}} + \frac{(-e_M^2)}{r_{a2}} + \frac{e_M^2}{r_{12}} \middle| u_a(1)u_b(2) \right\rangle \\ &= 2E_0 S^2 + \frac{e_M^2}{R_{ab}} S^2 + 2DS + E_{CE} \end{aligned} \tag{12.29a}$$

can be expressed in terms of the exchange integral D and the repulsive interaction computed with the use of exchange densities E_{CE}:

$$D = \left\langle u_b(1) \middle| \frac{-e_M^2}{r_{b1}} \middle| u_a(1) \right\rangle = \left\langle u_a(2) \middle| \frac{-e_M^2}{r_{a2}} \middle| u_b(2) \right\rangle, \tag{12.29b}$$

$$E_{CE} = \left\langle u_b(1)u_a(2) \middle| \frac{e_M^2}{r_{12}} \middle| u_a(1)u_b(2) \right\rangle, \tag{12.29c}$$

and the overlap integral S has already been defined in (12.25b).

The Bonding Energy Inserting (12.25) and (12.27)–(12.29) into (12.24), one finds after a straightforward rearrangements of terms,

$$E_{\pm} = 2E_0 + \frac{2C + E_{\mathrm{RI}}}{1 \pm S^2} \pm \frac{2DS + E_{CE}}{1 \pm S^2} + \frac{e_M^2}{R_{ab}} \tag{12.30}$$

The bonding of the H_2 molecule can now be explicitly discussed, based on (12.30). Since $0 < S < 1$ and $D < 0$ as discussed already, it is clear from (12.30) that E_+ is lower than E_-. The former is associated with the symmetric wavefunction φ_+ and therefore the singlet state χ_-, while the latter is associated with φ_- and the triplet state χ_+. Also, when the two protons are taken far apart from each other, with each carrying an electron, the corresponding energy consists of the total energy of two noninteracting hydrogen atoms, that is, $2E_0$. Therefore, the bonding can be specified by the difference between $E_\pm$ in (12.30) and $2E_0$, namely,

$$E_b \equiv E_\pm - 2E_0 = \frac{2C + E_{RI}}{1 \pm S^2} \pm \frac{2DS + E_{CE}}{1 \pm S^2} + \frac{e_M^2}{R_{ab}}. \tag{12.31}$$

The resulting bonding energy is plotted versus R_{ab} in Figure 12.8 for both singlet and triplet states. As is clear from the figure, for the singlet state, paired with the spatially symmetric wavefunction, the net effect of the various kinds of Coulomb interactions is to lower the energy level E_+ below $2E_0$. As a consequence, two hydrogen atoms are bonded into the H_2 molecule. In contrast, for the triplet state, which is paired with the spatially antisymmetric wavefunction, the net effect of the interactions is to raise E_- above $2E_0$, leading to the antibonding state.

The bonding versus antibonding can again be understood, based on the distribution of the charge clouds of two electrons. Plotted in Figure 12.9 are the probability densities of symmetric and antisymmetric wavefunctions $\varphi_\pm$. For the singlet state paired with φ_+, the probability density of two electrons between the two protons is high, as clearly seen from the figure. This suggests that the attractive interaction between the electron cloud and the two protons more than compensate for the repulsive interactions between the two protons or the two electrons, as indeed was the case with ionized hydrogen molecule.

The theoretical bonding energy of 3.14 eV, obtained from the Heitler–London theory is smaller than the measured dissociation energy of 4.48 eV. This disagreement between the theory and experiment can clearly be attributed to the inaccuracy of

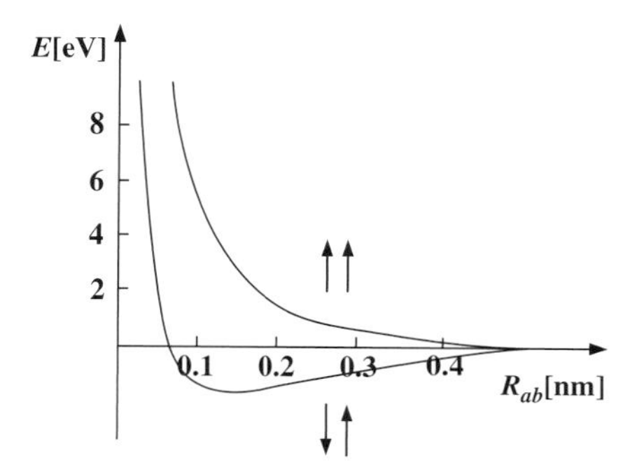

Figure 12.8 The bonding energy of hydrogen molecule versus the internuclear distance for both the triplet (antibonding) and singlet (bonding) states.

Figure 12.9 The probability densities of symmetrized and antisymmetrized ground-state wavefunctions $\varphi_{\pm}(\boldsymbol{r}_a, \boldsymbol{r}_b)$ in (12.22). The contrast between the two consists in differing probability densities between the two nuclei, as indicated by the degree of brightness.

the wavefunction used for describing the two electrons, as discussed earlier. However, the theory does provide the general mechanism where the homopolar molecules are formed, based purely upon the atomic orbitals and the Coulomb interactions.

12.3 Ionic Bond and van der Waals Attraction

The chemical bond underlying the formation of the H_2 molecule is called the covalent bond and constitutes a typical example of the interaction between two open-shell neutral atoms, in this case, two H-atoms. Moreover, there exist other kinds of chemical bonds, and some of these bonds are singled out for further discussion. First, the ionic bond is considered, which is based on the interaction between a closed-shell positive ion and a closed-shell negative ion. The discussion of this ionic bond is followed by the van der Waals attraction, which results from the interaction between two closed shells hosted by neutral atoms, for example, He–He.

12.3.1 The Ionic Bond

One of the well-known examples of the ionic bond is the sodium chloride, Na^+Cl^-. Here, the ionization of the Na atom requires the first ionization potential IP_1, while that of the Cl atom is accompanied by the electron capture and the affinity EA:

$$\mathrm{Na([Ne]3s)} \rightarrow \mathrm{Na^+([Ne])} + e^{-1}: \quad IP_1 = 5.14\,\mathrm{eV},$$

$$\mathrm{Cl\,([Ne]3s^2 3p^5)} + e^- \rightarrow \mathrm{Cl^-([Ar])}: \quad \mathrm{EA} = -3.65\,\mathrm{eV}.$$

Thus, the formation of these positive and negative ions at a distance $R \rightarrow \infty$ requires a net energy of 1.49 eV, namely,

$$\mathrm{Na} + \mathrm{Cl} \rightarrow \mathrm{Na^+} + \mathrm{Cl^-}: \quad \Delta E(\infty) = 1.49\,\mathrm{eV}.$$

As the two ions are brought together, an attractive Coulomb interaction begins to develop significantly and could make the net energy $\Delta E(R)$ of the two-ion system less than zero, thereby making it possible to form a stable molecule.

In the limit of small R, on the other hand, the electron clouds of the two ions begin to overlap with each other. When two closed-shell ions approach each other, the exclusion principle dictates that the overlapped electrons go into next higher lying states. This means that the energy of two ion pair rapidly increases with increasing interpenetration with shrinking R. Consequently, there ensues an effective repulsive interaction for $R \to 0$.

Thus, the energy of the two-ion system, $\Delta E(R)$, can generally be expressed as a function of R as

$$\Delta E(R) = Ae^{-\alpha R} - \frac{e_M^2}{R} + \Delta E(\infty), \tag{12.32}$$

where the first two terms represent the repulsive interaction arising from the exclusion principle and the attractive Coulomb interaction between the two ions. As can be clearly observed from Figure 12.10, $\Delta E(R)$ is dominated by repulsive and attractive interactions in the respective regions of R. As a consequence, there naturally appears the minimum level of $\Delta E(R)$ at the equilibrium distance R_e. Also, the parameters A and α can be specified with the use of electron wavefunctions, but those constants can be specified from the measured data of the molecule. For example, the equilibrium distance R_e and the vibrational frequency ω_c of the molecule can be measured accurately by the microwave spectroscopy and these data can be used for determining A and α.

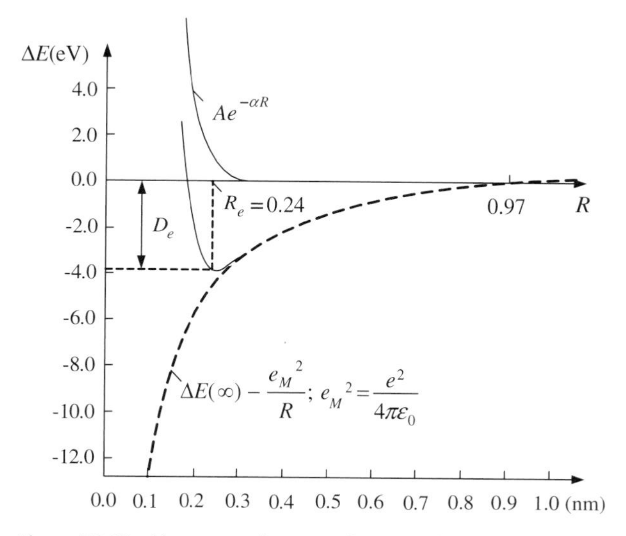

Figure 12.10 The potential energy of ionic molecule versus internuclear distance. The energy is contributed by repulsive and attractive interactions at small and large internuclear distances.

For this purpose, it is convenient to Taylor expand $\Delta E(R)$ centered at the minimum point, namely, $R = R_e$, thereby obtaining

$$\Delta E(R) = \Delta E(R_e) + \frac{1}{2}k(R-R_e)^2, \quad k \equiv \frac{\partial^2 \Delta E(R_e)}{\partial R^2}. \tag{12.33}$$

In (12.33), the first Taylor expansion coefficient is by definition zero because $\Delta E(R_e)$ has the minimum value at $R = R_e$. Also the expansion has been truncated after the second-order term, for simplicity of discussion. In this manner, $\Delta E(R)$ is naturally reduced to the potential energy of a harmonic oscillator and the measured vibrational frequency of the molecule can therefore be related to the effective spring constant k of the molecule and the reduced mass of the two ions, namely,

$$\omega_c^2 = \frac{k}{\mu}; \quad \frac{1}{\mu} = \frac{1}{m_{\text{Na}}} + \frac{1}{m_{\text{Cl}}}. \tag{12.34}$$

Now the fact that the first Taylor expansion coefficient vanishes at R_e is to be expressed from (12.32) as

$$0 = -\alpha A e^{-\alpha R_e} + \frac{e_M^2}{R_e^2}, \quad e_M^2 = \frac{e^2}{4\pi\varepsilon_0}, \tag{12.35a}$$

while the molecular spring constant is specified as

$$k \equiv \frac{\partial^2 \Delta E(R_e)}{\partial R^2} = \alpha^2 A e^{-\alpha R_e} - \frac{2e_M^2}{R_e^3}. \tag{12.35b}$$

Hence, from the measured data of R_e and ω_c, the parameters α and A are determined from (12.35) as

$$A = \frac{e_M^2 e^{\alpha R_e}}{\alpha R_e^2}, \quad \frac{e_M^2}{R_e^2}\left(\alpha - \frac{2}{R_e}\right) = k = \mu\omega_c^2. \tag{12.36}$$

Inserting the constants A and α thus determined into (12.32), one can estimate the strength of the ionic bond (see Figure 12.10). Thus, the ionic bond is to be specifically attributed to the Coulomb interaction between positive and negative ions.

12.3.2
van der Waals Attraction

It is interesting that the interaction between two closed-shell neutral atoms also leads to a chemical bond, called the van der Waals attraction. This kind of attractive interaction was first investigated by J. D. van der Waals and is shown to occur in nearly all atoms. The underlying force is also known as the London dispersion force after F. London who brought out its essential features using a simple model, sketched in Figure 12.11. In this model, two interacting atoms are simply represented by two one-dimensional negatively charged harmonic oscillators, bound to positive charge

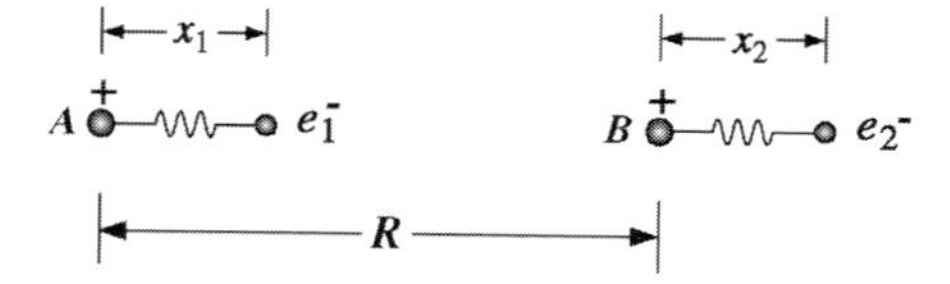

Figure 12.11 Two charged harmonic oscillators coupled via Coulomb interactions.

centers. Because the two oscillators are located in proximity with each other, there ensues additional coupling via the usual Coulomb interactions.

The Hamiltonian of the system is given by

$$H = H_1 + H_2 + V, \tag{12.37}$$

where

$$H_j = -\frac{\hbar^2}{2m}\frac{\partial^2}{\partial x_j^2} + \frac{1}{2}kx_j^2, \quad j = 1, 2, \tag{12.38}$$

represents the harmonic oscillator Hamiltonian with the spring constant k and

$$V = e_M^2\left(\frac{1}{R} - \frac{1}{R-x_1} + \frac{1}{R+x_2-x_1} - \frac{1}{R+x_2}\right), \quad e_M^2 = \frac{e^2}{4\pi\varepsilon_0} \tag{12.39}$$

is the coupling term, accounting for additional Coulomb interactions between two positively charged force centers and two negatively charged oscillators.

For $R \gg x_1, x_2$, one can simplify V by expanding the potential terms appearing therein in powers of x_j/R. For example, one can expand the second term in (12.39) as

$$\frac{1}{R-x_1} = \frac{1}{R}\left(1-\frac{x_1}{R}\right)^{-1} = \frac{1}{R}\left[1+\frac{x_1}{R}+\left(\frac{x_1}{R}\right)^2+\cdots\right]$$

and carry out similar expansions for other terms, thereby obtaining

$$V \simeq -\frac{2e_M^2 x_1 x_2}{R^3}. \tag{12.40}$$

It has thus become clear from (12.40) that additional Coulomb interactions provide a net attractive interaction between the two harmonic oscillator system.

Hence, inserting (12.38) and (12.40) into (12.37), the total Hamiltonian reads as

$$H = -\frac{\hbar^2}{2m}\frac{\partial^2}{\partial x_1^2} + \frac{1}{2}kx_1^2 - \frac{\hbar^2}{2m}\frac{\partial^2}{\partial x_2^2} + \frac{1}{2}kx_2^2 - \frac{2e_M^2 x_1 x_2}{R^3}. \tag{12.41}$$

The role of the coupling term in (12.41) is to be explicitly brought out by introducing a new set of variables

$$\xi = x_1 + x_2, \quad \eta = x_2 - x_1 \tag{12.42}$$

corresponding roughly to the center of mass and relative coordinates.

In terms of ξ and η, (12.41) is further simplified as

$$H = -\frac{\hbar^2}{2\mu}\frac{\partial^2}{\partial\xi^2} + \frac{1}{2}k_-\xi^2 - \frac{\hbar^2}{2\mu}\frac{\partial^2}{\partial\eta^2} + \frac{1}{2}k_+\eta^2, \quad k_\mp \equiv \left(\frac{k}{2} \mp \frac{e_M^2}{R^3}\right), \tag{12.43}$$

where the reduced mass μ,

$$\frac{1}{\mu} = \frac{1}{m} + \frac{1}{m} = \frac{1}{m/2} \tag{12.44}$$

becomes one half of the original mass m of the oscillators.

In this representation of H, the dynamics of the two coupled oscillator system has been transformed into that of two oscillators, which oscillate at two dressed frequencies independent of each other. Thus, the energy eigenfunctions and eigenvalues of the system can be obtained by a straightforward generalization of the results obtained in Chapter 8. The quantized energy level, for example, is now given by

$$E_{n_1,n_2} = \hbar\omega_+\left(n_1 + \frac{1}{2}\right) + \hbar\omega_-\left(n_2 + \frac{1}{2}\right), \tag{12.45}$$

where n_1, n_2 are the two quantum numbers. It is interesting to notice that the two dressed frequencies

$$\omega_\mp^2 \equiv \frac{k_\mp}{\mu} = \omega_c^2\left(1 \mp \frac{2e_M^2}{kR^3}\right); \quad \omega_c^2 \equiv \frac{k}{m} \tag{12.46}$$

are distinctly different from the original characteristic frequency ω_c of the two coupled oscillators, one is greater than ω_c and the other is less than ω_c.

These dressed frequencies can also be expressed in terms of the oscillator polarizability as follows. When an external field E is applied, the oscillator is pushed away from its equilibrium position, but in the process it is also subject to the restoring force provided by the spring. Thus, the total force acting on the oscillator is given by

$$F = -kx - eE,$$

but it vanishes at a position x_e such that

$$0 = -kx_e - eE.$$

The resulting dipole moment $\tilde{\mu}_{\text{ind}} \equiv -ex_e = (e^2/k)\text{E}$ can therefore be expressed in terms of the polarizability α, which connects $\tilde{\mu}_{\text{ind}}$ to its input field E via $\tilde{\mu}_{\text{ind}} \equiv 4\pi\varepsilon_0\alpha\text{E}$:

$$\alpha = \frac{(e^2/4\pi\varepsilon_0)}{k} = \frac{e_M^2}{k}. \tag{12.47}$$

With (12.47) inserted into (12.46), the dressed frequencies are specified as

$$\omega_\mp^2 \equiv \frac{k_\mp}{\mu} = \omega_c^2\left(1 \mp \frac{2\alpha}{R^3}\right), \quad \omega_c^2 \equiv \frac{k}{m}. \tag{12.48}$$

The ground-state energy of the system is thus given by

$$E_{0,0} = \frac{\hbar}{2}(\omega_- + \omega_+) \tag{12.49}$$

and the role of the coupling parameters can be brought out by making the expansion

$$\omega_\mp = \omega_c\left[1 \mp \frac{1}{2}\left(\frac{2\alpha}{R^3}\right) - \frac{1}{8}\left(\frac{2\alpha}{R^3}\right)^2 + \cdots\right], \quad \frac{2\alpha}{R^3} \ll 1 \tag{12.50}$$

Hence, inserting (12.50) into (12.49), one can write the zero-point energy of the system as

$$E_{00} = \hbar\omega_c - C\frac{1}{R^6}, \quad C = \frac{\hbar\omega_c\alpha^2}{2}. \tag{12.51}$$

In this expression, the first term obviously represents the total zero-point energy of two oscillators in the limit $R \to \infty$. In view of this fact, the second term evidently accounts for the net attractive interaction brought about by the interaction Hamiltonian $V(x)$ in (12.39). The resulting bonding, inversely proportional to R^6, is known as the van der Waals attraction.

Naturally, the simple model of two coupled oscillators used for illustrating the dispersion force can be generalized to two coupled neutral atomic systems. In this case, the power law dependence of the coupling strength, namely, $\propto 1/R^6$, is still preserved, but the oscillator polarizability is replaced by corresponding atomic polarizabilities and the binding parameter C is replaced by the first ionization potential IP_1.

The few chemical bonds discussed in this chapter are thus shown to originate from the attractive Coulomb interaction between the nuclear charge and electron charge clouds dominating over the repulsive Coulomb interactions.

12.4 Problems

12.1. The Heitler–London theory is mainly based on evaluating various matrix elements of the Hamiltonian of hydrogen molecule. Using the wavefunction given in (12.22), evaluate $E_\pm$ in (12.24) filling in algebraic steps and verify the final results of (12.30).

12.2. Take the trial wavefunction of the ionized hydrogen molecule H_2^+ to be

$$\varphi(\mathbf{r}_a, \mathbf{r}_b) = N(|u_a\rangle \pm |u_b\rangle),$$

with $|u_a\rangle, |u_b\rangle$ denoting the ground-state eigenfunctions of hydrogenic atom.

(a) Find the normalization constant N.

(b) Evaluate the expectation value of the Hamiltonian

$$\hat{H} = -\frac{\hbar^2}{2m}\nabla^2 - e_M^2\left(\frac{1}{r_a} + \frac{1}{r_b}\right), \quad e_M^2 = \frac{e^2}{4\pi\varepsilon_0},$$

and discuss the results obtained, in comparison with (12.13).

12.3. The interionic distance in NaCl molecule is 0.24 nm and the vibrational frequency of the molecule is $\nu_e = \omega_e/2\pi = 1.1 \times 10^{13}$ per second. Determine the parameters A and α in (12.32) and estimate the bonding energy using A and α thus determined together with $\Delta E(\infty) = 1.49$ eV.

12.4. Starting with the coupled Hamiltonian given in (12.41), use the new set of variables ξ and η introduced in (12.42) and obtain the decoupled Hamiltonian (12.43).

12.5. Two hydrogen atoms are located R distance away from each other along the z-axis (see Figure 12.7).

(a) Write down the interaction potentials between the volume charge elements of two electron charge clouds and two nuclear point charges.

(b) Assuming that $x_1, x_2, \ldots, z_1, z_2 \ll R$, show that the interaction Hamiltonian term in (a) reduces to

$$V \simeq \frac{e_M^2}{R}(x_1x_2 + y_1y_2 + z_1z_2).$$

(c) Model the hydrogen atom by an isotropic 3D harmonic oscillator and find the resulting van der Waals attractive energy.

12.6. A hydrogenic atom is placed in a uniform electric field E in the z-direction. The Hamiltonian is thus given by

$$H = -\frac{\hbar^2}{2m}\nabla^2 - \frac{e_M^2}{r} + \mathrm{E}er\cos\theta, \quad e_M^2 = \frac{e^2}{4\pi\varepsilon_0}.$$

(a) If the wavefunction is sought in the form

$$\varphi = c_1|u_{100}\rangle + c_2|u_{210}\rangle, \tag{12.52}$$

derive the coupled equation for c_1 and c_2 in a manner similar to (12.4)–(12.7).

(b) Solving the secular equation, find c_1, c_2, and the energy associated with the wavefunction given in (12.52).

(c) Evaluate the expectation value

$$\langle \boldsymbol{r} \rangle = \frac{\langle \varphi|\boldsymbol{r}|\varphi\rangle}{\langle \varphi|\varphi\rangle}$$

and the atom dipole $\mu_{\mathrm{ind}} = -e\langle \boldsymbol{r} \rangle$, and the polarizability α connecting the input field to the induced atom dipole as $\mu_{\mathrm{ind}} = -\alpha\mathrm{E}$.

Suggested Reading

1 Haken, H. and Wolf, H.C. (2004) *Molecular Physics and Elements of Quantum Chemistry: Introduction to Experiments and Theory*, 2nd edn, Springer.

2 Haken, H., Wolf, H.C., and Brewer, W.D. (2007) *The Physics of Atoms and Quanta: Introduction to Experiments and Theory*, 7th revised and enlarged edn, Springer.

3 Karplus, M. and Porter, R.N. (1970) *Atoms and Molecules: An Introduction for Students of Physical Chemistry*, Addison-Wesley Publishing Company.

4 Gasiorowicz, S. (2003) *Quantum Physics*, 3rd edn, John Wiley & Sons, Inc.

13
The Perturbation Theory

It is rather rare to be able to treat dynamical systems of practical interest in a closed analytical form. The rigorous analytical treatments of the hydrogen atom and the harmonic oscillator are admittedly notable exceptions. As a consequence, various perturbation and/or iteration schemes have been devised to deal with practical but complicated dynamic systems. These approximate schemes of analysis are still important in spite of rapid advances made in computational sciences. The approximate analysis is capable of shedding an overall insight of the problem and can also provide general guidelines for accessing the correctness of numerical outputs. More important, the time-dependent perturbation scheme provides a general framework for describing the coupling between different systems, such as light and matter. It is therefore important to discuss some of the perturbation schemes developed, in particular the harmonic perturbation in conjunction with Fermi's golden rule.

13.1
Time-Independent Perturbation Theory in Nondegenerate System

Given a dynamic system with the Hamiltonian $\hat{H}$, the problem at hand is to solve the energy eigenequation

$$\hat{H}\varphi = W\varphi. \tag{13.1}$$

As pointed out at the outset, it is not generally possible to solve the energy eigenequation in a closed analytic form. As a consequence, the perturbation schemes are required for solving such equations iteratively. A key point of the perturbation scheme is to partition the Hamiltonian into two parts in such a way that the following relation holds true:

$$\hat{H} = \hat{H}_0 + \lambda\hat{H}', \quad |\hat{H}_0| \gg |\hat{H}'|. \tag{13.2}$$

Introductory Quantum Mechanics for Semiconductor Nanotechnology. Dae Mann Kim

ISBN: 978-3-527-40975-4

Here, the first term $\hat{H}_0$ has to be chosen such that it lends itself to analytical treatment, while the remaining terms are lumped into the perturbation term $\hat{H}'$. The accuracy and the effectiveness of the scheme is obviously dependent on $\hat{H}'$ being much smaller than $\hat{H}_0$. The parameter λ is called the smallness or bookkeeping parameter and is introduced to keep track of the order of iterations.

Since the dominant term $\hat{H}_0$ is chosen such that it can be treated analytically, one can introduce a set of energy eigenfunctions

$$\hat{H}_0|u_n\rangle = E_n|u_n\rangle. \tag{13.3}$$

Here the discussion is confined to the nongenerate systems in which $E_m \neq E_n$, if $m \neq n$. Next, to solve (13.1), one may expand φ and W in powers of λ as

$$\varphi = \varphi_0 + \lambda\varphi_1 + \lambda^2\varphi_2 + \cdots, \tag{13.4}$$

$$W = W_0 + \lambda W_1 + \lambda^2 W_2 + \cdots. \tag{13.5}$$

Inserting (13.4) and (13.5) into (13.1) results in

$$\begin{aligned} &(\hat{H}_0 + \lambda\hat{H}')(\varphi_0 + \lambda\varphi_1 + \lambda^2\varphi_2 + \cdots) \\ &= (W_0 + \lambda W_1 + \lambda^2 W_2 + \cdots)(\varphi_0 + \lambda\varphi_1 + \lambda^2\varphi_2 + \cdots). \end{aligned} \tag{13.6}$$

Hence, equating the coefficients of the equal powers of λ from both sides of (13.6), one obtains a hierarchy of equations. One can, for example, write up to the second order in λ:

$$\hat{H}_0\varphi_0 = W_0\varphi_0, \tag{13.7}$$

$$\hat{H}'\varphi_0 + \hat{H}_0\varphi_1 = W_0\varphi_1 + W_1\varphi_0, \tag{13.8}$$

$$\hat{H}'\varphi_1 + \hat{H}_0\varphi_2 = W_0\varphi_2 + W_1\varphi_1 + W_2\varphi_0. \tag{13.9}$$

It is clear from the hierarchy of equations that once the zeroth-order eigenfunction φ_0 and eigenvalue W_0 are given, the true eigenfunction φ and eigenvalue W can be analyzed by incorporating iteratively the effect of the perturbing Hamiltonian $\hat{H}'$ to an arbitrary order of accuracy or equivalently to an arbitrary power of λ. Thus, choose for consideration the mth eigenstate of $\hat{H}_0$, for instance, and examine how u_m and E_m are modified due to the presence of $\hat{H}'$. Thus, the starting point of analysis is

$$\varphi_0 = |u_m\rangle \tag{13.10a}$$

and

$$W_0 = E_m. \tag{13.10b}$$

13.1.1
The First-Order Perturbation

To find the first-order corrections in the wavefunction φ_1 and the energy eigenvalue W_1 in (13.8), one can expand φ_1 in terms of the eigenfunctions of $\hat{H}_0$, that is,

$$\varphi_1 = \sum_n a_n^{(1)}|u_n\rangle, \tag{13.11}$$

and insert both (13.10) and (13.11) into (13.8), obtaining

$$\hat{H}'|u_m\rangle + \hat{H}_0\sum_n a_n^{(1)}|u_n\rangle = E_m\sum_n a_n^{(1)}|u_n\rangle + W_1|u_m\rangle. \tag{13.12}$$

The problem has thus been reduced to determining $\{a_n^{(1)}\}$ and W_1 from (13.12). To this end, one can take the inner product (13.12) with respect to $\langle u_k|$, obtaining

$$\hat{H}'_{km} + \sum_n a_n^{(1)}\langle u_k|\hat{H}_0|u_n\rangle = E_m\sum_n a_n^{(1)}\langle u_k|u_n\rangle + W_1\langle u_k|u_m\rangle, \tag{13.13}$$

where the matrix element of the perturbing Hamiltonian is defined as

$$\hat{H}'_{km} \equiv \langle u_k|\hat{H}'|u_m\rangle = \int d\mathbf{r}u_k^*(\mathbf{r})\hat{H}'u_m(\mathbf{r}). \tag{13.14}$$

Using the orthonormality of the eigenfunctions, namely, $\langle u_n|u_{n'}\rangle = \delta_{nn'}$, (3.13) simplifies as

$$\hat{H}'_{km} + a_k^{(1)}E_k = E_m a_k^{(1)} + W_1\delta_{km}. \tag{13.15}$$

Note that k is a dummy variable representing any of the eigenstates of $\hat{H}_0$, while m denotes the specific eigenstate chosen for investigation.

Since the analysis is confined to the nondegenerate system, if $k \neq m$, $E_k \neq E_m$, and $\delta_{km} = 0$, so that the first-order expansion coefficients are found from (13.15) with the exception of $a_m^{(1)}$:

$$a_k^{(1)} = \frac{\hat{H}'_{km}}{E_m - E_k}, \quad m \neq k. \tag{13.16}$$

For $k = m$, on the other hand, $E_k = E_m$ and $\delta_{km} = 1$, and therefore the first-order level shift W_1 due to $\hat{H}'$ is found from (13.15) as

$$W_1 \equiv \hat{H}'_{mm} = \langle u_m|\hat{H}'|u_m\rangle. \tag{13.17}$$

Thus, the results of the first-order perturbation analysis are summarized as

$$E_m^{(1)} = E_m + \hat{H}'_{mm}, \quad \hat{H}'_{mm} \equiv \langle u_m|\hat{H}'|u_m\rangle, \tag{13.18}$$

$$\varphi^{(1)} = |u_m\rangle(1+a_m^{(1)}) + \sum_{k\neq m} \frac{\hat{H}'_{km}}{E_m - E_k}|u_k\rangle, \quad \hat{H}'_{km} \equiv \langle u_k|\hat{H}'|u_m\rangle. \tag{13.19}$$

In this way, the modification in the zeroth order eigenfunction and the eigenvalue has been specified in terms of the matrix elements of $\hat{H}'$. The only unknown quantity $a_m^{(1)}$ can also be determined by imposing the normalization condition on $\varphi^{(1)}$. Again, in view of the orthonormality of $\{u_n\}$, the normalization condition up to the first order in λ is given from (13.19) by

$$\begin{aligned} 1 = \langle \varphi^{(1)}|\varphi^{(1)}\rangle &= \langle u_m(1+\lambda a_m^{(1)})|u_m(1+\lambda a_m^{(1)})\rangle + O(\lambda^2) \\ &= 1 + [a_m^{(1)}]^* + [a_m^{(1)}] + O(\lambda^2). \end{aligned} \tag{13.20}$$

Clearly, those terms proportional to λ^2 can be relegated to the second-order perturbation and one can put $a_m^{(1)} = 0$ to satisfy (13.20), thereby completing the first-order analysis.

13.1.2
The Second-Order Perturbation

One can likewise carry out the second-order perturbation analysis using (13.9). Here, φ_2 can again be expanded as

$$\varphi_2 = \sum_n a_n^{(2)}|u_n\rangle. \tag{13.21}$$

Inserting (13.21), (13.11), and (13.10b) into (13.9) results in

$$\hat{H}' \sum_n a_n^{(1)}|u_n\rangle + \hat{H}_0 \sum_n a_n^{(2)}|u_n\rangle = E_m \sum_n a_n^{(2)}|u_n\rangle + W_1 \sum_n a_n^{(1)}|u_n\rangle + W_2|u_m\rangle. \tag{13.22}$$

Thus, the second-order corrections in φ and W can be specified in terms of the zeroth order (u_m, E_m) and the first order (φ_1, W_1) results.

Again, taking the inner product of (13.22) with respect to $\langle u_k|$ and making use of the orthnormality, $\langle u_k|u_n\rangle = \delta_{kn}$, one finds

$$\sum_n a_n^{(1)} \hat{H}'_{kn} + a_k^{(2)} E_k = E_m a_k^{(2)} + W_1 a_k^{(1)} + W_2 \delta_{km}. \tag{13.23}$$

The second-order level shift is therefore obtained from (13.23) by putting $k = m$ and using the fact that $a_m^{(1)} = 0$ from the first-order analysis:

$$W_2 = \sum_n a_n^{(1)} \hat{H}'_{mn} - W_1 a_m^{(1)} = \sum_{n\neq m} a_n^{(1)} \hat{H}'_{mn}. \tag{13.24}$$

For $k \neq m$, (13.23) yields the seconds-order correction in the wavefunction, namely,

$$\begin{aligned} a_k^{(2)} &= \frac{1}{E_m - E_k}\left[\sum_n a_n^{(1)} \hat{H}'_{kn} - W_1 a_k^{(1)}\right] \\ &= \sum_{n \neq m} \frac{\hat{H}'_{nm}\hat{H}'_{kn}}{(E_m - E_n)(E_m - E_k)} - \frac{\hat{H}'_{mm}\hat{H}'_{km}}{(E_m - E_k)^2}, \quad k \neq m \end{aligned} \tag{13.25}$$

where the first-order results (13.16) and (13.17) have been used, together with $a_m^{(1)} = 0$.

Also, to find $a_m^{(2)}$, the normalization condition up to the second-order perturbation can be imposed, thus obtaining with the use of $\langle u_n | u_{n'} \rangle = \delta_{nn'}$,

$$1 = \left\langle u_m + \lambda \sum_{n \neq m} a_n^{(1)} |u_n\rangle + \lambda^2 \sum_n a_n^{(2)} |u_n\rangle | u_m + \lambda \sum_{n \neq m} a_n^{(1)} |u_n\rangle + \lambda^2 \sum_n a_n^{(2)} |u_n\rangle \right\rangle$$

$$= 1 + \sum_{n \neq m} \left|a_n^{(1)}\right|^2 + 2a_m^{(2)} + O(\lambda^3), \tag{13.26a}$$

so that

$$a_m^{(2)} = -\frac{1}{2}\sum_{n \neq m} \left|a_n^{(1)}\right|^2 = -\frac{1}{2}\sum_{n \neq m} \frac{\left|\hat{H}'_{nm}\right|^2}{(E_m - E_n)^2}. \tag{13.26b}$$

Hence, up to the second-order analysis, the results are summarized as

$$E_m^{(2)} = E_m + \hat{H}'_{mm} + \sum_{n \neq m} \frac{\left|\hat{H}'_{nm}\right|^2}{E_m - E_n}, \tag{13.27}$$

$$\begin{aligned} \varphi^{(2)} &= |u_m\rangle + \sum_{k \neq m} \frac{\hat{H}'_{km}}{E_m - E_k} |u_k\rangle \\ &+ \sum_{k \neq m} \left\{ \left[\sum_{n \neq m} \frac{\hat{H}'_{nm}\hat{H}'_{kn}}{(E_m - E_n)(E_m - E_k)} - \frac{\hat{H}'_{mm}\hat{H}'_{km}}{(E_m - E_k)^2} \right] |u_k\rangle - \frac{1}{2}\frac{\left|\hat{H}_{mk}\right|^2}{(E_m - E_k)^2} |u_m\rangle \right\}. \end{aligned} \tag{13.28}$$

13.1.3 The Stark Effect in Harmonic Oscillator

The validity of the perturbation analysis that has been carried out thus far is illustrated by considering the Stark effect in harmonic oscillator, as an example. Thus, consider a harmonic oscillator charged with q and placed in an external field E. The oscillator is then subject to additional force $q\mathrm{E}$ and the potential $-q\mathrm{E}x$. The Hamiltonian therefore reads as

$$\hat{H} = \hat{H}_0 + \hat{H}', \tag{13.29a}$$

where the two terms

$$\hat{H}_0 \equiv -\frac{\hbar^2}{2m}\frac{\partial^2}{\partial x^2} + \frac{1}{2}kx^2; \quad \hat{H}' \equiv -q\text{E}x \tag{13.29b}$$

denote, respectively, the unperturbed oscillator Hamiltonian and perturbing Hamiltonian arising from the external field.

To be specific, the lth eigenstate is chosen for examination. In this case, a simple parity consideration indicates that there is no first-order level shift, namely,

$$W_1 = \langle u_l | -q\text{E}x | u_l \rangle = -q\text{E}\langle u_l | x | u_l \rangle = 0 \tag{13.30}$$

(see (13.17)). Also, as detailed in Section 8.2, the matrix elements of x of the eigenfunctions connect only two nearest-neighbor states for given l and one can write from (8.51)

$$\hat{H}'_{l-1,l} = \langle u_{l-1} | -q\text{E}x | u_l \rangle = -q\text{E}\sqrt{\frac{\hbar}{2m\omega}}l^{1/2}, \tag{13.31a}$$

$$\hat{H}'_{l+1,l} = \langle u_{l+1} | -q\text{E}x | u_l \rangle = -q\text{E}\sqrt{\frac{\hbar}{2m\omega}}(l+1)^{1/2}, \tag{13.31b}$$

Hence, the first-order correction of the wavefunction reads from (13.19) as

$$\begin{aligned}\varphi_l^{(1)} &= |u_l\rangle + \frac{\hat{H}'_{l+1,l}}{E_l - E_{l+1}}|u_{l+1}\rangle + \frac{\hat{H}'_{l-1,l}}{E_l - E_{l-1}}|u_{l-1}\rangle \\ &= |u_l\rangle + \frac{q\text{E}}{\hbar\omega}\sqrt{\frac{\hbar}{2m\omega}}\left[(l+1)^{1/2}|u_{l+1}\rangle - (l)^{1/2}|u_{l-1}\rangle\right],\end{aligned} \tag{13.32}$$

where $E_l - E_{l\pm1} = \mp\hbar\omega$ (see (8.20)).

By the same token, it also follows from (13.24) and (8.51) that the second-order level shift is contributed solely by the two nearest-neighbor states and one can write using (13.27),

$$W_2 = \frac{\left|\hat{H}'_{l+1,l}\right|^2}{E_l - E_{l+1}} + \frac{\left|\hat{H}'_{l-1,l}\right|^2}{E_l - E_{l-1}} = \frac{1}{\hbar\omega}q^2\text{E}^2\frac{\hbar}{2m\omega}[-(l+1)+l] = -\frac{q^2\text{E}^2}{2m\omega^2}. \tag{13.33}$$

And, the level shift up to the second order perturbation analysis is given by

$$E_l^{(2)} = E_l^{(0)} + W_1 + W_2 = \hbar\omega\left(l+\frac{1}{2}\right) - \frac{q^2\mathrm{E}^2}{2m\omega^2}. \tag{13.34}$$

The Stark effect of the harmonic oscillator can in fact be treated analytically and the exact result can therefore be compared with approximate results obtained in (13.34). The analytical treatment is made possible by first completing the square in the Hamiltonian, namely,

$$\hat{H} = -\frac{\hbar^2}{2m}\frac{\partial^2}{\partial x^2} + \frac{1}{2}kx^2 - q\mathrm{E}x = -\frac{\hbar^2}{2m}\frac{\partial^2}{\partial x^2} + \frac{1}{2}k\left(x-\frac{q\mathrm{E}}{k}\right)^2 - \frac{q^2\mathrm{E}^2}{2k}. \tag{13.35}$$

The energy eigenequation

$$\hat{H}u(x) = Eu(x) \tag{13.36a}$$

can then be expressed in terms of the shifted x-coordinate $x' = x - q\mathrm{E}/k$ as

$$\left[-\frac{\hbar^2}{2m}\frac{\partial^2}{\partial x'^2} + \frac{1}{2}kx'^2\right]u(x') = E_{\mathrm{eff}}\,u(x'), \tag{13.36b}$$

where

$$E_{\mathrm{eff}} \equiv E + \frac{q^2\mathrm{E}^2}{2k} = E + \frac{q^2\mathrm{E}^2}{2m\omega^2}, \quad k \equiv m\omega^2, \tag{13.36c}$$

consists of the unknown energy eigenvalue E and an additional term resulting from completing the square.

Now, the energy eigenequation (13.36b) is identical in form to the original unperturbed energy eigenequation of an oscillator, provided E_{eff} is formally viewed as the effective energy eigenvalue (see (8.2)). One can therefore use the same eigenfunction as derived in (8.40) with the proviso that x is replaced by x'. Moreover, the quantized energy level can be written from (8.20) as

$$E_{\mathrm{eff}\,l} \equiv E_l + \frac{q^2\mathrm{E}^2}{2m\omega^2} = \hbar\omega\left(l+\frac{1}{2}\right). \tag{13.37a}$$

Hence, the true energy eigenvalue E_l of the charged harmonic oscillator in the presence of the external field is given exactly by

$$E_l = \hbar\omega\left(l+\frac{1}{2}\right) - \frac{q^2\mathrm{E}^2}{2m\omega^2} \tag{13.37b}$$

and this exact result agrees with the approximate result obtained from the second-order perturbation analysis, that is, (13.34). The precise agreement between the exact treatment and the second-order perturbation calculation is fortuitous. Nevertheless, the agreement points to the fact that the perturbation scheme works.

13.2 Time-Dependent Perturbation Theory

13.2.1 The Formulation

The Hamiltonian of a dynamic system often depends explicitly on time. Also, the dynamic system is often coupled to other systems via a time-dependent Hamiltonian term. It is therefore important to analyze the system with time-dependent Hamiltonian in conjunction with the time-dependent perturbation theory. Thus, consider a dynamic system with the Hamiltonian

$$\hat{H} = \hat{H}_0 + \lambda \hat{H}'(t), \tag{13.38}$$

where λ denotes again the smallness or bookkeeping parameter. Then, the Schrödinger equation reads as

$$i\hbar \frac{\partial}{\partial t} |\psi(\boldsymbol{r}, t)\rangle = \left[\hat{H}_0 + \lambda \hat{H}'(t)\right] |\psi(\boldsymbol{r}, t)\rangle. \tag{13.39}$$

A convenient way to analyze (13.39) is to expand the wavefunction in terms of the eigenfunctions of $\hat{H}_0$:

$$|\psi\rangle = \sum_n a_n(t) e^{-i(E_n/\hbar)t} |u_n\rangle, \tag{13.40}$$

where

$$\hat{H}_0 |u_n\rangle = E_n |u_n\rangle. \tag{13.41}$$

It should be noted in (13.40) that the expansion coefficient $a_n(t)$ is taken to be time dependent. Inasmuch as $|a_n|^2$ represents the probability of finding the system in the nth eigenstate, the change in time of $a_n(t)$ entails the possibility of transition from one state to another or the coupling between eigenstates driven by $\hat{H}'(t)$.

Inserting (13.40) into (13.39) and using (13.41), one finds

$$i\hbar \textstyle\sum_n \dot{a}_n(t) e^{-i(E_n/\hbar)t} |u_n\rangle + \displaystyle\sum_n E_n a_n(t) e^{-i(E_n/\hbar)t} |u_n\rangle$$

$$= \sum_n E_n a_n(t) e^{-i(E_n/\hbar)t} |u_n\rangle + \lambda \hat{H}'(t) \sum_n a_n(t) e^{-i(E_n/\hbar)t} |u_n\rangle. \tag{13.42}$$

In (13.42), the second term on the left-hand side and the first term on the right-hand side cancel each other out, leaving only two terms to consider. The expansion coefficient can be singled out from the summation by taking the inner product of (13.42) with respect to, say, $\langle u_k|$ and making use of the orthnormality of the eigenfunctions. Thus, one finds

$$\dot{a}_k = -\frac{i}{\hbar}\lambda \sum_n \hat{H}'_{kn} a_n e^{i\omega_{kn}t}, \tag{13.43a}$$

where

$$\hat{H}'_{kn} \equiv \langle u_k | \hat{H}'(t) | u_n \rangle = \int_{-\infty}^{\infty} d\boldsymbol{r} u_k^*(\boldsymbol{r}) \hat{H}'(t) u_n(\boldsymbol{r}) \tag{13.43b}$$

is called the transition matrix element and

$$\omega_{kn} \equiv \frac{E_k - E_n}{\hbar} \tag{13.44}$$

is the atomic transition frequency between u_n and u_k states.

The problem of solving the Schrödinger equation(13.39) has thus been reduced to solving (13.43a). The solution can again be obtained by expanding $a_k(t)$ in powers of λ, that is,

$$a_k = a_k^{(0)} + \lambda a_k^{(1)} + \lambda^2 a_k^{(2)} + \cdots . \tag{13.45}$$

Inserting (13.45) into (13.43a), one can write

$$\dot{a}_k^{(0)} + \lambda \dot{a}_k^{(1)} + \lambda^2 \dot{a}_k^{(2)} + \cdots = \lambda \frac{(-i)}{\hbar} \sum_n \hat{H}'_{kn} e^{i\omega_{kn}t} (a_n^{(0)} + \lambda a_n^{(1)} + \lambda^2 a_n^{(2)} + \cdots). \tag{13.46}$$

Again, equating the coefficients of equal powers of λ from both sides results in

$$\dot{a}_k^{(0)} = 0 \tag{13.47}$$

and the time rate of change of higher order coefficients is given in terms of lower order ones by

$$\dot{a}_k^{(j)} = -\frac{i}{\hbar} \sum_n \hat{H}'_{kn} e^{i\omega_{kn}t} a_n^{(j-1)}, \quad j = 1, 2, 3, \ldots . \tag{13.48}$$

It has thus become clear from (13.47) and (13.48) that given the initial values of the expansion coefficients, $\{a_n^{(0)}\}$, the evolution in time of higher order coefficients can be followed in time iteratively via (13.48) to an arbitrary order in λ. Specifically, if the system is initially in the mth eigenstate, namely,

$$a_n^{(0)} = \delta_{nm}, \tag{13.49}$$

then (13.48) simplifies as

$$\dot{a}_k^{(1)} = -\frac{i}{\hbar} \hat{H}'_{km} e^{i\omega_{km}t}. \tag{13.50}$$

And the growth in time of the first-order expansion coefficients is readily obtained form (13.50) and these coefficients in turn serve as the source for higher order analysis. In this way, one can follow the change in time of the probability of finding the system in each eigenstate and examine the electron making transitions from one state to another.

13.2.2 Harmonic Perturbation and Fermi's Golden Rule

The perturbation scheme presented in the previous section is next applied to the problem of a harmonic perturbation, which carries a wide range of practical applications. Thus, consider a system, initially prepared in the mth eigenstate and driven by a harmonic perturbation Hamiltonian

$$\hat{H}'(t) = \hat{H}'e^{-i\omega t} + \hat{H}'^{*}e^{+i\omega t}, \tag{13.51}$$

where ω is the driving frequency.

Inserting (13.51) into (13.50) and performing the time integration, one finds

$$\begin{aligned} a_k^{(1)}(t) &= \frac{-i}{\hbar}\int_0^t dt\hat{H}'_{km}(t)e^{i\omega_{km}t} \\ &= \frac{-i}{\hbar}\left[\hat{H}'_{km}\frac{e^{i(\omega_{km}-\omega)t}-1}{i(\omega_{km}-\omega)} + \hat{H}'^{*}_{km}\frac{e^{i(\omega_{km}+\omega)t}-1}{i(\omega_{km}+\omega)}\right]. \end{aligned} \tag{13.52}$$

A case of great practical interest is the resonant interaction, in which the atomic transition frequency ω_{km} is approximately equal to the driving frequency ω, so that $|(|\omega_{km}|-\omega)| \ll |\omega_{km}|, \omega$. In this case, either the first or second term in (13.52) becomes dominant, depending on whether $E_k - E_m > 0$ or $E_k - E_m < 0$. Thus, under such a condition, one can choose the dominant term in the form

$$a_k^{(1)}(t) = \pm\frac{i}{\hbar}\hat{H}'_{km}\frac{e^{\pm i(\omega_a-\omega)t}-1}{i(\omega_a-\omega)} \equiv \pm\frac{i\hat{H}'_{km}e^{\pm i(\omega_a-\omega)t/2}}{\hbar}\frac{\sin[(\omega_a-\omega)t/2]}{(\omega_a-\omega)/2}, \tag{13.53a}$$

where ω_a denotes the magnitude of the transition frequency,

$$\omega_a \equiv |\omega_{km}|, \tag{13.53b}$$

and the identity $\sin x = [\exp(ix) - \exp-(ix)]/2i$ has been used. Hence, the probability of the atomic system making a transition from the initial m to a final k state is given by

$$\left|a_k^{(1)}(t)\right|^2 = \frac{|\hat{H}'_{km}|^2}{\hbar^2}\frac{\sin^2[(\omega_a-\omega)t/2]}{[(\omega_a-\omega)/2]^2}. \tag{13.54}$$

Fermi's Golden Rule When the transition involves a quasi-continuous final states, as sketched in Figure 13.1, the total transition probability is obtained by

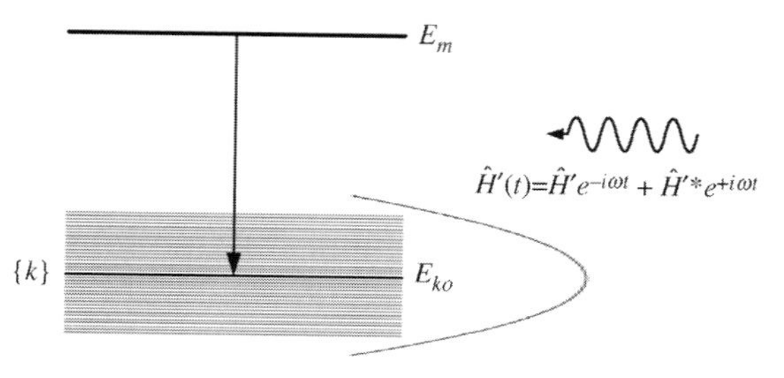

Figure 13.1 A two level atom driven by a harmonic perturbation.The electron can make the transition between two atomic levels. The energy levels are broadened due to the finite electron lifetime therein.

summing (13.54) over these final states:

$$\left|a_{\{k\}}^{(1)}\right|^2 \equiv \sum_k \left|a_k^{(1)}(t)\right|^2 = \frac{1}{\hbar^2}\int_{-\infty}^{\infty} \left|\hat{H}'_{km}\right|^2 \frac{\sin^2[(\omega_a-\omega)t/2]}{[(\omega_a-\omega)/2]^2}\varrho(\omega_a)d\omega_a, \qquad (13.55)$$

where $\varrho(\omega_a)$ is the density of final states between ω_a and $\omega_a + d\omega_a$. In the long time limit, the integrand of (13.55) is primarily dictated by the square of the sinc function $(\sin\xi/\xi)^2$ with $\xi = (\omega_a-\omega)t/2$, as clearly shown in Figure 13.2. Under this condition, the integration can be accurately carried out by fixing the rest of the factors in the integrand at the peak of the sinc function, namely, at $\omega_a = \omega$. Thus, one finds

$$\left|a_{\{k\}}^{(1)}\right|^2 = \frac{2\pi}{\hbar^2}\left|\hat{H}'_{km}\right|^2 \varrho(\omega)t \qquad (13.56)$$

using the formula

$$\int_{-\infty}^{\infty} d\xi(\sin\xi/\xi)^2 = \pi. \qquad (13.57)$$

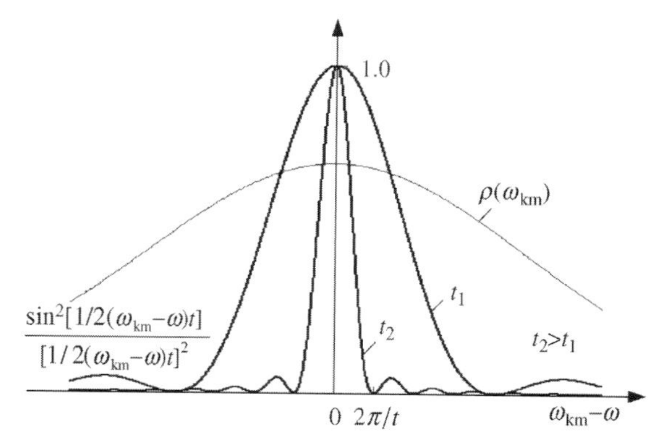

Figure 13.2 The density of states, ϱ, and sinc function versus $\omega_{km}-\omega$. The sinc function dictates the behavior of the integrand in (13.55) in the long time limit.

It is interesting to observe that the transition probability from the initial state to the quasi-continuous final states grows linearly in time, so that one can introduce the transition rate as

$$\frac{d}{dt}\left|a_{\{k\}}^{(1)}(t)\right|^2 = W_{m\to k}, \tag{13.58a}$$

with

$$W_{m\to k} = \frac{2\pi}{\hbar^2}\left|\hat{H}'_{km}\right|^2 \varrho(\omega) = \frac{2\pi}{\hbar}\left|\hat{H}'_{km}\right|^2 \varrho(|E_m - E_k| = \hbar\omega). \tag{13.58b}$$

Here, the two densities of states, ϱ, entering in the transition rate in (13.58b) are related by definition to each other via $\varrho(\omega)d\omega \equiv \varrho(E)dE$, that is, $\varrho(\omega) = \hbar\varrho(E)$. Equation (13.58) is the celebrated Fermi's golden rule.

The golden rule is often expressed in terms of the δ-function by using the representation

$$\delta(\omega_a) = \lim_{t\to\infty}\frac{2}{\pi}\frac{\sin^2(\omega_a t/2)}{t\omega_a^2} = \lim_{t\to\infty}\frac{t}{2\pi}\left[\frac{\sin(\omega_a t/2)}{\omega_a t/2}\right]^2. \tag{13.59}$$

This representation of the delta function is readily seen to satisfy the basic properties of the δ-function, namely, that (i) it goes to infinity for $t\to\infty$ at $\omega_a\to 0$, since $\operatorname{sinc} x\to 1$ for $x\to 0$ and (ii) it yields unity upon integrating over ω_a from $-\infty$ to $+\infty$, as clearly seen from (13.57). Thus, combining (13.54) and (13.59), one can write

$$\left|a_k^{(1)}(t)\right|^2 = \frac{2\pi t\left|\hat{H}'_{km}\right|^2}{\hbar^2}\delta(\omega_a - \omega) = \frac{2\pi t\left|\hat{H}'_{km}\right|^2}{\hbar}\delta(|E_k - E_m| - \hbar\omega), \tag{13.60}$$

with the relationship $\delta(\omega) = \delta(E)/\hbar$. Hence, the transition rate can also be expressed in terms of the δ-function as

$$W_{m\to k} = \frac{2\pi\left|\hat{H}'_{km}\right|^2}{\hbar^2}\delta(\omega_a - \omega) = \frac{2\pi\left|\hat{H}'_{km}\right|^2}{\hbar}\delta(|E_k - E_m| - \hbar\omega). \tag{13.61}$$

The Fermi's golden rule is extensively utilized in a number of interesting problems and has thus been derived in detail. In the derivation, a few assumptions were made and it thus behooves to point out the limits of its validity. First, the integration over the quasi-continuous final states in (13.55) was carried out in the long time limit so that

$$2\pi/t \ll \Delta\omega_a = 2\pi\Delta\nu_a, \tag{13.62}$$

where $\Delta\omega_a$ is the atomic line width shown in Figure 13.2. Also, the transition rate was derived, using the results of the first-order perturbation analysis. Hence, in order for the result to be valid, $\left|a_k^{(1)}\right| \ll 1$, and this condition can be explicitly stated from (13.54) as

$$\left|a_k^{(1)}(t)\right| \simeq \frac{\hat{H}'_{km}}{\hbar}\frac{\sin[(\omega_a - \omega)t/2]}{(\omega_a - \omega)/2} \approx \frac{\hat{H}'_{km}t}{\hbar} \ll 1. \tag{13.63}$$

Combining (13.62) and (13.63), the limits of validity are shown to be

$$\frac{\left|\hat{H}'_{km}\right|}{\hbar} \ll \frac{1}{t} \ll \Delta\nu_a. \tag{13.64}$$

13.3 Problems

13.1. The Hamiltonian of one-dimensional anharmonic oscillator is given by

$$\hat{H} = -\frac{1}{2m}\frac{\partial^2}{\partial x^2} + \frac{1}{2}k_1x^2 + k_2x^3 + k_3x^4, \quad k_1 \gg k_2, k_3.$$

Find the corrections in ground and first excited state energy level and corresponding wavefunctions to the first order of perturbation analysis.

13.2. Consider an isotropic two-dimensional harmonic oscillator, weakly coupled via a perturbing Hamiltonian:

$$\hat{H} = -\frac{1}{2m}\frac{\partial^2}{\partial x^2} + \frac{1}{2}kx^2 + -\frac{1}{2m}\frac{\partial^2}{\partial y^2} + \frac{1}{2}ky^2 + \hat{H}', \quad \hat{H}' = Cxy.$$

(a) Write down the eigenfunctions and energy eigenvalues of this 2D oscillator in the absence of the perturbing Hamiltonian $\hat{H}'$

(b) Evaluate the corrections in the total energy level of this oscillator for the ground and first excited states, arising from the perturbing term $\hat{H}'$ up to the second-order perturbation analysis.

(c) Introduce a set of new coordinates

$$\xi = x + y, \quad \eta = x - y$$

and express the total Hamiltonian in terms of these new variables.

(d) Discuss the motion of the oscillator in ξ and η and compare the result with those obtained from the perturbation analysis.

13.3. Consider an electron in infinite square well potential of width W in which an electric field E is applied, as shown in Figure 13.3.

(a) Find the potential acting on the electron in the well.

(b) Calculate the first- and second-order corrections in energy levels in the ground and first excited states.

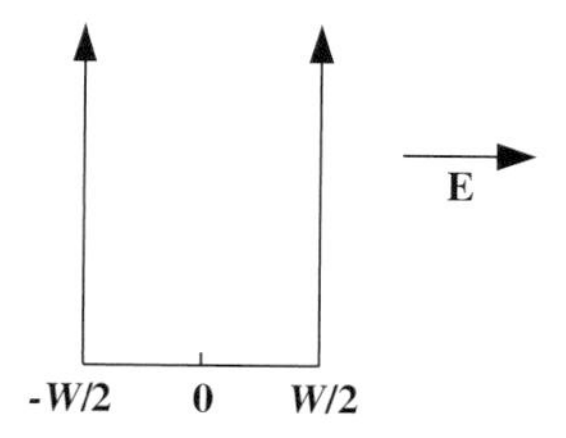

Figure 13.3 An electron in an infinite square well potential in the presence of electric field.

13.4. A hydrogenic atom is placed in an electric field varying harmonically in time, so that the perturbing Hamiltonian is given by

$$\hat{H}' = ez\mathrm{E}_0 \cos \omega t,$$

where $z = r \cos \theta$ is the z-component of the displacement $\boldsymbol{r}$ of the electron from the nucleus.

(a) Given an eigenfunction u_{nlm}, find eigenfunctions that are connected to u_{nlm} via $\hat{H}'$, that is,

$$\left\langle u_{nlm} \middle| \hat{H}' \middle|, u_{n'l'm'} \right\rangle \neq 0. \tag{13.65}$$

The conditions required for quantum numbers l' m' in $u_{n'l'm'}$ to satisfy (13.65) is called the selection rule.

(b) Determine the selection rule for the perturbing Hamiltonian, assuming $|E_n - E_{n'}| \approx \hbar\omega$.

$$\hat{H}' = exE_0 \cos \omega t.$$

13.5. Consider a charged one-dimensional oscillator with the charge to mass ratio q/m. The oscillator is prepared at $t = 0$ in nth eigenstate. A harmonic electric field

$$\mathrm{E}(t) = \mathrm{E}_0 \cos \omega_0 t$$

is applied at $t = 0$.

(a) Write down the perturbing Hamiltonian.

(b) Evaluate the matrix element

$$\left\langle u_n \middle| \hat{H}' \middle| u_{n'} \right\rangle$$

and specify the final states connected to the initial state.

(c) Find the probability that the oscillator makes the transition to those connected final states at $t = \pi/\omega_0$ for $|E_n - E_{n'}| \approx \hbar\omega_0$.

13.6. Consider a circularly polarized light

$$\mathbf{E} = \mathrm{E}_0(\hat{x} \cos \omega t + \hat{y} \sin \omega t)$$

interacting with hydrogen atom initially prepared in u_{nl0}. Find the states $u_{n'l'm'}$ that can be connected to the initial state for both $n' > n$ and $n' < n$, assuming $|E_n - E_{n'}| \approx \hbar\omega$.

Suggested Reading

1 Singh, J. (1996) *Quantum Mechanics, Fundamentals & Applications to Technology*, John Wiley & Sons, Inc.

2 Yariv, A. (1982) *An Introduction to Theory and Applications of Quantum Mechanics*, John Wiley & Sons, Inc.

3 Gasiorowicz, S. (2003) *Quantum Physics*, 3rd edn, John Wiley & Sons, Inc.

4 Liboff, R.L. (2002) *Introductory Quantum Mechanics*, 4th edn, Addison-Wesley Publishing Company, Reading, MA.

14
Atom–Field Interaction

The interaction between the field and matter is one of the most important processes occurring in nature and is discussed herein by first considering a single atom interacting with the field. Both the semiclassical and quantum treatments of the interaction are considered. In the former, the atom is described quantum mechanically, while the field is treated classically. In the latter, both the atom and the field are treated quantum mechanically. The concept of the stimulated versus spontaneous emission of radiation is highlighted in conjunction with the Einstein A coefficient and field quantization. Also, the driven two-level atom is discussed, together with the Rabi flopping formula.

14.1
Field Quantization

Field Energy Density As discussed in Chapter 1, the traveling electromagnetic (EM) wave is described by

$$\mathbf{E}(\boldsymbol{r}, t) = \mathbf{E}_0 e^{-i(\omega t - \boldsymbol{k}\cdot\boldsymbol{r})}, \tag{14.1}$$

where the frequency and the wave vector satisfy the dispersion relation (see (1.50))

$$k = \frac{2\pi}{\lambda} = \omega\sqrt{\mu\varepsilon} = \frac{\omega}{c/n}. \tag{14.2}$$

In a medium free of charge, the Coulomb's law reads as

$$\nabla \cdot \boldsymbol{D} = \varepsilon \nabla \cdot \mathbf{E} = 0$$

(see (1.42)) and when combined with (14.1), it leads to

$$\nabla \cdot \mathbf{E} = i\boldsymbol{k} \cdot \mathrm{E} = 0. \tag{14.3}$$

Thus, $\boldsymbol{k}$ is shown perpendicular to $\mathbf{E}$.

Also, the EM wave is inherently associated with the power, which is built into the coupled electric and magnetic fields in Maxwell's equation. For the sourceless case,

Introductory Quantum Mechanics for Semiconductor Nanotechnology. Dae Mann Kim

ISBN: 978-3-527-40975-4

the coupled equations read as

$$\nabla \times \mathbf{E} = -\mu \dot{\boldsymbol{H}}, \tag{14.4a}$$

$$\nabla \times \boldsymbol{H} = \varepsilon \dot{\mathbf{E}} \tag{14.4b}$$

(see (1.40) and (1.41)). Inserting (14.1) into (14.4a) and performing the curl operation on the left-hand side and the time integration on the right-hand side, $\boldsymbol{H}$ is found as

$$\boldsymbol{H} = \frac{i\boldsymbol{k} \times \mathbf{E}}{i\omega\mu} = \frac{1}{\eta}\hat{k} \times \mathbf{E}, \quad \eta \equiv \sqrt{\frac{\mu}{\varepsilon}}, \ \hat{k} = \left(\frac{\boldsymbol{k}}{k}\right), \tag{14.5}$$

where $\hat{k}$ is the dimensionless unit vector in the direction of the wave vector and the dispersion relation (14.2) has been used and η denotes the impedance of the medium. It is therefore clear from (14.3), (14.5), and the definition of the vector product that $\mathbf{E}$, $\boldsymbol{k}$, and $\boldsymbol{H}$ are mutually orthogonal.

The power in the EM wave is obtained by performing the dot product of (14.4b) and (14.4a) with $\mathbf{E}$ and $\boldsymbol{H}$, respectively, and by subtracting the latter from the former:

$$\mathbf{E} \cdot \nabla \times \boldsymbol{H} - \boldsymbol{H} \cdot \nabla \times \mathbf{E} = \varepsilon \mathbf{E} \cdot \dot{\mathbf{E}} + \mu \boldsymbol{H} \cdot \dot{\boldsymbol{H}}. \tag{14.6}$$

Thus, using the identities

$$\mathbf{E} \cdot \nabla \times \boldsymbol{H} - \boldsymbol{H} \cdot \nabla \times \mathbf{E} \equiv -\nabla \cdot \mathbf{E} \times \boldsymbol{H},$$

$$\varepsilon \mathbf{E} \cdot \dot{\mathbf{E}} + \mu \boldsymbol{H} \cdot \dot{\boldsymbol{H}} \equiv \frac{d}{dt}\left[\frac{\varepsilon \mathbf{E} \cdot \mathbf{E}}{2} + \frac{\mu \boldsymbol{H} \cdot \boldsymbol{H}}{2}\right],$$

(14.6) is recast into a form representing the conservation of the field energy:

$$-\nabla \cdot (\mathbf{E} \times \boldsymbol{H}) \equiv -\nabla \cdot \boldsymbol{P} = \frac{d}{dt}\left[\frac{\varepsilon \mathbf{E} \cdot \mathbf{E}}{2} + \frac{\mu \boldsymbol{H} \cdot \boldsymbol{H}}{2}\right], \quad \boldsymbol{P} \equiv \mathbf{E} \times \boldsymbol{H}. \tag{14.7a}$$

Here, the right-hand side represents the energy density residing in electric and magnetic fields and the Poynting vector $\boldsymbol{P}$ denotes the power crossing per unit area. It is therefore clear that (14.7a) represents the conservation of energy, namely, that the field energy in a volume element grows in time according to the net input power flowing into the volume element. Since $\varepsilon \mathrm{E}^2 = \mu H^2$ from (14.5), the electric and magnetic energy densities are the same and the Poynting vector

$$P \equiv \mathrm{E}^2/\sqrt{\mu/\varepsilon} = \varepsilon \mathrm{E}^2/\sqrt{\varepsilon\mu} = \varepsilon \mathrm{E}^2 (c/n) \tag{14.7b}$$

can therefore be represented by the total EM field energy density propagating with the velocity of light in the medium, c/n.

14.1.1
One-Dimensional Resonator and Its Eigenmodes

The one-dimensional cavity resonator consists of two parallel metal plates, L distance apart, say in the z-direction (see Figure 14.1), and EM waves are traversing back and forth in the resonator. In this case, the electric field can be taken, for example, as

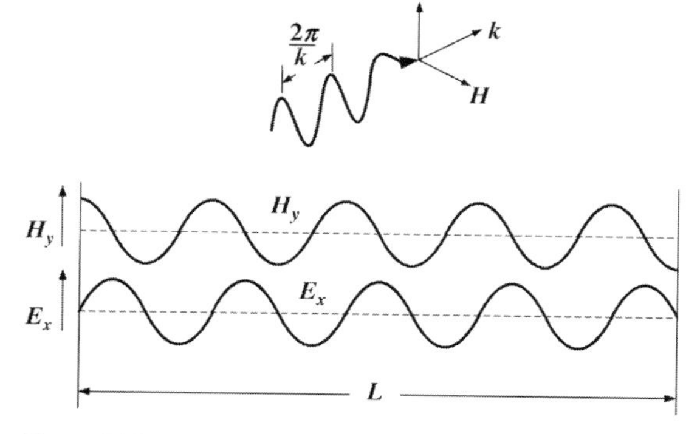

Figure 14.1 A cavity resonator in which **E** and ***H*** forming the Poynting vector (power/unit area) bounces back and forth between two metal plates.

y-polarized standing wave, represented by

$$\mathbf{E}_l = \hat{y}\sqrt{\frac{2}{V\varepsilon}}p_l(t)\sin k_l z, \tag{14.8}$$

where V is the volume of the resonator and l is a positive integer denoting the lth standing wave mode. The boundary condition of the standing wave mode is given by $k_l L = l\pi$, so that $\mathbf{E}_l$ vanishes at the metal surfaces at $z = 0, L$. The time dependence of $\mathbf{E}_l$ is accounted for by $p_l(t)$. With $\mathbf{E}_l$ thus specified, $\boldsymbol{H}_l$ can be represented by

$$\boldsymbol{H}_l = \hat{x}\sqrt{\frac{2}{V\mu}}q_l(t)\omega_l \cos k_l z, \tag{14.9}$$

with $q_l(t)$ accounting for time dependence of $\boldsymbol{H}_l$.

The consistency in representing $\mathbf{E}_l$ and $\boldsymbol{H}_l$ by (14.8) and (14.9), respectively, is seen by inserting them into (14.4a) and performing the operations entailed:

$$\begin{aligned} \nabla \times \mathbf{E}_l &= -\hat{x}\sqrt{\frac{2}{V\varepsilon}}k_l p_l(t)\cos k_l z \\ &= -\hat{x}\mu\sqrt{\frac{2}{V\mu}}\omega_l \dot{q}_l \cos k_l z. \end{aligned} \tag{14.10}$$

Hence, in view of the dispersion relation of (14.2), namely, $k = \omega\sqrt{\mu\varepsilon}$, one finds

$$p_l(t) = \dot{q}_l(t). \tag{14.11}$$

By the same token, when (14.8) and (14.9) are substituted into (14.4b), it yields

$$\begin{aligned} \nabla \times \boldsymbol{H}_l &= \hat{y}(-k_l)\sqrt{\frac{2}{V\mu}}\omega_l q_l(t)\sin k_l z \\ &= \hat{y}\varepsilon\sqrt{\frac{2}{V\varepsilon}}\dot{p}_l(t)\sin k_l z \end{aligned} \tag{14.12}$$

and again in view of the dispersion relation of (14.2), there results

$$\dot{p}_l(t) = -\omega_l^2 q_l(t). \tag{14.13}$$

It thus follows from (14.11) and (14.13) that p_l and q_l are uncoupled as

$$\ddot{q}_l(t) = \dot{p}_l = -\omega_l^2 q_l(t), \tag{14.14a}$$

$$\ddot{p}_l(t) = -\omega_l^2 \dot{q}_l = -\omega_l^2 p_l(t), \tag{14.14b}$$

and are shown to obey the harmonic oscillator equation. The analogy existing between the lth mode field and the harmonic oscillator becomes more apparent, when the energy associated with the lth mode field in the cavity is considered:

$$\hat{H}_l = \int dr \left(\frac{\varepsilon \mathbf{E}_l \cdot \mathbf{E}_l}{2} + \frac{\varepsilon \boldsymbol{H}_l \cdot \boldsymbol{H}_l}{2} \right). \tag{14.15a}$$

Inserting (14.8) and (14.9) into (14.15a) and performing the integration, one obtains

$$\begin{aligned} \hat{H}_l &= \int_0^L Adz \left[\frac{1}{AL} \sin^2(k_l z) p_l^2(t) + \frac{1}{AL} \omega_l^2 \cos^2(k_l z) q_l^2(t) \right] \\ &= \frac{1}{2} p_l^2(t) + \frac{1}{2} \omega_l^2 q_l^2(t), \end{aligned} \tag{14.15b}$$

where A is the cross-sectional area of the resonator and $V = AL$. In performing the integration, the boundary condition $k_l L = l\pi$ has been used, together with the identities $\cos^2 x = (1+\cos 2x)/2$ and $\sin^2 x = (1-\cos 2x)/2$.

Thus, the energy residing in the lth mode, $\hat{H}_l$, in (14.15) is identical to the Hamiltonian of the harmonic oscillator with unit mass, that is, $m = 1$, and the characteristic frequency ω_l. Moreover, $p_l(t)$ and $q_l(t)$ are canonically conjugate variables obeying Hamilton's equation of motion, namely,

$$\dot{q}_l = \frac{\partial}{\partial p_l} \hat{H}_l = p_l, \tag{14.16a}$$

$$\dot{p}_l = -\frac{\partial}{\partial q_l} \hat{H}_l = -\omega_l^2 q_l. \tag{14.16b}$$

Field Quantization The quantum treatment of the EM field, namely, the quantization of the electromagnetic field, simply consists of imposing the basic assumption that

$$[q_l, p_{l'}] = i\hbar \delta_{ll'}, \tag{14.17}$$

while all other combinations of q_l and p_l are taken to commute. Implicit in (14.17) is the fact that p_l and q_l are taken as operators.

It has become clear that there is an apparent similarity existing between a resonator standing wave mode of frequency ω_l and the harmonic oscillator having the same frequency. One can therefore introduce the raising and lowering operators and carry out the operator treatment of the field by repeating exactly the same procedures as was done in the analysis of the harmonic oscillator (see (8.57) and (8.58)):

$$\begin{pmatrix} a_l \\ a_l^+ \end{pmatrix} \equiv \frac{\alpha_l}{\sqrt{2}} q_l \pm i \frac{1}{\sqrt{2}\hbar\alpha_l} p_l$$
$$= \left(\frac{1}{2\hbar\omega_l}\right)^{1/2} (\omega_l q_l \pm i p_l), \quad \alpha_l \equiv \left(\frac{\omega_l}{\hbar}\right)^{1/2}. \tag{14.18}$$

Thus, in view of (8.59), these operators satisfy the commutation relation

$$[a_l, a_{l'}^+] = \delta_{ll'}. \tag{14.19}$$

Also, inverting (14.18), one can write

$$q_l = \left(\frac{\hbar}{2\omega_l}\right)^{1/2} (a_l^+ + a_l), \tag{14.20a}$$

$$p_l = i\left(\frac{\hbar\omega_l}{2}\right)^{1/2} (a_l^+ - a_l), \tag{14.20b}$$

and express $\hat{H}_l$ in terms of the number operator $a_l^+ a_l$ as

$$\hat{H}_l = \hbar\omega_l \left(a_l^+ a_l + \frac{1}{2} \right). \tag{14.21}$$

In (14.21), the commutation relation (14.19) has been used (see (8.62)).

Since it is evident from (14.21) that $\hat{H}_l$ commutes with $a_l^+ a_l$, the eigenfunction u_l of the harmonic oscillator can also be used as the eigenfunction of $a_l^+ a_l$, and one can write from (8.65)

$$a_l{}^+ a_l |n_l\rangle = n|n_l\rangle, \quad |n_l\rangle \equiv |u_l\rangle. \tag{14.22}$$

When (14.22) and (14.21) are combined together, it becomes clear that the quantum number n carries the obvious meaning of the number of photons, each carrying the basic quantum of energy $\hbar\omega_l$. Moreover, the eigenstate $|u_l\rangle$ is to be interpreted as the quantum state, consisting of n such photons. In such context, the role of the raising operator as derived in (8.64), namely,

$$a_l^+ |n_l\rangle = \sqrt{n+1}|n_l + 1\rangle, \tag{14.23}$$

simply consists of creating one photon, thereby transforming the n photon state u_n to $(n+1)$ photon state u_{n+1}. Likewise, the role of the lowering operator in (8.63),

$$a_l|n_l\rangle = \sqrt{n}|n_l - 1\rangle, \tag{14.24}$$

is to annihilate one photon, transforming u_n to u_{n-1}. Hence, the operators a_l^+ and a_l are also called the creation and annihilation operators.

Furthermore, it is obvious from (14.15) that the total Hamiltonian of the field in the resonator is given by summing over standing wave modes therein:

$$\hat{H} = \sum_{l=1}^{\infty} \hat{H}_l. \tag{14.25}$$

Therefore, the wavefunction of the total field in the cavity is given by the product of the oscillator eigenfunctions, namely,

$$|\psi_f\rangle = |n_1, n_2, \ldots n_l, \ldots\rangle, \quad |n_j\rangle = |u_j(\boldsymbol{r})\rangle \tag{14.26}$$

so that

$$\hat{H}|\psi_f\rangle = \sum_{l=1}^{\infty} \hbar\omega_l(n_l + 1/2)|\psi_f\rangle. \tag{14.27}$$

Thus, the energy residing in the field is represented by a denumerable, infinite set of harmonic oscillator eigenenergies.

Operator Representation of the Field To describe the field quantum mechanically, it is necessary to consider the evolution in time of the operators a_l^+ and a_l. For the former, one can write from (14.18) and (14.16)

$$\begin{aligned} \dot{a}_l^+ &= \left(\frac{1}{2\hbar\omega_l}\right)^{1/2} (\omega_l \dot{q}_l - i\dot{p}_l) \\ &= i\left(\frac{1}{2\hbar\omega_l}\right)^{1/2} \omega_l(\omega_l q_l - ip_l) \equiv i\omega_l a_l^+ . \end{aligned} \tag{14.28}$$

Hence, a simple time integration of (14.28) yields

$$a_l^+(t) = a_l^+(0)e^{i\omega_l t}. \tag{14.29}$$

One can similarly find

$$a_l(t) = a_l(0)e^{-i\omega_l t}. \tag{14.30}$$

Hence, inserting (14.20) into (14.8) and (14.9), together with (14.29) and (14.30), one can represent the lth standing wave mode in the resonator as

$$\mathbf{E}_l = \hat{y}i\sqrt{\frac{\hbar\omega_l}{V\varepsilon}}\left[a_l^+(t) - a_l(t)\right]\sin k_l z, \tag{14.31a}$$

$$\boldsymbol{H}_l = \hat{x}\sqrt{\frac{\hbar\omega_l}{V\mu}}\left[a_l^+(t) + a_l(t)\right]\cos k_l z. \tag{14.31b}$$

By the same token, one can represent the traveling wave modes by combining the factors $\exp \pm i\omega t$ and $\exp \pm i\boldsymbol{k}\cdot\boldsymbol{r}$ in (14.31) and selecting the appropriate mode function:

$$\mathbf{E}_{k\lambda} = ie_{k\lambda}\sqrt{\frac{\hbar\omega_k}{2V\varepsilon}}\left[a_{k\lambda}^{+}(0)e^{i(\omega_k t-\boldsymbol{k}\cdot\boldsymbol{r})} - a_{k\lambda}(0)e^{-i(\omega_k t-\boldsymbol{k}\cdot\boldsymbol{r})}\right], \tag{14.32a}$$

$$\boldsymbol{H}_{k\lambda} = \left(\boldsymbol{e}_{k\lambda}\times\frac{\boldsymbol{k}}{k}\right)\sqrt{\frac{\hbar\omega_k}{2V\mu}}\left[a_{k\lambda}^{+}(0)e^{i(\omega_k t-\boldsymbol{k}\cdot\boldsymbol{r})} + a_{k\lambda}(0)e^{-i(\omega_k t-\boldsymbol{k}\cdot\boldsymbol{r})}\right], \tag{14.32b}$$

where $\boldsymbol{e}_{k\lambda}$ is the polarization vector of the electric field.

14.1.2 The Blackbody Radiation Revisited

It is an opportune time to revisit Planck's blackbody radiation theory in the light of quantized electromagnetic field. As already discussed, the energy density of the field in thermodynamic equilibrium consists of the number of the standing wave modes in the cavity in the frequency interval ν and $\nu + d\nu$, multiplied by the energy residing in those standing waves. In a cubic box of length L, for example, the standing wave modes are given, as discussed in Section 4.2, by

$$\mathrm{E}_{nlm} \propto \sin k_x x \sin k_y y \sin k_z z, \tag{14.33}$$

where $\boldsymbol{k}$ satisfies the usual standing wave boundary conditions, namely,

$$k_j L = l_j\pi, \quad j = x, y, z, \tag{14.34}$$

with l_j denoting a positive integer (see (4.17)). The dispersion relation is then given by

$$k^2 = \omega^2\mu\varepsilon = [2\pi\nu/(c/n)]^2 = \frac{\pi^2}{L^2}(l_x^2 + l_y^2 + l_z^2). \tag{14.35}$$

In the k-space, each lattice point representing a single standing wave mode occupies the volume element $(\pi/L)^3$, as shown in Figure 14.2. Hence, in analogy with the density of states analysis discussed in Section 4.3, one can find the total number of modes in the spherical volume having the radius k by

$$N = 2\frac{[(4\pi k^3/3)/8]}{(\pi/L)^3} = \frac{k^3 V}{3\pi^2} = \frac{8\pi\nu^3 n^3 V}{3c^3}. \tag{14.36}$$

Here, the factor 2 is introduced to account for the two polarizations that every wave vector k can possess. Also, the spherical volume in k-space is divided by 8 due to l_x, l_y, l_z being confined to the positive integers. Finally, (14.35) was used for transcribing k into ν. Thus, the mode density representing the number of modes between k and $k + dk$ per unit volume is given by

$$\varrho_f(k) \equiv \frac{1}{V}\frac{dN}{dk} = \frac{k^2}{\pi^2} \tag{14.37a}$$

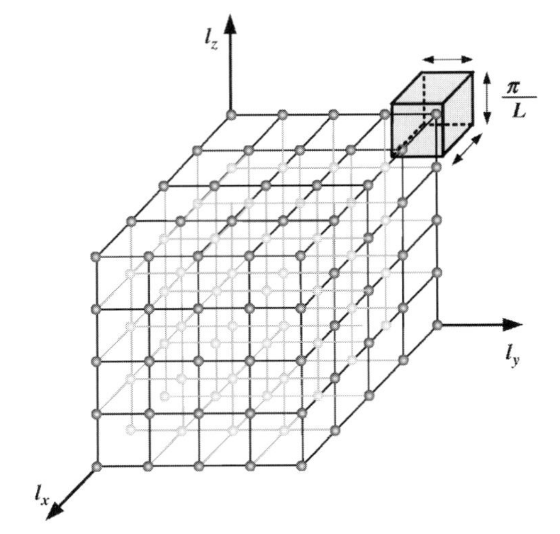

Figure 14.2 The k-space lattice points representing standing wave modes in the cavity.

and the corresponding mode density in frequency domain between ν and $\nu + d\nu$ is given by

$$\varrho_f(\nu) \equiv \frac{1}{V}\frac{dN}{d\nu} = \frac{8\pi n^3 \nu^2}{c^3}. \tag{14.37b}$$

The expressions of the mode density, given in (14.37), are not necessarily limited to the cubic box but are valid for other cavities as well.

Next, the energy residing in the field at the frequency ν is given as in (2.5) by

$$\langle \varepsilon(\nu) \rangle = \sum_{n=0}^{\infty} \varepsilon_n p(n), \tag{14.38a}$$

where

$$p(n) = \frac{e^{-\varepsilon_n/k_B T}}{Z}, \quad Z \equiv \sum_{s=0}^{\infty} e^{-\varepsilon_s/k_B T}, \tag{14.38b}$$

is the Boltzmann probability factor with Z denoting the normalization constant or the partition function and

$$\varepsilon_n = \hbar\omega(n + 1/2) = h\nu(n + 1/2) \tag{14.38c}$$

is the quantized field energy, which consists of n photons, oscillating with frequency ν, and the zero-point energy inherent in the quantum oscillator. In evaluating the average energy of the field, the zero-point energy, $h\nu/2$, should be deleted. This is because it originates purely from the uncertainty principle, as noted earlier, and is not related to the number of photons representing the EM field intensity. In this case, the average energy density given in (14.38a) becomes identical to (2.7) and the spectral energy density is therefore obtained from (14.37) and (2.7) as

$$\varrho(\nu) \equiv \langle \varepsilon(\nu) \rangle \varrho_f(\nu) = \frac{8\pi n^3 h \nu^3}{c^3 (e^{h\nu/k_B T} - 1)} \tag{14.39}$$

in complete agreement with Planck's radiation theory. It has thus become clear that Planck's concept of the quantum of energy is built into the field quantization, which further incorporates the photons as the carriers of the basic quantum of energy.

14.2 Atom–Field Interaction

Semiclassical Treatment of the Field Consider an atom interacting with an electromagnetic field of frequency ω. Given this driving frequency, the atom can interact efficiently with the field via the resonant interaction, if the two energy levels involved are such that $E_2 - E_1 \approx \hbar\omega$. In this case, the coupling of the field with other atomic levels can be neglected and the atom is well approximated by two-level atom, as sketched in Figure 14.3.

In the classical treatment, the electric field is represented by

$$\mathbf{E}(t) = \hat{e}_f \mathrm{E}_0 \cos \omega t = \hat{e}_f \frac{\mathrm{E}_0}{2} (e^{i\omega t} + e^{-i\omega t}), \tag{14.40}$$

where $\hat{e}_f$ is the polarization vector and E_0 is the field amplitude. The wavelength of the field is typically much larger than the atomic dimension and the spatial dependence of the field does not enter in the discussion. The coupling between the atom and the field is well described by the dipole interaction, namely,

$$V = -e\mathbf{E} \cdot \boldsymbol{r}, \tag{14.41}$$

where $-e$ is the electron charge and $\boldsymbol{r}$ is the electron displacement from the nucleus. Hence, the interaction potential can be expressed from (14.40) and (14.41) as

$$\hat{H}' = -\hat{\mu} \frac{\mathrm{E}_0}{2} (e^{i\omega t} + e^{-i\omega t}), \quad \hat{\mu} \equiv e(\hat{e}_f \cdot \boldsymbol{r}). \tag{14.42}$$

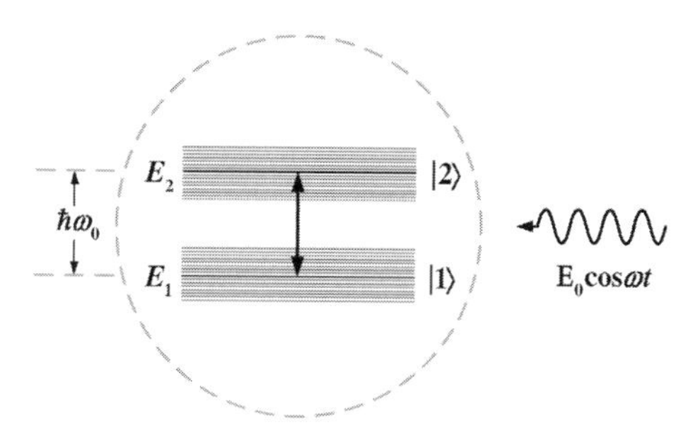

Figure 14.3 A two-level atom driven by harmonic field with frequency ω. The two coupled levels are broadened due to the finite lifetime of the electron therein.

When this dipole interaction Hamiltonian term is used in the Fermi's golden rule, the transition rate between the upper (u_2) and lower (u_1) levels is given by (see (13.61))

$$W_{1\to 2} = W_{2\to 1} = W_i = \frac{2\pi}{\hbar}\left|\hat{H}'_{12}\right|^2\delta(E_2-E_1-\hbar\omega), \quad \left|\hat{H}'_{12}\right|^2 = \frac{\tilde{\mu}^2 E_0^2}{4}, \tag{14.43}$$

where the transition matrix element

$$\tilde{\mu} \equiv e\langle u_1|\hat{e}_f\cdot \boldsymbol{r}|u_2\rangle \tag{14.44}$$

is determined by the energy eigenfunction of the atom under consideration and is also called the atomic dipole matrix. It is important to notice that the rate is the same whether the transition occurs from the upper to the lower level or vice versa.

In a real atomic system, the atomic energy levels E_1 and E_2 are not to be sharply defined but are diffused due to the finite lifetime τ of the electron in each level and the uncertainty relation $\Delta E \approx \hbar/\tau$ (see Figure 14.3). The lifetime of the electron in each level is generally short because of the collisions or the nonradiative transitions. As a result, the difference in energy levels, E_2-E_1, is to be viewed as a random variable and a lineshape factor $g(E_2-E_1)$ has to be introduced to account for the transition from an initial state to the quasi-continuous final manifolds, that is,

$$\begin{aligned} W_i &= \frac{\pi\tilde{\mu}^2 E_0^2}{2\hbar}\int_{-\infty}^{\infty} d(E_2-E_1)g(E_2-E_1)\delta(E_2-E_1-\hbar\omega) \\ &= \frac{\pi\tilde{\mu}^2 E_0^2}{2\hbar}g(\hbar\omega) = \frac{\tilde{\mu}^2 E_0^2}{4\hbar^2}g(\nu). \end{aligned} \tag{14.45}$$

In (14.45), $g(E)dE = g(\nu)d\nu$ has been used.

14.2.1 Stimulated and Spontaneous Transitions

Next, the atom–field interaction is extended to the electromagnetic field interacting with the ensemble of two-level atoms in thermodynamic equilibrium. The thermal equilibrium is generally characterized by a few basic facts. First, the probability of the atom being in E_1 or E_2 level is dictated by the Boltzmann probability factor, as discussed, namely,

$$N_j \propto e^{-E_j/k_B T}, \quad j = 1, 2, \tag{14.46a}$$

so that

$$\frac{N_2}{N_1} = e^{-(E_2-E_1)/k_B T}. \tag{14.46b}$$

Also, the physical quantities are time invariant, due to the fact that every process is balanced by its inverse process. This suggests that the number of atoms making the transition from the upper (2) to lower (1) level, that is, $N_2 W_i$, should be equal to its inverse process, namely, $N_1 W_i$, so that $N_2 W_i = N_1 W_i$. But, this equality is in obvious contradiction with the Boltzmann probability factor, which states that $N_1 > N_2$.

Einstein A Coefficient This apparent but fundamental inconsistency was resolved by Einstein, who introduced a new mode of transition from upper to lower level:

$$W_{2\to 1} = B\varrho(\nu) + A, \tag{14.47}$$

where the first term, proportional to the field energy density $\varrho(\nu)$ (see (14.39), represents the transition that is driven by the field, while the second term A accounts for additional field-independent transition. The inverse transition is given from (14.43) by

$$W_{1\to 2} = B\varrho(\nu). \tag{14.48}$$

Hence, the detailed balancing between the two opposing processes should read as

$$N_2[B\varrho(\nu) + A] = N_1 B\varrho(\nu). \tag{14.49}$$

Equivalently,

$$\frac{N_2}{N_1} = \frac{B\varrho(\nu)}{[B\varrho(\nu) + A]}. \tag{14.50}$$

Substituting (14.46b) for N_2/N_1 into (14.50), together with (14.39) for $\varrho(\nu)$, results in

$$\frac{1}{e^{h\nu/k_B T}} = \frac{1}{1 + \frac{A}{B\varrho(\nu)}} = \frac{1}{1 + \frac{A}{B}\frac{c^3}{8\pi n^3 h\nu^3}\left(e^{h\nu/k_B T} - 1\right)}, \quad E_2 - E_1 = h\nu \tag{14.51}$$

Therefore, the detailed balancing is realized if and only if

$$\frac{A}{B} = \left(\frac{c^3}{8\pi n^3 h\nu^3}\right)^{-1}, \tag{14.52}$$

in which case both sides of (14.51) become identical. The quantity A thus introduced and specified is called the Einstein A coefficient and represents a new mode of transition, namely, the spontaneous transition.

The significance of this A coefficient is clearly seen by considering an ensemble of atoms all prepared in the upper level. In this case, the number of atoms decreases in time even in the absence of the external driving field by this spontaneous transition process:

$$\frac{\partial N_2}{\partial t} = -AN_2. \tag{14.53}$$

Hence, N_2 exponentially falls off in time as

$$N_2(t) = N_2(0)e^{-At}. \tag{14.54a}$$

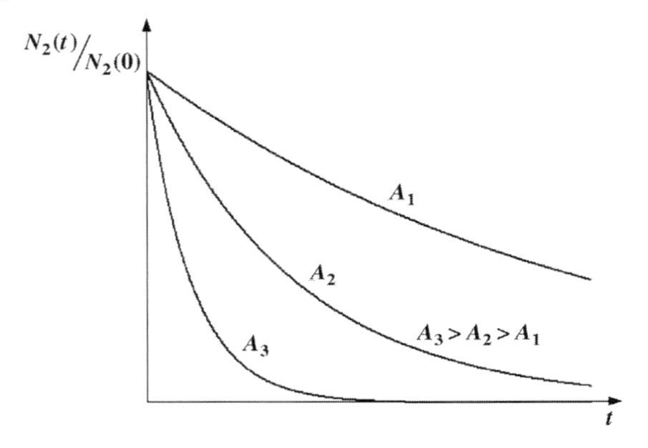

Figure 14.4 The spontaneous decay of the atomic population initially prepared in the upper level. The spontaneous emission rate $\propto \tilde{\mu}^2$ is specified by the atom dipole matrix element and it varies widely, depending on the atomic species. The spontaneous lifetime is given by A^{-1}.

The average time the electron spends in the upper level or equivalently the lifetime of the electron,

$$\tau_{sp} \equiv \frac{\int_0^\infty dt t N_2(t)}{\int_0^\infty dt N_2(t)} = \frac{1}{A}, \tag{14.54b}$$

is therefore determined by A^{-1} (Figure 14.4). Apparently, the spontaneous transition rate A is proportional to B (see (14.52)) and hence to the atomic dipole matrix element $\tilde{\mu}^2$, which is characteristic of the atomic species. Thus, the decay rate due to the spontaneous transition varies sensitively, depending on the atomic or molecular species.

Quantum Treatment of Spontaneous Transition The spontaneous transition as foreseen by Einstein from purely thermodynamic considerations is shown a posteriori a basic and inherent property of the quantized electromagnetic field. This can be seen by considering the dipole interaction between the quantized field and atom, namely,

$$\hat{H}' = -e\mathbf{E}\cdot\boldsymbol{r} = -ie(\hat{e}_{l\lambda}\cdot\boldsymbol{r})\sqrt{\frac{\hbar\omega_l}{2V\varepsilon}}(a_{l\lambda}^{+}(t)e^{-i\boldsymbol{k}\cdot\boldsymbol{r}} - a^{l\lambda}(t)e^{i\boldsymbol{k}\cdot\boldsymbol{r}}), \tag{14.55}$$

where the field has been represented by (14.32a). In this quantum treatment of the field, the two-level atom is coupled, in essence, to a harmonic oscillator with the frequency ω_l and the eigenstate u_l comprised of l photons, as sketched in Figure 14.5.

The transition rate in Fermi's golden rule is then described from (14.43) by

$$W = \frac{2\pi}{\hbar}\frac{e^2\hbar\omega_l}{2V\varepsilon}\left|\langle 1, n_l+1)(\hat{e}_{l\lambda}\cdot\boldsymbol{r})(a_{k\lambda}^{+}e^{-i\boldsymbol{k}\cdot\boldsymbol{r}} - a_{k\lambda}e^{i\boldsymbol{k}\cdot\boldsymbol{r}})|2, n_l\rangle\right|^2 \delta(E_2 - E_1 - \hbar\omega_l). \tag{14.56}$$

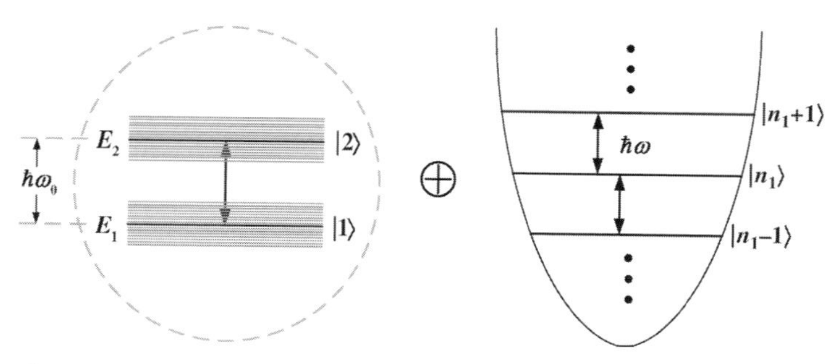

Figure 14.5 The sketch of two-level atom, driven by quantized field. The atom is coupled in essence to a harmonic oscillator.

Here the transition has been confined to the energy-conserving processes. That is, between two composite states, one in which the atom is in the lower level and the field in the state with $(l+1)$ photons, namely, $|1,\ u_{l+1}\rangle$, and the other in which the atom is in the upper level while the field is in the l photon state, namely, $|2,\ u_l\rangle$. The total energies involved are therefore approximately the same, that is, $E_2 + n\hbar\omega_l \approx E_1 + (n+1)\hbar\omega_l$. Using (14.23), (14.24) and the orthonormality of the oscillator eigenfunctions in (14.56), one finds

$$W = \frac{2\pi\omega_l}{2V\varepsilon}\sum_{\lambda=1}^{2}\tilde{\mu}_\lambda^2(n_l+1)\delta(E_2-E_1-\hbar\omega) \equiv W_{\text{ind}} + W_{\text{sp}}, \quad \tilde{\mu}_\lambda \equiv e\langle 1|\hat{e}_{l\lambda}\cdot \boldsymbol{r}|2\rangle \tag{14.57}$$

and the total transition rate W is naturally shown to consist of two terms.

The first term W_{ind} is proportional to the number of photons n_l and, therefore, represents the induced transition. The second term W_{sp} is independent of n_l and accounts for the spontaneous transition. The ratio of these two transitions, $W_{\text{ind}}/W_{\text{sp}} = n_l$, is commensurate with the number of photons present, that is, the field intensity. Since the spontaneous transition does not depend on n_l, one has to sum over the entire modes in the resonator to obtain the total spontaneous transition rate:

$$\begin{aligned} W_{\text{sp}} &\equiv \int_0^\infty d\nu_l\, W_{\text{sp}}^{(l)}\varrho_f(\nu_l)V \\ &= \frac{2\pi\tilde{\mu}^2}{2V\varepsilon}\int_0^\infty d\nu_l\omega_l\frac{8\pi\nu_l^2 n^3}{c^3}V\delta(E_2-E_1-h\nu_l) \\ &= \frac{16\pi^3\tilde{\mu}^2\nu_0^3 n^3}{\varepsilon h c^3}, \quad h\nu_0 = E_2 - E_1. \end{aligned} \tag{14.58}$$

In summing over the radiation modes in (14.58), the mode density given in (14.37) has been used, together with the identity $\varrho(E)hd\nu = \varrho(\nu)d\nu$.

Evidently, this spontaneous transition rate that arises naturally from the field quantization is the quantum mechanical analog of the Einstein A coefficient. The physical equivalence between W_{sp} in (14.58) and the Einstein A coefficient can be seen as follows. First, formally identify W_{sp} in (14.58) with A coefficient in (14.52) and find B, obtaining $B = \tilde{\mu}^2/2\varepsilon\hbar^2$. The induced transition rate in (14.47) then reads as

$$W_{\mathrm{ind}} \equiv B\varrho(\nu) = \frac{\tilde{\mu}^2}{2\varepsilon\hbar^2}\varrho(\nu).$$

The physical significance of this expression can be seen by comparing it with the transition rate W_i in (14.45), which was derived based on the Fermi's golden rule in a harmonically driven atom. From these two transition rates, one finds a relationship between two corresponding quantities involved, namely,

$$\frac{\varepsilon E_0^2}{2} g(\nu) \leftrightarrow \varrho(\nu).$$

Evidently, the left-hand side represents the field energy density at the driving frequency, while the right-hand side denotes the energy density of the radiation field in thermodynamic equilibrium at the same frequency. Obviously, these two energy densities should enter into the respective induced transitions and the resulting relationship is thus entirely consistent. In this way, the A coefficient is shown to be equivalent to W_{sp} and the concept of the spontaneous emission of radiation, which was introduced based purely on the thermodynamic equilibrium considerations is shown as an integral part of the quantized field.

14.3 Driven, Damped Two-Level Atom

Next, the two-level atom interacting with the radiation field is further discussed. There are in general two regimes of interaction between the field and atom, namely, the collisionless and the collision-dominated regimes. In the former, the transition time is much shorter than the mean collision time, while in the latter, the mean collision time is much shorter instead. In this section, a simple two-level atom is singled out for consideration, which is driven by a harmonic field and at the same time is subject to collisions or other nonradiative transitions, as sketched in Figure 14.6. The

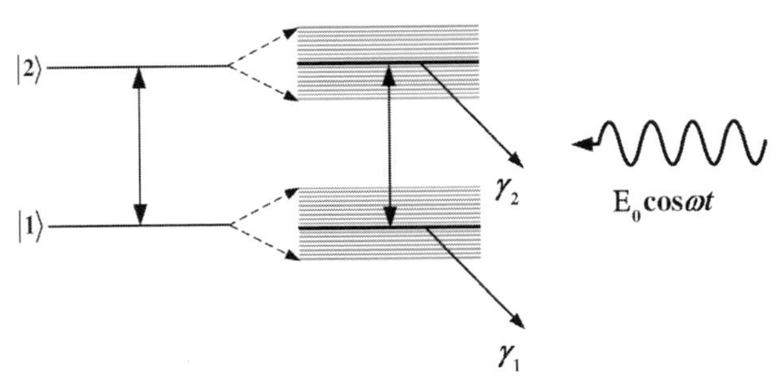

Figure 14.6 A sketch of a two-level atom interacting harmonically with the classical field, while decaying at the same time.

Hamiltonian of the driven two-level atom is given by

$$\hat{H} = \hat{H}_0 + \hat{H}'(t), \tag{14.59a}$$

where

$$\hat{H}'(t) = -\hat{\mu}\mathrm{E}(t), \quad \hat{\mu} \equiv e(\hat{e}_f \cdot \boldsymbol{r}), \quad \mathrm{E}(t) = \mathrm{E}_0 \cos \omega t \tag{14.59b}$$

represents the dipole interaction between the atom and the field (see (14.42)) and the field is treated classically, for simplicity.

The coupling between the atom and the field can be accounted for in a manner similar to the time-dependent perturbation theory, but is simpler in this case because only two levels are involved. The wavefunction of the atom evolves in time via the Schrödinger equation, namely,

$$i\hbar \frac{\partial}{\partial t} \psi(\boldsymbol{r}, t) = \left[\hat{H}_0 + \hat{H}'(t)\right] \psi(\boldsymbol{r}, t), \tag{14.60}$$

and the wavefunction is in this case spanned by two states, that is,

$$\psi(\boldsymbol{r}, t) = a_1(t) e^{-i\omega_1 t} |1\rangle + a_2(t) e^{-i\omega_2 t} |2\rangle, \quad \omega_j = E_j/\hbar, \tag{14.61a}$$

where each state is characterized by

$$\hat{H}_0 |j\rangle = E_j |j\rangle, \quad |j\rangle \equiv |u_j(\boldsymbol{r})\rangle, j = 1, 2. \tag{14.61b}$$

It is important to notice in (14.61) that the expansion coefficients $a_1(t)$ and $a_2(t)$ have been taken time dependent. This is again to account for the interaction Hamiltonian $\hat{H}'(t)$ giving rise to transitions from one level to another, as detailed in Section 13.2. Inserting (14.61) into (14.60) and taking the inner product with respect to $|u_j(\boldsymbol{r})\rangle$, $j = 1, 2$, on both sides, one finds in a manner similar to Section 13.2 (see (13.43)),

$$\dot{a}_1 = i \frac{\tilde{\mu}\mathrm{E}(t)}{\hbar} a_2 e^{-i\omega_0 t}, \quad \omega_0 \equiv \omega_2 - \omega_1, \tag{14.62a}$$

$$\dot{a}_2 = i \frac{\tilde{\mu}\mathrm{E}(t)}{\hbar} a_1 e^{i\omega_0 t}, \tag{14.62b}$$

where ω_0 is the atomic transition frequency. In obtaining (14.62), the orthonormality of the eigenfunctions has been used, and the fact that these eigenfunctions are of the opposite parity, so that the transition matrix element $\tilde{\mu}$ does not vanish, that is,

$$\mu_{jj} \equiv \langle j|\hat{H}'(t)|j\rangle = 0, \quad j = 1, \ 2 \tag{14.62c}$$

$$\tilde{\mu} \equiv \langle 1|\hat{\mu}|2\rangle = e\langle \hat{e}_f \cdot \boldsymbol{r}\rangle. \tag{14.62d}$$

Next, in view of $E(t)$ varying in time as

$$\cos\omega t = (\exp i\omega t + \exp - i\omega t)/2,$$

the time variation of the coupling term consists of two types, namely, $\exp \pm i(\omega \pm \omega_0)t$. In the rotating wave approximation, the rapidly oscillating terms $\exp \pm i(\omega + \omega_0)t$ can be discarded without losing too much accuracy. One could also account for the finite lifetime of the electron in each level by introducing the decay rates γ_1 and γ_2 phenomenologically. Thus, the coupled equations read as

$$\dot{a}_1 = i\Omega a_2 e^{i\Delta t} - \gamma_1 a_1, \tag{14.63a}$$

$$\dot{a}_2 = i\Omega a_1 e^{-i\Delta t} - \gamma_2 a_2, \tag{14.63b}$$

where the quantities

$$\Omega \equiv \frac{\tilde{\mu} E_0}{2\hbar}; \quad \Delta \equiv \omega - \omega_0 \tag{14.63c}$$

denote the transition frequency and the frequency detuning factor, respectively. These coupled equations can be further compacted by introducing a'_j such that

$$a_j = a'_j e^{-\gamma_j t}, \quad j = 1, 2, \tag{14.64}$$

in which case (14.63) reduces to

$$\dot{a}'_1 = i\Omega a'_2 e^{i\Lambda t}; \quad \Lambda \equiv \Delta - i(\gamma_1 - \gamma_2), \tag{14.65a}$$

$$\dot{a}'_2 = i\Omega a'_1 e^{-i\Lambda t}. \tag{14.65b}$$

This coupled equation can be solved in a straightforward manner. For instance, one may look for the solution in the form

$$a'_1 = e^{iSt} \tag{14.66}$$

and insert it into (14.65a), obtaining

$$a'_2 = (S/\Omega) e^{i(S-\Lambda)t}. \tag{14.67}$$

When (14.66) and (14.67) are inserted into (14.65b), a quadratic equation for the unknown S ensues, namely,

$$S^2 - \Lambda S - \Omega^2 = 0,$$

and the resulting two roots

$$S_{\pm} = \frac{1}{2}\left(\Lambda \pm \sqrt{\Lambda^2 + 4\Omega^2}\right) \tag{14.68a}$$

provide the solution for a'_1 and a'_2:

$$a'_1 = A_+ e^{iS_+ t} + A_- e^{iS_- t}, \tag{14.68b}$$

$$a'_2 = \left(\frac{S_+}{\Omega} A_+ e^{iS_+ t} + \frac{S_-}{\Omega} A_- e^{iS_- t} \right) e^{-i\Lambda t}. \tag{14.68c}$$

Here, the two constants of integration, A_+ and A_-, are determined by the given initial conditions. Thus, consider the case in which the atom is prepared initially in the lower level, that is,

$$a'_1(t = 0) = 1, \quad a'_2(t = 0) = 0. \tag{14.69}$$

Also, for simplicity, a nonessential simplification is made, namely, that the two decay rates are the same, that is, $\gamma_1 = \gamma_2 = \gamma$, so that $\Lambda = \Delta$ (see (14.65a). When the initial condition is used in (14.68) together with the definition (14.64), one finds the Rabi flopping formula

$$a_1(t) = e^{-\gamma t} e^{i\Delta t/2} \left[\cos(\omega' t/2) - i \frac{\Delta}{\omega'} \sin(\omega' t/2) \right], \tag{14.70a}$$

$$a_2(t) = e^{-\gamma t} e^{-i\Delta t/2} i \frac{2\Omega}{\omega'} \sin(\omega' t/2), \tag{14.70b}$$

where

$$\omega' = \sqrt{\Delta^2 + 4\Omega^2}. \tag{14.70c}$$

Rabi Flopping Formula In the absence of decay, that is, $\gamma = 0$, and for the resonant interaction $\Delta = 0$, the solutions (14.70) reduce to

$$a_1(t) = \cos \Omega t, \tag{14.71a}$$

$$a_2(t) = i \sin \Omega t \tag{14.71b}$$

and are shown to oscillate with the transition frequency Ω. Also, for $\gamma = 0$, regardless of the values of the detuning factor Δ, it clearly follows from (14.70) or (14.71) that the probability is conserved, namely,

$$|a_1(t)|^2 + |a_2(t)|^2 = 1. \tag{14.72}$$

At resonance, these two probabilities swing completely between 0 and 1 in time, as the coupling between the atom and field becomes maximal (see Figure 14.7). In the process, a photon is exchanged between the atom and the field to conserve energy. But at off-resonance, in which $\Delta \neq 0$, the coupling strength between the atom and field degrades and the amplitude of swing for two probabilities is reduced, while the oscillation frequency of the swing increases (see Figure 14.7).

In the presence of decay ($\gamma \neq 0$), these two probabilities, $|a_1(t)|^2$ and $|a_2(t)|^2$ decay in time and, consequently, the flopping oscillation is damped. In the limit where the

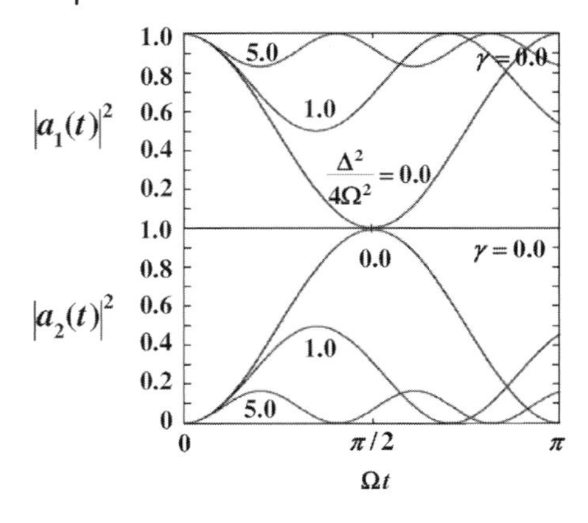

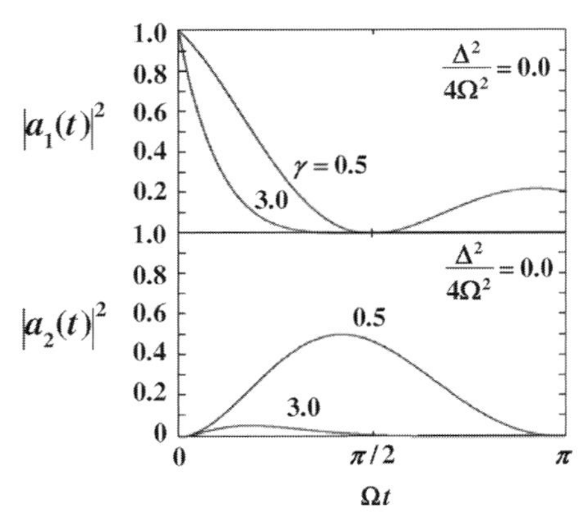

Figure 14.7 The Rabi flopping curves at resonance and off-resonance versus time Ωt without decay (left). With increasing detuning, the flopping amplitude decreases, while the frequency of flopping increases. The flopping curve at resonance versus time Ωt for different decay rates (right). The interaction time in effect is determined by the decay rates.

decay time $1/\gamma$ is much shorter than the flipping period, the change in time of a_1 and a_2 due to flipping becomes small and the atom–field interaction simply yields the probability of a photon being absorbed or emitted (Figure 14.7). For example, for the case under consideration, the probability of the photon being absorbed is given by $|a_2(\gamma^{-1})|^2 \approx \Omega^2/\gamma^2$.

Atom Dipole The Rabi flopping formula (14.70) or (14.71) provides a convenient means by which to understand the atom dipole being induced by the driving field. When the atom is either in upper or lower level, the wavefunction is given by

$$|\psi\rangle = e^{-i\omega_j t}|u_j(\mathbf{r})\rangle, \quad j = 1, 2. \tag{14.73}$$

In this case, the atom does not possess the dipole moment, as clearly follows from the parity considerations, that is,

$$\langle\psi|\hat{\mu}|\psi\rangle = 0, \quad \hat{\mu} \equiv e\hat{e}_f \cdot \mathbf{r}, \tag{14.74}$$

regardless of whether $u_j(\mathbf{r})$ is even or odd.

However, when the atom is driven by the harmonic field, the wavefunction evolves into a linear superposition of upper and lower states, that is,

$$\psi(\mathbf{r}, t) = a_1(t)e^{-i\omega_1 t}|u_1(\mathbf{r})\rangle + a_2(t)e^{-i\omega_2 t}(t)|u_2(\mathbf{r})\rangle, \quad \omega_j = E_j/\hbar. \tag{14.75}$$

Or equivalently one may write

$$\psi(\mathbf{r}, t) = a_{1S}(t)|u_1(\mathbf{r})\rangle + a_{2S}(t)|u_2(\mathbf{r})\rangle, \quad a_{jS}(t) \equiv a_j(t)e^{-i\omega_j t}. \tag{14.76}$$

In (14.75), the time dependence of the wavefunction is divided into the usual built-in component, $\exp -i\omega_j t$, for each level and additional dependence $a_j(t)$ is introduced to account for the interaction Hamiltonian. This representation is known as the interaction picture. In (14.76), the time dependence is relegated entirely to the expansion coefficients and is known as the Schrödinger picture. In either case, the induced atom dipole is described by

$$\langle\psi|\hat{\mu}|\psi\rangle = -\tilde{\mu}[a_1^*(t)a_2(t)e^{-i\omega_0 t} + a_1(t)a_2^*(t)e^{i\omega_0 t}], \quad \tilde{\mu} = \langle u_1(\boldsymbol{r})|\hat{\mu}|u_2(\boldsymbol{r})\rangle \tag{14.77a}$$

or

$$\langle\psi|\,\hat{\mu}|\psi\rangle = -\,\tilde{\mu}[a_{1S}^*(t)a_{2S}(t) + a_{1S}(t)a_{2S}^*(t)] \tag{14.77b}$$

The atom dipole being induced due to the wavefunction evolving into a linear superposition of the two states is illustrated in Figure 14.8. The magnitude of the atom dipole is determined by both the dipole matrix element and $a_1(t)$, $a_2(t)$ and oscillates with a frequency close to ω (see (14.70), (14.71) and (14.75)). Naturally, an oscillating dipole either emits or absorbs the radiation, thereby providing the mechanism for exchanging a photon between the atom and field.

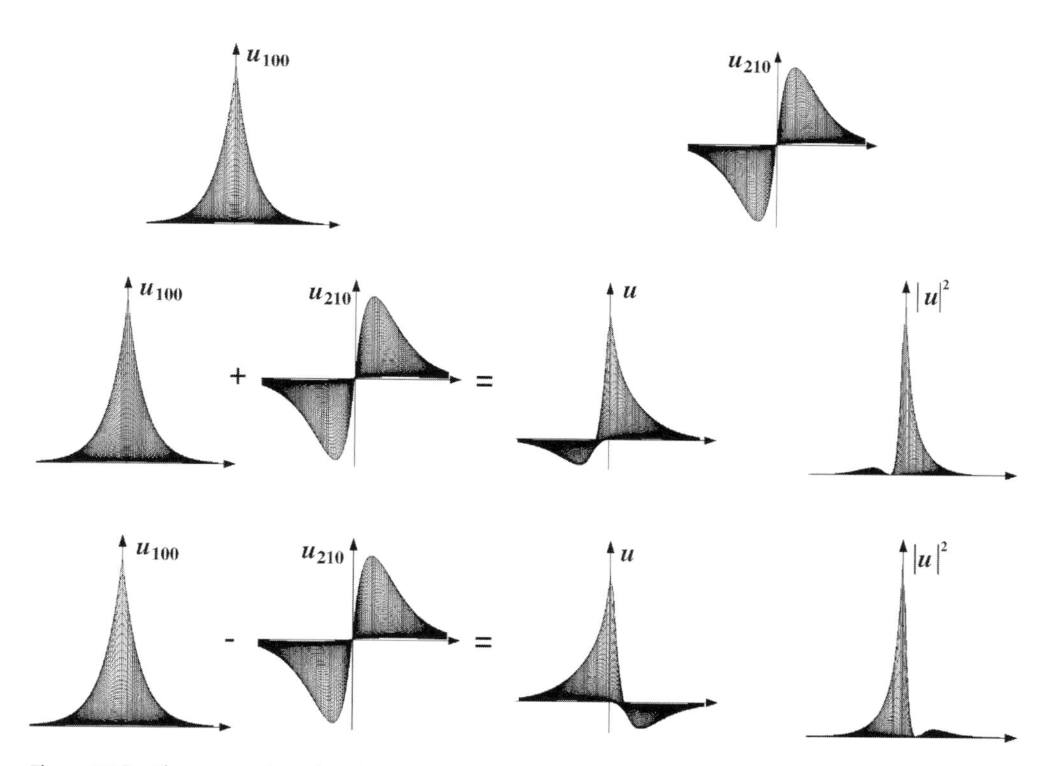

Figure 14.8 The symmetric and anti-symmetric wavefunctions of the ground (u_{100}) and first excited (u_{210}) states of hydrogen atom (top). When superposed linearly with equal amplitudes, the wavefunction varies between two limits ($u_{100} + u_{210}$) (middle) and ($u_{100} - u_{210}$) (bottom), thereby exhibiting oscillating atom dipole.

14.4 Problems

14.1. The spontaneous transition rate W_{sp} or the Einstein A coefficient given in (14.58) depends on the transition matrix element, namely, $\tilde{\mu}^2 = |e\langle 1|\hat{e}_{l\lambda} \cdot \boldsymbol{r}|2\rangle|^2$. The random orientations of atoms in the ensemble or different electric field polarizations make it necessary to average the transition matrix element over x, y, and z directions:

$$\tilde{\mu}^2 = e^2 \frac{1}{3}[\langle 1|x|2\rangle^2 + \langle 1|y|2\rangle^2 + \langle 1|z|2\rangle^2]$$

(a) Estimate the spontaneous lifetime $1/W_{sp}$ for the transition from u_{32m} to u_{21m} in hydrogen atom.

(b) Calculate the spontaneous lifetimes for $|p_x\rangle$, $|p_y\rangle$, and $|p_z\rangle$ states in (10.55) and (10.56), for $n = 2$ and $l = 1$ in hydrogen atom decaying to ground state and compare the results.

14.2. Consider a two-level atom driven by a harmonic field $\mathrm{E}(t) = \mathrm{E}_0 \cos \omega t$. The Hamiltonian is given by

$$\hat{H} = \hat{H}_0 + \hat{H}'(t); \quad \hat{H}'(t) = -\hat{\mu}\mathrm{E}(t), \quad \hat{\mu} \equiv e(\hat{e}_f \cdot \boldsymbol{r}).$$

(a) Starting from the time-dependent Schrödinger equation and using the wavefunction in the interaction picture, that is,

$$\psi(\boldsymbol{r}, t) = a_1(t)e^{-i\omega_1 t}|1\rangle + a_2(t)e^{-i\omega_2 t}|2\rangle, \quad \omega_j = E_j/\hbar,$$

reproduce the coupled equation of motion (14.62) for the expansion coefficients $a_1(t)$ and $a_2(t)$.

(b) Derive the corresponding coupled equation of motion in Schrödinger picture by representing the wavefunction as

$$\psi(\boldsymbol{r}, t) = a_{1S}(t)|1\rangle + a_{2S}(t)|2\rangle.$$

(c) Compare the two coupled equations in (a) and (b) and discuss whether the two sets of coupled equations could describe the same probabilities in each level and the atom dipole moment.

14.3. Consider the coupled equations for a'_1 and a'_2, derived for the driven, damped two-level atom in (14.65a) and (14.65b).

(a) For $\gamma_1 \neq \gamma_2$, show that the roots $S_\pm$ corresponding to (14.68) are given by

$$S_\pm = \frac{1}{2}\{\Delta + i\gamma_{12} \pm [\Delta + i(\gamma_1 - \gamma_2)]^2 + 4\Omega^2\}^{1/2}, \quad \gamma_{12} = \frac{1}{2}(\gamma_1 + \gamma_2)$$

(b) Given the initial conditions $a_1(t = 0) = 1$ and $a_2(t = 0) = 0$, find the solution for $a_1(t)$ and $a_2(t)$.

(c) Compare the results with those given in (14.70), including the case $\gamma_2 \gg \gamma_1$ and $\gamma_2 \ll \gamma_1$.

14.4. The coupled equation for driven two-level atom is given for $\gamma_1 = \gamma_2 = 0$ by

$$\dot{a}_1 = i\Omega a_2 e^{i(\omega-\omega_0)t}, \quad \Omega \equiv \frac{\tilde{\mu} \mathrm{E}_0}{2\hbar}, \ \omega_0 = (E_2 - E_1)/\hbar,$$

$$\dot{a}_2 = i\Omega a_1 e^{-i(\omega-\omega_0)t},$$

where E_0 and ω are the amplitude and frequency of the field (see (14.63).

(a) When the atom is prepared initially in the lower level so that $a_1(0) = 1$ and $a_2(0) = 0$, show that $a_2(t)$ found to the first-order iteration, that is, by putting $a_1 = 1$, is given by

$$a_2(t) = i\Omega e^{i(\omega-\omega_0)t/2} \frac{\sin(\omega-\omega_0)t/2}{(\omega-\omega_0)/2}.$$

(b) Take $E_2 - E_1$ as a distributed quantity arising from the finite lifetimes of two levels and derive the Fermi's golden rule by finding the ensemble averaged transition probability,

$$\int d\omega \varrho(\omega) |a_2(t)|^2; \quad \hbar\omega_0 = E_2 - E_1,$$

with $\varrho(\omega)$ denoting the lineshape factor centered at ω_0.

(c) Using (b), derive the attenuation coefficient of light traversing through an optical medium.

14.5. Consider a lossless transmission line in which the voltage and current are coupled as

$$\frac{\partial V}{\partial z} = -L\frac{\partial I}{\partial t}; \quad \frac{\partial I}{\partial z} = -C\frac{\partial V}{\partial t},$$

where L and C are the inductance and capacitance per unit length along the z-direction.

(a) Show that the equations can be decoupled as

$$\frac{\partial^2 V}{\partial z^2} = \frac{1}{c^2}\frac{\partial^2 V}{\partial t^2}; \quad \frac{\partial^2 I}{\partial z^2} = \frac{1}{c^2}\frac{\partial^2 I}{\partial t^2},$$

with $c^2 \equiv 1/LC$ denoting the velocity of propagation.

(b) Sow that

$$V(z,t) = Z_0 I(z,t) = \left(\frac{\hbar\omega}{2Cz}\right)^{1/2} \left[a e^{-i(\omega t - kz)} + a^+ e^{i(\omega t - kz)}\right], \quad Z_0 \equiv \sqrt{L/C},$$

is the solution with Z_0 representing the characteristic impedance, and $k = \omega/c = \omega\sqrt{LC} = 2\pi/\lambda$ is the propagation constant or the wave vector. Also, z is the length of the transmission line, a and its complex conjugate a^+ are arbitrary constants, and ω is a given frequency.

(c) Assume that the line length is an integer multiple of wavelengths, namely, $z = n\lambda$ and show that the energy contained in the line is expressed as

$$H = \int_0^z dz \left[\frac{1}{2}CV^2 + \frac{1}{2}LI^2\right] = \hbar\omega a^* a,$$

(d) Show that if a and a^+ are defined such that

$$a = \frac{1}{\sqrt{2\hbar\omega}}(\omega q + ip); \quad a^+ = \frac{1}{\sqrt{2\hbar\omega}}(\omega q - ip),$$

the energy can be formally expressed as

$$H = \frac{1}{2}(p^2 + \omega^2 q^2).$$

(e) Show that q and p are canonically conjugate variables.

(f) If it is assumed that $[q, p] = i\hbar$, show that $[a, a^+] = 1$. This is how the LC circuit is quantized.

Suggested Reading

1 Yariv, A. (1982) *An Introduction to Theory and Applications of Quantum Mechanics*, John Wiley & Sons, Inc.

2 Sargent, M.I., Scully, M.O., and Lamb, W.E., Jr. (1978) *Laser Physics*, Westview Press.

15
Interaction Between EM Waves and Optical Media

The atom–field interaction discussed in Chapter 14 is next extended to the general case of the EM waves interacting with ensemble of atoms or molecules. In particular, the absorption and/or gain and the concomitant dispersion of the wave as it propagates in the medium are discussed. This kind of interaction is a most fundamental and important phenomenon and serves as a powerful tool for basic studies in various disciplines. Also, an outstanding example of the utilization of the atom–field interaction is the invention of laser devices. The device is based upon a fundamental concepts in quantum mechanics, for example, the spectroscopy, the population inversion and controlled emission of radiation, and so on. The operational principles of the device are briefly discussed in this chapter. In addition, the density matrix formulation bridging the microscopic world to macroscopic world is discussed.

15.1
Attenuation and Dispersion of Waves

Attenuation In the collision-dominated regime, the interaction time between the atom and the field is much shorter than the transition time. Hence, the atom in this regime is not allowed to undergo the full Rabi flopping but simply ends up with making a transition from one level to another with a certain probability, as discussed in Chapter 14 (see Figure 14.7). Thus, given an ensemble of two-level atoms with N_1, N_2 atoms per unit volume in each level, the number of the induced transitions involved is given by

$$N_{1\to 2} = N_1 W_i, \quad N_{2\to 1} = N_2 W_i, \tag{15.1}$$

where the induced transition rate is given from Fermi's golden rule (14.45) as

$$W_i = \frac{\tilde{\mu}^2 E_0^2 g(\nu)}{4\hbar^2} = \frac{\tilde{\mu}^2 n g(\nu)}{2\hbar^2 c\varepsilon} I_\nu \propto I_\nu. \tag{15.2}$$

Introductory Quantum Mechanics for Semiconductor Nanotechnology. Dae Mann Kim

ISBN: 978-3-527-40975-4

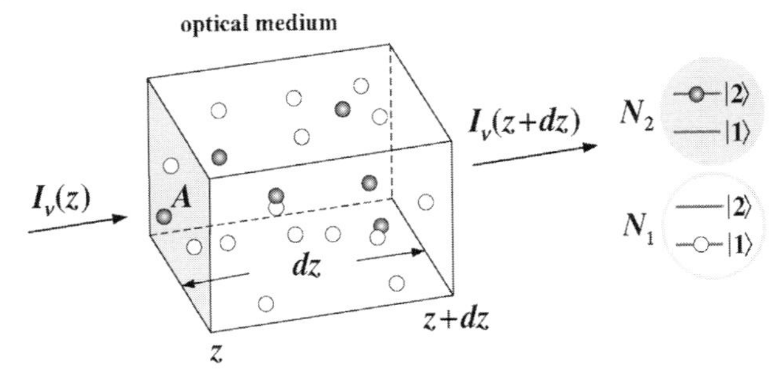

Figure 15.1 The power conservation in a differential volume element. The input power at z is absorbed in part and exits at the output plane at $z + dz$.

Here, W_i has been re-expressed in terms of light intensity

$$I_\nu \equiv (\varepsilon E_0^2/2)(c/n) = cn\varepsilon_0 E_0^2/2$$

and the spontaneous transition is neglected, compared with the induced one.

Then the light incident on a slab at z having the unit cross-sectional area and thickness dz (Figure 15.1) is absorbed in part via the net transition of atoms from lower to upper level, while the remainder exits at the output plane:

$$I_\nu(z + dz) - I_\nu(z) = -(N_1 - N_2)W_i h\nu dz. \tag{15.3}$$

One may recast (15.3) into a differential form by Taylor expanding $I_\nu(z + dz)$ at z and retaining only the first-order term in dz, obtaining with the use of (15.2) for W_i

$$\frac{dI_\nu}{dz} = -\alpha I_\nu, \tag{15.4a}$$

where

$$\alpha \equiv (N_1 - N_2)\frac{\tilde{\mu}^2 n g(\nu)}{2\hbar^2 c\varepsilon} h\nu \tag{15.4b}$$

is called the linear attenuation coefficient and is positive in the absorbing medium in which $N_1 > N_2$. One can readily integrate (15.4), obtaining

$$I_\nu = I_\nu(0)e^{-\alpha z}. \tag{15.5}$$

The intensity of light is thus attenuated exponentially as it traverses through an absorbing medium (Figure 15.2), with α depending sensitively on the dipole matrix elements of the atoms or molecules constituting the medium.

However, when the atomic population in the medium is inverted, that is, $N_1 < N_2$, α becomes negative and is converted into a gain factor and the input light is then amplified, as shown in Figure 15.2:

$$\gamma \propto (N_2 - N_1) = -\alpha. \tag{15.6}$$

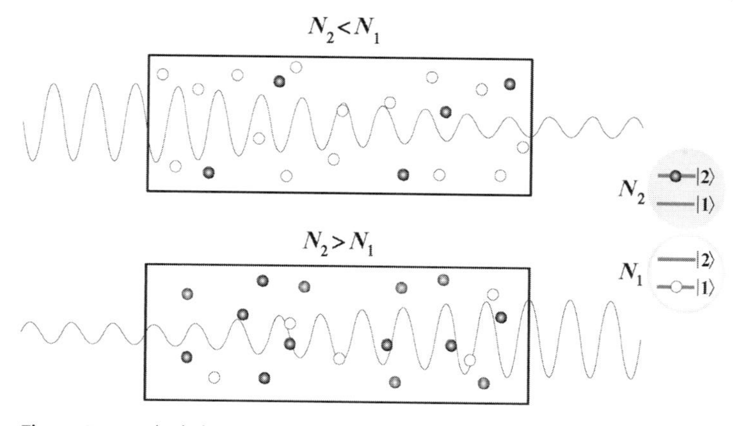

Figure 15.2 The light attenuation and amplification in passive and active media. When $N_1 > N_2$, light is attenuated while for $N_1 < N_2$ it is amplified as more emission ensues than absorption.

Dispersion Concomitant with the absorption or gain, the EM wave undergoes the dispersion as it propagates in the medium. The dispersion occurs because the incident light induces a microscopic dipole out of each atom, as detailed in Section 14.3. These atom dipoles collectively give rise to the macroscopic polarization vector $\boldsymbol{P}$, which in turn acts as the source for the field:

$$\mathbf{E} \to \langle \mu_j \rangle \to \sum_j \langle \mu_j \rangle \to \boldsymbol{P} \to \mathbf{E}.$$

This feedback process can be quantified by tracing the input field as it interacts with the medium. Thus, consider the input field in phasor notation,

$$\mathbf{E} = \mathrm{Re}\mathbf{E}_0 e^{i\omega t}. \tag{15.7}$$

The field induces the polarization vector

$$\boldsymbol{P} = \mathrm{Re}\boldsymbol{P}_0 e^{i\omega t}, \quad \boldsymbol{P}_0 \propto \mathbf{E}_0 \equiv \varepsilon_0 \chi_a \mathbf{E}_0, \tag{15.8a}$$

where the atomic susceptibility as a response function, connecting $\boldsymbol{P}$ to the input field $\mathbf{E}$, is contributed by the ensemble of atom dipoles per unit volume and is in general complex, that is,

$$\chi_a \equiv \chi_a' - i\chi_a''. \tag{15.8b}$$

Hence, one can write

$$\boldsymbol{P} = \mathrm{Re}\varepsilon_0[(\chi_a' - i\chi_a'')\mathbf{E}_0 e^{i\omega t}] = \varepsilon_0 \chi_a' \mathbf{E}_0 \cos \omega t + \varepsilon_0 \chi_a'' \mathbf{E}_0 \sin \omega t. \tag{15.9}$$

Now, the polarization vector is well known to be an inherent part of the displacement vector, that is,

$$\boldsymbol{D} \equiv \varepsilon \mathbf{E} \equiv \varepsilon_0 \mathbf{E} + \boldsymbol{P}, \tag{15.10}$$

so that

$$D_0 \propto \mathbf{E}_0 = \varepsilon_0 \mathbf{E}_0 + \varepsilon_0 \chi_a \mathbf{E}_0 \equiv \varepsilon \mathbf{E}_0. \tag{15.11}$$

And, the resulting permittivity ε or the dielectric constant $\varepsilon_r = \varepsilon/\varepsilon_0$ with ε_0 denoting the vacuum permittivity can be partitioned into the background and the resonant interaction terms, namely,

$$\varepsilon \equiv \varepsilon_0(1+\chi) = \varepsilon_0(1+\chi_b+\chi_a). \tag{15.12}$$

In practice, $1+\chi_b \gg \chi_a$ and (15.12) can further be regrouped in terms of the background susceptibility as

$$\begin{aligned} \varepsilon &\equiv \varepsilon_0(1+\chi_b)\left[1+\frac{\varepsilon_0\chi_a}{\varepsilon_0(1+\chi_b)}\right] \\ &= \varepsilon_b\left\{1+\frac{[\chi_a'(\omega)-i\chi_a''(\omega)]}{n^2}\right\}, \quad \varepsilon_b \equiv \varepsilon_0(1+\chi_b),\ n^2 = \varepsilon_b/\varepsilon_0. \end{aligned} \tag{15.13}$$

Hence, the input wave concomitantly undergoes the absorption and/or gain and the dispersion, which is described by the complex wave vector, given by

$$\begin{aligned} k &= \omega\sqrt{\mu\varepsilon} = \omega\sqrt{\mu\varepsilon_b\{1+[\chi_a'(\omega)-i\chi_a''(\omega)]/n^2\}} \\ &= k_b[1+\chi_a'(\omega)/2n^2]-ik_b\chi_a''(\omega)/2n^2, \quad k_b \equiv \omega\sqrt{\mu\varepsilon_b}. \end{aligned} \tag{15.14}$$

In (15.14), use has been made of

$$(1+x)^{1/2} \simeq 1+x/2, \quad x \equiv (\chi_a'(\omega)-i\chi_a''(\omega))/n^2 \ll 1.$$

It is thus clear from (15.14) that the input wave is attenuated or amplified depending on whether $N_1 > N_2$ or $N_1 < N_2$ and dispersed concomitantly:

$$\begin{aligned} E(z,t) &= \mathrm{Re}E_0 e^{i(\omega t - kz)} \\ &= \mathrm{Re}E_0 e^{i\{\omega t - k_b z[1+\chi_a'(\omega)/2n^2]\} - k_b z\chi_a''(\omega)/2n^2}. \end{aligned} \tag{15.15}$$

It is clear from (15.15) that the real part of the susceptibility affects the phase dispersion, while the imaginary part accounts for the power exchange between the field and medium.

15.2 Density Matrix and Ensemble Averaging

Now that the phenomenology of the absorption, amplification, and dispersion of the wave has been discussed, the ensemble of atom dipoles giving rise to the macroscopic quantities such as the permittivity is addressed to. To this end, the convenient starting point is to recall that the dynamics of driven two-level atoms are conveniently characterized by the expansion coefficients of the two levels involved in the resonant interaction. Specifically, the quantities

$$a_j^*(t)a_j(t) \equiv a_{sj}^*(t)a_{sj}(t); \quad a_j e^{-i\omega_j t} \equiv a_{sj}, \quad j = 1, 2, \tag{15.16}$$

represent the probabilities of finding the electron in the jth level at t and a_j and a_{sj} denote the respective expansion coefficients in the interaction and Schrödinger pictures (see (14.75) and (14.76)).

Also, the atom dipole has been shown to be described by the real part of

$$a_1^*(t)a_2(t)e^{-i\omega_0 t}; \quad \omega_0 = \omega_2 - \omega_1; \quad \omega_j = E_j/\hbar \tag{15.17a}$$

in the interaction picture or

$$a_{s1}^*(t)a_{s2}(t) \tag{15.17b}$$

in the Schrödinger picture (see (14.77)).

These physical quantities of interest are to be compacted into a 2×2 matrix:

$$\varrho = \begin{pmatrix} \varrho_{11} & \varrho_{12} \\ \varrho_{21} & \varrho_{22} \end{pmatrix} \equiv \begin{pmatrix} a_{s1}a_{s1}^* & a_{s1}a_{s2}^* \\ a_{s2}a_{s1}^* & a_{s2}a_{s2}^* \end{pmatrix}. \tag{15.18}$$

The matrix thus introduced in (15.18) is called the density matrix and has been represented in Schrödinger picture, for convenience. In this matrix, the diagonal elements carry the probability information, while the off-diagonal elements provide the information of the atom dipole.

The equations of motion of these matrix elements readily follow from the coupled equations derived earlier for $a_1(t)$ and $a_2(t)$ in (14.62), which can be transcribed into $a_{s1}(t)$ and $a_{s2}(t)$, using the relation $a_j \exp{-i\omega_j t} = a_{sj}$ (see (15.16)):

$$\dot{a}_{s1} = -i\omega_1 a_{s1} + i\frac{\tilde{\mu}E(t)}{\hbar}a_{s2}, \quad \omega_1 = E_1/\hbar, \tag{15.19a}$$

$$\dot{a}_{s2} = -i\omega_2 a_{s2} + i\frac{\tilde{\mu}E(t)}{\hbar}a_{s1}, \quad \omega_2 = E_2/\hbar. \tag{15.19b}$$

Thus, one can write

$$\begin{aligned} \frac{d}{dt}\varrho_{11} &= \dot{a}_{s1}a_{s1}^* + a_{s1}\dot{a}_{s1}^* \\ &= i\frac{\tilde{\mu}E(t)}{\hbar}a_{s2}a_{s1}^* - i\frac{\tilde{\mu}E(t)}{\hbar}a_{s1}a_{s2}^* \\ &= i\frac{\tilde{\mu}E(t)}{\hbar}(\varrho_{21} - \varrho_{12}) \end{aligned} \tag{15.20a}$$

and

$$\frac{d}{dt}\varrho_{22} = -\frac{i\tilde{\mu}E(t)}{\hbar}(\varrho_{21} - \varrho_{12}), \tag{15.20b}$$

$$\frac{d}{dt}\varrho_{21} = -i\omega_0\varrho_{21} + i\frac{\tilde{\mu}E(t)}{\hbar}(\varrho_{11} - \varrho_{22}) = \frac{d}{dt}\varrho_{12}^{*}. \tag{15.20c}$$

Ensemble Averaging and Relaxation Approach It is clear from (15.20) that the diagonal elements of the density matrix are coupled to the off-diagonal elements in a manner indicative of the atom being driven into a microscopic dipole. To proceed further, it is convenient to regroup (15.20) and simultaneously introduce the ensemble averaging procedure. Thus, first subtract (15.20b) from (15.20a), thereby forming the equation of motion of the difference in probabilities or the population difference $\varrho_{11} - \varrho_{22}$, and incorporate the relaxation term at the same time:

$$\frac{d}{dt}(\varrho_{11} - \varrho_{22}) = \frac{2i\tilde{\mu}E(t)}{\hbar}(\varrho_{21} - \varrho_{12}) - \frac{(\varrho_{11} - \varrho_{22}) - (\varrho_{11}^{(0)} - \varrho_{22}^{(0)})}{\tau}. \tag{15.21a}$$

Here, τ is the longitudinal relaxation time and $\varrho_{11}^{(0)} - \varrho_{22}^{(0)}$ denotes the population difference at equilibrium. Likewise, the transverse relaxation time T_2 can be introduced for the off-diagonal element as

$$\frac{d}{dt}\varrho_{21} = -i\omega_0\varrho_{21} + i\frac{\tilde{\mu}E(t)}{\hbar}(\varrho_{11} - \varrho_{22}) - \frac{\varrho_{21}}{T_2}. \tag{15.21b}$$

In the absence of the driving field, that is, $E(t) = 0$, two equations in (15.21) are compacted with the introduction of integration factors as

$$\frac{d}{dt}\left[(\varrho_{11} - \varrho_{22})e^{t/\tau}\right] = \frac{(\varrho_{11}^{(0)} - \varrho_{22}^{(0)})}{\tau}e^{t/\tau},$$

$$\frac{d}{dt}\left[\varrho_{21}e^{i\omega_0 t + t/T_2}\right] = 0,$$

the time integration of which readily yields

$$[\varrho_{11}(t) - \varrho_{22}(t)] = [\varrho_{11}(0) - \varrho_{22}(0)]e^{-t/\tau} + [\varrho_{11}^{(0)} - \varrho_{22}^{(0)}](1 - e^{-t/\tau}), \tag{15.22a}$$

$$\varrho_{21}(t) = \varrho_{21}(0)e^{-t/T_2 - i\omega_0 t}. \tag{15.22b}$$

It is thus clear from (15.22) that when the perturbing field is turned off, the population difference relaxes back to its equilibrium value in a few τ's, regardless of the initial value that decays away in the same time frame (see Figure 15.3). The off-diagonal element ϱ_{21} also relaxes back to the equilibrium level in a few T_2's and T_2 is generally shorter than τ.

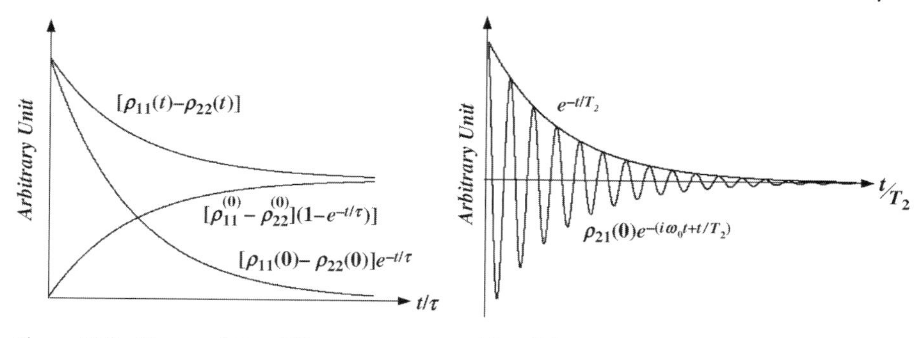

Figure 15.3 The population difference $\varrho_{11} - \varrho_{22}$ and the off-diagonal element ϱ_{21} describing the atom dipole versus time. The two quantities relax back to the equilibrium values in respective relaxation times, while the initial nonequilibrium values decay away in the same time frame.

The faster decay of ϱ_{21} is due to the rapid dephasing of the atom dipoles, which is caused by collisions or other nonradiative decay. When two atoms undergo a collision, for example, one perturbs the other, thereby providing a burst of perturbing Hamiltonian and inducing a shift in energy level ΔE, as detailed in Section 13.1. This gives rise to additional phase term in the off-diagonal element, namely,

$$\varrho_{21}(t) \equiv \frac{1}{N}\sum_{j=1}^{N} a_{s2j}a_{s1j}^{*} \propto \sum_{j} e^{-i(E_2-E_1)t/\hbar - i\Delta E_j t/\hbar} \simeq \sum_{j} e^{-i\omega_0 t - i\Delta\varphi_j}.$$

Inasmuch as the collision is a random process, the resulting phase shift is also a random variable and, therefore, these colliding atom dipoles add up to zero in equilibrium.

It has thus become clear that with the incorporation of the relaxation terms in the equation of motion, the density matrix elements are to be viewed as ensemble averaged atomic parameters, capable of describing the macroscopic quantities.

Steady-State Analysis When the driving field oscillates with frequency ω near the atomic transition frequency ω_0, namely,

$$\mathrm{E}(t) = \frac{\mathrm{E}_0}{2}(e^{i\omega t} + e^{-i\omega t}), \tag{15.23}$$

ϱ_{21} describing the atom dipole is forced to oscillate with the same driving frequency, as discussed in Section 14.3. One can thus put

$$\varrho_{21} = \sigma_{21}e^{-i\omega t}. \tag{15.24}$$

Inserting (15.23) and (15.24) into (15.21b) results in

$$\left(\frac{d}{dt}\sigma_{21}\right)e^{-i\omega t} = i(\omega - \omega_0)\sigma_{21}e^{-i\omega t} + i\frac{\tilde{\mu}\mathrm{E}_0}{2\hbar}(e^{i\omega t} + e^{-i\omega t})(\varrho_{11} - \varrho_{22}) - \frac{\sigma_{21}e^{-i\omega t}}{T_2}. \tag{15.25}$$

The terms on the right-hand side of (15.25) divide into two groups, one oscillating synchronously with the driving term $\exp -i\omega t$ and the other oscillating fast at the second harmonic frequency $\exp \pm 2i\omega t$. Obviously, the harmonic terms become negligible in time frame greater than $2\pi/\omega$ via the time averaging and can be disregarded. Thus, one can single out the synchronous terms from (15.25) and write

$$\frac{d}{dt}\sigma_{21} = i(\omega - \omega_0)\sigma_{21} + i\frac{\tilde{\mu}E_0}{2\hbar}(\varrho_{11} - \varrho_{22}) - \frac{\sigma_{21}}{T_2}. \tag{15.26a}$$

One can also insert (15.23) and (15.24) into (15.21a) and single out the DC components from both sides, obtaining

$$\frac{d}{dt}(\varrho_{11} - \varrho_{22}) = \frac{i\tilde{\mu}E_0}{\hbar}(\sigma_{21} - \sigma_{21}^*) - \frac{(\varrho_{11} - \varrho_{22}) - (\varrho_{11}^{(0)} - \varrho_{22}^{(0)})}{\tau}. \tag{15.26b}$$

In (15.26b), the harmonic components oscillating with $\exp \pm i2\omega t$ have likewise been discarded.

The problem has thus been reduced to finding σ_{21} and $\varrho_{11} - \varrho_{22}$ in (15.26). At this point, the discussion is confined to the steady state, in which the physical quantities are time invariant. Hence, one can put the time derivatives in (15.26) to zero. Also, the real and imaginary parts of σ_{21}, namely, $\sigma_{21} = \sigma_{21}^{(r)} + i\sigma_{21}^{(i)}$ can be singled out by colleting real and imaginary parts from both sides of (15.26) for separate analysis. Hence, there results

$$(\omega - \omega_0)\sigma_{21}^{(i)} + \frac{\sigma_{21}^{(r)}}{T_2} = 0, \tag{15.27a}$$

$$-\frac{\sigma_{21}^{(i)}}{T_2} + (\omega - \omega_0)\sigma_{21}^{(r)} + \frac{\tilde{\mu}E_0}{2\hbar}(\varrho_{11} - \varrho_{22}) = 0, \tag{15.27b}$$

$$-\frac{2\tilde{\mu}E_0}{\hbar}\sigma_{21}^{(i)} - \frac{(\varrho_{11} - \varrho_{22})}{\tau} = -\frac{(\varrho_{11}^{(0)} - \varrho_{22}^{(0)})}{\tau}. \tag{15.27c}$$

Thus, the density matrix equations have been reduced to three coupled equations for three unknowns. One can look for the solution by first finding $\sigma_{21}^{(r)}$ and $\sigma_{21}^{(i)}$ in terms of $\varrho_{11} - \varrho_{22}$ from (15.27a) and (15.27b) and then inserting the result in (15.27c), obtaining

$$\varrho_{11} - \varrho_{22} = (\varrho_{11}^{(0)} - \varrho_{22}^{(0)})\frac{1 + (\omega - \omega_0)^2 T_2^2}{1 + (\omega - \omega_0)^2 T_2^2 + 4\Omega^2 T_2 \tau}, \quad \Omega \equiv \frac{\tilde{\mu}E_0}{2\hbar}. \tag{15.28}$$

The solution (15.28) can in turn be used for specifying $\sigma_{21}^{(i)}$ and $\sigma_{21}^{(r)}$, namely,

$$\sigma_{21}^{(i)} = (\varrho_{11}^{(0)} - \varrho_{22}^{(0)})\frac{T_2\Omega}{1 + (\omega - \omega_0)^2 T_2^2 + 4\Omega^2 T_2 \tau}, \tag{15.29}$$

$$\sigma_{21}^{(r)} = (\varrho_{11}^{(0)} - \varrho_{22}^{(0)})\frac{-(\omega - \omega_0)T_2^2\Omega}{1 + (\omega - \omega_0)^2 T_2^2 + 4\Omega^2 T_2 \tau}. \tag{15.30}$$

Atomic Susceptibility and Dispersion Using these density matrix elements thus found at steady state, the macroscopic physical quantities of interest can be specified in terms of the basic atomic or molecular parameters. For example, the ensemble averaged atom dipole is given by

$$\begin{aligned}\langle\hat{\mu}\rangle &\equiv \tilde{\mu}(\varrho_{21}+\varrho_{12}) = 2\mathrm{Re}\tilde{\mu}\sigma_{21}e^{-i\omega t}\\ &= 2\tilde{\mu}[\sigma_{21}^{(r)}\cos\omega t+\sigma_{21}^{(i)}\sin\omega t].\end{aligned} \tag{15.31}$$

Hence, the polarization vector P is obtained by multiplying this ensemble averaged atom dipole with the number of atoms per unit volume N, namely,

$$P(t) \equiv N\langle\mu(t)\rangle = \frac{\tilde{\mu}^2(N_1^{(0)}-N_2^{(0)})T_2}{\hbar}\cdot\frac{-(\omega-\omega_0)T_2\mathrm{E}_0\cos\omega t+\mathrm{E}_0\sin\omega t}{1+(\omega_0-\omega)^2T_2^2+4\Omega^2T_2\tau}, \tag{15.32a}$$

where the density of atoms in the upper and lower levels in equilibrium are naturally specified by

$$N_j^{(0)} = N\varrho_{jj}^{(0)}, \quad j=1,2, \tag{15.32b}$$

and Ω appearing in the numerator of (15.29) and (15.30) has been spelled out, namely, $\Omega \equiv \tilde{\mu}\mathrm{E}_0/2\hbar$.

The atomic susceptibility connecting P to E_0 via (15.9) can therefore be extracted from (15.32) as

$$\chi_a'(\omega) = \frac{\tilde{\mu}^2T_2(N_1^{(0)}-N_2^{(0)})}{\hbar\varepsilon_0}\frac{-(\omega-\omega_0)T_2}{1+(\omega_0-\omega)^2T_2^2+4\Omega^2T_2\tau}, \tag{15.33a}$$

$$\chi_a''(\omega) = \frac{\tilde{\mu}^2T_2(N_1^{(0)}-N_2^{(0)})}{\hbar\varepsilon_0}\frac{1}{1+(\omega_0-\omega)^2T_2^2+4\Omega^2T_2\tau}. \tag{15.33b}$$

The frequency response of the atomic susceptibility is plotted in Figure 15.4.

With χ_a thus specified, the microscopic description of the wave dispersion is completed by combining (15.33) with (15.14). As was pointed out, the real part of χ_a affects the phase velocity as a function of ω, namely,

$$v_p(\omega) = \frac{\omega}{k_b(1+\chi_a'(\omega)/2n^2)} = \frac{c/n}{(1+\chi_a'(\omega)/2n^2)}. \tag{15.34}$$

However, the imaginary part of χ_a accounts for the power exchange between the field and medium as a function of ω. Thus, the output intensity at z with respect to the input intensity at $z=0$ is given from (15.15) by $I(z)/I(0) = |E(z)/E(0)|^2 = e^{-\alpha z}$ with α denoting the linear attenuation coefficient. Putting aside the power-broadening term proportional to Ω^2 in the denominator of (15.33), one can specify α as

$$\alpha(\omega) \equiv k_b\frac{\chi_a''(\omega)}{n^2} = \frac{\pi\tilde{\mu}^2(N_1^{(0)}-N_2^{(0)})}{\lambda n^2\hbar\varepsilon_0}g(\nu), \tag{15.35}$$

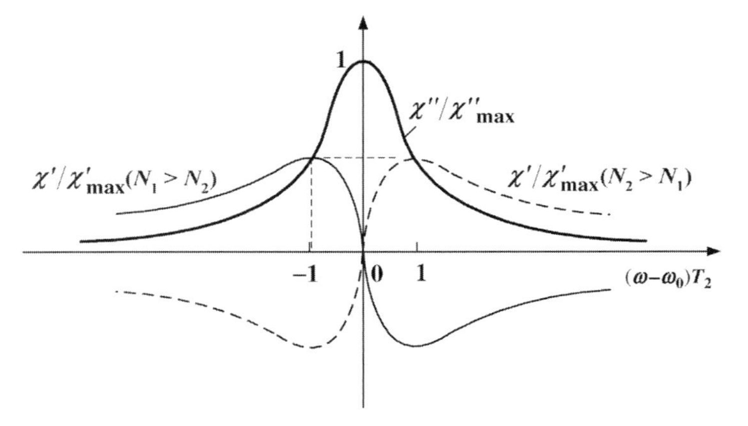

Figure 15.4 The real and imaginary parts of atomic susceptibility versus frequency detuning, $\omega - \omega_0$. The dispersion curve $(\chi'_a(\omega))$ flips its sign depending on whether $N_1 > N_2$ or $N_1 < N_2$, while the power exchange curve $(\chi''_a(\omega))$peaks at $\omega - \omega_0$.

where $g(\nu)$ is the normalized Lorentzian lineshape function,

$$g(\nu) = \frac{2T_2}{1 + (\omega - \omega_0)^2 T_2^2} = \frac{2T_2}{1 + 4\pi^2(\nu - \nu_0)^2 T_2^2}. \tag{15.36}$$

It is interesting that the attenuation coefficient α derived in terms of χ''_a in (15.35) is in complete agreement with the result obtained from the Fermi's golden rule in (15.4) with the identification $\lambda\nu = c/n$. The lineshape factor $g(\nu)$ introduced in (15.4) has now been explicitly specified by the homogeneous line broadening factor, namely Lorentzian function resulting from the transverse relaxation time T_2. Moreover, the relaxation has in turn been attributed to the nonradiative transitions and the collision-induced dephasing of the atomic dipoles.

When the translational motion of the atoms is taken into account, the resulting Doppler shift gives rise to the inhomogeneous Gaussian lineshape function (see (10.66)). Also, in equilibrium, where the Boltzmann probability factor prevails and $N_1 > N_2$, the light is attenuated. But when the atomic population is inverted, so that $N_1 < N_2$, the wave is amplified, and this is further amplified upon in Section 15.3.

15.3
Laser Device

The laser device is one of the premier inventions of the twentieth century and is based upon the innovative applications of the basic concepts inherent in quantum mechanics. Specifically, the population inversion and the controlled emission of radiation are based upon the utilization of the spectroscopy. In addition, the device banks on the Bose–Einstein statistics and capitalizes upon the feedback mechanism, where the photons with given frequency and propagation direction are regenerated via the induced emission of radiation.

Moreover, the device can be miniaturized into the micrometer regime in the form of laser diode. The device can therefore be conveniently interfaced with optical fibers and semiconductor devices for wideband optical communications and a variety of optoelectronic applications. The output of the continuous wave (cw) operation of device provides near-ideal and tunable monochromatic plane waves over the range of frequencies from microwave to X-ray regimes and is utilized as a powerful optical tool for probing atomic and molecular structures. The laser output can also be operated in the form of ultrafast optical pulses of femtosecond time duration or even shorter and can be used for exploring fast chemical and physical processes in femtosecond time regimes. The list of the laser applications is fast increasing in this era of nanotechnology. The operating principles of the device are briefly described in the following sections.

Axial or Longitudinal Cavity Modes Consider a Fabry–Perot type cavity, consisting of two parallel mirrors L distance apart (see Figure 15.5). A longitudinal standing wave mode therein satisfies the boundary condition

$$(\lambda_l/2)l = L, \quad l = 1, 2, 3, \ldots, \tag{15.37a}$$

or in terms of frequency

$$\nu_l = \frac{c/n}{\lambda_l} = l\frac{1}{2L/(c/n)}. \tag{15.37b}$$

Hence, the axial modes are separated in frequency by a constant value

$$\Delta\nu = \nu_{l+1} - \nu_l = \frac{1}{2L/(c/n)} \tag{15.38}$$

given by the inverse round-trip time of the wave inside the cavity.

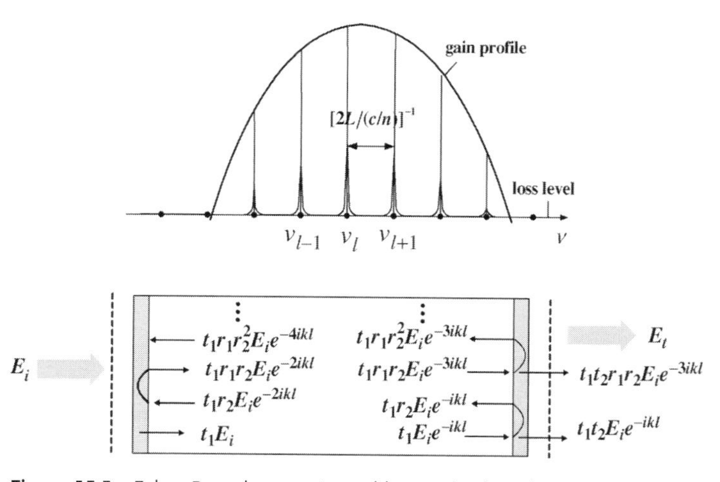

Figure 15.5 Fabry–Perot laser cavity and longitudinal modes, separated by the same frequency spacing. The input beam gives rise to a string of output waves, with each succeeding beam having undergone one more cavity round-trip.

The Laser Oscillator The cavity when filled with an active medium works as an oscillator and light amplifier. That is, an input wave E_i is amplified and sustained at a constant level and is out-coupled through the output mirror (Figure 15.5). When the population inversion is achieved to a sufficient extent, the oscillator is self-sustained and provides the output power without the need of the input wave.

The input wave is amplified as it traverses the active medium, but at the same time, it is subject to loss due to imperfect reflectivity and transmittance of the mirrors and the scattering loss, as illustrated in Figure 15.5. The output wave is comprised of a string of transmitted beams, with each succeeding beam having undergone one more cavity round-trip. The transmitted beam is thus given by the sum of these transmitted waves:

$$\begin{aligned} E_t &= t_1 t_2 E_i e^{-ikL}[1+s+s^2+\cdots], \quad s = r_1 r_2 e^{-2ikL}, \\ &= \frac{t_1 t_2 E_i e^{-ikL}}{1-s}, \end{aligned} \tag{15.39}$$

where t_j and r_j denote the transmission and reflection coefficients of the two mirrors and s is the net gain in one cavity round-trip. The infinite geometric series in (15.39) can be readily summed up, provided $s \leq 1$. As already discussed, the amplification and dispersion of the wave in one cavity round-trip is to be incorporated into the effective wave vector

$$k = k_b + \Delta k + i\frac{1}{2}(\gamma - \alpha_s), \tag{15.40a}$$

where k_b is the background wave vector of the cavity,

$$\Delta k = k_b \frac{\chi_a'(\omega)}{2n^2} \tag{15.40b}$$

is the dispersion term contributed by the real part of the atomic susceptibility of the active medium (see (15.15)),

$$\gamma \equiv -\alpha = -k_b \frac{\chi_a''(\omega)}{n^2} \tag{15.40c}$$

represents the gain resulting from the imaginary part of the susceptibility (see (15.15)), and α_s is the phenomenological scattering loss term in the cavity.

15.3.1 Population Inversion

The condition for gain, that is, the condition for γ to be positive, is that $N_2 > N_1$ (see (15.33b) and 15.40c)). This condition can only be met by driving the medium away from the equilibrium. Specifically, the population should be inverted by means of a pumping process, promoting the electrons to the upper lasing level, so that more atoms or molecules are in the upper level than in the lower level.

Thus, consider a two-level atom, which is pumped, driven, and damped at the same time, as sketched in Figure 15.6. The rate equation of the system of these two-level

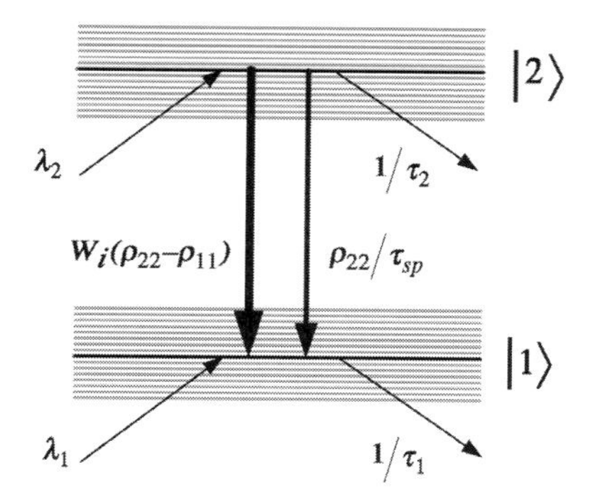

Figure 15.6 Two lasing levels driven, pumped, and damped simultaneously. The two levels are coupled via induced as well as spontaneous transitions.

atoms is given by

$$\dot{\varrho}_{22} = \lambda_2 - \left(\frac{1}{\tau_2} + \frac{1}{\tau_{sp}}\right)\varrho_{22} - W_i(\varrho_{22} - \varrho_{11}), \tag{15.41}$$

$$\dot{\varrho}_{11} = \lambda_1 - \frac{1}{\tau_1}\varrho_{11} + \frac{1}{\tau_{sp}}\varrho_{22} + W_i(\varrho_{22} - \varrho_{11}), \tag{15.42}$$

where λ_j and $1/\tau_j$ are the respective pumping and decay rates with τ_j denoting the lifetime of the jth level and $1/\tau_{sp}$ is the spontaneous transition rate, discussed in Section 14.4. These two levels are coupled by the driving field and the electron undergoes the induced transition between the two, as described by the term proportional to W_i (see (15.2)). At steady state in which the time derivatives of ϱ_{11} and ϱ_{22} are zero, one readily finds,

$$\varrho_{22} - \varrho_{11} = \frac{\lambda_2\tau_2 - \lambda_1\tau_1}{1 + W_i(\tau_1 + \tau_2)}. \tag{15.43}$$

In (15.43), a nonessential simplification has been made, namely, $\tau_2 \ll \tau_{sp}$ and $\tau_1 \ll \tau_{sp}$.

It thus follows from (15.43) that to achieve the population inversion, the pumping rate at the upper level should be strong enough and the electron lifetime therein long enough, so that the condition $\lambda_2\tau_2 > \lambda_1\tau_1$ prevails. The number of atoms in the upper and lower levels per unit volume is then given by multiplying (15.43) by the density of atoms, N, and expressing the transition rate W_i explicitly from (15.2), namely,

$$N_2 - N_1 \equiv N(\varrho_{22} - \varrho_{11}) = \frac{(N_2 - N_1)_0}{1 + I_\nu g(\nu)/I_s}, \tag{15.44a}$$

where

$$(N_2 - N_1)_0 \equiv N(\lambda_2\tau_2 - \lambda_1\tau_1) \tag{15.44b}$$

represents the bare population inversion, given strictly by the pumping rate in the absence of light intensity and

$$I_s^{-1} = \frac{\tilde{\mu}^2(\tau_1 + \tau_2)}{2\hbar^2 cn\varepsilon_0} \tag{15.44c}$$

is the saturation intensity and $g(\nu)$ is the Lorentzian lineshape function (see (15.36)).

Oscillation Condition The gain coefficient γ for the given bare population inversion (15.44b) can obviously be expressed by transcribing the expression of the attenuation coefficient α in (15.35). The only modification needed is to replace $N_1^{(0)} - N_2^{(0)}$ operative in equilibrium by the bare population inversion $(N_2 - N_1)_0$ given in (15.44b). Thus, one can write

$$\gamma \equiv -\alpha = \frac{\pi\tilde{\mu}^2(N_2 - N_1)_0}{\lambda n^2 \hbar\varepsilon_0} g(\nu). \tag{15.45}$$

When γ is sufficient to balance the net loss in one cavity round-trip, then $s = 1$ in the transfer function (15.39). In this case, the denominator of the transfer function vanishes and, consequently, the transmitted field amplitude E_t diverges. This simply means that a finite E_t is to be generated with infinitesimal input E_i. That is, it signals the onset of laser oscillation.

The condition for the steady-state laser oscillation is therefore given by

$$s \equiv r_1 r_2 e^{-2i(k_b + \Delta k)L} e^{(\gamma - \alpha_s)L} = 1 \equiv e^{-2\pi i l}. \tag{15.46}$$

This complex equality can be split into the amplitude and phase parts:

$$r_1 r_2 e^{(\gamma - \alpha_s)L} = 1, \tag{15.47}$$

$$2(k_b + \Delta k)L = 2\pi l. \tag{15.48}$$

The amplitude equation (15.47), when combined with (15.45) for γ, determines the threshold value of the bare population inversion for laser oscillation, namely,

$$\frac{\pi\tilde{\mu}^2(N_2 - N_1)_{T0}}{\lambda n^2 \hbar\varepsilon_0} g(\nu) = \alpha_s - \frac{1}{L}\ln(r_1 r_2). \tag{15.49}$$

Here, the loss consists of both the scattering loss and the imperfect mirror reflectivities.

It is interesting to notice that with increased detuning of the laser oscillation frequency ν from the atomic transition frequency ν_0, the Lorentzian lineshape function $g(\nu)$ decreases (see (15.36)). Consequently, the required population inversion should be increased by raising the pumping level. This is understandable, since with detuning, the coupling between the field and the medium degrades, as discussed in Section 14.3, and therefore more atoms should be pumped into the upper lasing level to compensate for the given loss in the cavity.

Operating Laser Intensity Once the pumping level exceeds the threshold value, so that $N_2 - N_1 > (N_2 - N_1)_{T0}$, the oscillation sets in. However, it is important to notice that this inequality does not mean that there is more gain than loss in one cavity round-trip and the string of transmitted field amplitudes grows without any upper bound. Rather, the oscillation condition (15.47) should still prevail, regardless of the pumping level if the oscillator is to be operated at a steady state.

It is at this point that the physical meaning of the saturated population inversion (15.44a) becomes apparent. At the onset of oscillation, the laser intensity is still at zero level. With the population inversion larger than the threshold value (19.49), the intensity of the laser builds up in the cavity, so that the gain coefficient γ becomes intensity dependent (see (15.44a)). But, the same condition (15.49) has to be met for the laser oscillator to sustain the steady-state operation, however with $I_\nu \neq 0$ namely,

$$\frac{\pi\tilde{\mu}^2 g(\nu)}{\lambda n^2 \hbar\varepsilon_0} \frac{(N_2 - N_1)_0}{1 + I_\nu g(\nu)/I_s} = \alpha_s - \frac{1}{L}\ln(r_1 r_2). \tag{15.50}$$

Hence, combining (15.49) and (15.50), one can write

$$\frac{(N_2 - N_1)_0}{(N_2 - N_1)_{T_0}} \frac{1}{1 + I_\nu g(\nu)/I_s} = 1. \tag{15.51}$$

Clearly, (15.51) describes the same oscillation condition as (15.47) but at a pumping level above the bare threshold value and operating at a finite laser intensity. In fact, it is from this dressed oscillation condition that the operating laser intensity I_ν is obtained, that is,

$$I_\nu = [I_s/g(\nu)]\left[\frac{(N_2 - N_1)_0}{(N_2 - N_1)_{T_0}} - 1\right]. \tag{15.52}$$

Hence, it is clear from (15.52) that the output intensity of the laser increases with increasing pumping level, as it should. Also, it increases with increasing detuning and decreased Lorentzian lineshape factor, however, at the expense of increased pumping level (see (15.49)). Moreover, the lasing intensity depends on the inherent property of the active medium as represented in the saturated intensity I_s such as the atomic dipole moment $\tilde{\mu}$ and electron lifetimes (see (15.44c)).

Operation Frequency When the phase part of the oscillation condition (15.48) is spelled out explicitly using (15.40b), (15.48) and the relation $k_b = 2\pi\nu_l/(c/n)$, it reads as

$$\frac{2\pi\nu_l L}{c/n}\left[1 + \frac{\chi_a'(\omega)}{2n^2}\right] = l\pi, \quad l = 1, 2, 3, \ldots. \tag{15.53}$$

Or in terms of the bare longitudinal cavity mode given in (15.37b),

$$\nu_l^0 \equiv \frac{l}{2L/(c/n)},$$

the operating laser frequency (15.53) is given by

$$\nu_l = \nu_l^0 \frac{1}{1+\chi_a'(\omega)/2n^2}. \tag{15.54}$$

It is therefore clear that the dispersion occurring concomitantly with the amplification shifts the lasing frequency from ν_l^0. In fact ν_l is pulled or pushed from ν_l^0 depending on the intensity of the laser.

Operation Modes The frequency of the laser spans from microwaves to optical, UV, EUV, and X-ray regimes. The active medium is provided by gases (helium, neon, argon ion, carbon dioxide, free-electron ring), liquids (dye), and solid state materials (ruby, GaAs).

The device operates in both continuous wave (cw) and pulsed modes. The single-mode cw operation provides near-ideal monochromatic optical beams as the powerful tool for spectroscopic studies, investigating the structure and dynamics of atoms and molecules. The longitudinal modes within the broad gain profile (see Figure 15.4) can be excited simultaneously as a free running wave for generating powerful light sources. Furthermore, the phases of these excited waves can be locked together to generate powerful light pulses. The time duration of these optical pulses ranges from microseconds to femtoseconds, approaching attosecond time regime. These laser pulses have opened up a new horizon for investigating fast natural phenomena in nano regimes. More important, the laser diode as a miniaturized coherent light source is instrumental for fiber optical communications as well as other optoelectronic applications in conjunction with semiconductor devices.

15.4 Problems

15.1. The equation of motion of the 2×2 density matrix for two-level atoms was derived in the Schrödinger picture (see (15.20)).

(a) Derive the equation of motion of the density matrix in the interaction picture in which the coefficients are related by $a_j e^{-i(E_j/\hbar)t} = a_{sj}$, $j = 1, 2$ (see (14.75), (14.76), (15.16), and (15.17)).

(b) Compare the two equations of motion and discuss whether or not the matrix elements in both cases lead to identical descriptions of physical quantities involved.

15.2. The equation of motion of the density matrix in (15.20) is often solved iteratively to an arbitrary order of coupling strength.

(a) Given the initial condition that the atom is in the lower level, so that $a_{s1}(t=0) = 1$ and $a_{s2}(0) = 0$, find to the first order of iteration, $\varrho_{21}(t)$ and $\varrho_{12}(t)$ by taking $\varrho_{11} = 1$ and $\varrho_{22} = 0$ in (15.20c).

(b) Using $\varrho_{21}(t)$ and $\varrho_{12}(t)$ thus found, specify the atom dipole

$$\langle\hat{\mu}\rangle = \tilde{\mu}(\varrho_{21} + \varrho_{12}); \quad \hat{\mu} = e\hat{e}_f \cdot \boldsymbol{r}, \quad \tilde{\mu} = \langle u_1|\hat{\mu}|u_2\rangle.$$

(c) Repeat (a) and (b) using the equation of motion in the interaction picture found in 15.1 and compare the two results.

15.3. Solve the coupled equations (15.27) for three unknowns, $\varrho_{11} - \varrho_{22}$, $\sigma_{21}^{(r)}$, and $\sigma_{21}^{(i)}$ in the manner described in the text and reproduce the results in (15.28)–(15.30).

15.4. Solve the rate equation in (15.41) and (15.42) at steady state without neglecting the spontaneous transition rate and specify the conditions by which the expression

$$\varrho_{22} - \varrho_{11} = \frac{\lambda_2\tau_2 - \lambda_1\tau_1}{1 + W_i(\tau_1 + \tau_2)}$$

in (15.43) holds true.

15.5. A scheme by which to achieve the population inversion is sketched in Figure 15.7. In this scheme, electrons are optically pumped into the upper lasing level $|3\rangle$ from $|1\rangle$, while the lower lasing level $|2\rangle$ is strongly coupled to $|1\rangle$, spontaneously or nonradiatively so that the lifetime of electron therein is short. Consider the three lowest levels in hydrogen atom, corresponding to $n = 1, 2, 3$.

(a) If the perturbing Hamiltonian for the optical pumping is given by

$$\hat{H}' = \mathbf{E}_0 \cdot \boldsymbol{r} \cos\omega t,$$

find the eigenstates with $n = 3$ satisfying the selection rule required for the optical pumping.

(b) Find the eigenstates with $n = 2$ that is optimally coupled for efficient lasing with the upper state found in (a).

(c) Find the ratio N_3/N_2 with N_j denoting the density of electrons in the jth state at equilibrium.

(d) Find the pumping field intensity required to satisfy $N_3/N_2 \geq 1$ in terms of the given lifetime in each level, τ_3 and τ_2.

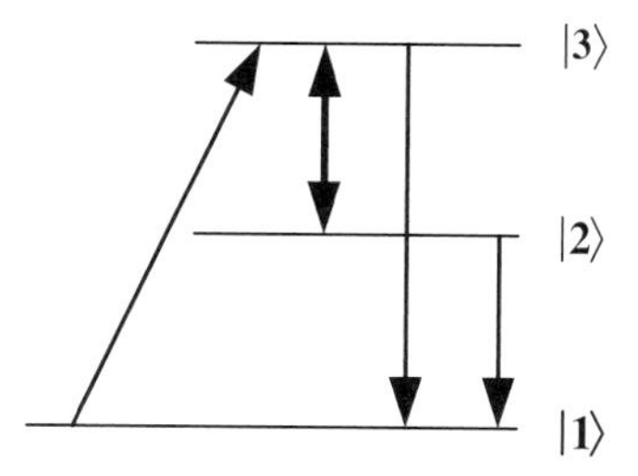

Figure 15.7 Three level lasing scheme. The population inversion is attained by pumping electrons from $|1\rangle$ to $|3\rangle$ and fast draining electrons from $|2\rangle$ nonradiatively.

15.6. Consider a passive Fabry–Perot etalon in which $\gamma = \alpha = 0$.

(a) Calculate the net transmission and reflection coefficients, namely, $|E_t/E_i|$ and $|E_r/E_i|$ as a function of input frequency for given cavity length L.

(b) Find the standing wave modes and the mode spacing in frequency unit for the cavity lengths 1 m, 1 cm, 100 μm, and 10 μm.

(c) Repeat the calculations in (a) and (b) for the case of Fabry–Perot etalon, filled with active lasing medium, and compare the results obtained with those in (a) and (b).

Suggested Reading

1 Yariv, A. (1982) *An Introduction to Theory and Applications of Quantum Mechanics*, John Wiley & Sons, Inc.

2 Sargent, M.I., Scully, M.O., and Lamb, W.E., Jr. (1978) *Laser Physics*, Westview Press.

16
Quantum Statistics

The microscopic world of electrons, atoms, and molecules are connected to the macroscopic world via the cumulative effects of a large number of such particles. This has been amply illustrated in Chapter 15, dealing with the interaction of radiation with the ensemble of atoms. The statistical description of these microscopic objects is therefore important and the quantum statistics is an essential part of such description. For modeling the semiconductor devices, for example, the quantum and semiconductor statistics are the prerequisites.

16.1
General Background and Three Kinds of Particles

Given a system of N identical particles, it evolves in time according to the Schrödinger equation,

$$i\hbar\frac{\partial}{\partial t}\psi = \hat{H}(\boldsymbol{q}_1, \boldsymbol{p}_1, \ldots, \boldsymbol{q}_N, \boldsymbol{p}_N, t)\psi. \tag{16.1}$$

However, solving the equation is difficult even for small N, as has indeed been shown to be the case even for simple molecular systems considered in Chapter 12. When a large number of particles are involved, solving the Schrödinger equation becomes practically impossible even with the use of a massive computational power. Even if such solutions are attained, the management of the information entailed in such a large number of particles would be near impossible. In view of these facts, the statistical approach provides the only viable and meaningful description for such systems.

Thus, consider a system of N identical and weakly interacting particles, in which each particle is associated with eigenfunction u_s, with eigenenergy ε_s, and the degeneracy g_s. When the energy levels are quasi-continuous, ε_s may be replaced by the energy bin ranging from ε_s to $\varepsilon_s + d\varepsilon_s$ with g_s denoting the number of quantum states in the range. The total number of particles and the energy of the system are then given by

$$N = \sum_s n_s; \quad E = \sum_s n_s \varepsilon_s. \tag{16.2}$$

Introductory Quantum Mechanics for Semiconductor Nanotechnology. Dae Mann Kim

ISBN: 978-3-527-40975-4

For the given N and E, there generally exist different sets of $\{n_s\}$, with each set satisfying the same condition (16.2). That is, there is more than one way of distributing the particles among the eigenstates so as to satisfy the given constraints (16.2). Furthermore, for each set of $\{n_s\}$, there exists more than one microscopically distinguishable arrangement, with each arrangement possessing a unique wavefunction.

A fundamental postulate of quantum statistics states that in thermodynamic equilibrium, each microscopically distinguishable arrangement having the same total energy E is equally likely to occur.

Thus, an essential task of the quantum statistics is to find the most probable distribution $(n_1, n_2, \ldots, n_s, \ldots)$ of the particles under the given constraints of (16.2). Since there exist more than one microscopic arrangement and since each arrangement is equally likely to occur, the problem reduces to finding the distribution $(n_1, n_2, \ldots, n_s, \ldots)$ with the largest number of microscopic arrangements.

The analysis of obtaining the number of microscopic arrangements is generally different, depending on the kinds of particles considered. As mentioned, there are three different kinds of particles: (i) identical but distinguishable, for example, identical atoms and molecules in a volume, (ii) fermions with half-odd integer spins, $\hbar/2, 3\hbar/2, \ldots$ and obeying the Pauli exclusion principle, for example, electrons, holes, protons, neutrons, and so on, and (iii) bosons with integer spins, $\hbar, 2\hbar, \ldots$, for example, photons, phonons, α-particles, and so on.

The fact that the microscopic arrangements are different, depending on the types of particles considered, is specifically illustrated in Figure 16.1. In this figure, two identical particles are assigned to two possible quantum states. For the case of two distinguishable particles, there are four distinct arrangements, one with the total energy $2E_1$, one with $2E_2$, and two with $E_1 + E_2$. For two fermions, however, because of the Pauli exclusion principle, there is only one arrangement with the total energy

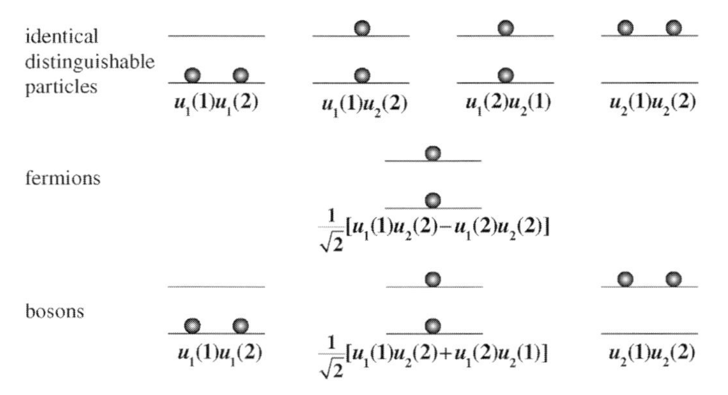

Figure 16.1 The graphical illustration of microscopically distinguishable arrangements of two identical particles in the presence of two quantum states for (i) distinguishable particles, (ii) fermions, and (iii) bosons, respectively. A distinguishable arrangement is represented by a unique wavefunction.

$E_1 + E_2$. For two bosons, there exist three arrangements, with each arrangement having the total energies E_1, $E_1 + E_2$, and E_2, respectively, as clearly shown in the figure.

16.2
Statistics for Distinguishable Particles

Consider first the distribution $(n_1, n_2, \ldots, n_s, \ldots)$ of distinguishable particles, subject to the constraints (16.2). The number of finding the distinguishable arrangements in the first bin is as follows. First, there are

$$N(N-1)\ldots(N-n_1+1) = N!/(N-n_1)!$$

ways for choosing n_1 particles out of the total number N. But the permutations among the chosen particles $n_1!$ do not provide distinguishable arrangements. Moreover, any one particle chosen can be in any of g_1 states in the bin. The number of distinguishable states is therefore given by

$$P_1 = \frac{N!}{(N-n_1)!n_1!}g_1^{n_1}. \tag{16.3}$$

One can repeat the same argument for assigning n_2 particles to the second bin out of the remaining $N-n_1$ particles, obtaining

$$P_2 = \frac{(N-n_1)!}{(N-n_1-n_2)!n_2!}g_2^{n_2}. \tag{16.4}$$

Repeating the similar arguments in succession, the total number of distinguishable arrangements for the distribution $(n_1, n_2, \ldots, n_s, \ldots)$ is obtained as

$$\begin{aligned} P(n_1, n_2 \ldots, n_s, \ldots) &= \frac{N!}{(N-n_1)!n_1!}g_1^{n_1}\frac{(N-n_1)!}{(N-n_1-n_2)!n_2!}g_2^{n_2}\ldots \\ &= N!\prod_{s=1}^{\infty}\frac{g_s^{n_s}}{n_s!}. \end{aligned} \tag{16.5}$$

In (16.5), the factorials of the type $(N-n_j)!$ cancel out in the process of multiplication.

Lagrange Undetermined Multiplier Technique Next, the distribution $(n_1, n_2, \ldots, n_s, \ldots)$ possessing the maximum number of microscopically distinguishable arrangements is considered, subject to the condition (16.2). This can be done using Lagrange undetermined multiplier technique, and a brief digression is needed for discussing the method. Given a function $f(x, y)$, its maximum value is routinely found by finding x_0 and y_0 such that

$$\frac{\partial}{\partial x}f(x_0, y_0) = 0; \quad \frac{\partial}{\partial y}f(x_0, y_0) = 0, \tag{16.6}$$

provided there is no restriction. However, in the presence of a constraint

$$y = y_a(x)$$

the maximum value is found instead by determining x_0 such that

$$\frac{d}{dx}f(x_0, y_a(x_0)) = \frac{\partial}{\partial x}f(x_0) + \frac{\partial f}{\partial y_a}\frac{\partial y_a(x_0)}{\partial x} = 0. \tag{16.7}$$

This procedure is equivalent to introducing a function

$$F(x, y) \equiv f(x, y) - \lambda g(x, y); \quad g(x, y) \equiv y - y_a(x) = 0 \tag{16.8}$$

and regarding x, y, and λ as independent variables free of any constraints and finding the maximum value of $F(x, y)$, namely,

$$\frac{\partial}{\partial x}F = \frac{\partial}{\partial y}F = \frac{\partial}{\partial \lambda}F = 0. \tag{16.9}$$

The equivalence of (16.9) to (16.7) can be shown as follows. First, note from (16.8) that

$$\frac{\partial F}{\partial x} = 0 = \frac{\partial f}{\partial x} - \lambda\frac{\partial g}{\partial x} = \frac{\partial f}{\partial x} + \lambda\frac{\partial y_a(x)}{\partial x}, \tag{16.10a}$$

$$\frac{\partial F}{\partial y} = 0 = \frac{\partial f}{\partial y} - \lambda, \tag{16.10b}$$

$$\frac{\partial F}{\partial \lambda} = g \equiv 0. \tag{16.10c}$$

Here, the condition (16.7c) is automatically satisfied by the very definition of g (see (16.8)) and (16.10a) and (16.10b) are combined to yield

$$\frac{\partial f}{\partial x} + \lambda\frac{\partial y_a}{\partial x} = \frac{\partial f}{\partial x} + \frac{\partial f}{\partial y}\frac{\partial y_a}{\partial x} = 0. \tag{16.11}$$

Therefore, (16.9) is identical to (16.7). The constant λ thus introduced is known as the Lagrange undetermined multiplier. The technique can be readily generalized to the case of multiple constraints.

16.2.1 Boltzmann Distribution Function

Using the undetermined multiplier technique, the maximum value of $P(\{n_s\})$ in (16.5) for distinguishable particles can be found. In so doing, however, the maximum value of $\ln P(\{n_s\})$ is found instead, for convenience. Thus, the F-function incorporating the constraints (16.2) should read as

$$F(\{n_s\}, \alpha, \beta) = \ln P(n_1, n_2 \ldots, n_s, \ldots) - \alpha\left[\sum_s n_s - N\right] - \beta\left[\sum_s n_s\varepsilon_s - E\right]. \tag{16.12}$$

Inserting (16.5) into (16.12) results in

$$F(\{n_s\},\alpha,\beta) = \ln N! + \sum_{s=1}^{\infty}[n_s \ln g_s - \ln(n_s!)] - \alpha\left[\sum_s n_s - N\right] - \beta\left[\sum_s n_s\varepsilon_s - E\right]. \tag{16.13}$$

Since a large number of particles are involved, $n_s \gg 1$ and one can therefore use the Stirling's formula for simplicity of analysis, namely,

$$\ln n_s! = n_s \ln n_s - n_s. \tag{16.14}$$

Inserting (16.14) into (16.13) and differentiating F with respect to n_s, the condition for the maximum value of $\ln P(\{n_s\})$ is obtained, namely,

$$\frac{\partial}{\partial n_s} F(\{n_s\},\alpha,\beta) = \ln g_s - \ln n_s - \alpha - \beta\varepsilon_s = 0. \tag{16.15}$$

Hence, one can readily find from (16.15) the set of n_s with the largest number of microscopic arrangements as

$$n_s = g_s e^{-\alpha-\beta\varepsilon_s}. \tag{16.16}$$

***Determination of* α *and* β** The specification of the most probable distribution of $\{n_s\}$ derived in (16.16) is thus completed, when Lagrange multipliers α and β are determined explicitly in terms of the constraints (16.2). Using (16.16), the constraints now read as

$$N = \sum_s n_s = e^{-\alpha}\sum_s g_s e^{-\beta\varepsilon_s}, \tag{16.17}$$

$$E = \sum_s n_s\varepsilon_s = e^{-\alpha}\sum_s g_s\varepsilon_s e^{-\beta\varepsilon_s}. \tag{16.18}$$

To perform the summations in (16.17) and (16.18), it is convenient to consider the density of states, $g(\varepsilon)$, of distinguishable particles in a cubic box of volume $V(=L^3)$. The procedure for finding $g(\varepsilon)$ is exactly the same as that of finding the density of states of electrons, except that there is in this case no spin factor of 2 involved. Thus, it follows from (4.27) that

$$g(\varepsilon) = \frac{m^{3/2}}{\sqrt{2}\pi^2\hbar^3}\varepsilon^{1/2}. \tag{16.19}$$

Using (16.19), one can replace the summation in (16.17) by an integral, obtaining the total number N in the volume V as

$$N \equiv Ve^{-\alpha}\int_0^{\infty} d\varepsilon g(\varepsilon)e^{-\beta\varepsilon} = \frac{Vm^{3/2}}{\sqrt{2}\pi^2\hbar^3}\frac{e^{-\alpha}}{\beta^{3/2}}\int_0^{\infty} dx x^{1/2}e^{-x} = \frac{Vm^{3/2}}{\sqrt{2}\pi^2\hbar^3}\frac{e^{-\alpha}}{\beta^{3/2}}\Gamma\left(\frac{3}{2}\right), \tag{16.20}$$

where the integral has been put into the gamma function, namely,

$$\Gamma(z) \equiv \int_0^\infty dx x^{z-1} e^{-x}. \tag{16.21}$$

Using the well-known properties of the gamma function, namely, $\Gamma(z+1) = z\Gamma(z)$ and $\Gamma(1/2) = \sqrt{\pi}$, one can rewrite (16.20) as

$$N = \frac{Vm^{3/2}}{2\sqrt{2}\pi^{3/2}\hbar^3} \frac{e^{-\alpha}}{\beta^{3/2}}. \tag{16.22}$$

One can evaluate E in (16.18) in a similar manner, obtaining

$$E \equiv Ve^{-\alpha} \int_0^\infty d\varepsilon g(\varepsilon) \varepsilon e^{-\beta\varepsilon} = \frac{Vm^{3/2}}{\sqrt{2}\pi^2\hbar^3} \frac{e^{-\alpha}}{\beta^{5/2}} \int_0^\infty dx x^{3/2} e^{-x} = \frac{Vm^{3/2}}{\sqrt{2}\pi^2\hbar^3} \frac{e^{-\alpha}}{\beta^{5/2}} \Gamma\left(\frac{5}{2}\right). \tag{16.23a}$$

Using $\Gamma(5/2) = (3/2)(1/2)\sqrt{\pi}$, one can rewrite (16.23a) as

$$E = \frac{3Vm^{3/2}}{4\sqrt{2}\pi^{3/2}\hbar^3} \frac{e^{-\alpha}}{\beta^{5/2}}. \tag{16.23b}$$

Hence, the average energy of the particle can be obtained from (16.22) and (16.23b) as

$$\frac{E}{N} = \frac{3}{2\beta}. \tag{16.24}$$

It is interesting to note that the same average energy of a free particle in thermodynamic equilibrium was found in (1.26) from the equipartition theorem, namely,

$$\frac{E}{N} = \frac{3k_B T}{2}. \tag{16.25}$$

It therefore follows from (16.24) and (16.25) that

$$\beta \equiv \frac{1}{k_B T}. \tag{16.26}$$

When β thus identified is inserted into (16.22), α is specified as

$$e^{-\alpha} = \frac{N}{V} \left(\frac{h^2}{2\pi m k_B T} \right)^{3/2}. \tag{16.27}$$

Hence, the ratio of n_s with respect to the total number N is now specified from (16.16) and (16.17) as

$$f_i = \frac{n_i}{N} \equiv \frac{g_i e^{-\varepsilon_i/k_B T}}{\sum_s g_s e^{-\varepsilon_s/k_B T}} \tag{16.28}$$

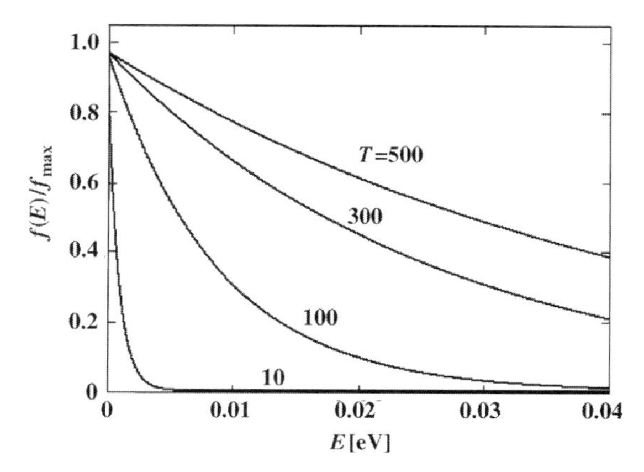

Figure 16.2 The Boltzmann probability factor versus energy at different temperatures.

and is shown independent of α. In addition, the ratio of n_i, n_j associated with the quantum states with energies ε_i, ε_j,

$$\frac{f_i}{f_j} = \frac{g_i}{g_j} e^{-(\varepsilon_i - \varepsilon_j)/k_B T}, \tag{16.29}$$

is described by the Boltzmann probability factor at equilibrium. Combining (16.16), (16.26), and (16.27), one can write

$$\frac{n_s}{g_s} = e^{-\alpha - \beta \varepsilon_s} = \frac{N}{V} \left(\frac{h^2}{2\pi m k_B T} \right)^{3/2} e^{-\varepsilon_s / k_B T}. \tag{16.30}$$

Clearly, (16.30) states that the ratio of the average number of particles per quantum state is dictated by the Boltzmann probability factor. In this way, the quantum statistics when applied to distinguishable particles is shown to yield the same Boltzmann distribution function as derived from the transport equation at equilibrium. The distribution function is plotted in Figure 16.2 as a function of energy at different temperatures. As clearly shown in the figure, the distribution function exponentially decreases with increasing energy E and decreasing temperature T, as expected.

16.3
Statistics for Fermions and Fermi-Dirac Distribution Function

Fermi-Dirac Distribution Function For fermions, the number of distinguishable states P_s in the sth bin with n_s particles is found as follows. The first particle chosen out of n_s particles has g_s possible ways for occupying one state in the bin. The second particle however has $g_s - 1$ options, because of the exclusion principle. The process continues until the last particle, which has $g_s - n_s + 1$ options of occupying the states therein. However, these particles can be permuted among themselves without giving rise to distinguishable states. Hence, P_s is given by

$$P_s = \frac{g_s(g_s-1)\dots(g_s-n_s+1)}{n_s!} = \frac{g_s!}{(g_s-n_s)!n_s!}. \tag{16.31}$$

Therefore, by repeating the same arguments for different bins, the total number of distinguishable states for the N fermion system can be expressed as

$$P(n_1,\dots,n_s,\dots) = \prod_{s=1}^{\infty} \frac{g_s!}{(g_s-n_s)!n_s!}. \tag{16.32}$$

Given the total number of distinguishable arrangements, one can find the most probable distribution $\{n_s\}$ again by finding the maximum value of $\ln P(\{n_s\})$, subject to the constraints (16.2) in a manner identical to the analysis of distinguishable particles. The F-function in this case reads as

$$F(\{n_s\},\alpha,\beta) = \sum_{s=1}^{\infty} \ln\left[\frac{g_s!}{(g_s-n_s)!n_s!}\right] - \alpha\left[\sum_s n_s - N\right] - \beta\left[\sum_s n_s\varepsilon_s - E\right] \tag{16.33}$$

and the maximum distribution of $\{n_s\}$ is readily found by using the Stirling's formula (16.14) for simplifying the expressions $\ln(g_s-n_s)!$ and $\ln n_s!$:

$$\frac{\partial}{\partial n_s} F(\{n_s\},\alpha,\beta) = \ln(g_s-n_s) - \ln n_s - \alpha - \beta\varepsilon_s = \ln\left(\frac{g_s-n_s}{n_s}\right) - \alpha - \beta\varepsilon_s = 0. \tag{16.34}$$

Thus, one obtains from (16.34)

$$n_s = \frac{g_s}{e^{\alpha+\beta\varepsilon_s}+1}. \tag{16.35}$$

With β identified as $\beta = 1/k_BT$ in (16.26) and by expressing the undetermined multiplier α in terms of the chemical potential μ or the Fermi level E_F, namely,

$$\alpha \equiv -\frac{\mu}{k_BT} \equiv -\frac{E_F}{k_BT}, \tag{16.36}$$

n_s is expressed as

$$n_s = \frac{g_s}{1+e^{(\varepsilon_s-\mu)/k_BT}} = \frac{g_s}{1+e^{(\varepsilon_s-E_F)/k_BT}}. \tag{16.37}$$

The distribution per state,

$$f(\varepsilon_s) \equiv \frac{n_s}{g_s} = \frac{1}{1+e^{(\varepsilon_s-E_F)/k_BT}}, \tag{16.38}$$

is the celebrated Fermi-Dirac distribution function. It is also called Fermi occupation factor.

Figure 16.3 plots the Fermi occupation factor versus energy. At $T = 0, f(\varepsilon)$ is a step function and is equal to unity for $\varepsilon < E_F$, representing 100% probability of occupation, while it is zero for $\varepsilon > E_F$, showing the zero probability. For $T \neq 0$, the shape of

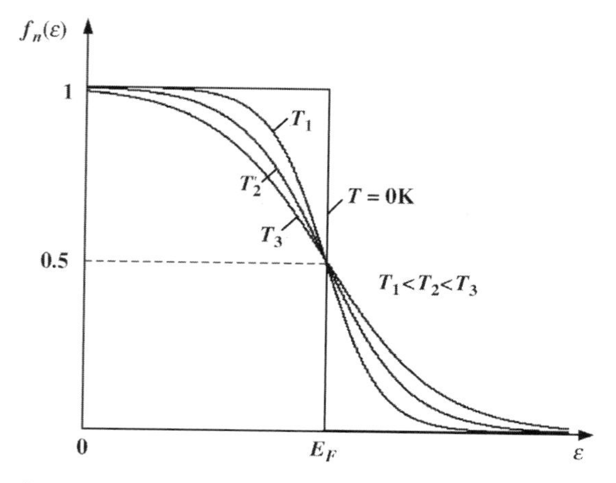

Figure 16.3 The Fermi-Dirac distribution function or occupation factor versus energy at different temperatures.

$f(\varepsilon)$ is generally preserved except that the probability curve is rounded off near E_F. Specifically, $f(\varepsilon)$ is less than unity in the range of a few $k_B T$ below E_F and tails out exponentially a few $k_B T$ beyond E_F, thereby transferring the occupation probability from below E_F to above E_F. This tailing of the occupation probability above E_F is called the Boltzmann tail. With increasing temperature, the round-off energy range is naturally broadened and the Boltzmann tail becomes progressively pronounced.

16.3.1 3D Electrons

A fundamental difference between the distinguishable particles and fermions can be shown by considering a metal. As noted earlier, the metal is represented by a sea of free electrons confined therein by a potential barrier at the surface, called the work function, as sketched in Figure 16.4. The electrons in the metal start filling up the quantum states from the bottom of the energy level to higher lying states

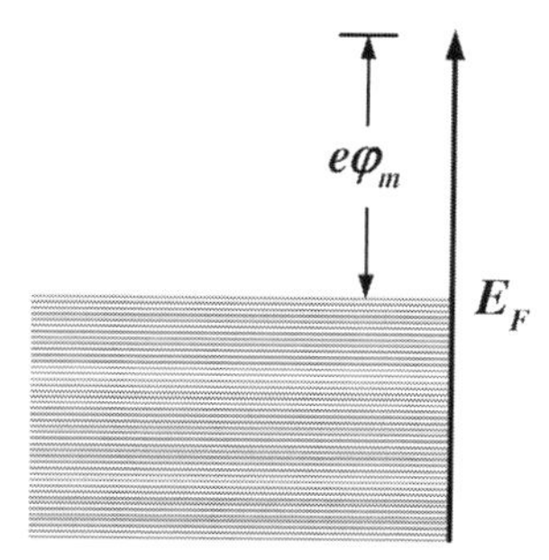

Figure 16.4 A simple model of metal, consisting of sea of electrons filled up to Fermi level E_F and confined by a potential barrier at the surface, called the work function.

according to Pauli exclusion principle. Thus, the total number of electrons N in the metal is given by

$$N = \int_0^\infty d\varepsilon\, Vg(\varepsilon)f(\varepsilon), \tag{16.39}$$

where V is the volume, $g(\varepsilon)$ is the density of states, and $f(\varepsilon)$ is the Fermi occupation factor given in (16.38).

To be specific, one can model the metal to consist of electrons confined in 3D box, in which case (16.39) reads as

$$\frac{N}{V} = \frac{\sqrt{2}m^{3/2}}{\pi^2\hbar^3}\int_0^\infty d\varepsilon \frac{\varepsilon^{1/2}}{1+e^{(\varepsilon-E_F)/k_BT}}; \quad g(\varepsilon) = \frac{\sqrt{2}m^{3/2}\varepsilon^{1/2}}{\pi^2\hbar^3}, \tag{16.40a}$$

where the 3D density of states of electrons (4.27) has been used. The integral in (16.40a) can be compacted by introducing a dimensionless variable of integration, $\eta = \varepsilon/k_BT$, in which case the electron density N/V can be expressed as

$$\frac{N}{V} \equiv N_C \frac{2}{\sqrt{\pi}} F_{1/2}(\eta_f), \quad \eta_f \equiv E_F/k_BT, \tag{16.40b}$$

with

$$N_C \equiv 2\left(\frac{mk_BT}{2\pi\hbar^2}\right)^{3/2} = 2\left(\frac{2\pi mk_BT}{h^2}\right)^{3/2}, \tag{16.40c}$$

and the Fermi 1/2 integral is defined as

$$F_{1/2}(\eta_f) \equiv \int_0^\infty \frac{d\eta\eta^{1/2}}{1+e^{\eta-\eta_f}}, \quad \eta_f \equiv \frac{E_F}{k_BT}, \tag{16.40d}$$

and has been extensively tabulated. Thus, the electron density in the metal has been specified in terms of basic physical parameters and the Fermi integral.

In the limit of zero temperature, $f(\varepsilon)$ reduces to the step function, as discussed (see Figure 16.3), and therefore the Fermi 1/2 integral is readily evaluated as

$$F_{1/2} \equiv \int_0^{E_F(0)/k_BT} d\eta\eta^{1/2} = \frac{2}{3}\left(\frac{E_F(0)}{k_BT}\right)^{3/2}. \tag{16.41}$$

Hence, combining (16.41) with (16.40), one can find the Fermi level as a function of the electron density at $T = 0$ as

$$E_F(0) = \frac{h^2}{8m}\left(\frac{3}{\pi}\frac{N}{V}\right)^{2/3}. \tag{16.42}$$

Figure 16.5 shows $E_F(0)$ as a function of the electron density N/V. As is clear from this figure, the Fermi level increases monotonously with increasing density. For N/V

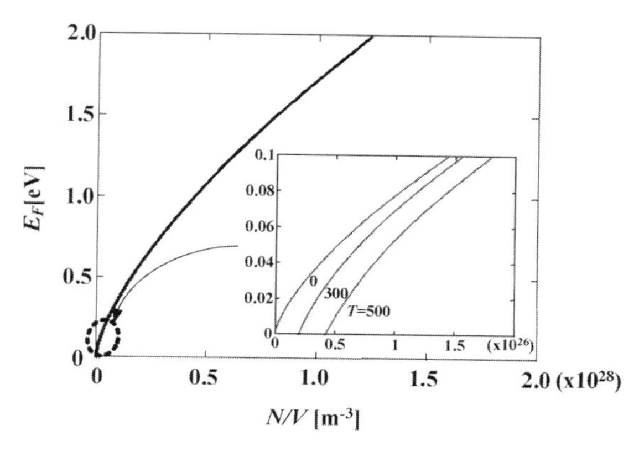

Figure 16.5 The Fermi level versus 3D electron concentration $n = N/L^3$ for $T = 0$ and $T \neq 0$.

of the order of Avogadro's number, that is, $N/V = 10^{29}/m^3$, for example, $E_F(0) \approx 7.9$ eV and the Fermi velocity of the electrons on top of $E_F(0)$, namely,

$$mv_F^2/2 = E_F(0),$$

is as large as $\simeq 1.7 \times 10^6$ ms^{-1} at $T = 0$. This is in marked contrast with the classical distinguishable particles, which should be completely at rest at $T = 0$. Also shown for comparison are Fermi level versus N/V curves, obtained at different temperatures by a simple numerical analysis using (16.40). Clearly, $E_F(T)$ also increases with increasing N/V in the range $0 \leq E_F \leq 0.1\, eV$ but at fixed N/V E_F(T) decreases with increasing T (see inset), as more electrons are in the Boltzmann tail with kinetic energies larger than E_F. However, these E_F versus N/V curves converge essentially into one for $E_F \approx 0.2\, eV$ or larger, in which region E_F is mainly dictated by N/V.

16.3.2
2D Electrons

One can also analyze 2D electron gas system by using (16.39). The number of electrons per unit area is thus given from (16.39) by

$$n_S \equiv \frac{N}{L^2} = \frac{m}{\pi\hbar^2}\int_0^\infty \frac{d\varepsilon}{1 + e^{(\varepsilon - E_F)/k_B T}}, \quad g_{2D} = \frac{m}{\pi\hbar^2}, \tag{16.43}$$

where the 2D density of states has been used from (4.29). For $T = 0$, the Fermi level is simply obtained by

$$n_S = \frac{m}{\pi\hbar^2}\int_0^{E_F(0)} d\varepsilon = \frac{m}{\pi\hbar^2} E_F(0) \tag{16.44}$$

and $E_F(0)$ increases linearly with the 2D electron density.

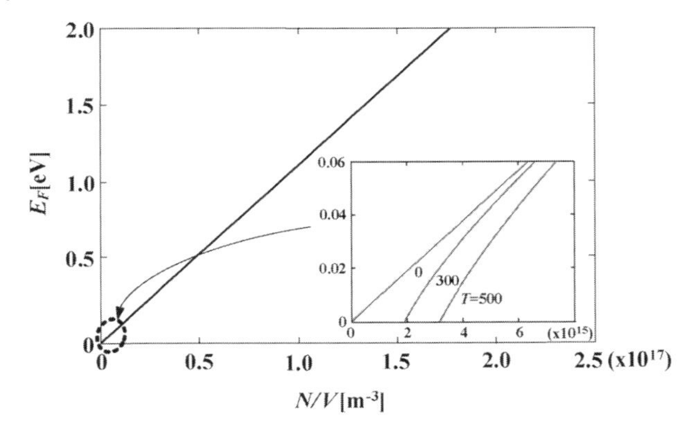

Figure 16.6 The Fermi level versus 2D electron concentration $n_s = N/L^2$ for $T = 0$ and $T \neq 0$.

One can perform the integration in (16.43) exactly, obtaining

$$n_S = \frac{mk_BT}{\pi\hbar^2}\ln(1+e^{\eta(T)}); \quad \eta(T) \equiv \frac{E_F(T)}{k_BT}, \tag{16.45}$$

and correlate $E_F(T)$ with $E_F(0)$ by dividing (16.44) by (16.45):

$$\eta(0) = \ln(1+e^{\eta(T)}); \quad \eta(0) \equiv \frac{E_F(0)}{k_BT}. \tag{16.46}$$

Or by inverting (16.46), one can find $E_F(T)$ as a function of T and in terms of n_S or $E_F(0)$:

$$E_F(T)/k_BT = \ln(e^{E_F(0)/k_BT} - 1). \tag{16.47}$$

Figure 16.6 shows E_F versus n_S curves both for $T = 0$ and $T \neq 0$, obtained from (16.44) and (16.45), respectively. Again, these curves show that the Fermi level increases with increasing n_S, as it should. At fixed n_S, the Fermi level decreases considerably with increasing temperature for $E_F < 0.1$ eV but E_F versus n_s curves again converge into one curve for $E_F \approx 0.2$ eV or higher.

16.4 Statistics for Bosons and Bose–Einstein Distribution Function

Bose–Einstein Distribution Function For bosons, the number of ways for assigning n_s bosons to the sth bin with g_s states can be conveniently analyzed with the aid of Figure 16.7. In this figure, the dots represent assigned bosons and the horizontal axis is divided into g_s quantum states using $g_s - 1$ partitions drawn in vertical lines. Bosons are free of the exclusion principle and there is no restriction on the number of bosons occupying a given quantum state. This means that the total number of distinguishable arrangements or states is to be obtained by $(n_s + g_s - 1)!$, provided the dots and

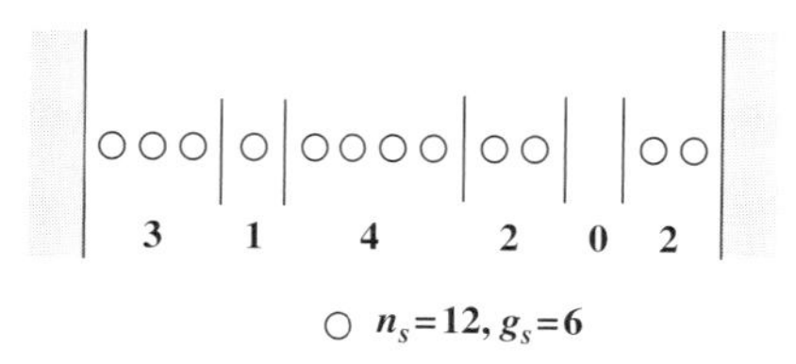

Figure 16.7 A graphical scheme devised for finding the number of microscopically distinguishable arrangements of bosons (white dots) in g_s states bounded by partitions or by a partition and the wall.

partitions are distinguishable. But the dots and partitions are in fact not distinguishable and therefore $n_s!$ permutations among the dots and $(g_s-1)!$ permutations among the partitions do not lead to distinguishable arrangements. Hence, the number of distinguishable arrangements of the bosons in the sth bin is given by

$$P_s = \frac{(n_s+g_s-1)!}{n_s!(g_s-1)!}. \tag{16.48}$$

And the total number of distinguishable arrangements for $\{n_s\}$ set therefore reads as

$$P(n_1,\ldots,n_s,\ldots) = \prod_{s=1}^{\infty} \frac{(n_s+g_s-1)!}{n_s!(g_s-1)!}. \tag{16.49}$$

The most probable distribution $\{n_s\}$ can again be obtained by finding the maximum value of $\ln P(\{n_s\})$, subject to the constraints (16.2). Thus, one may introduce F-function as

$$F(\{n_s\},\alpha,\beta) = \sum_{s=1}^{\infty} \ln\left[\frac{(n_s+g_s-1)!}{n_s!(g_s-1)!}\right] - \alpha\left[\sum_s n_s - N\right] - \beta\left[\sum_s n_s\varepsilon_s - E\right] \tag{16.50}$$

and employ a similar analysis as was used for fermions, for example, including the use of Stirling's formula for $\ln(n_s+g_s-1)!$, $\ln n_s!$, and $\ln\,(g_s-1)!$, obtaining

$$\frac{\partial}{\partial n_s}F(\{n_s\},\alpha,\beta) = \ln\,(n_s+g_s-1) - \ln n_s - \alpha - \beta\varepsilon_s = \ln\frac{n_s+g_s-1}{n_s} - \alpha - \beta\varepsilon_s = 0. \tag{16.51}$$

Hence, one finds from (16.51),

$$n_s = \frac{g_s}{e^{\alpha+\beta\varepsilon_s}-1}. \tag{16.52}$$

With the identification $\beta = 1/k_BT$ in (16.52), the resulting n_s per state

$$f(\varepsilon_s) \equiv \frac{n_s}{g_s} = \frac{1}{e^{\alpha}e^{\varepsilon_s/k_BT}-1} \tag{16.53}$$

is known as the Bose–Einstein distribution function. In a system of boson gas, α approaches zero value as the temperature is lowered. This is because the bosons tend to condense into the lowest energy state as T approaches zero irrespective of the number of bosons involved, so that $f(\varepsilon_s)$ should be sharply peaked at $\varepsilon_s \simeq 0$. This phenomenon is known as Bose condensation and is a general characteristic of a boson gas system. With α put to zero and for $T \to 0$, $f(\varepsilon_s)$ diverges with $\varepsilon_s \to 0$, thereby describing Bose condensation.

Photons as Bosons The photons constitute a typical boson system. When enclosed in a container, the photons are absorbed or emitted by the atoms on the wall. Thus, there is no constraint for the number of photons to be conserved. Therefore, one can put $\alpha = 0$ (see (16.2)) and write the photon distribution function as

$$f(\varepsilon_s) \equiv \frac{n_s}{g_s} = \frac{1}{e^{h\nu_s/k_B T} - 1}, \qquad \varepsilon_s = h\nu_s, \tag{16.54}$$

where ν_s represents the frequency of the sth mode or bin. Also, the total number of states g_s is given by the mode density therein (see (14.37b)) times the volume V, namely,

$$g_s = \frac{8\pi n^3 \nu_s^2}{c^3} V.$$

Hence, the field energy density at the thermodynamic equilibrium is obtained by multiplying n_s by the photon energy, that is,

$$\varrho(\nu_s) \equiv \frac{n_s h\nu_s}{V} = \frac{8\pi n^3 \nu_s^2}{c^3} \frac{h\nu_s}{(e^{h\nu_s/k_B T} - 1)} \tag{16.55}$$

in complete agreement with Planck's theory of radiation. The distribution function of photons (16.54) is plotted versus energy in Figure 16.8 at different temperatures. Also plotted in the figure are corresponding Boltzmann distributions, for comparison. The photon distribution function is clearly shown to exhibit the Bose condensation at low temperatures. Although the Boltzmann distribution function itself is sharply

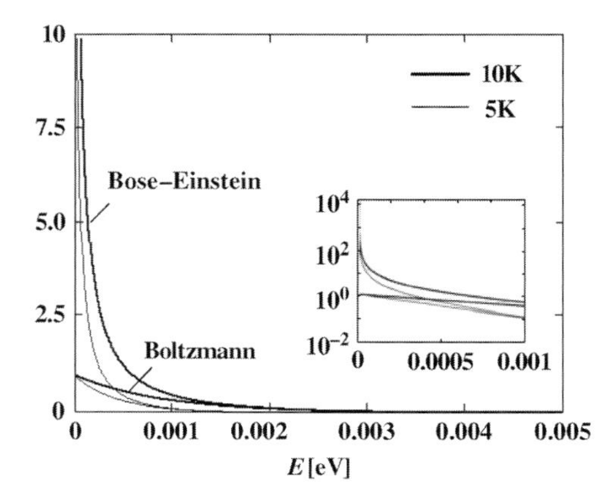

Figure 16.8 The Bose–Einstein distribution function versus energy at different temperatures. Also shown are the corresponding Boltzmann distribution functions for comparison. The inset amplifies the Bose condensation of the photon distribution function in finer energy units.

peaked at zero energy in low temperatures, as discussed in Figure 16.2, it is practically constant in the energy range from 0 to 0.001 eV, while the photon distribution function decreases in the same energy range by more than two orders of magnitude.

16.5 Problems

16.1. An ideal 3D gas system is described by the Boltzmann distribution function

$$f(v_x, v_y, v_z) = Ke^{-m(v_x^2+v_y^2+v_z^2)/2k_BT},$$

with K denoting the normalization constant.

(a) Find the normalization constant K.

(b) Show that the number of particles having the thermal speed between v and $v + dv$ is described by

$$N(v)dv = N\left(\frac{m}{2\pi k_B T}\right)^{3/2} e^{-v^2/k_BT} 4\pi v^2 dv,$$

where N is the total number of particles per volume.

(c) Find the most probable speed, that is, the maximum value of $N(v)$ versus T.

(d) Show that

$$\langle v \rangle = \frac{\int_0^\infty vN(v)dv}{\int_0^\infty N(v)dv} = \left(\frac{8k_BT}{\pi m}\right)^{1/2}.$$

(e) Show that the number of particles having v_x and $v_x + dv_x$ regardless of v_y and v_z is given by

$$N(v_x)dv_x = N\left(\frac{m}{2\pi k_B T}\right)^{1/2} e^{-mv_x^2/k_BT} dv_x.$$

16.2. (a) Show that the average kinetic energy of conduction electrons in a metal at zero temperature is given by

$$\langle E \rangle = \frac{\int_0^\infty dEEg(E)f(E)}{\int_0^\infty dEg(E)f(E)} = 3E_F/5,$$

where $g(E)$ and $f(E)$ are the density of states and Fermi occupation factor, respectively.

(b) Calculate numerically the same average energy versus T and discuss the result, in comparison with the result for $T = 0$.

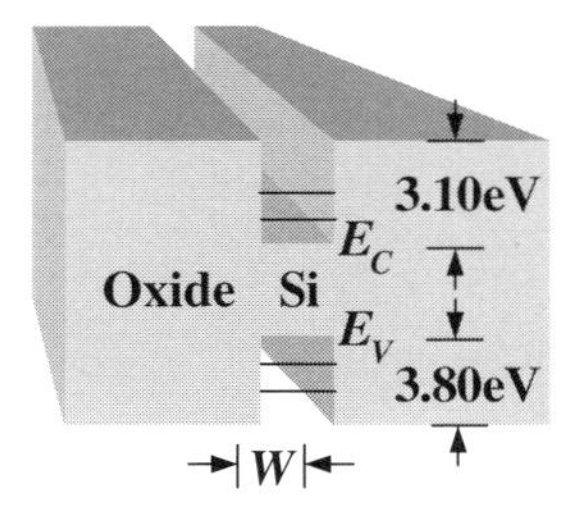

Figure 16.9 Quantum well for electrons and holes provided by silicon in between two silicon dioxide layers.

16.3. The Fermi energy of copper at 0 K is 7.05 eV. Find the average kinetic energy of electrons and the equivalent temperature in an ideal classical gas system having the same kinetic energy.

16.4. Using the distribution function for bosons, find numerically the average energy per photon in equilibrium as a function of the frequency and cavity temperature and discuss the result in comparison with that of electrons.

16.5. Consider the intrinsic silicon sandwiched between two silicon dioxide layers, as shown in Figure 16.9.

(a) Find the sublevels of 2D electrons in eV unit for $W = 500, 100, 10, 1$ nm, respectively.
(b) Calculate the corresponding electron densities in the conduction band.
(c) Repeat the analysis of (a) and (b) for holes in the valence band.

You may approximate the quantum well by infinite square well potentials and use $m_n/m_0 = 0.98$ and $m_p/m_0 = 0.49$ for the effective mass of electrons and holes with m_0 denoting electron rest mass.

16.6. Consider the same intrinsic silicon quantum wire surrounded by silicon dioxide layers, as sketched in Figure 16.10.

(a) Find and discuss the subbands and degeneracy of 1D electrons for $W = 100, 10, 5, 1$ nm, respectively.
(b) Find the corresponding densities of electrons and holes in conduction and valence bands. Use the same m_n, m_p as in 16.5.

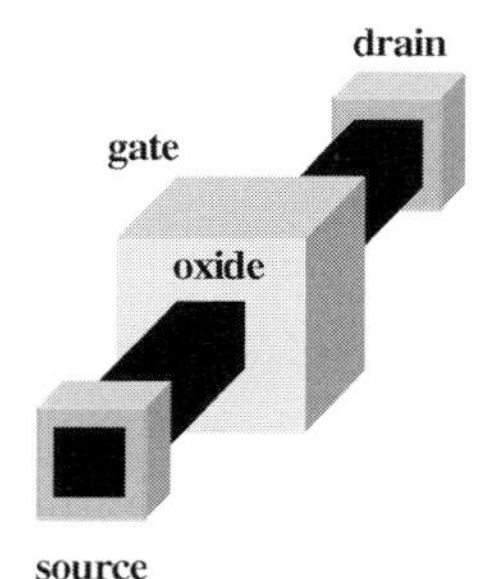

Figure 16.10 Silicon quantum wire surrounded by SiO_2 layer.

Suggested Reading

1 Yariv, A. (1982) *An Introduction to Theory and Applications of Quantum Mechanics*, John Wiley & Sons, Inc.

2 McKelvey, J.P. (1982) *Solid State and Semiconductor Physics*, Krieger Publishing Company.

17
Semiconductor Statistics

The invention of semiconductor transistors is one of the epoch-making achievements made in the twentieth century. The semiconductor devices constitute the flagship technology for digitalization and information technology and are firmly rooted in the concepts inherent in quantum mechanics. The basic principles of the device operation and the underlying concepts are compactly discussed in a few chapters to follow. The first of which will be the semiconductor statistics as a natural follow-up of Chapter 16. It is in essence the quantum statistics at work in semiconducting materials. As well known, the "charge control" is the keyword underpinning the operation of the semiconductor devices currently at work. A factor important for charge control is the carrier density, as determined by doping level, temperature, and other electronic properties of the semiconductor. The carrier densities are quantified as a function of such physical and material parameters.

17.1
Carrier Densities in Intrinsic Semiconductors

An important application of quantum mechanics consists of describing the motion of electrons in solids. As discussed in Kronig–Penny model in Chapter 7, a solid is generally characterized by allowed energy bands separated by forbidden gaps. In allowed bands, an electron moves as a free particle with an effective mass as determined by the dispersion relation operative in the energy band. Of particular interest are the valence and conduction bands, as illustrated in Figure 17.1. The valence band consists of the highest energy states occupied by the valence electrons in the outermost shell of host atoms. The conduction band is the next higher lying allowed energy band.

Conductors, Insulators, and Semiconductors As well known, solids are classified into metals, insulators, and semiconductors. These are in turn characterized by differing configurations of the valence and conduction bands. In metals, the valence electrons constitute a sea of free electrons and the valence and conduction bands overlap and there is no forbidden gap in between (Figure 17.1). This means that the valence electrons can move up to the conduction band upon acquiring kinetic energy or contribute to the current under bias.

Introductory Quantum Mechanics for Semiconductor Nanotechnology. Dae Mann Kim

ISBN: 978-3-527-40975-4

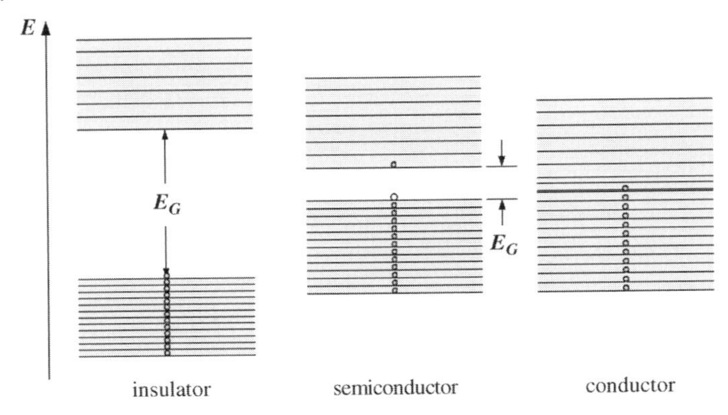

Figure 17.1 The valence (bottom) and conduction (top) energy bands in conductors, insulators, and semiconductors.

An insulator such as silicon dioxide is characterized, on the other hand, by the conduction and valence bands with relatively narrow widths and separated by a large bandgap, typically 10 eV or more (Figure 17.1). In an insulator, the valence electrons form strong bonds with neighboring atoms and these bonds are difficult to break, thereby giving rise to a large bandgap. As a consequence, there are essentially very few electrons in the conduction band to carry currents under bias.

The semiconductors are characterized by the conduction and valence band configurations in between those of metals and insulators. These two bands are separated by the bandgap ranging from about 0.5 eV to a few eV, as depicted in Figure 17.1. Also, the bonds between neighboring atoms are moderately strong, giving rise to narrower bandgap, compared with the insulator. Thus, the bonds are relatively easy to be broken at room temperature and an appreciable number of electrons are promoted into the conduction band via the band-to-band thermal excitation to conduct the current under bias. The holes left behind the valence band are also capable of conducting current under bias.

Electrons and Holes The fact that electrons in the filled valence band are thermally excited into the conduction band unambiguously implies that they leave behind the lack of electrons, that is, holes in the valence band (see Figure 17.1). These holes are to be viewed as the positive charge carriers just as electrons are the negative charge carriers. Also, the motion of these holes can be described in the same way as electrons using the dispersion curves operative in the valence band including the effective mass.

In intrinsic semiconductors, the concentrations of electrons and holes are the same by definition, that is, $n = p \equiv n_i$, and holes can be roughly taken as the mirror image of electrons in many respects, when viewed from the midgap. For example, an electron going up the conduction band by acquiring kinetic energy is matched by a hole going down the valence band upon acquiring the kinetic energy.

Thermal Equilibrium The semiconductor statistics at equilibrium is considered next. In so doing, the properties of the equilibrium are briefly described from different points of view. The thermodynamic equilibrium is generally characterized by a few basic facts: (i) the physical quantities are time invariant, and this is closely correlated with, (ii) every process is balanced by its inverse process (detailed balancing), (iii) the electron and hole concentrations, n and p, are quantified by a single Fermi level E_F, (iv) the Fermi level is spatially flat and also lines up in composite semiconductor system, and finally, (v) the law of mass action holds true, that is, $np = n_i^2$ with n_i denoting the intrinsic concentration.

17.1.1 Electron Concentration

The electron density in the conduction band is specified by a few basic factors:

$$n = \int_{E_C}^{E_C+\Delta E_c} dE g_n(E) f_n(E). \tag{17.1a}$$

Here, ΔE_C is the conduction band width and

$$g_n(E) = \frac{1}{2\pi^2}\left(\frac{2m_n}{\hbar^2}\right)^{3/2}(E-E_C)^{1/2} \tag{17.1b}$$

is the 3D density of states of electrons in the bulk conduction band (see (4.27)). The bottom of the conduction band E_C serves as the reference level from which to define the kinetic energy of electrons moving with the effective mass m_n and also for $g_n(E)$, as shown in Figure 17.2. Thus, $g_n(E)dE$ represents the total number of states per unit

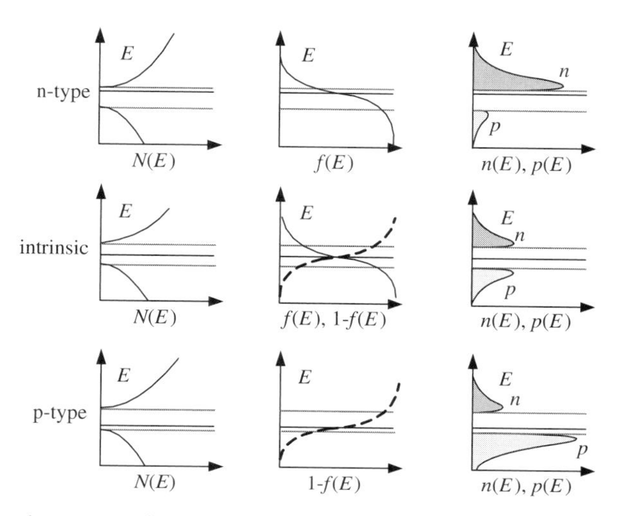

Figure 17.2 The graphical description of electron and hole concentrations in terms of the 3D density of states, Fermi occupation factors for electrons (solid line) and holes (broken line), and the location of Fermi level. n, p are commensurate with the shaded area.

volume between E and $E+dE$. Also, the Fermi–Dirac distribution function

$$f_n(E) = \frac{1}{1+e^{(E-E_F)/k_BT}} \tag{17.1c}$$

represents the probability that an electron occupies the quantum states at E or, more simply, the electron occupation factor in equilibrium (Figure 17.2).

Inserting (17.1b) and (17.1c) into (17.1a) and introducing a dimensionless variable of integration, $\eta \equiv (E-E_C)/k_BT$, and making a further approximation, namely, $\Delta E_C/k_BT \to \infty$, one can obtain in exactly identical manner as in (16.40) the electron concentration as

$$n = \frac{2}{\sqrt{\pi}} N_c F_{1/2}(\eta_{Fn}), \quad \eta_{Fn} \equiv (E_F - E_c)/k_BT, \tag{17.2a}$$

where the quantity

$$N_C \equiv 2\left(\frac{2\pi m_n k_B T}{h^2}\right)^{3/2} \tag{17.2b}$$

is known as the effective density of states at the conduction band and

$$F_{1/2}(\eta_F) = \int_0^\infty \frac{\eta^{1/2} d\eta}{1+e^{\eta-\eta_F}} \tag{17.2c}$$

is the Fermi 1/2 integral as introduced in (16.40c). It should be noted that the approximation, namely, $\Delta E_C/k_BT \to \infty$ is reasonable, since ΔE_C is typically few eV, while $k_BT \simeq 25$ meV at room temperature. Moreover, the Fermi occupation factor cuts off the contribution coming from those states a few k_BT above E_C, as clear from in Figure 17.2.

For the nondegenerate case in which the Fermi level E_F ranges in the energy gap and stays below E_C by a few k_BT so that $\exp -\eta_F \gg 1$, the Fermi 1/2 integral simplifies as

$$F_{1/2}(\eta_{Fn}) \simeq e^{\eta_{Fn}} \int_0^\infty d\eta e^{-\eta} \eta^{1/2} = e^{\eta_{Fn}} \frac{\sqrt{\pi}}{2}. \tag{17.3}$$

Hence, inserting the Boltzmann approximation (17.3) into (17.2a), one can express the nondegenerate electron concentration in a simple analytic form as

$$n = N_C e^{-(E_C-E_F)/k_BT}. \tag{17.4}$$

Thus, n is shown to increase exponentially with temperature following the power law $1/T$, and becomes equal to N_C when E_F coincides with E_C. Plotted in Figure 17.3 are both the Fermi 1/2 integral and its Boltzmann approximation given in (17.3) versus η_{Fn}. These two curves practically coincide with each other for $\eta_{Fn} \equiv (E_F - E_C)/k_BT \leq -2$, thereby indicating the range of E_F for the nondegenerate statistics to be valid. For $\eta_{Fn} \geq -2$, the two curves start to depart appreciably from

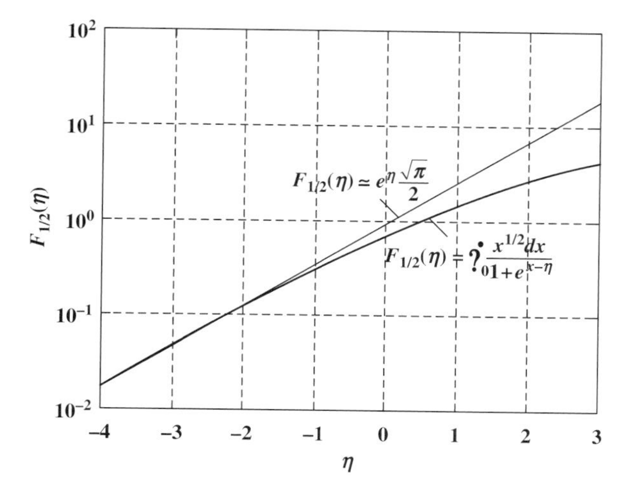

Figure 17.3 The Fermi 1/2 integral $F_{1/2}(\eta)$ and its Boltzmann approximation $(\sqrt{\pi}/2)\exp\eta$ versus $\eta[=(E_F-E_C)/k_BT]$.

each other and the gap between the two widens progressively as E_F is further raised. Consequently, the increase of n is shown to slow down considerably from the projected value of Boltzmann approximation as E_F is raised above E_C.

17.1.2 Hole Concentration

One can find the hole concentration in a similar manner. The hole concentration is given by

$$p = \int_{E_V-\Delta E_V}^{E_V} dE g_p(E) f_p(E), \tag{17.5a}$$

where ΔE_V is the valence band width and

$$g_p(E) = \frac{1}{2\pi^2}\left(\frac{2m_p}{\hbar^2}\right)^{3/2}(E_V-E)^{1/2} \tag{17.5b}$$

is the hole density of states in the valence band with E_V serving as the reference level of the kinetic energy of holes moving with the effective mass m_p (Figure 17.2). Then, $g_p(E)dE$ represents the total number of hole states per unit volume between E and $E+dE$. Naturally, the hole occupation factor represents, by definition, the probability that the state is not occupied by the electron, namely,

$$f_p(E) \equiv 1-\frac{1}{1+e^{(E-E_F)/k_BT}} = \frac{1}{1+e^{(E_F-E)/k_BT}}. \tag{17.5c}$$

Thus, inserting (17.5b) and (17.5c) into (17.5a) and introducing the variable of integration, $\eta \equiv (E_V - E)/k_B T$, and again making the approximation $\Delta E_V/k_B T \to \infty$ (see Figure 17.2), one obtains

$$p = \frac{2}{\sqrt{\pi}} N_V F_{1/2}(\eta_{Fp}), \quad \eta_{Fp} \equiv (E_V - E_F)/k_B T, \tag{17.6a}$$

where the effective density of states at the valence band is given by

$$N_V \equiv 2\left(\frac{2\pi m_p k_B T}{h^2}\right)^{3/2}. \tag{17.6b}$$

For the nondegenerate case in which E_F ranges in the energy gap, a few $k_B T$ above E_V, the Fermi 1/2 integral simplifies as

$$F_{1/2}(\eta_{Fp}) \simeq e^{\eta_{Fp}} \int_0^\infty d\eta e^{-\eta} \eta^{1/2} = e^{\eta_{Fp}} \frac{\sqrt{\pi}}{2}, \tag{17.7}$$

and by inserting (17.7) into (17.6a), one obtains a simple analytic expression of the nondegenerate hole concentration:

$$p = N_V e^{-(E_F - E_V)/k_B T}. \tag{17.8}$$

Again, p increases exponentially with increasing T following the power law $1/T$.

In intrinsic semiconductors, electrons in the valence band are continually excited into the conduction band at a finite temperature, leaving behind the same number of holes in the valence band. This excitation process is balanced by its inverse process of the electron–hole recombination in equilibrium. Thus, the electron and hole concentrations are identical, that is, $n = p = n_i$. The intrinsic concentration is therefore given from (17.4) and (17.8) by

$$\begin{aligned} n_i &\equiv \sqrt{np} = \sqrt{N_C N_V} e^{-[(E_C - E_F) + (E_F - E_V)]/2k_B T} \\ &= \sqrt{N_C N_V} e^{-E_G/2k_B T}, \quad E_G \equiv E_C - E_V. \end{aligned} \tag{17.9}$$

Figure 17.4 shows the intrinsic carrier concentrations in silicon, germanium, and gallium arsenide versus the inverse temperature $1/T$. Indeed, n_i varies exponentially with $1/T$ and this temperature variation is drastically accentuated with increasing bandgap of the material. Also, n_i exponentially increases with decreased bandgap at given T, as more electron–hole pairs are thermally excited across the smaller bandgap. For instance, in Si with the bandgap of about 1.12 eV at room temperature, $n_i = 1.45 \times 10^{10}\,\text{cm}^{-3}$, while in GaAs with the bandgap of 1.424 eV, $n_i = 1.79 \times 10^{6}\,\text{cm}^{-3}$. Thus, n_i varies by nearly four orders of magnitude due to the difference in bandgap of about 0.3 eV.

Intrinsic Fermi Level, E_{Fi} The location of the Fermi level is generally determined from the charge neutrality condition. In intrinsic semiconductors, the charge

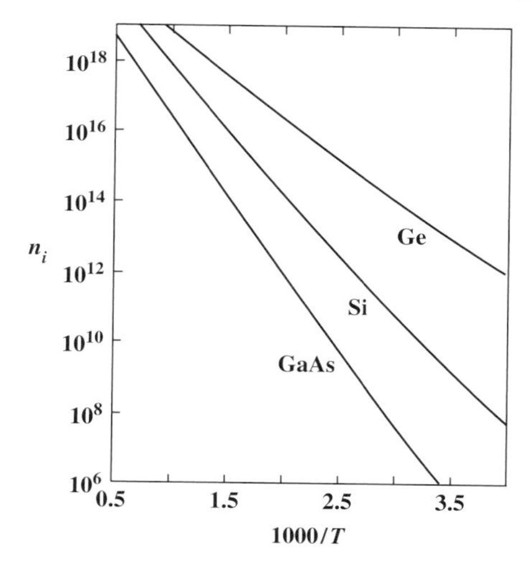

Figure 17.4 The intrinsic carrier concentration versus $1000/T$ in germanium, silicon, and gallium arsenide.

neutrality condition is simply given by $n = p$. This in turn can be explicitly specified from (17.4) and (17.8) by

$$N_C e^{-(E_C - E_{Fi})/k_B T} = N_V e^{-(E_{Fi} - E_V)/k_B T} \tag{17.10a}$$

and one can therefore specify E_{Fi} from (17.10a) as

$$E_{Fi} = \frac{1}{2}(E_C + E_V) + \frac{3k_B T}{4}\ln\frac{m_p}{m_n}; \quad \ln\frac{N_V}{N_C} = \frac{3}{2}\ln\frac{m_p}{m_n}. \tag{17.10b}$$

Clearly, E_{Fi} is located very close to the midgap and the departure from the midgap arises from the difference in the effective masses of electron and hole and amounts to a fraction of the thermal energy $k_B T$ which is $\approx 25meV$ at room temperature.

17.2
Carrier Densities in Extrinsic Semiconductors

The control of n and p is naturally one of the key factors for controlling the charge contents and is done by doping impurity atoms via, for example, the ion implantation technique. The statistics of n and p in doped extrinsic semiconductors is therefore important and analyzed in this section. In so doing, however, a brief discussion of the physics of doping the semiconductor with impurity atoms is in order. For silicon, for example, whose electron configuration is specified by $[\mathrm{Ne}]3\mathrm{s}^2 3\mathrm{p}^2$, there are four valence electrons outside the closed neon core and Si atoms are covalently bonded with its four neighbors by sharing one valence electron with each other, so that the subshell is filled up, as sketched in Figure 17.5. Doping consists of incorporating

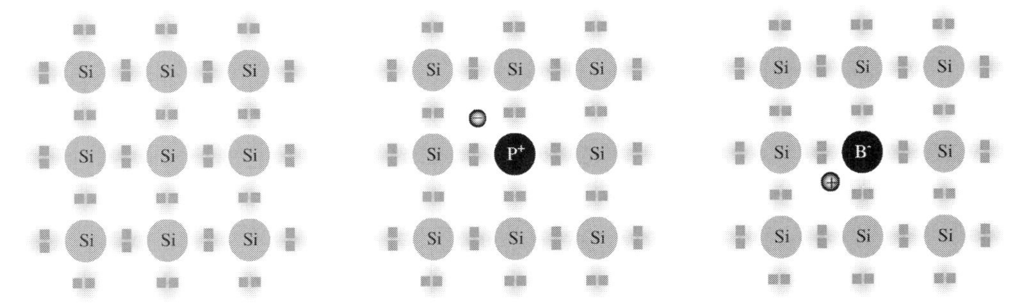

Figure 17.5 The donor and acceptor atoms in tetrahedrally bonded silicon, forming loosely bound hydrogenic atoms, consisting of positively charged nucleus and electron in the former and negatively charged nucleus and hole in the latter.

impurity atoms at substitutional sites (Figure 17.5) and these impurity atoms generally divide into two groups, namely, donors and acceptors.

Donors and Acceptors The donor atoms are from Column V in periodic table, for example, phosphorus ([Ne]$3s^2 3p^3$) or arsenic ([Ar]$4s^2 3d^{10} 4p^3$) atoms, which have five valence electrons instead of four as in silicon atoms. Of these, four valence electrons are used up in the tetrahedral bonding, thereby replacing the role of four valence electrons in Si atom. The remaining fifth electron forms a hydrogenic atom with P^+ or As^+ ion core (Figure 17.5).

On the other hand, the acceptor atoms are from the Column III, for instance, boron ([He]$2s^2 2p$) with three valence electrons. These atoms cannot complete the required tetrahedral bonding but readily accept an electron from other Si–Si bonds, thereby completing its bonding, but in the process creates a hole in the valence band (Figure 17.5). The resulting negative ion and hole again form a hydrogenic atom.

The ionization energy of the donor atom, for example, can be estimated by transcribing the ionization energy of the hydrogen atom as follows. As detailed earlier, the ionization energy in hydrogen atom is the energy required to release the electron from the ground state and is given from (10.43) or (2.35) by

$$E_D = -\frac{e_M^2}{2a_0}, \quad a_0 \equiv \frac{\hbar^2}{m_n e_M^2}, \; e_M^2 = \frac{e^2}{4\pi\varepsilon_S}, \tag{17.11}$$

where the effective electron mass m_n and the permittivity of the semiconductor, ε_S, have replaced the rest mass of the electron and the vacuum permittivity appearing in the ionization energy of the hydrogen atom. In this manner, the effective radius of the donor atom, a_0, can be scaled in units of basic Bohr radius in H-atom, namely, 0.05 nm:

$$a_0 \equiv \frac{\hbar^2 4\pi\varepsilon_S}{m_n e^2} = \left(\frac{\hbar^2 4\pi\varepsilon_0}{m_0 e^2}\right)\left(\frac{m_0}{m_n}\right)\left(\frac{\varepsilon_S}{\varepsilon_0}\right) = 0.05\left(\frac{m_0}{m_n}\right)\left(\frac{\varepsilon_S}{\varepsilon_0}\right)\text{nm}. \tag{17.12}$$

Likewise, the ionization energy E_D can be scaled in units of the ionization energy of H-atom, namely, 13.64 eV, as

$$E_D = \frac{e^4 m_n}{2\hbar^2(4\pi\varepsilon_S)^2} = \frac{e^4 m_0}{2\hbar^2(4\pi\varepsilon_0)^2}\left(\frac{m_n}{m_0}\right)\left(\frac{\varepsilon_0}{\varepsilon_S}\right)^2 = 13.64\left(\frac{m_n}{m_0}\right)\left(\frac{\varepsilon_0}{\varepsilon_S}\right)^2 \text{ eV}. \tag{17.13}$$

Thus, with $m_n/m_0 \approx 0.98, 0.2$, depending on the crystallographic directions, and $\varepsilon_S/\varepsilon_0 \simeq 12$ in silicon, for example, the effective radius of the donor atom is larger than the radius of the hydrogen atom by the factor of about 60–12, amounting approximately to 2.9–0.5 nm. Also, the ionization energy is reduced to about 20–100 meV, a few thermal energies at room temperature. It is thus clear that the fifth valence electron in the donor atom is loosely bound to the donor ion and can be readily promoted into the conduction band to become freely mobile, hence the name the donor. This has been illustrated in Figure 17.6.

The similar estimations can also be carried out for acceptor atoms and the ionization energy of the hole can also be shown to be about the same as that of electrons in donor atoms. Thus, the acceptor atoms can readily accept electrons from the valence band, creating holes in the valence band, hence the name the acceptor (Figure 17.6).

17.2.1
Donor and Acceptor Statistics

The statistics of donor and acceptor atoms can be analyzed in a way similar to the Fermi–Dirac statistics at equilibrium, except that the valence requirement of the degenerate states has to be accounted for. Thus, consider the donor impurity atoms incorporated into the semiconductor. For the given donor level, ε_s and g_s quantum states therein, the number of distinct arrangements of n_s electrons can be analyzed in exactly the same way as was done for the statistics of fermions and is given by

$$P_s = \frac{g_s(g_s-g_D)\ldots(g_s-g_D(n_s-1))}{n_s!} = \frac{g_D^{n_s}(g_s/g_D)!}{n_s!(g_s/g_D-n_s)!}. \tag{17.14}$$

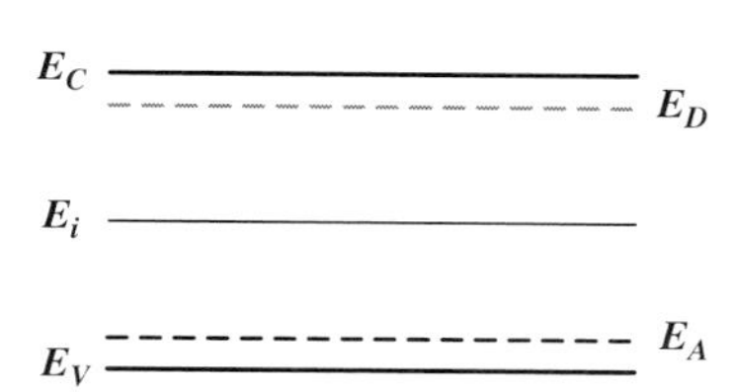

Figure 17.6 Donor and acceptor levels *vis-à-vis* E_C and E_V. The solid line represents the extended nature of the electron states in conduction and valence bands, while the broken lines denote the localized donor or acceptor levels.

Here, the only difference existing between (17.14) and (16.31) is the introduction of the degeneracy factor g_D arising from the valence requirement of the impurity atom to be satisfied by one electron only. That is, when one state is occupied, the occupancy of other degenerate states is precluded. Hence, repeating the same argument as in Section 16.1, one can write for the total number of distinguishable arrangements as

$$P(n_1, \ldots, n_s, \ldots) = \prod_{s=1}^{\infty} \frac{g_D^{n_s} (g_s/g_D)!}{n_s!(g_s/g_D - n_s)!}. \tag{17.15}$$

Next, to find the maximum number of arrangements one can introduce the F-function in the usual way as

$$F(\{n_s\}, \alpha, \beta) = \ln P(\{n_s\}) - \alpha\left[\sum_s n_s - N\right] - \beta\left[\sum_s n_s \varepsilon_s - E\right], \tag{17.16}$$

where the constraints used are the same as in (16.2). Thus, the maximum value of F-function is found by using the Stirling's formula (16.14) for factorials in (17.15) and putting the derivative of F with respect to n_s to zero:

$$\frac{\partial}{\partial n_s} F(\{n_s\}, \alpha, \beta) = \ln g_D - \ln n_s + \ln (g_s/g_D - n_s) - \alpha - \beta\varepsilon_s = 0. \tag{17.17}$$

Hence, the most probable distribution of electrons is readily found from (17.17):

$$n_s = \frac{g_s/g_D}{1 + (1/g_D)e^{\alpha + \beta\varepsilon_s}}. \tag{17.18}$$

The Lagrange undetermined multipliers α and β can again be identified and treated in a way similar to (16.26) and (16.36), and one can thus write

$$n_s = \frac{g_s/g_D}{1 + (1/g_D)e^{(\varepsilon_s - E_F)/k_B T}}. \tag{17.19}$$

Now the density of donor atoms N_D incorporated into the semiconductor is to be identified as g_s/g_D, while ε_s is the ground-state energy of the donor atom E_D lying below E_C by a few thermal energies at room temperature, as shown in Figure 17.6. Therefore, with the identifications $\varepsilon_s \equiv E_D$ and $n_s \equiv n_D$, the distribution of electrons in donor atoms is given by

$$n_D = \frac{N_D}{1 + (1/g_D)e^{(E_D - E_F)/k_B T}}. \tag{17.20}$$

It therefore follows from (17.20) that the ionized donor atoms are given by

$$N_D^+ \equiv N_D - n_D = \frac{N_D}{1 + g_D e^{(E_F - E_D)/k_B T}}. \tag{17.21}$$

One can find the distribution of the holes in acceptor atoms in similar manner, obtaining

$$p_A = \frac{N_A}{1+(1/g_A)e^{(E_F-E_A)/k_BT}} \tag{17.22}$$

and

$$N_A^- \equiv N_A - p_A = \frac{N_A}{1+g_A e^{(E_A-E_F)/k_BT}}, \tag{17.23}$$

where N_A and E_A are the acceptor concentration and energy level, respectively (Figure 17.6).

17.3 Fermi Level in Extrinsic Semiconductors

Now that the donor and acceptor ions incorporated into the semiconductor have been quantified in terms of the pertinent semiconductor parameters, the Fermi energy E_F in extrinsic semiconductors in equilibrium can be found by imposing the charge neutrality condition, namely,

$$n + N_A^- = p + N_D^+, \tag{17.24}$$

where n, p, N_D^+, and N_A^- are specified in terms of E_F from (17.2a), (17.6a),(17.21), and (17.23), respectively. Hence, the charge neutrality condition (17.24) prevailing at every point of the extrinsic semiconductor uniquely determines the location of the Fermi level in terms of the doping level, temperature, the bandgap of the material, and so on.

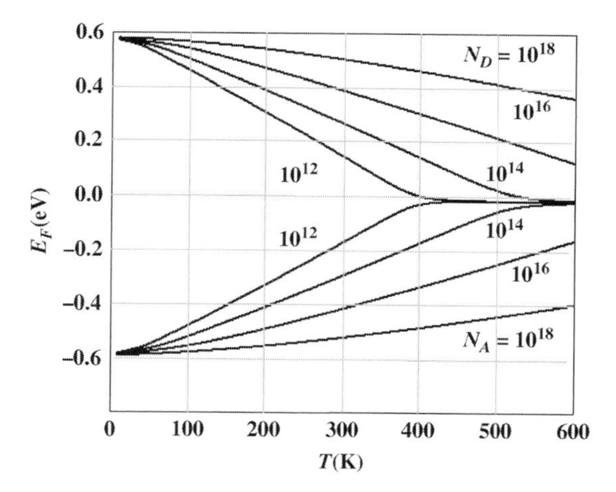

Figure 17.7 The Fermi level versus T in silicon at different doping levels of donor and acceptor atoms. The bandgap weakly depends on T given by $E_G(T) = 1.17 - \alpha T^2/(T+\beta)$ eV, $\alpha = 4.73\times 10^{-4}$, $\beta = 636$, and $m_n = 0.98m_0$ and $m_p = 0.49m_0$.

Figure 17.7 shows the location of the Fermi level found numerically from (17.24) as a function of temperature in both n- and p-type silicon at different doping levels. In n-type silicon, where $N_A = 0$, E_F level is monotonously raised above the midgap with increasing N_D, as it should. Also, for the given N_D, E_F level is lowered with increasing temperature to approach asymptotically the intrinsic Fermi level near the midgap. This is because the electron concentration at high temperature is primarily determined by the thermal excitation, regardless of the doping level. In p-type silicon, where $N_D = 0$, the behavior of E_F versus N_A and T essentially mirrors the behavior of E_F in n-type silicon.

To better understand the behavior of E_F in Figure 17.7 in correlation with n in n-type semiconductor, both n and N_D^+/N_D are plotted in Figure 17.8 versus the inverse temperature $1000/T$ at different doping levels N_D. In this figure, one can notice the three regimes of the electron concentration. The first of which is the intrinsic regime at high T, in which n is primarily determined by temperature regardless of N_D, as more electron–hole pairs are thermally generated via the band-to-band excitation. In this regime, E_F is located close to the midgap. Hence, the probability of electrons to occupy E_D level well above E_F is vanishingly small and nearly all of the donor atoms are ionized, thereby donating the valence electrons to the conduction band. However, the number of electrons donated is understandably smaller than n_i at high T, as is clearly shown in the figure.

With decreasing temperature, however, the intrinsic region is followed by the saturation regime, in which n is practically determined by N_D and nearly all of the donor atoms are ionized as E_D is still well above E_F. It is in this regime that the electron concentration is mainly controlled by the doping level. With further decrease of temperature, the saturation range is followed by the freeze-out regime in which n is decreased nearly exponentially with decreasing T. This is attributed to two factors, namely, no appreciable thermal excitation of electron–hole pairs and the electrons donated by the dopant atoms are captured back by the ionized atoms. This is made explicitly clear by E_F approaching and surpassing E_D level with decreasing

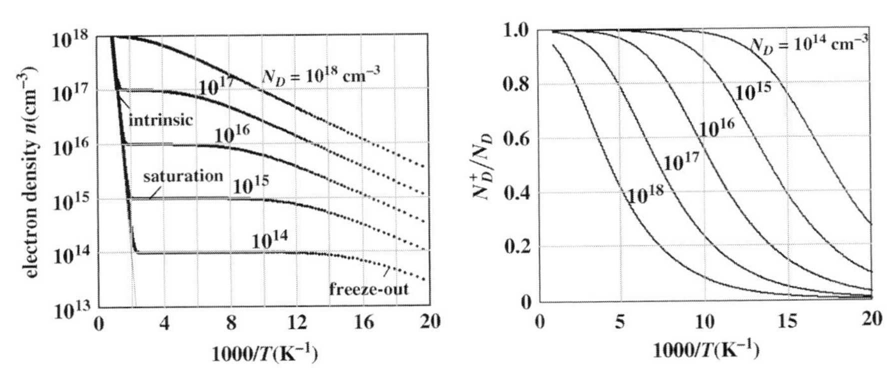

Figure 17.8 The electron concentration and ionized donors N_D^+/N_D in silicon versus $1000/T$ at different doping levels.

T (Figure 17.7) and the fraction N_D^+/N_D rapidly decreasing with decreasing T (Figure 17.8). One can understand the behavior of the hole concentrations as a function of temperature for various doping levels in similar contexts.

Fermi Potentials In the nondegenerate regime, one can also specify n and p in terms of the intrinsic carrier concentration n_i by

$$n = N_C e^{-(E_C - E_{Fi} + E_{Fi} - E_F)/k_B T} = n_i e^{(E_F - E_{Fi})/k_B T}, \tag{17.25}$$

$$p = N_V e^{-(E_F - E_{Fi} + E_{Fi} - E_V)/k_B T} = n_i e^{(E_{Fi} - E_F)/k_B T}, \tag{17.26}$$

where the intrinsic carrier concentration is given in terms of the intrinsic Fermi level $E_{Fi} \approx E_i$ as

$$n_i = N_C e^{-(E_C - E_{Fi})/k_B T} = N_V e^{-(E_{Fi} - E_V)/k_B T}. \tag{17.27}$$

Also, in the saturation regime in which nearly all the dopant atoms are ionized, n can be well approximated by

$$n \simeq N_D = n_i e^{(E_F - E_{Fi})/k_B T}, \tag{17.28}$$

so that the Fermi potential,

$$q\varphi_{Fn} \equiv E_F - E_{Fi} \simeq E_F - E_i,$$

defined as the difference between E_F and the midgap E_{Fi}, is found explicitly as

$$\varphi_{Fn} = (k_B T/q)\ln(N_D/n_i). \tag{17.29}$$

One can likewise specify the Fermi potential $q\varphi_{Fp} \equiv E_i - E_F$ in the p-type semiconductor as

$$\varphi_{Fp} = (k_B T/q)\ln(N_A/n_i). \tag{17.30}$$

It is thus clear from (17.29) and (17.30) that the Fermi level in n-type semiconductor is raised above that in p-type by the sum of these Fermi potentials, as is clear from Figure 17.9.

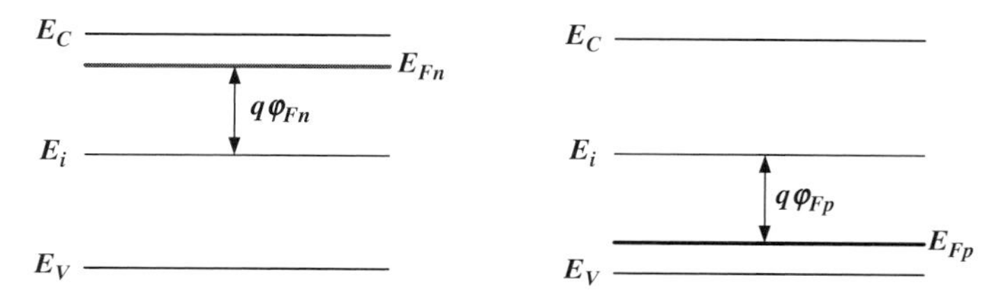

Figure 17.9 The Fermi potentials in n- and p-type semiconductors.

17.4
Problems

17.1. Derive the distribution of holes in p-type semiconductor (17.22), together with ionized acceptor atoms (17.23) filling in algebraic steps.

17.2. The minority hole concentrations in n-type silicon is found to be $10, 10^2, 10^5, 10^8$ per cm^3, respectively.

(a) Find the corresponding electron concentrations at $T = 10, 100, 300, 500$ K.

(b) Calculate the corresponding doping level and ionized dopant atoms.

(c) Discuss whether or not the analytical nondegenerate statistics can be used in (a) and (b).

17.3. In the limit of $T \rightarrow 0$ K, E_F is raised above E_D and approaches E_C regardless of the donor concentrations in n-type silicon. Likewise, with $T \rightarrow 0$ K, E_F is lowered below E_A and approaches E_V in p-type silicon (see Figure 17.7). Are these behaviors to be expected? Discuss the reasons for such behavior.

17.4. The fabrication of laser diode requires heavily doped n- and p-type semiconductors.

(a) The Fermi level in n-type GaAs is located above E_C by 0.1 eV. Find the corresponding electron concentration in the conduction band.

(b) Calculate the doping level required and the fraction of ionized donor atoms.

(c) Repeat the calculations in (a) and (b) for the case where the Fermi level in p-type GaAs is below E_V by 0.15 eV. The bandgap in GaAs is 1.424 eV at room temperature and $m_n/m_0 = 0.067$ and $m_p/m_0 = 0.45$, respectively with m_0 denoting the electron rest mass.

17.5. Plot the hole concentrations and N_A^-/N_A in p-type silicon versus $1000/T$ for $N_A = 10^{14}, 10^{16}, 10^{18}, 10^{20}$ per cm^3.

17.6. Calculate the Fermi potentials in n- and p-type silicon and gallium arsenide for N_D and N_A ranging from 10^{17}–10^{21} per cm^3 for $T = 10, 100, 300, 500$ K. Plot $q\varphi_{Fn} - q\varphi_{Fp}$ versus the common doping level $N(= N_D = N_A)$ in the same doping range for $T = 10, 100, 300, 500$ K.

17.7. Consider an intrinsic silicon sandwiched between two silicon dioxide layers W distance apart, as sketched in Figure 16.9.

(a) Discuss the ground and first excited subbands versus W.

(b) Find the electron concentrations in the conduction band for $W = 500, 100, 10$ nm for $T = 10, 100, 300, 500$ K.

(c) Compare the results obtained in (b) with n_i in bulk silicon at same temperatures.

(d) Specify the charge neutrality condition for finding E_F when the intrinsic silicon in (a) is doped with donor atoms and discuss how n is affected by the quantum effects.

For simplicity of analysis, you may approximate the quantum well of electrons by infinite square well potential and use the analytical energy eigenvalue expression. Take $m_n/m_0 = 0.98$.

17.8. Consider a quantum wire consisting of intrinsic silicon of cross-sectional area W^2. The quantum wire is surrounded by silicon dioxide, as sketched in Figure 16.10.

(a) Discuss ground and first excited subbands versus W^2.
(b) Find the electron concentrations in the conduction band for $W = 100$, 10 nm for $T = 70, 300, 500$ K.
(c) Compare the results obtained in (b) with electron concentrations resulting from classical analysis.

For simplicity of analysis, you may approximate the quantum wire by an infinite square potentials. Take $m_n/m_0 = 0.98$.

Suggested Reading

1 Blakemore, J.S. (2002) *Semiconductor Statistics*, Dover Publications.
2 McKelvey, J.P. (1982) *Solid State and Semiconductor Physics*, Krieger Publishing Company.
3 Sze, S.M. and Ng, K.K. (2006) *Physics of Semiconductor Devices*, 3rd edn, Wiley-Interscience.
4 Pierret, R.F. (1988) *Semiconductor Fundamentals, Modular Series on Solid State Devices*, 2nd edn, vol. I, Prentice Hall.
5 Pierret, R.F. (2002) *Advanced Semiconductor Fundamentals, Modular Series on Solid State Devices*, 2nd edn, vol. VI, Prentice Hall.
6 Muller, R.S., Kamins, T.I., and Chan, M. (2002) *Device Electronics for Integrated Circuits*, 3rd edn, John Wiley & Sons, Inc.
7 Streetman, B.G. and Banerjee, S. (2005) *Solid State Electronic Devices*, 6th edn, Prentice Hall.
8 Grove, A.S. (1967) *Physics and Technology of Semiconductor Devices*, John Wiley & Sons, Inc.
9 Yariv, A. (1982) *An Introduction to Theory and Applications of Quantum Mechanics*, John Wiley & Sons, Inc.
10 Pierret, R.F. (2002) *Advanced Semiconductor Fundamentals, Modular Series on Solid State Devices*, 2nd edn, vol. VI, Prentice Hall.

18
Charge Transport in Semiconductors

In addition to the carrier densities n and p, the transport of charge carriers constitutes another key factor for controlling the charge. In conventional semiconductor devices, the charge transport is done primarily via drift and diffusion. The former is based upon the drift the carriers undergo in between collisions when driven by the external electric field. The latter comes about due to the diffusion of carriers, driven by the concentration gradient. These two basic transport processes give rise to the drift and diffusion currents, respectively.

Equally important, however, are the recombination and generation currents that are brought about by injection or extraction of charge carriers above or below the equilibrium level. The recombination or generation processes are drastically enhanced in the presence of traps in the energy gap and the trap-assisted recombination and generation are considered in some detail. Also the roles of Fermi level and quasi-Fermi level are discussed in equilibrium and away from equilibrium, respectively.

18.1
Drift and Diffusion Currents

Continuity Equation Consider the one-dimensional volume element of unit cross-sectional area at x, as shown in Figure 18.1. The time rate of change of the electron charge therein is determined by the net input current density:

$$\frac{\partial}{\partial t}(-qndx) = J_n(x) - J_n(x+dx) = -\frac{\partial J_n(x)}{\partial x}dx. \tag{18.1}$$

From here on the basic electronic charge $-e$ is also denoted by $-q$. In (18.1), the current density $J_n(x+dx)$ at the output surface has been Taylor expanded at x and truncated after the term linear in dx. Thus, (18.1) reduces to

$$\frac{\partial n}{\partial t} = \frac{1}{q}\frac{\partial J_n(x)}{\partial x}. \tag{18.2}$$

Now, the electron current density consists of both drift and diffusion components, namely,

Introductory Quantum Mechanics for Semiconductor Nanotechnology. Dae Mann Kim

ISBN: 978-3-527-40975-4

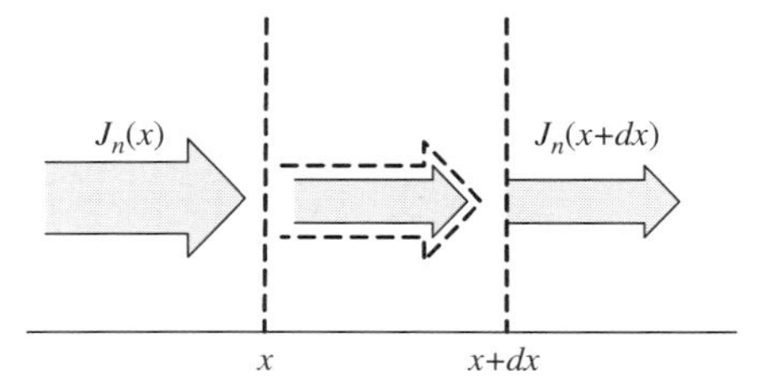

Figure 18.1 A sketch for the charge conservation in 1D differential volume element.

$$J_n = J_{n\,\text{drift}} + J_{n\,\text{diff}}. \tag{18.3}$$

As discussed in Chapter 1, the drift current is contributed by the drift velocity v_{dn} the electrons acquire on the average in between collisions, when driven by external electric field E, that is,

$$J_{n\,\text{drift}} \equiv (-q) n v_{dn} = q n \mu_n \mathrm{E}, \quad v_{dn} = -\mu_n \mathrm{E}, \tag{18.4}$$

with μ_n denoting the electron mobility, a key transport coefficient connecting the input field to the output drift velocity.

On the other hand, the diffusion current is induced by the spatial gradient of the electron density profile, namely,

$$J_{n\,\text{diff}} \propto -\frac{dn}{dx} \equiv (-q) D_n \left(-\frac{dn}{dx}\right), \tag{18.5}$$

where D_n is the electron diffusion coefficient, another key transport coefficient accounting for the net diffusion of electrons from higher to lower concentrations.

Hence, the total electron current density consists of the drift and diffusion components:

$$J_n = q n \mu_n \mathrm{E} + q D_n \frac{dn}{dx} \equiv q F_n. \tag{18.6}$$

The quantity F_n introduced in (18.6) represents the electron flux per unit cross-sectional area per unit time. Combining (18.2) and (18.6), one obtains the well-known continuity equation for electrons, which also represents the conservation of charge and/or matter, namely,

$$\frac{\partial n}{\partial t} = \frac{d}{dx}\left(n \mu_n \mathrm{E} + D_n \frac{dn}{dx}\right). \tag{18.7}$$

One can similarly write down the continuity equation for holes by changing the polarity of the charge, namely,

$$J_p = q p \mu_p \mathrm{E} + q D_n \left(-\frac{dp}{dx}\right) \equiv q F_p, \quad v_{dp} \equiv \mu_p \mathrm{E}, \tag{18.8}$$

and

$$\frac{\partial p}{\partial t} = -\frac{d}{dx}\left(p \mu_p \mathrm{E} - D_p \frac{dp}{dx}\right). \tag{18.9}$$

Generalizing these two continuity equations from one to three dimensions is straightforward and one can write

$$\frac{\partial n}{\partial t} = \frac{1}{q}\nabla \cdot \boldsymbol{J}_n, \quad \boldsymbol{J}_n = q\mu_n n\mathbf{E} + qD_n\nabla n, \tag{18.10a}$$

$$\frac{\partial p}{\partial t} = -\frac{1}{q}\nabla \cdot \boldsymbol{J}_p, \quad \boldsymbol{J}_p = q\mu_p p\mathbf{E} - qD_p\nabla p. \tag{18.10b}$$

18.2 Transport Coefficients

Drift Velocity and Mobility As noted earlier, the drift velocity v_{dn} of electrons is driven by the external electric field E in between collisions. Thus, given the mean collision time τ_n, the drift velocity can be roughly specified by the product of the acceleration and the collision time, that is,

$$\boldsymbol{v}_{dn} \simeq \frac{(-q)\mathbf{E}}{m_n}\tau_n \equiv -\mu_n\mathbf{E}, \tag{18.11}$$

where

$$\mu_n = \frac{q\tau_n}{m_n} \tag{18.12}$$

is the electron mobility, connecting the output $\boldsymbol{v}_{dn}$ to the input $\mathbf{E}$. The total velocity of an electron therefore consists of two components, namely, the random thermal velocity and the uniform drift velocity. Obviously, only the latter component contributes to the current, as illustrated in Figure 18.2. One can likewise express the hole drift velocity and mobility as

$$\boldsymbol{v}_{dp} \simeq \frac{q\mathbf{E}}{m_p}\tau_p \equiv \mu_p\mathbf{E}, \tag{18.13}$$

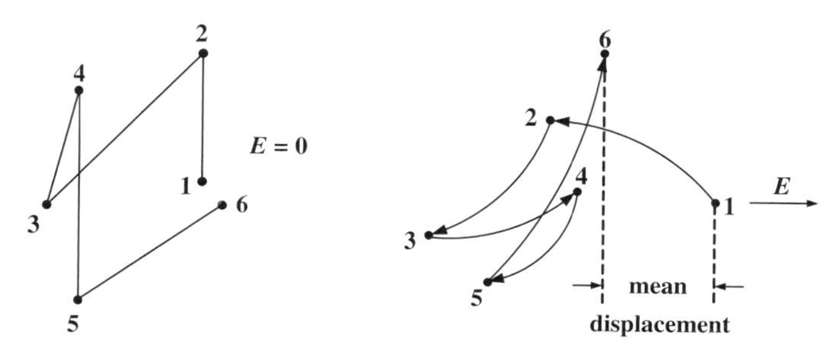

Figure 18.2 The graphical illustration of random thermal motion of electrons (left) and the random motion superposed with field-induced uniform drift velocity (right).

where

$$\mu_p = \frac{q\tau_p}{m_p}. \quad (18.14)$$

The key parameters specifying the mobility are the effective masses m_n and m_p and the mean collision times τ_n and τ_p. The effective mass has already been discussed in conjunction with the dispersion curves in energy bands in Chapter 7. The mean collision time is rooted in the quantum transport processes, and is briefly discussed here.

Quantum Transport Equation Consider a system of electrons, spatially homogeneous and free, such as those electrons in the conduction band in bulk semiconductors. When an electric field is applied along, say, z-direction, the transport equation at steady state reads in relaxation approach as

$$-\frac{q\mathrm{E}}{m_n}\frac{\partial f}{\partial v_z} = -\frac{f - f_0}{\tau_n}, \quad (18.15)$$

where τ_n is the longitudinal relaxation time and the equilibrium distribution is now specified by the Fermi–Dirac distribution function, namely,

$$f_0(E) = \frac{1}{1 + e^{(E-E_F)/k_BT}}, \quad E = \frac{1}{2}m_n(v_x^2 + v_y^2 + v_z^2). \quad (18.16)$$

Note that the transport equation (18.15) is identical to the Boltzmann transport equation (1.31), as applied to a homogeneous system where $\nabla f = 0$ and in steady state in which $\partial f/\partial t = 0$. The only difference consists of the effective mass of electron m_n used in (18.15) instead of the rest mass and the Boltzmann distribution function (1.21) or (16.30) replaced by the Fermi–Dirac distribution function (16.38) or (17.1c). Naturally, the electron energy consists entirely of the kinetic energy for the case under consideration.

Again, for simplicity, the departure of f from f_0 is taken small, in which case f on the left-hand side of (18.15) can be replaced by f_0 to the first-order iteration, yielding

$$f = f_0 + \frac{q\mathrm{E}}{m_n}\tau_n\frac{\partial f_0}{\partial v_z}. \quad (18.17)$$

Now, using (18.16), the derivative of f_0 can be expressed as

$$\frac{\partial f_0}{\partial v_z} = \frac{\partial f_0}{\partial E}\frac{\partial E}{\partial v_z} = -\frac{e^{(E-E_F)/k_BT}}{(1 + e^{(E-E_F)/k_BT})^2}\frac{m_n v_z}{k_BT} \equiv -\frac{m_n v_z}{k_BT}f_0(1 - f_0). \quad (18.18)$$

Thus, combining (18.17) and (18.18), one finds f to the first order of approximation as

$$f = f_0 - \frac{q\mathbf{E}v_z\tau_n}{k_BT}f_0(1 - f_0). \quad (18.19)$$

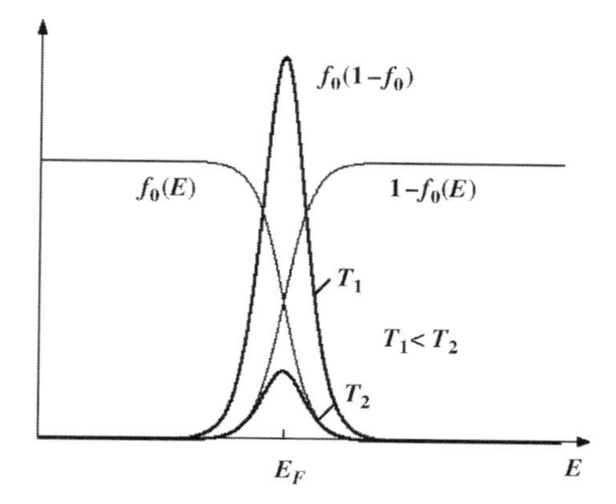

Figure 18.3 The plot of $f_0(E)(1-f_0(E))$ versus energy, with $f_0(E)$ denoting the Fermi distribution function.

Here, the product of f_0 and $(1-f_0)$ is sharply peaked at the Fermi level E_F, as shown in Figure 18.3, and it can therefore be well approximated by a delta function:

$$f_0(1-f_0) \approx k_B T \delta(E - E_F). \tag{18.20}$$

The validity of (18.20) is assured by the fact that both sides are sharply peaked at E_F, as they should, and also lead to the same result when integrated with respect to E from $-\infty$ to $+\infty$.

Mobility Now that the steady-state distribution function f has been found in (18.19), the average drift velocity of electrons along the direction of the field, that is, the z-direction, on top of their random thermal motion can be obtained by

$$\begin{aligned}\langle v_z \rangle &= \frac{\int d\mathbf{v} v_z f}{\int d\mathbf{v} f} \\ &= \frac{-\frac{q\mathrm{E}}{k_B T}\int_0^{2\pi} d\varphi \int_{-1}^{1} d\mu \int_0^{\infty} \tau(v) v^2 dv v^2 \mu^2 f_0(1-f_0)}{\int_0^{2\pi} d\varphi \int_{-1}^{1} d\mu \int_0^{\infty} v^2 dv f_0}, \quad v_z = v\mu, \ \mu = \cos\theta.\end{aligned} \tag{18.21}$$

In (18.21), the integrations involving the first term of f in the numerator and the second term of f in the denominator are readily seen to be zero in view of odd parities of the integrands involved. Also, the integral has been cast in the spherical coordinates in velocity space and the longitudinal relaxation time τ_n is taken a function of v. Carrying out the angular integrations φ and μ with the use of (18.20) and $dE = m_n v dv$ results in

$$\langle v_z \rangle = -\frac{2q\mathrm{E}}{3m_n} \frac{\int_0^\infty dE E^{3/2} \tau(E) \delta(E - E_F)}{\int_0^\infty dE E^{1/2} f_0(E)}, \qquad E = m_n v^2/2. \tag{18.22}$$

In evaluating (18.22), the integration in the denominator can be simplified without losing too much accuracy by considering the limiting case of $T = 0$, so that $f_0 = 1$ for $E < E_F$, while $f_0 = 0$ for $E > E_F$ (see Figure 16.3). Thus, one obtains

$$\int_0^\infty dE E^{1/2} f_0(E) = \int_0^{E_F} dE E^{1/2} = \frac{2}{3} E_F^{3/2}. \tag{18.23}$$

Hence, performing the delta function integration in the numerator and using (18.23), one finds the drift velocity of electrons from (18.22) as

$$v_{dn} \equiv \langle v_z \rangle = -\frac{q\tau_n(E_F)}{m_n} \mathrm{E} = -\mu_n \mathrm{E} \tag{18.24}$$

and the resulting mobility,

$$\mu_n = \frac{q\tau_n(v_F)}{m_n}; \qquad \frac{m_n v_F^2}{2} = E_F, \tag{18.25}$$

is of the same form as in (18.12), except that the mean collision or relaxation time is specified at the Fermi velocity v_F.

Diffusion Coefficient Another important transport coefficient is the diffusion constant, which specifies the flux of particles arising from the concentration gradient. Thus, consider a nonuniform electron density, as sketched in Figure 18.4. The electron flux at x can be conveniently analyzed by introducing the mean free path of electrons l_n on both sides of x, as sketched in Figure 18.4. In this way, electrons can be treated simply as free particles in the volume elements from $x - l_n$ to x and from x to $x + l_n$. Thus, the net number of electrons crossing x per unit area during the mean collision time τ_n from left to right is given by

$$N = N_{LR} - N_{RL} = \frac{1}{2} n(x - l_n) l_n - \frac{1}{2} n(x + l_n) l_n, \tag{18.26}$$

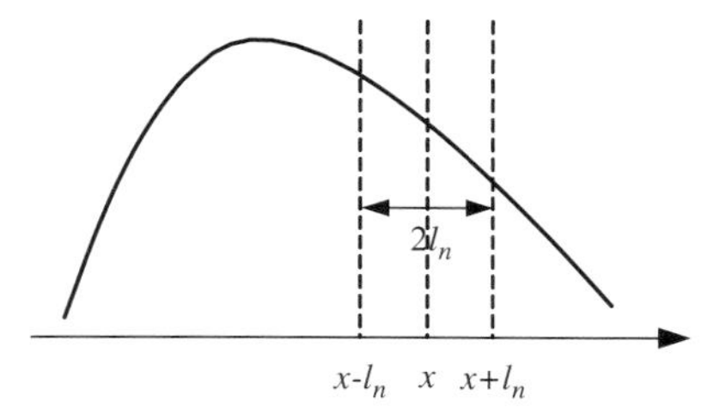

Figure 18.4 The electron concentration profile and a differential volume element, consisting of two planes at $x \pm l_n$ and centered at x, with l_n denoting the electron mean free path.

where 1/2 factor entering in (18.26) accounts for the fact that due to the random motion in equilibrium, only one-half of the electrons are moving from left to right or vice versa. Since l_n can be taken small, compared with the spatial range over which appreciable changes in concentration occur, one can Taylor expand the electron concentrations as

$$n(x \pm l_n) \simeq n(x) \pm \frac{\partial n}{\partial x} l_n. \tag{18.27}$$

Inserting (18.27) into (18.26) results in

$$N = -l_n^2 \frac{\partial n(x)}{\partial x}. \tag{18.28}$$

The flux of electrons from left to right at x is therefore found by dividing (18.28) by the mean collision time τ_n, obtaining

$$F_n \equiv \frac{N}{\tau_n} = -\frac{l_n^2}{\tau_n}\frac{\partial n(x)}{\partial x} \equiv -D_n \frac{\partial n(x)}{\partial x}. \tag{18.29}$$

The resulting diffusion coefficient of electrons

$$D_n \equiv \frac{l_n^2}{\tau_n} \tag{18.30}$$

connects the flux of electrons to the concentration gradient and is specified via the mean free path l_n and the collision time τ_n. These two physical parameters are related by

$$l_n = v_T \tau_n, \tag{18.31}$$

where the thermal speed of electrons v_T is taken much larger than v_{dn} and is given from the equipartition theorem (1.25) by

$$\frac{m_n v_T^2}{2} = \frac{k_B T}{2}. \tag{18.32}$$

Einstein Relation Hence, combining (18.30)–(18.32) for D_n and comparing with mobility in (18.12) or (18.25), one finds Einstein relation between μ_n and D_n, namely,

$$\frac{D_n}{\mu_n} = \frac{l_n^2/\tau_n}{q\tau_n/m_n} = \frac{k_B T}{q}. \tag{18.33}$$

One can similarly carry out the analysis for holes and write

$$\mu_p = \frac{e\tau_p}{m_p}, \tag{18.34}$$

$$D_p \equiv \frac{l_p^2}{\tau_p}, \tag{18.35}$$

and the Einstein relation for holes, thus, also reads as

$$\frac{D_p}{\mu_p} = \frac{l_p^2/\tau_p}{q\tau_p/m_p} = \frac{k_B T}{q}. \tag{18.36}$$

18.3
Equilibrium and Nonequilibrium

18.3.1
Equilibrium and Fermi Level

Single Semiconductor System One of the basic properties of the thermal equilibrium is that the electron and hole concentrations therein are characterized by Fermi distribution function and a single Fermi level E_F. In addition, the charge transport is inextricably related to the Fermi level and this is discussed next. For this purpose, consider the 1D current density of electrons, which is given from (18.6) by

$$J_n = q\mu_n n\mathrm{E} + qD_n \frac{dn}{\partial x}. \tag{18.37}$$

Here, the electric field E driving the drift current component is related to the electrostatic potential φ via $\mathrm{E} = -\partial\varphi/\partial x$. Also, φ in turn represents the electron potential energy when multiplied by $-q$, and $-q\varphi$ carries identical physical contents as E_C, E_V, or the midgap energy E_i, that is,

$$-q\varphi \equiv E_i (= E_c = E_v),$$

so that one can write

$$\mathrm{E} = -\partial\varphi/\partial x \equiv \frac{1}{q}\frac{\partial E_i}{\partial x}. \tag{18.38}$$

Also, the nondegenerate electron concentration in equilibrium is analytically expressed in terms of E_i and E_F by (17.25) with $E_i = E_{Fi}$. Hence, with the use of the Einstein relation (18.33), the electron current density (18.37) can be compacted as

$$J_n = \mu_n n\left[\frac{dE_i}{dx} + q\left(\frac{k_B T}{q}\right)\frac{d}{dx}\left(\frac{E_F - E_i}{k_B T}\right)\right] = \mu_n n \frac{dE_F}{dx}. \tag{18.39}$$

Similarly, one can write the hole current density as

$$\begin{aligned} J_p &= q\mu_p p\mathrm{E} - qD_p \frac{dp}{\partial x} \\ &= \mu_p p\left[\frac{dE_i}{dx} - q\left(\frac{k_B T}{q}\right)\frac{d}{dx}\left(\frac{E_i - E_F}{k_B T}\right)\right] \\ &= \mu_p p \frac{dE_F}{dx}. \end{aligned} \tag{18.40}$$

Now, in equilibrium no current flows, that is,

$$J_n = J_p = 0,$$

so that it follows from (18.39) or (18.40) that

$$\frac{dE_F}{dx} = 0. \qquad (18.41)$$

Therefore, the Fermi level in a single semiconductor system should be flat and independent of x in equilibrium.

Composite Semiconductor System Consider next a composite system consisting of two semiconductors interfaced at $z = 0$ (Figure 18.5). Once the thermodynamic equilibrium is established between the two, the electron flux from left to right, F_{LR}, should be balanced by the reverse flux from right to left, F_{RL}. Now, because of the Pauli exclusion principle, F_{LR} at given E requires filled quantum states on the left and vacant states on the right for the electrons to fill in, that is,

$$F_{LR} = Mn_L(E)\nu_R(E), \qquad (18.42)$$

where M is the transfer matrix and the electron density on the left,

$$n_L = g_L(E)dEf_L(E),$$

is specified by the density of states at E times the Fermi occupation factor. Likewise, the density of vacant states on the right,

$$\nu_R = g_R(E)dE[1 - f_R(E)],$$

is given by the density of states times the probability that the state is empty of electron.

Hence, one can explicitly specify (18.42) by

$$F_{LR} = Mg_L(E)dEf_L(E)g_R(E)dE[1 - f_R(E)]. \qquad (18.43)$$

By the same token, the flux from right to left is given by

$$F_{RL} = Mg_R(E)dEf_R(E)g_L(E)dE[1 - f_L(E)]. \qquad (18.44)$$

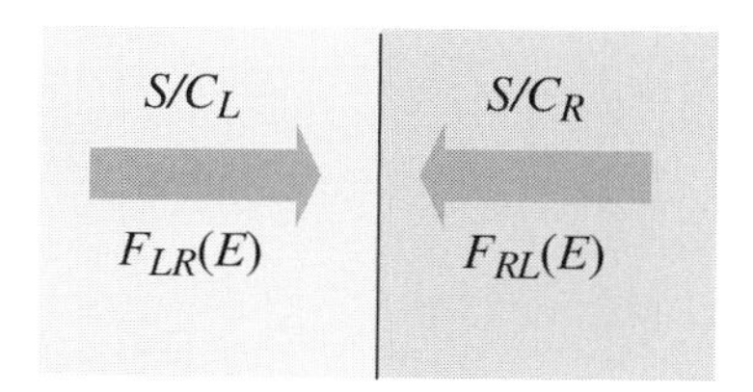

Figure 18.5 A composite semiconductor system, consisting of two semiconductors interfaced. The electron flux from left to right is balanced by its inverse flux from right to left, in equilibrium.

Since these two opposing fluxes should balance each other out in equilibrium, that is, $F_{LR} = F_{RL}$, it follows from (18.43) and (18.44) that

$$f_L(E) = f_R(E). \quad (18.45a)$$

That is,

$$\frac{1}{1 + e^{(E - E_{FL})/k_B T}} = \frac{1}{1 + e^{(E - E_{FR})/k_B T}}. \quad (18.45b)$$

Therefore, it follows from (18.45b) that

$$E_{FL} = E_{FR} \quad (18.45c)$$

and Fermi levels in composite semiconductor system should therefore line up. This, together with (18.41) leads to the general conclusion, namely, that the Fermi level in equilibrium should line up and be flat. The conclusion holds true for any number of semiconductor layers in equilibrium contact.

18.3.2
Nonequilibrium and Quasi-Fermi Level

A system, when subjected to irradiation or bias, is driven away from equilibrium into nonequilibrium. In nonequilibrium, it is not possible to specify both electron and hole concentrations by means of a single Fermi level E_F. Rather, two quasi-Fermi levels, one for electrons and one for holes, are required.

This can be clearly shown by considering a slab of a semiconductor under irradiation, as sketched in Figure 18.6. Under illumination, the electron–hole pairs are generated via the band-to-band excitation and at the same time the electrons and holes are constantly recombined. Hence, the rate equations of the electron and hole concentrations read as

$$\frac{\partial n}{\partial t} = g - \frac{n}{\tau_n}, \quad (18.46a)$$

$$\frac{\partial p}{\partial t} = g - \frac{p}{\tau_p}, \quad (18.46b)$$

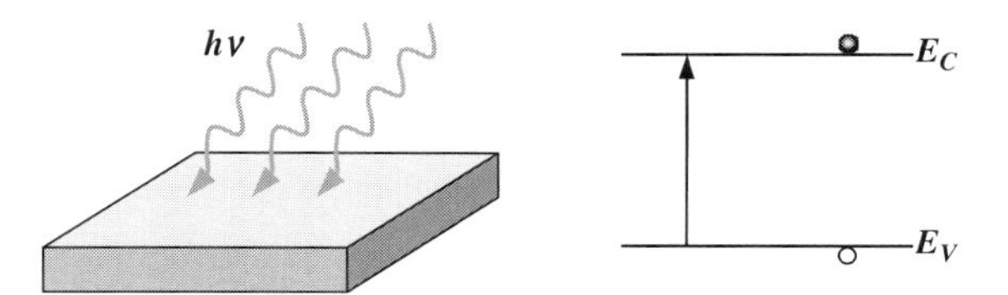

Figure 18.6 A semiconductor sample under uniform irradiation and the generation of electron–hole pairs via band-to-band excitation.

where τ_n and τ_p are the respective recombination lifetimes and the band-to-band generation rate of electron–hole pairs,

$$g = \alpha I/h\nu, \tag{18.46c}$$

is specified by the linear attenuation coefficient α and the flux of photons incident on the semiconductor, $I/h\nu$, as discussed.

At steady state in which the time rate of change in n and p is zero, the photogenerated electron and hole concentrations are readily found from (18.46). Thus, the total electron and hole concentrations, n and p, consist of the intrinsic as well as photogenerated concentrations:

$$n = n_i + n_{\text{ph}} = n_i + g\tau_n, \tag{18.47a}$$

$$p = n_i + p_{\text{ph}} = n_i + g\tau_p. \tag{18.47b}$$

Now, consider the case in which the light intensity is high, so that $n_{\text{ph}} \gg n_i$ and $p_{\text{ph}} \gg p_i$. The former inequality in (18.47a) requires that E_F should be above E_i just as in n-type semiconductor (see (17.25)). The latter inequality in (18.47b) requires, on the other hand, that E_F should be below E_i as in p-type semiconductor (see (17.26)).

Clearly, these two requirements cannot be met with a single Fermi level simultaneously. The only way to satisfy these two requirements and to quantify n and p simultaneously is to introduce two quasi-Fermi levels, one for electron E_{Fn} and one for hole E_{Fp}, and express the carrier concentrations in the usual manner as

$$n = n_i e^{(E_{Fn} - E_i)/k_B T}, \tag{18.48a}$$

$$p = n_i e^{(E_i - E_{Fp})/k_B T}. \tag{18.48b}$$

The quasi-Fermi levels E_{Fn} and E_{Fp} are often called imrefs.

The splitting of these two imrefs under illumination is obtained by inverting the respective quasi-Fermi levels from (18.48) in combination with (18.47):

$$E_{Fn} - E_{Fp} = k_B T \ln\left[\left(1 + \frac{g\tau_n}{n_i}\right)\left(1 + \frac{g\tau_p}{n_i}\right)\right]. \tag{18.49}$$

Indeed, the splitting of these imrefs increases with increasing light intensity and/or the generation rate, as shown in Figure 18.7. In the absence of illumination, that is,

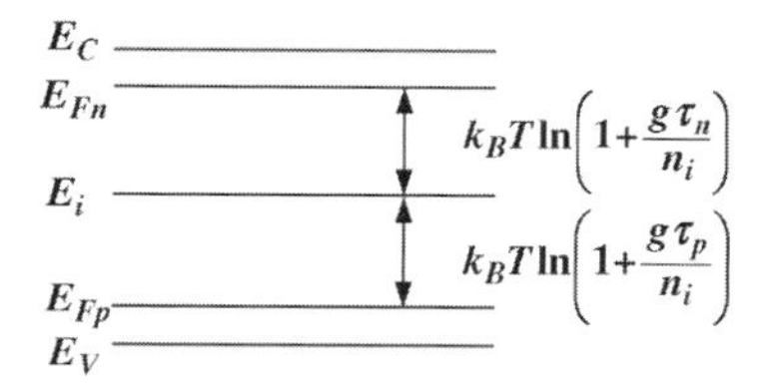

Figure 18.7 Quasi-Fermi levels of electrons and holes in irradiated semiconductor sample. The splitting of two imrefs is determined by the illumination intensity.

$g = 0$, these two quasi-Fermi levels collapse into a common Fermi level, that is, $E_{Fn} = E_{Fp} = E_F$, as the system relaxes back to equilibrium.

The roles of these two quasi-Fermi levels are identical to those of the single Fermi level in equilibrium. For example, the carrier concentrations in nonequilibrium are generally specified by

$$n = N_C \frac{2}{\sqrt{\pi}} F_{1/2}(\eta_{Fn}), \quad \eta_{Fn} = \frac{E_{Fn} - E_C}{k_B T}, \tag{18.50a}$$

$$p = N_V \frac{2}{\sqrt{\pi}} F_{1/2}(\eta_{Fp}), \quad \eta_{Fp} = \frac{E_V - E_{Fp}}{k_B T} \tag{18.50b}$$

(see (17.2) and (17.6)). Also, the total current densities of electrons and holes consisting of the drift and diffusion components are described by the slope of these imrefs in exactly the same way as in (18.39) and (18.40), that is,

$$J_n = \mu_n n \frac{d}{dx} E_{Fn}, \tag{18.51a}$$

$$J_p = \mu_p p \frac{d}{dx} E_{Fp}. \tag{18.51b}$$

Unlike the flat Fermi level at equilibrium leading to zero current, the quasi-Fermi levels generally vary as a function of position and account for the local current densities.

18.4 Generation and Recombination Currents

The drift and diffusion currents considered in previous sections are based on the motion of electrons and holes in the conduction and valence bands, respectively. In addition to these currents, there exist other kinds of currents, namely, the generation and recombination currents. These two currents come about because of the law of mass action, $np = n_i^2$, being broken due to external perturbation. Specifically, when the charges are injected in excess of the equilibrium level, so that $np > n_i^2$, the recombination current ensues driven by the reactive force pushing the system back to the equilibrium. If charge is extracted, on the other hand, so that $np < n_i^2$, there arises the generation current again to drive the system back to equilibrium. These generation and recombination currents play an important role in semiconductor devices just like the drift and diffusion currents and will be considered in this section.

18.4.1 Band-to-Band Excitation or Recombination

Carrier Emission and Capture As noted earlier, some of the electrons in the valence band are promoted into the conduction band via the band-to-band thermal excitation, leaving behind the same number of holes in the valence band, as sketched in

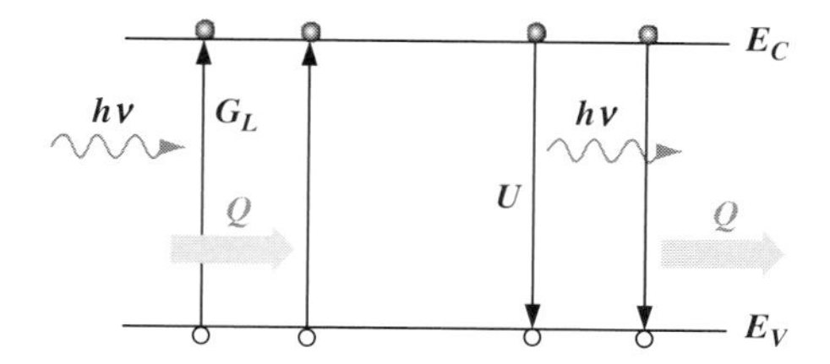

Figure 18.8 The band-to-band generation of electron–hole pairs via thermal and optical excitations and the inverse process of radiative and nonradiative recombinations of electrons and holes.

Figure 18.8. In addition, when the semiconductor is irradiated with photons with energy larger than the bandgap, the electron–hole pairs are photogenerated via the band-to-band excitation (Figure 18.8).

The electrons and holes in conduction and valence bands also undergo the band-to-band recombination. In such recombination processes, the electron has to lose energy amounting approximately to the bandgap energy. This energy is emitted either in the form of light or consumed via the heat dissipated. This kind of band-to-band recombination of electron–hole pairs is obviously proportional to the electron and hole concentrations in conduction and valence bands, respectively. In n-type semiconductors, for example, one can therefore express the recombination rate by

$$R \propto n_n p_n = \Gamma n_n p_n. \tag{18.52}$$

In equilibrium in which the detailed balancing between recombination and generation prevails, the thermal excitation can be expressed as

$$G_{\text{th}} \equiv R_{\text{eq}} = \Gamma n_{n0} p_{n0}, \tag{18.53}$$

with n_{n0} and p_{n0} denoting the majority and minority carrier concentrations in equilibrium, respectively. Thus, for the case of charge injection, the net recombination rate is given by

$$\begin{aligned} U &= R - G_{\text{th}} \\ &= \Gamma(n_n p_n - n_{n0} p_{n0}) \simeq \frac{1}{\tau_p}(p_n - p_{n0}), \quad \frac{1}{\tau_p} \equiv \Gamma n_{n0}. \end{aligned} \tag{18.54}$$

In (18.54), an approximation has been made, namely, that $n_n \simeq n_{n0}$ for the low-level injection and τ_p thus introduced is the hole lifetime as minority carrier.

18.4.2 Trap-Assisted Recombination and Generation

The recombination (r) and generation (g) processes are drastically enhanced when trap sites are present in the bandgap. This trap-assisted r/g processes are discussed, based on the theories of Shockley and Read and also of Hall. Thus, consider the simple case of a single trap level, as sketched in Figure 18.9. The general case of

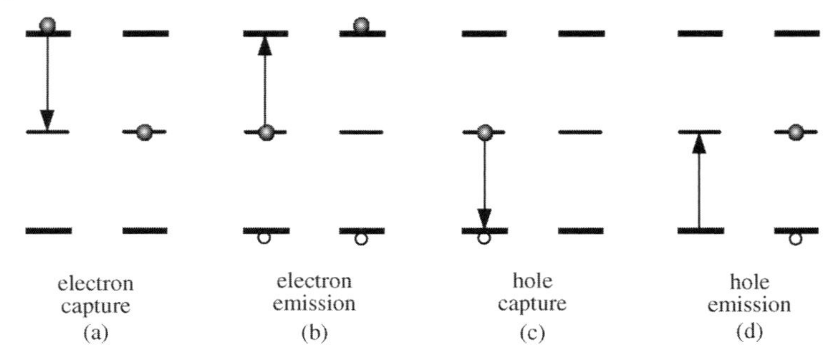

Figure 18.9 The schematics for the single-level trap-assisted emission and capture of electrons and holes: (a) electron capture, (b) electron emission, (c) hole capture, and (d) hole emission.

multiple trap levels can be described by a straightforward generalization of this single trap model.

There are in general four possible processes, as sketched in Figure 18.9: (i) the capture of an electron from the conduction band by an empty trap site, (ii) its inverse process of the trapped electron being emitted from the trap level to the conduction band, or (iii) the trapped electron capturing a hole in the valence band, and (iv) its inverse process of an electron in valence band being captured by the trap site, thereby inducing a hole emission.

As a reference for the discussions to follow, it is noted that the number of electrons trapped in equilibrium is specified by the Fermi occupation factor, that is,

$$n_t = N_t f_0(E_t) = \frac{N_t}{1 + e^{(E_t - E_F)/k_B T}}, \tag{18.55}$$

where N_t is the trap density and E_t the trap level.

Now, the electron capture rate should obviously be proportional to the number of electrons in the conduction band and the empty trap sites available, that is,

$$r_{ec} \propto n N_t (1 - f) = v_T \sigma_n n N_t (1 - f), \tag{18.56}$$

where the proportionality constant is specified in terms of the thermal speed v_T of the electron and the electron capture cross section σ_n of the trap, $\approx 10^{-15}$ cm^2. The capture cross section is to be roughly represented by the cross-sectional area of the impact parameter within which an electron when incident on the trap site with thermal speed v_T is captured, as illustrated in Figure 18.10. Also, it is important to

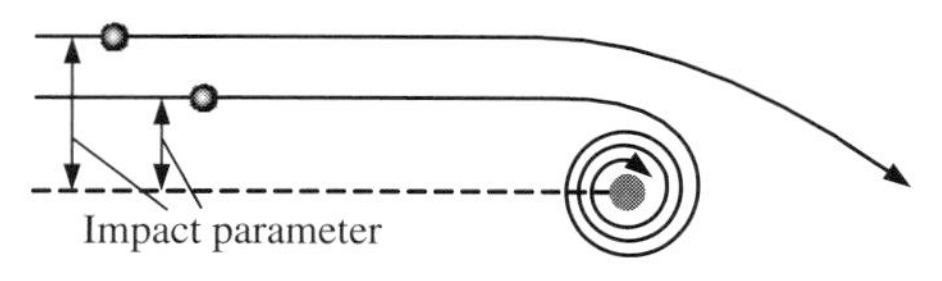

Figure 18.10 The graphical illustration of electron capture by a trap. The capture cross section is determined by the type of attractive force, impact parameter, and incident speed.

recognize that the distribution function f away from equilibrium is generally different from the Fermi distribution function in equilibrium.

The inverse process of the electron emission from the trap into the conduction band should obviously be proportional to the filled trap sites, namely,

$$r_{ee} \propto N_t f = e_n N_t f, \tag{18.57}$$

where the proportionality constant e_n thus introduced is called the electron emission probability.

The hole capture and emission processes can be described in a similar fashion. The hole capture is done by the trapped electron recombining with a hole in the valence band (Figure 18.9). Hence, the corresponding capture rate should be proportional to the number of holes in the valence band and the number of trapped electrons, that is,

$$r_{hc} \propto pN_t f = v_T \sigma_p p N_t f. \tag{18.58}$$

Here again, the proportionality constant is given in terms of the thermal speed and the hole capture cross section σ_p. Likewise, the hole emission should be proportional to the number of empty traps into which electrons in the valence band can be captured and is given by

$$r_{he} \propto N_t(1-f) = e_p N_t(1-f), \tag{18.59}$$

with the proportionality constant provided by the hole emission probability e_p.

Steady State Versus Equilibrium Consider the recombination and generation processes at steady state, as distinct from equilibrium. The difference between the equilibrium and steady state can be best brought out by considering a semiconductor under uniform irradiation. The rate equation of the electron and hole concentrations are then given by

$$\dot{n} = g_L - (r_{ec} - r_{ee}), \quad g_L = \alpha I / h\nu, \tag{18.60a}$$

$$\dot{p} = g_L - (r_{hc} - r_{he}). \tag{18.60b}$$

Here, the only difference between (18.46) and (18.60) consists of explicitly specifying the net recombination rate in terms of respective emission and capture rates.

At steady state, the time rate of change of both n and p is zero, so that it follows from (18.60) that

$$r_{ec} - r_{ee} = r_{hc} - r_{he}, \tag{18.61}$$

thereby indicating that the net recombinations of both electrons and holes are the same, as it should. At equilibrium, on the other hand, the respective capture processes are balanced by their inverse emission processes, namely,

$$r_{ec} = r_{ee}, \tag{18.62a}$$

$$r_{hc} = r_{he}. \tag{18.62b}$$

It is therefore clear from the comparison of (18.61) and (18.62) that the condition for equilibrium is more stringent than that of the steady state in that the equilibrium condition automatically satisfies the steady-state condition.

Emission Probabilities and Distribution Function To proceed further, a reasonable approximation is made at this point, namely, that the electron emission probability, e_n, is taken to remain unchanged in both equilibrium and nonequilibrium conditions. With this assumption, e_n can be specified explicitly from (18.62). For example, combining (18.62a), (18.56), and (18.57) and using the equilibrium Fermi distribution function

$$f_0(E_t) = \frac{1}{1+e^{(E_t - E_F)/k_B T}}$$

in equilibrium, one can write

$$v_T \sigma_n n N_t \left(1 - \frac{1}{1+e^{(E_t - E_F)/k_B T}}\right) = e_n N_t \frac{1}{1+e^{(E_t - E_F)/k_B T}}, \quad n = n_i e^{(E_F - E_i)/k_B T}. \tag{18.63}$$

Therefore, e_n can be explicitly specified from (18.63) in terms of the basic physical parameters:

$$e_n = v_T \sigma_n n_i e^{(E_t - E_i)/k_B T}. \tag{18.64}$$

Similarly, the hole emission probability can also be obtained as

$$e_p = v_T \sigma_p n_i e^{(E_i - E_t)/k_B T}. \tag{18.65}$$

With the emission probabilities thus found, the distribution function at steady state can next be found from the steady-state condition (18.61), namely,

$$\begin{aligned} v_T \sigma_n n N_t (1-f) &- v_T \sigma_n n_i e^{(E_t - E_i)/k_B T} N_t f \\ &= v_T \sigma_p p N_t f - v_T \sigma_p n_i e^{(E_i - E_t)/k_B T} N_t (1-f). \end{aligned} \tag{18.66}$$

In (18.66), the emission and capture processes (18.56)–(18.59) and the emission probabilities of electrons and holes, (18.64) and (18.65), have all been put together. It is now a simple matter to find f at steady state from (18.66) as

$$f = \frac{\sigma_n n + \sigma_p n_i e^{(E_i - E_t)/k_B T}}{\sigma_n (n + n_i e^{(E_t - E_i)/k_B T}) + \sigma_p (p + n_i e^{(E_i - E_t)/k_B T})}. \tag{18.67}$$

The resulting f is shown distinctly different from the equilibrium Fermi distribution function f_0. The comparison of f with f_0 can be made much more explicit by taking $\sigma_n = \sigma_p$, in which case (18.67) simplifies as

$$f = \frac{n + n_i e^{(E_i - E_t)/k_B T}}{n + p + 2 n_i \cosh(E_t - E_i)/k_B T}. \tag{18.68}$$

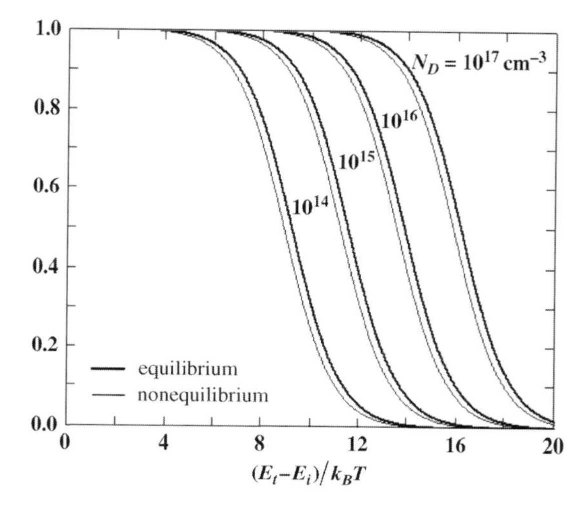

Figure 18.11 The equilibrium (17.1c) and steady-state distribution functions (18.68) versus the trap level $(E - E_t)/k_BT$ at different doping levels.

Figure 18.11 shows the distribution function thus found in (18.68) plotted versus the trap level, $(E_t - E_i)/k_BT$, for various electron concentrations or doping levels. Also shown in the figure are the corresponding Fermi distribution functions f_0 in equilibrium, for comparison. Indeed the trap occupation factors f and f_0 are near unity for $E_t < E_F$ but decrease rapidly as E_t is raised above E_F. Although the expression of the nonequilibrium distribution function f in (18.68) appears rather different from that of f_0, the two curves do not depart very much from each other but in fact generally look alike. But f is consistently and systematically different from f_0 and this small difference is responsible for net recombination or generation rates.

Recombination Rate Now that the distribution function f has been found, the net recombination rate, $r_{ec} - r_{ee}$, for example, is readily obtained by combining (18.61), (18.68), (18.56) and (18.57):

$$
\begin{aligned}
U &= v_T \sigma_n n N_t (1 - f) - v_T \sigma_n n_i e^{(E_t - E_i)/k_BT} N_t f \\
&= \frac{\sigma_n \sigma_p v_T N_t (pn - n_i^2)}{\sigma_n (n + n_i e^{(E_t - E_i)/k_BT}) + \sigma_p (p + n_i e^{(E_i - E_t)/k_BT})}.
\end{aligned}
\tag{18.69}
$$

The recombination rate for electrons given in (18.69) also applies to holes as evident from (18.61).

It should be noted that the recombination rate (18.69) can also account for generation rate. At equilibrium, in which the law of mass action holds true, namely, $np = n_i^2$, $U = 0$, as it should. In this case, there is a detailed balancing between the recombination and the generation. However, if the system is driven away from equilibrium and charge is injected, so that $np > n_i^2$, then $U > 0$ and there ensues the net recombination of the excess charge to restore the equilibrium. When the charge is

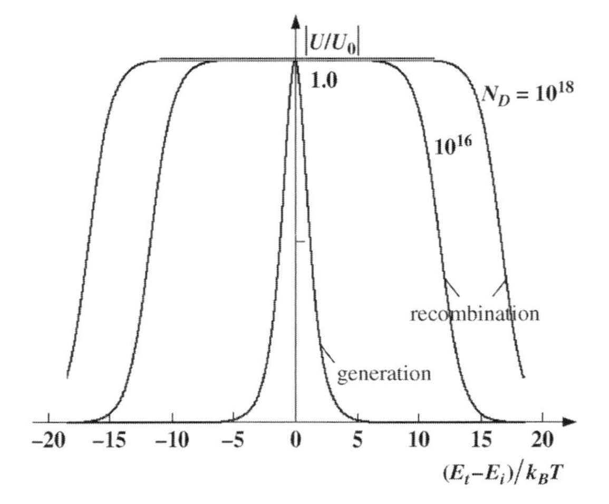

Figure 18.12 The recombination and generation rates (18.70) versus the trap level $(E-E_t)/k_BT$ for different doping levels.

extracted on the other hand, so that $np < n_i^2$ and $U < 0$, there naturally follows the generation of electrons and holes with the traps playing the role of stepping sites for generation again to restore the equilibrium condition.

The recombination rate can be simplified by assuming that $\sigma_n = \sigma_p = \sigma$, in which case (18.69) reduces to

$$U = \frac{1}{\tau}\frac{(pn-n_i^2)}{n+p+2n_i\cosh(E_t-E_i)/k_BT}, \quad \frac{1}{\tau} \equiv \sigma v_T N_t, \tag{18.70}$$

where τ thus defined is the recombination lifetime. Clearly, U depends sensitively on the location of the trap level E_t with respect to the midgap E_i. Specifically, U attains the maximum value when E_t coincides with E_i. That is, the recombination or generation of electron–hole pairs is maximally enhanced when the trap level is located near the midgap, so that the differences in levels, E_C-E_t and E_t-E_V, are approximately the same. Figure 18.12 shows both the recombination and generation rates for different n_n or N_D. In n-type semiconductor, in which $n_n \gg p_n, n_i$, the recombination rate U is mainly dictated by n_n and is pinned at its maximum level corresponding to $E_t = E_i$ over a wide range, in which $n_n \gg n_i\cosh(E_t-E_i)/k_BT$. On the other hand, when carriers are depleted so that $np \ll n_i^2$, the generation rate falls off exponentially as a function of the trap level from its maximum value at $E_t = E_i$.

In the presence of a shallow trap near E_C, the electrons can be easily emitted to the conduction band from this trap site. However, the large energy difference existing between the trap level and valence band slows down the hole emission process, so that the overall efficiency of the trap-assisted generation of electron–hole pair is low. As a consequence, the electron emission is more likely to be accompanied by its inverse process of electron capture for shallow traps near the conduction band edge. By the same token for traps near the valence band edge, the hole capture and emission constitute the dominant processes.

The Minority Carrier Lifetime The recombination rate (18.70) lends itself to another useful expression when it is applied to, say, n-type semiconductor. In this case, $n_n \gg p_n, n_i$ and n_n can be taken to be the same as the equilibrium value $n_n \approx n_{n0}$. Thus, one can write

$$pn - n_i^2 \approx n_{n0}(p_{n0} + p_n - p_{n0}) - n_i^2 = n_{n0}(p_n - p_{n0}), \tag{18.71}$$

where n_{n0} and p_{n0} denote the respective equilibrium concentrations and the law of mass action $n_{n0}p_{n0} = n_i^2$ has been used.

Thus, inserting (18.71) into (18.70) and using the fact that $n_n \gg p_n, n_i$, the recombination rate (18.69) simplifies as

$$U = \sigma_p v_T N_t (p_n - p_{n0}) = \frac{p_n - p_{n0}}{\tau_p}, \tag{18.72}$$

where τ_p defined as

$$\frac{1}{\tau_p} \equiv \sigma_p v_T N_t \tag{18.73}$$

is the lifetime of holes as minority carriers.

One can similarly analyze the recombination rate in the p-type semiconductor, obtaining

$$U = \frac{n_p - n_{p0}}{\tau_n}, \tag{18.74}$$

with τ_n defined as

$$\frac{1}{\tau_n} = \sigma_n v_T N_t \tag{18.75}$$

denoting the lifetime of electrons as minority carriers.

Recombination Velocity Moreover, near the surface of an n-type semiconductor, for example, the recombination occurring near the surface can be described with the aid of (18.72) by introducing the effective thickness of the surface t_s and one may write

$$U = v_R(p_n(0) - p_{n0}) \tag{18.76}$$

with

$$v_R = \frac{\sigma_p v_T N_t t_s}{\tau_p} \tag{18.77}$$

representing the surface recombination velocity of holes as minority carriers.

18.5 Problems

18.1. The conductivity $\sigma = qn\mu$ is specified in terms of the charge, concentration, and mobility, while the resistivity is the inverse conductivity $\varrho = 1/\sigma$. An n-type silicon with resistivity of 10 Ω cm is uniformly illuminated and 10^{21} e–h pairs are generated per cm^3 s. The carrier lifetime in the sample is 1 μs.

(a) Calculate the dark and photoconductivity.

(b) Calculate the contributions made by electrons and holes to the total conductivity.

Take $\mu_n = 800\ \mathrm{cm^2/V\,s}$ and $\mu_p = 400\ \mathrm{cm^2/V\,s}$.

18.2. An n-type semiconductor is irradiated uniformly with light, having the wavelength of $\lambda = 500$ nm, attenuation coefficient of $\alpha = 10\ \mathrm{cm^{-1}}$, and irradiation intensity of $10\ \mathrm{W/cm^2}$.

(a) Find the photon flux (the number of photons crossing per area and time).

(b) Calculate the concentration of electron–hole pairs generated per second.

(c) The rate equation for the minority hole carrier concentration p_n is given by

$$\frac{\partial p_n}{\partial t} = g_L - \frac{p_n - p_{n0}}{\tau_p}, \quad g_L = \alpha I/h\nu.$$

Find and plot $p_n(t)$ versus time.

(d) When the light is turned off after the steady state has reached, find and plot $p_n(t)$ versus time.

18.3. A silicon sample doped with $10^{17}/\mathrm{cm}^3$ donor atoms contains $10^{15}/\mathrm{cm}^3$ recombination and generation centers with the capture cross sections, $\sigma_n = \sigma_p = 10^{-15}\ \mathrm{cm}^2$, and $m_n/m_0 \simeq 0.98$ and $m_p/m_0 = 0.45$ with m_0 denoting electron rest mass.

(a) Find the generation rate in the depletion region of the sample where $n_n, p_n \ll n_i$ when the trap level is at the midgap.

(b) Find the generation rates if $E_t = E_C - 0.25$ eV and $E_t = E_V + 0.25$ eV.

18.4. An n-type semiconductor is under uniform irradiation, as sketched in Figure 18.13. The rate equation of the minority hole concentration near the surface is given by

$$\frac{\partial p_n}{\partial t} = D_p \frac{\partial^2 p_n}{\partial x^2} + g_L - \frac{p_n - p_{n0}}{\tau_p}, \quad g_L = \alpha I/h\nu.$$

The boundary conditions for p_n deep in the bulk and at the surface are given by

$$p_n(\infty) = g_L \tau_p + p_{n0}; \quad D_p \frac{\partial p_n(0)}{\partial x} = v_R[p_n(0) - p_{n0}],$$

where v_R is the surface recombination velocity.

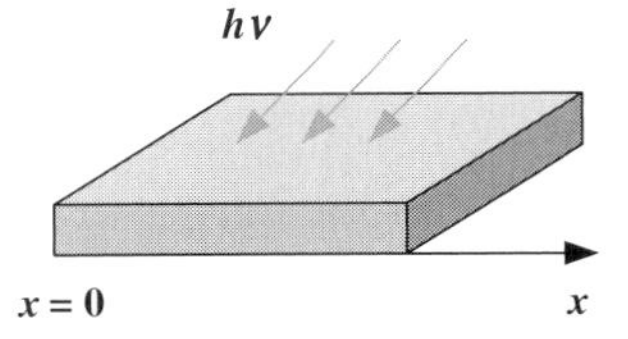

Figure 18.13 N-type semiconductor under uniform irradiation.

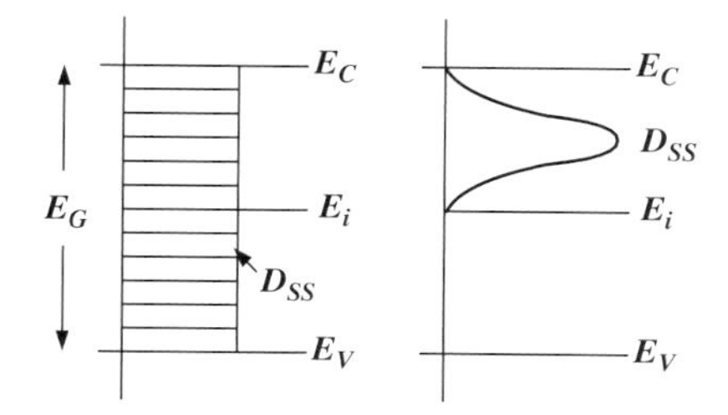

Figure 18.14 The energy gap in which trap centers are uniformly distributed at a constant level D_{SS} (left) and Gaussian distributed with its center half way between E_i and E_C (right).

(a) Show that at steady state where $\partial p_n/\partial t = 0$, the solution is given by

$$p_n(x) = p_{n0} + g_L\tau_p - g_L\tau_p \frac{v_R\tau_p/L_p}{1+v_R\tau_p/L_p} e^{-x/L_p},$$

with $L_p = (D_p\tau_p)^{1/2}$ denoting the hole diffusion length.

(b) Plot $p_n(x)$ for different v_R and discuss the effect of v_R.

18.5. A 0.6 Ω cm n-type silicon is uniformly irradiated by light and 10^{18} e–h pairs are generated per cm^3 s. Take the hole lifetime of 10 μs and the surface recombination velocity of 100 cm/s.

(a) Calculate the number of photogenerated holes recombined at the surface per unit area and time.

(b) Estimate the total number of holes recombined near the surface per area and time.

18.6. Consider the recombination–generation centers uniformly distributed across the energy gap, as sketched in Figure 18.14.

(a) Derive the recombination rates by generalizing the single-level trap model and assuming that $\sigma_n = \sigma_p = \sigma$, for simplicity.

(b) Derive the expression for minority carrier lifetime.

(c) Repeat the analysis of (a) and (b) for the case of traps Gaussian distributed, $N_t \exp-(E_t - E_{tc})^2/2\sigma^2$ centered at E_{tc} half-way between midgap and E_C.

Suggested Reading

1. Muller, R.S., Kamins, T.I., and Chan, M. (2002) *Device Electronics for Integrated Circuits*, 3rd edn, John Wiley & Sons, Inc.
2. Pierret, R.F. (2002) *Advanced Semiconductor Fundamentals, Modular Series on Solid State Devices*, 2nd edn, vol. VI, Prentice Hall.
3. Streetman, B.G. and Banerjee, S. (2005) *Solid State Electronic Devices*, 6th edn, Prentice Hall.
4. Grove, A.S. (1967) *Physics and Technology of Semiconductor Devices*, John Wiley & Sons, Inc.
5. Yariv, A. (1982) *An Introduction to Theory and Applications of Quantum Mechanics*, John Wiley & Sons, Inc.

19
p–n Junction Diode

The p–n junction diode is one of the simplest two-terminal, solid-state devices but is rich in physics and applications. The device is based upon the innovative utilization of basic concepts inherent in quantum mechanics, in particular, the interface physics and equilibrium and nonequilibrium statistics. It also provides platforms for a number of novel devices, such as photodetectors, light-emitting or laser diodes, solar cells, tunnel diodes, and so on. Furthermore, when two p–n junctions are placed together in interacting vicinity, a bipolar junction transistor is formed. As well known, BJT was one of the first driving forces for the digitalization and information technology. In addition, the p–n junction is a key element in metal oxide silicon field effect transistors, and as such plays an important role in the overall performance of the transistor. In addition, the junction diode covers a lot of ground in device physics and modeling, the understanding of which is essential for comprehending other active devices.

19.1
The Junction Interface Physics in Equilibrium

Overview The p–n junction consists of n- and p-type semiconductors in equilibrium contact, as sketched in Figure 19.1. There are two kinds of junctions, namely, homo and heterojunctions, depending on whether the two bandgaps in contact are the same or different. In this chapter, the discussion is confined to the former, for simplicity.

A positive voltage, when applied to p side with respect to n side is called the forward bias, while a negative voltage applied therein is called the reverse bias (Figure 19.1). Under forward bias, a large amount of current flows from p to n, while under reverse bias, a minimal amount of current flows in opposite direction. Hence, the device works as a rectifier and/or switch. In an ideal rectifier, a positive bias should induce a large amount of current with a minimal resistance, while the negative bias should draw essentially no current.

Figure 19.1 also shows the general features of the current–voltage (I–V) characteristics of the diode. For $V > 0$, I increases exponentially with V, while for $V < 0$, I quickly saturates at a very low level, as expected from a rectifier. With further increase in positive bias, the increase of the forward current slows down. Also, with increasing magnitude of negative bias beyond a certain critical value, the reverse current

Introductory Quantum Mechanics for Semiconductor Nanotechnology. Dae Mann Kim

ISBN: 978-3-527-40975-4

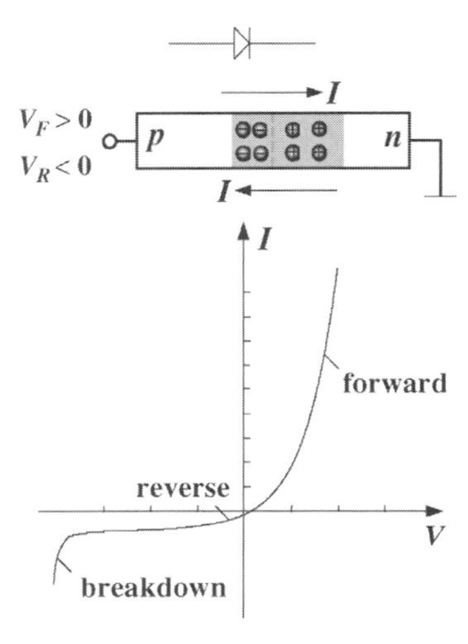

Figure 19.1 The p–n junction diode and diode I–V curve, showing forward, reverse, and breakdown currents.

suddenly starts to grow exponentially. This sudden and explosive increase of reverse current is known as the junction breakdown.

The diode operation is based upon the innovative utilization of the interface physics and the interplay between equilibrium and nonequilibrium conditions. The basic features of the equilibrium properties of the interface are therefore first considered.

19.1.1
Junction in Equilibrium Contact

Band Bending When p and n bulk regions of a given semiconductor are brought together into equilibrium contact, a junction is formed, accompanied by a series of physical processes occurring at the interface. Before the contact, E_F in the n bulk lies above E_F in the p bulk, as is clear from Figure 19.2. Specifically, E_F in the n region lies above E_i while E_F in the p region lies below E_i and evidently the difference between these two Fermi levels is given by the sum of respective Fermi potentials, namely, $q\varphi_{Fn} + q\phi_{Fp}$.

Upon contact, however, the equilibrium condition requires that the Fermi level should line up and be flat, as detailed in Chapter 18. Otherwise, the current would flow in violation of the equilibrium condition (see (18.41)). However, making the Fermi level flat throughout the entire junction region clearly necessitates the lowering of n side with respect to p side, thereby inducing the band bending by an amount $q\varphi_{Fn} + q\phi_{Fp}$. With this band bending, E_F can indeed be made flat and at the same time the relative position of E_F *vis-à-vis* E_i or E_C, E_V is maintained in both n and p regions, thereby preserving the electronic properties therein.

Space Charge, Field, and Potential A question then naturally arises as to what physical mechanisms are responsible for such a band bending. The answer to this

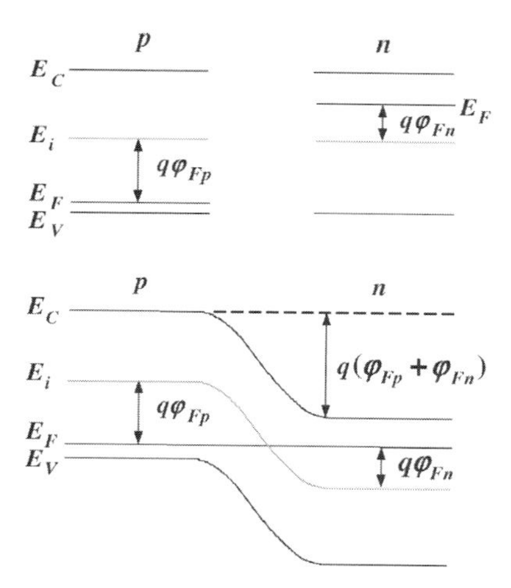

Figure 19.2 The p–n junction energy bands before and after contact and the equilibrium junction band bending.

question can be sought from the space charge ϱ, which is developed near the junction interface. The space charge ϱ results from the diffusion of electrons from the high-concentration region n to the low-concentration region p, and likewise, the diffusion of holes from p to n regions. These carrier diffusions leave behind the uncompensated donor and acceptor ions in n and p regions, giving rise to a fixed dipole charge layer of heights qN_D^+ and $-qN_A^-$ in the interface, as shown in Figure 19.3. Once the space charge is formed, it acts as the source for the space charge field E and potential φ. The potential when multiplied by the basic electron charge $-q$, namely, $-q\varphi$, connects the misaligned E_C and E_V in the junction interface, accounting for the band bending.

It should be noted that the space charge field E induced from the carrier diffusion acts in turn as the source for the opposite flow of electrons and holes via drift. These drift fluxes balance the diffusion fluxes, so that the net flux of electrons or holes becomes zero. To sum it up then, the carrier diffusion driven by the concentration gradient induces ϱ, giving rise to E and φ, and the band bending $-q\varphi$, that is, $\varrho \Rightarrow \mathrm{E} \Rightarrow \varphi \Rightarrow -q\varphi$. The accompanying drift of carriers completes the detailed balancing of two opposite fluxes required for equilibrium.

The qualitative discussion given above can be quantified by a simple electrostatic analysis. For this purpose, consider the 1D Coulomb's law that reads as

$$\frac{\partial}{\partial x}\mathrm{E} = \frac{\varrho}{\varepsilon_S}, \quad \varrho = \begin{cases} qN_D, & 0 \le x \le x_n, \\ -qN_A, & -x_p \le x \le 0, \end{cases} \tag{19.1}$$

where ε_S is the permittivity of the semiconductor and ϱ is taken a step function in completely depleted approximation, as depicted in Figure 19.3. Note that x_n and $-x_p$ demarcate the junction interface region from n and p bulk regions. One can readily

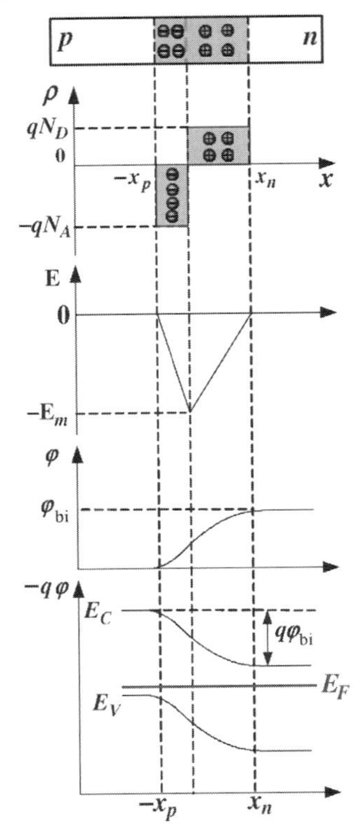

Figure 19.3 The dipole space charge ϱ inducing the space charge field E and potential φ. The resulting electron potential energy $-q\varphi$ accounts for the band bending, connecting E_C and E_V on both sides of the junction.

integrate (19.1), obtaining

$$\mathrm{E}(x) = \begin{cases} (qN_D/\varepsilon_S)(x - x_n), & 0 \le x \le x_n, \\ -(qN_A/\varepsilon_S)(x + x_p), & -x_p \le x \le 0. \end{cases} \tag{19.2}$$

Here, E does not penetrate into the n and p bulk regions, that is, $\mathrm{E}(x_n) = \mathrm{E}(-x_p) = 0$. Also, E should be continuous everywhere, and this condition when applied at the interface, namely, at $x = 0$, yields

$$qN_D x_n = qN_A x_p. \tag{19.3}$$

Obviously, (19.3) represents the global charge neutrality, namely, that the total number of electrons and holes spilled over from n to p and from p to n regions, respectively, is the same. Moreover, (19.3) provides in conjunction with (19.2) an important parameter, namely, the maximum space charge field (see (Figure 19.3)),

$$|\mathrm{E}_{\max}| = qN_D x_n/\varepsilon_S = qN_A x_p/\varepsilon_S. \tag{19.4}$$

Now that the space charge field $E(x)$ has been found, the corresponding potential can be obtained from

$$E(x) = -\frac{\partial \varphi}{\partial x}. \tag{19.5}$$

Inserting (19.2) for $E(x)$ in (19.5) and performing the x-integration, $\varphi(x)$ can be found as a quadratic function of x, leading to another key parameter of interest, namely, the built-in potential φ_{bi}. Here, φ_{bi} is defined as the potential difference across the interface region, that is, between $-x_p$ and x_n. This built-in potential φ_{bi} can be simply obtained without performing the integration but by considering the triangular area under the $E(x)$ curve in Figure 19.3:

$$\begin{aligned}\varphi_{bi} &\equiv \varphi(x_n) - \varphi(-x_p) \\ &= \frac{1}{2}\frac{qN_D x_n}{\varepsilon_S}x_n + \frac{1}{2}\frac{qN_A x_p}{\varepsilon_S}x_p.\end{aligned} \tag{19.6}$$

Depletion Depth and Built-in Potential The junction depletion depth W is defined as the interface region from $x = -x_p$ to $x = x_n$, namely,

$$\begin{aligned}W &\equiv x_n + x_p \\ &= x_n(1 + N_D/N_A) = x_p(1 + N_A/N_D)\end{aligned} \tag{19.7}$$

where (19.4) has been used. Hence, combining (19.7), (19.6), and (19.4), one can also express E and φ_{bi} in terms of W, namely,

$$E_{max} = \frac{qN_A N_D}{\varepsilon_S(N_A + N_D)}W, \tag{19.8a}$$

$$\varphi_{bi} \equiv \frac{1}{2}E_{max}W = \frac{q}{2\varepsilon_S}\frac{N_A N_D}{N_A + N_D}W^2. \tag{19.8b}$$

Now, φ_{bi} represents precisely the amount of band bending required for the equilibrium contact and is given by the sum of two Fermi potentials (see Figure 19.2):

$$\varphi_{bi} \equiv \varphi_{Fn} + \varphi_{Fp} = \frac{k_B T}{q}\ln\left(\frac{N_A N_D}{n_i^2}\right), \tag{19.9}$$

where (17.29) and (17.30) have been used for the respective Fermi potentials. Hence, for the given doping levels N_D and N_A, the junction parameters such as W, E_{max}, x_n, x_p and φ_{bi} are analytically specified in the completely depleted approximation that has been introduced for treating the dipole charge layer.

Equilibrium Carrier Profiles With the junction band bending in equilibrium thus quantified in terms of doping levels N_D and N_A, it is clear from Figure 19.3 that the law of mass action $np = n_i^2 \propto \exp - E_G/k_B T$ holds true everywhere, including the depletion region, as it should. In the n region, the equilibrium electron concentration as the majority carrier therein is given by

$$n_{n0} = n_i e^{(E_F - E_i)/k_B T} = n_i e^{q\varphi_{Fn}/k_B T}, \tag{19.10}$$

while the electron concentration in the p region as the minority carrier is given by

$$n_{p0} = n_i e^{(E_F - E_i)/k_B T} = n_i e^{-q\varphi_{Fp}/k_B T}. \tag{19.11}$$

Hence, these two concentrations are correlated via φ_{bi} as

$$n_{p0} = n_{n0} e^{-q\varphi_{\text{bi}}/k_B T}. \tag{19.12}$$

By the same token, one can also write

$$p_{n0} = p_{p0} e^{-q\varphi_{\text{bi}}/k_B T}. \tag{19.13}$$

19.2
The Junction Interface Under Bias

The Nonequilibrium and Quasi-Fermi Levels Under a bias, the junction is pushed away from equilibrium to nonequilibrium conditions and the resulting modifications of the junction interface are discussed. Specifically, when the junction is forward biased by applying a positive voltage V on the p side, E_C and/or E_V therein is lowered with respect to the n bulk by an amount qV, as sketched in Figure 19.4. This indicates that the junction band bending has to be reduced from the equilibrium value, $q\varphi_{\text{bi}}$ to $q(\varphi_{\text{bi}} - V)$. Concomitantly, the location of E_F with respect to E_i should be kept the same in both n and p regions, so that the respective bulk properties are preserved.

Obviously, these two requirements cannot be met with the single E_F operative in equilibrium. Rather, two quasi-Fermi levels or imrefs, one for electrons E_{Fn} and the other for holes E_{Fp} are required to satisfy the new boundary conditions under bias, as shown in Figure 19.4. It has therefore become clear that the introduction of two Fermi levels are essential in nonequilibrium conditions, whether the system is pushed away from equilibrium via irradiation (see (18.49)) or by applying the voltages.

Moreover, the reduced junction band bending by an amount qV should be accompanied by reduced depletion depth W and the corresponding reduction of the maximum field $\mathrm{E}_{\max}$ at the interface, as shown in Figure 19.4. It should also be noted that the quasi-Fermi levels E_{Fn} and E_{Fp} are split in W but merge in both n and p bulk regions to become flat therein, so that the equilibrium properties still prevail in the bulk. Note further that E_{Fn} remains approximately flat in W in quasi-equilibrium approximation and gradually merges with E_{Fp} in the p region. Likewise, E_{Fp} remains flat in W and merges with E_{Fn} in the n bulk. In the process, a loop is formed by these two imrefs in the junction interface.

When the junction is reverse biased by applying negative voltage $-V$ on the p side, the band therein is raised and consequently the band bending should increase from $q\varphi_{\text{bi}}$ to $q[(\varphi_{\text{bi}} - (-V)]$. Concomitantly, W is broadened and $\mathrm{E}_{\max}$ is increased, as shown in Figure 19.5. The changes in W and $\mathrm{E}_{\max}$ resulting from the forward ($V > 0$) or reverse ($V < 0$) bias are described from (19.8) by

$$W(V) = \left[\frac{2\varepsilon_S(N_A + N_D)}{qN_A N_D}(\varphi_{\text{bi}} - V)\right]^{1/2}, \tag{19.14}$$

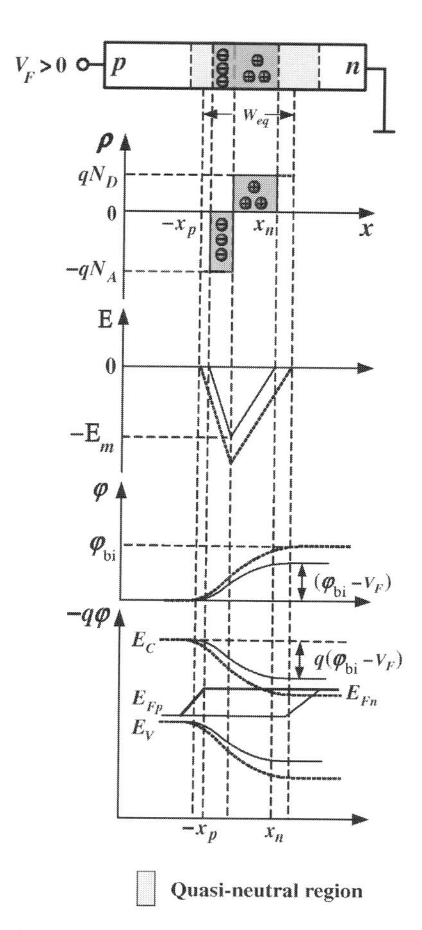

Figure 19.4 The p–n junction under forward bias: the band bending and W, E, and φ are reduced from the equilibrium values. The two quasi-Fermi levels, E_{Fn} and E_{Fp} account for n, p in the junction region and accommodate bias via splitting, namely, $E_{Fn} - E_{Fp} = qV$.

$$\mathrm{E}_{\max}(V) = \frac{qN_D N_A}{\varepsilon_S(N_A + N_D)} W(V). \tag{19.15}$$

Charge Injection and Extraction The voltage-controlled modulation of charge carrier concentrations is a key factor underpinning the operation of the junction diode and, for that matter, of any other active devices. The overall picture of the carrier concentrations can be clearly observed from Figure 19.6, which summarizes the band bending both in equilibrium and nonequilibrium under forward and reverse biases. Note in particular the splitting of quasi-Fermi levels or the imrefs in the junction depletion region to accommodate the decreased or increased band bending due to bias,

$$E_{Fn} - E_{Fp} = qV. \tag{19.16}$$

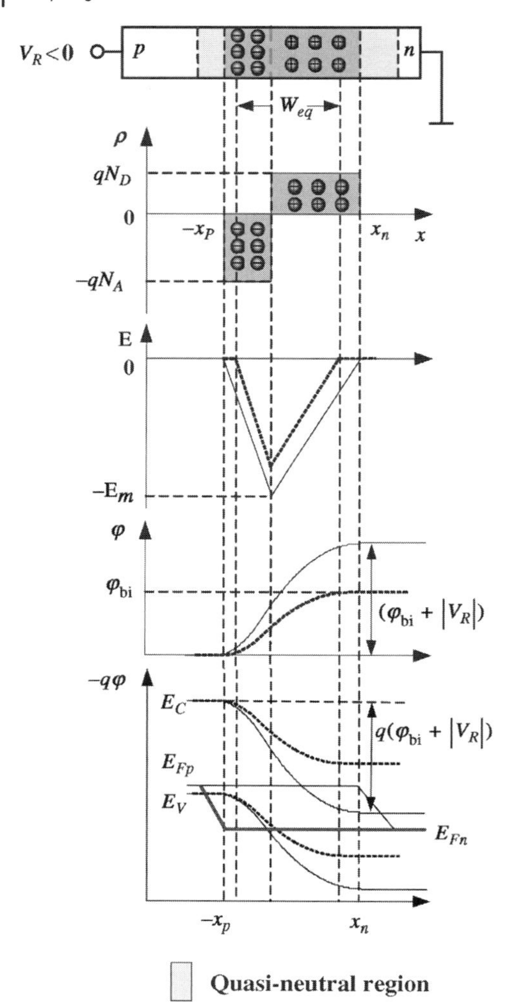

Figure 19.5 The p–n junction under reverse bias: the band bending is increased, together with W, E, and φ and quasi-Fermi levels split to accommodate the bias $E_{Fn}-E_{Fp} = -qV_R$.

Also, these two imrefs merge gradually just outside the x_n and $-x_p$ and the regions in which these two imrefs merge are called the quasi-neutral regions.

Since E_C and/or E_V varies as a function of x in the interface region, n and p should also depend on x therein. For the nondegenerate case and under bias, one can write

$$n(x) = N_C e^{-[E_C(x)-E_{Fn}(x)]/k_B T}, \tag{19.17}$$

$$p(x) = N_V e^{-[E_{Fp}(x)-E_V(x)]/k_B T}, \tag{19.18}$$

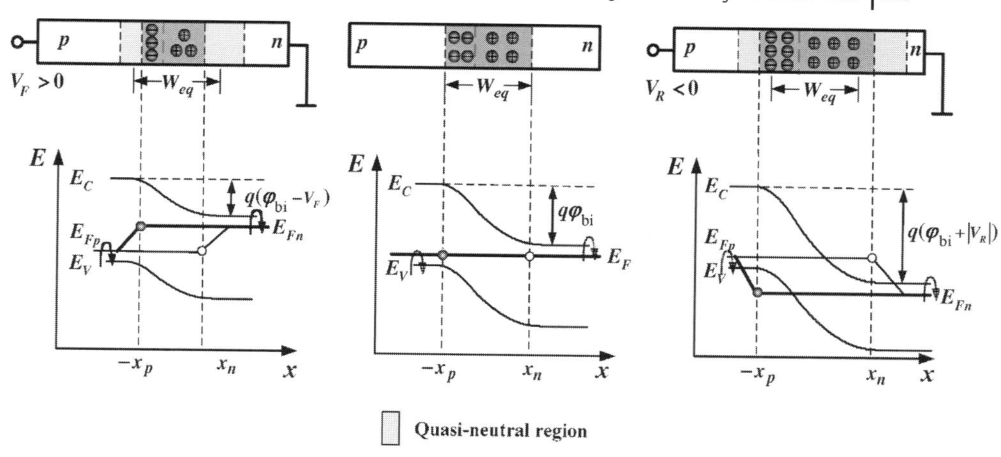

Figure 19.6 The junction band bending in equilibrium and nonequilibrium under forward (left) and reverse (right) biases, respectively. E_F in equilibrium splits under bias into E_{Fn} and E_{Fp} in the junction region, specifying changes in minority carrier concentrations therein, and these two imrefs merge in quasi-neutral regions.

so that

$$n(x)p(x) = N_C N_V e^{-[E_C(x)-E_V(x)]/k_B T} e^{[E_{Fn}(x)-E_{Fp}(x)]/k_B T} = n_i^2 e^{qV/k_B T}. \tag{19.19}$$

As noted earlier, the role of imrefs is identical to that of E_F in specifying n and p and it is therefore clear from (19.19) that the charge is injected in the interface region under forward bias, while it is extracted under reverse bias, namely,

$$n(x)p(x) > n_i^2, \quad V > 0, \tag{19.20}$$

$$n(x)p(x) < n_i^2, \quad V < 0. \tag{19.21}$$

The voltage-controlled injection and extraction of minority carriers are graphically summarized in Figure 19.7.

Equations 19.20 and 19.21 are the key driving forces responsible for the diode operation. When the junction is driven away from equilibrium via bias, there ensues a reactive process to drive the system back to equilibrium. Specifically, the charge injection is countered by the recombination of electrons and holes, giving rise to the recombination current I_R. By the same token, the charge extraction is accompanied by the inverse process of the generation of electron–hole (e–h) pairs and the generation current I_G. These two current components are the main contributors to the diode current and the diode operation is thus based upon the interplay between the equilibrium and nonequilibrium. That is, the operation is based upon driving the system away from the equilibrium and utilizing the reactive force pushing it back to equilibrium.

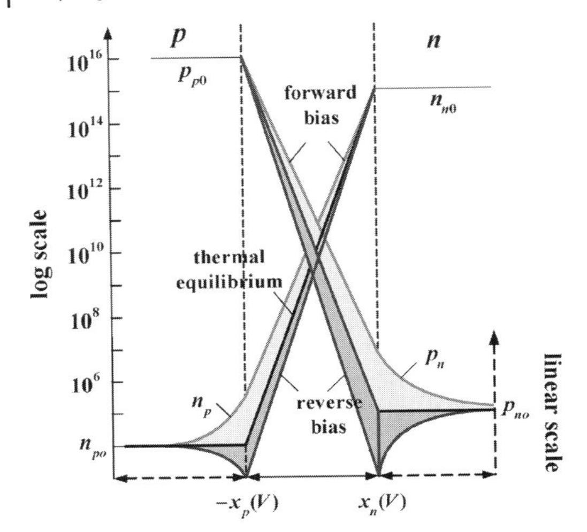

Figure 19.7 The charge injection and extraction in the junction depletion depth W and quasi-neutral regions under forward and reverse biases. The logarithmic and linear scales were used in W and quasi-neutral regions, respectively.

19.3
The Diode *I–V*

Overview In equilibrium, there is no net flux of electrons or holes across the junction interface. Specifically, the diffusion flux of electrons, for example, from n to p regions is balanced by the drift flux driven by the space charge field in the opposite direction, as pointed out. In spilling over from n to p regions via diffusion, electrons have to overcome the potential barrier of height $q\varphi_{bi}$, while they can freely roll down the barrier from p to n regions via drift (Figure 19.2). Likewise diffusing holes from p to n must overcome the barrier φ_{bi} while drifting holes from n to p freely rollup the potential hill. This detailed balancing between diffusion and drift is broken when the junction is biased and as a consequence the current flows.

The output current of the diode is next analyzed, using the *I–V* model by Shockley. In his model, Shockley introduced a few nonessential assumptions, for simplicity: (i) the abrupt deletion approximation has been used, in which the dipole charge layer supporting the band bending is taken constant at the level given by N_D and N_A and ending abruptly at x_n and $-x_p$ (see Figure 19.3), (ii) the charge injection is confined to the low level in which n_p and p_n injected in p and n regions are taken much smaller than the equilibrium majority carrier concentrations therein, namely, p_{p0} and n_{n0}, (iii) the carrier concentrations are confined to the nondegenerate regime, so that n and p are specified by the Boltzmann distribution function, and finally (iv) the junction interface is taken ideal, so that the recombination or generation currents in the depletion depth W are neglected.

In *I–V* modeling, the p–n junction is divided into three regions, as shown in Figure 19.6: (a) the depletion region W that is comprised of the dipole layer supporting the band bending, (b) the quasi-neutral regions on both sides of the depletion region in which the two imrefs gradually merge into one, and (c) n and p bulk regions.

19.3.1
Ideal I–V Behavior

Diffusion Equation Consider first a positive bias V applied to the p side. In this case, the detailed balancing between drift and diffusion fluxes of the charge carriers is broken. Consequently excess electrons and holes spill in from n and p regions into the interface region, thereby providing excess charge carriers in depletion and quasi-neutral regions (Figure 19.7). The hole concentration p_n injected, for instance, into the quasi-neutral region on the n side (Figures 19.6 and 19.7) evolves in time via the equation

$$\frac{dp_n}{dt} = -\frac{d}{dx}\left(p_n \mu_p \mathrm{E} - D_p \frac{dp_n}{dx}\right) - \frac{p_n - p_{n0}}{\tau_p}. \tag{19.22}$$

Here, the rate equation is comprised of the usual continuity equation (see (18.10)) and the recombination term for minority carriers, as discussed in Section 18.3 (see (18.72)).

In the steady state, p_n is the time invariant and the small electric field E in the quasi-neutral region can be neglected, so that (19.22) reduces to diffusion equation:

$$\frac{d^2 p_n}{dx^2} - \frac{p_n - p_{n0}}{L_p^2} = 0, \quad L_p \equiv (D_p \tau_p)^{1/2} \quad \text{for } x \geq x_n, \tag{19.23}$$

with L_p denoting the hole diffusion length.

The solution to this equation is of the form $\exp \pm x/L_p$, of which the positive branch is not suitable for the region under consideration. Also, the boundary conditions are specified by

$$p_n(x_n) = p_{n0} e^{qV/k_B T}, \tag{19.24a}$$

$$p_n(x \to \infty) = p_{n0} = n_i^2 / N_D, \tag{19.24b}$$

where (19.24a) represents the charge injection or extraction, depending on the polarity of V (see (19.17) or (19.18) and Figures 19.6 and 19.7), while (19.24b) reiterates the bulk property in the n region.

Incorporating the boundary conditions (19.24) in (19.23), p_n is given by

$$p_n(x - x_n) = p_{n0}(e^{qV/k_B T} - 1) e^{-(x - x_n)/L_p} + p_{n0}, \quad x \geq x_n. \tag{19.25}$$

One can carry out a similar analysis for n_p in the quasi-neutral region on p side, obtaining

$$n_p(x + x_p) = n_{p0}(e^{eV/k_B T} - 1) e^{(x + x_p)/L_n} + n_{p0}, \quad L_n \equiv (D_n \tau_n)^{1/2}, \; x \leq -x_p, \tag{19.26}$$

where L_n is the electron diffusion length.

Diffusion Current Having found both $p_n(x)$ and $n_p(x)$ in two quasi-neutral regions, the respective diffusion current densities are obtained as

$$J_p(x) \equiv qD_p\left(-\frac{dp_n}{dx}\right) = \frac{qD_p p_{n0}}{L_p}\left[e^{qV/k_BT}-1\right]e^{-(x-x_n)/L_p}, \quad x \geq x_n, \tag{19.27}$$

and

$$J_n(x) \equiv -qD_n\left(-\frac{dn_p}{dx}\right) = \frac{qD_n n_{p0}}{L_n}\left[e^{qV/k_BT}-1\right]e^{(x+x_p)/L_p}, \quad x \leq -x_p. \tag{19.28}$$

In the ideal I–V model, the total current density is taken to be the sum of these two diffusion currents, evaluated at x_n and $-x_p$ respectively, namely,

$$I_{\text{ideal}} = I_n(-x_p) + I_p(x_n) = I_S(e^{qV/k_BT}-1), \tag{19.29a}$$

where

$$I_S = \left(\frac{qD_n n_{p0}}{L_n} + \frac{qD_p p_{n0}}{L_p}\right)A_J = qn_i^2\left(\frac{D_n}{L_n N_A} + \frac{D_p}{L_p N_D}\right)A_J \tag{19.29b}$$

is the saturation current and A_J is the cross-sectional area of the diode. Equation 19.29 is the ideal diode I–V model.

As is clear from (19.29), the diode current is contributed by both electrons and holes. Under the forward bias $V > 0$, the current exponentially increases, driven by exponentially enhanced minority carrier injection. Also, electrons and holes diffuse in opposite directions, but the two current components add up obviously due to the opposite polarity of the charge. The forward current flows from p to n direction.

It is essential to note that the electron and hole diffusion currents in (19.27) and (19.28) are sensitively dependent on x in their respective quasi-neutral regions. However, the majority carrier drift currents in the same regions also vary, in such a manner that the total current is maintained at a constant level, as specified by (19.29). This is illustrated in Figure 19.8. Without satisfying this basic condition of a constant current level at every point in x, it is not possible to maintain a steady-state current. The voltage required to induce the required drift currents takes up a minute fraction of the total voltage applied because of the large majority carrier concentrations involved in the drift.

Under a reverse bias $V < 0$, p_n and n_p are now depleted in their respective quasi-neutral regions and electrons and holes diffuse in the reverse direction, as shown in Figure 19.8. The ideal I–V behavior of the p–n junction diode is shown in Figure 19.9. The forward current is indeed exponentially increased with V, while the reverse current reaches the saturation level I_S at a small reverse bias, typically a few thermal voltages, k_BT/q.

Finally, the average distance an excess hole, for example, traverses in the quasi-neutral n region before recombination can be found from (19.25) by

$$\langle x-x_n\rangle = \frac{\int_{x_n}^{\infty} dx e^{-(x-x_n)/L_p}(x-x_n)}{\int_{x_n}^{\infty} dx e^{-(x-x_n)/L_p}} = L_p. \tag{19.30}$$

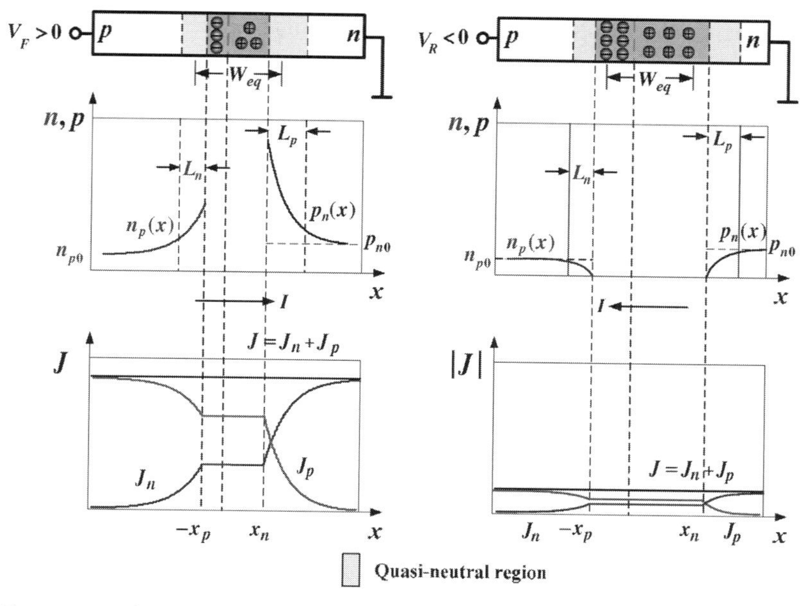

Figure 19.8 The injected or extracted minority carrier profiles in quasi-neutral regions. The resulting minority carrier diffusion current profiles are shown together with majority carrier drift currents accompanied to make the total current constant.

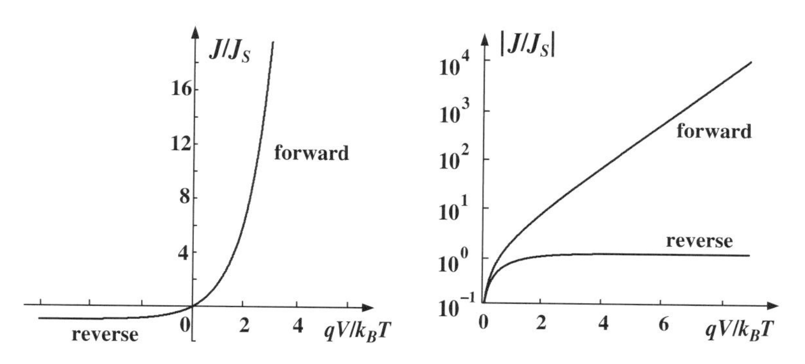

Figure 19.9 The ideal diode I–V behavior, consisting of forward and reverse currents.

The diffusion length therefore represents the average distance an excess minority carrier traverses before recombination.

19.3.2
Nonideal I–V Behavior

Generation and Recombination The nonideal I–V behavior of the junction diode is next discussed briefly. In the ideal I–V model, the junction interface has been put to play a passive role, accommodating merely the charge injection or extraction, depending on the polarity of the applied voltage. In practice, however, there constantly occurs in the depletion depth W the recombination and/or generation of electron–hole pairs, thereby providing additional current components.

This is conveniently described by the recombination rate derived in (18.70), namely,

$$U = \frac{1}{\tau} \frac{n_i^2 (e^{qV/k_B T} - 1)}{n + p + 2n_i \cosh(E_t - E_i)/k_B T}, \tag{19.31}$$

where the product of the two carrier concentrations pn in the numerator has been put into an explicit function of V(see (19.19)). It is clear from (19.31) that the maximum recombination rate ensues when the trap level coincides with the midgap, $E_t = E_i$, and also for the minimum value of $n + p$, that is,

$$d(n+p) = 0,$$

subject to the condition that the product pn is constant, namely,

$$pn = n_i^2 e^{qV/k_B T}.$$

Combining these two conditions, one readily finds

$$dn = -dp = -d \frac{n_i^2 e^{qV/k_B T}}{n} = \frac{p}{n} dn$$

or

$$n = p = n_i e^{qV/2k_B T}. \tag{19.32}$$

Thus, inserting (19.32) in (19.31), the maximum recombination rate is given by

$$U_{\max} = \frac{1}{\tau} \frac{n_i (e^{qV/k_B T} - 1)}{2(e^{qV/2k_B T} + 1)} \approx \frac{1}{2\tau} n_i e^{qV/2k_B T} \tag{19.33}$$

and the resulting recombination current can therefore be estimated as

$$I_R \approx q U_{\max} W A_J \approx \frac{q}{2\tau} n_i e^{qV/2k_B T} W A_J. \tag{19.34}$$

Under a reverse bias, in which $\exp -qV/k_B T \ll 1$ and $n, p \ll n_i$ in the depletion region, the maximum generation rate for $E_t = E_i$ is to be found from (19.31) as

$$U_G \equiv |U_G| \approx \frac{1}{2\tau} n_i. \tag{19.35}$$

Hence, the generation current in W can be approximately given by

$$I_G \approx \frac{q}{2\tau} n_i W A_J. \tag{19.36}$$

In summary, the I–V behavior of the p–n junction diode, including the generation and recombination currents in W, is given by

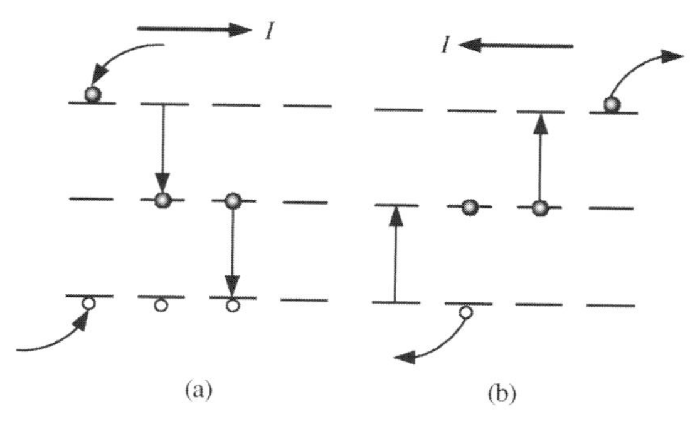

Figure 19.10 The schematic description of the cyclic trap-assisted recombination and generation of electron–hole pairs in W, giving rise to recombination (a) and generation (b) currents.

$$I_{\text{total}} = \begin{cases} I_{\text{ideal}} + I_R, & V > 0, \\ I_{\text{ideal}} + I_G, & V < 0, \end{cases} \tag{19.37}$$

where each term in (19.37) has been specified in (19.29), (19.34), and (19.36), respectively.

A few comments are due regarding the general features of I–V characteristics. First, I_R and I_G can be contributed by the band-to-band generation or recombination of electron–hole pairs. But in practice, I_R and I_G are primarily due to the trap-assisted recombination or generation. Thus, the mechanism responsible for I_R can be illustrated with the aid of Figure 19.10. Under the forward bias, the excess electrons and holes are constantly supplied from both n and p bulk regions into the junction interface and are recombined in two steps in succession, namely, the electron capture followed by the hole capture. This two-step recombination when repeated in cyclic fashion closes the current loop for I_R.

Similarly, I_G can be understood in terms of the alternating emissions of holes and electrons in succession (see Figure 19.10). The electron–hole pairs thus generated are swept across the depletion region, holes to the p region and electrons to the n region, by strong junction electric field to close the loop for I_G.

The nonideal current components are often incorporated empirically into an analytical form as

$$I = I_S\left[\exp\left(\frac{qV}{mk_BT}\right) - 1\right], \quad I_S \approx A_J q n_i^2\left(\frac{D_n}{L_n N_A} + \frac{D_p}{L_p N_D}\right) + A_J\frac{qn_i}{2\tau}W. \tag{19.38}$$

In this expression, I_G adds to I_S while I_R is embedded into the ideality factor m that varies from 1 to about 2. The ideality factor empirically accounts for the different V-dependence of the diffusion and recombination components of the total forward current (see (19.29) and (19.34)). Figure 19.11 shows the typical I–V characteristics of the diode. For small forward voltage V, the recombination current $I_R \propto \exp qV/2k_BT$ dominates (a) and at high-level injection for large V, the exponential growth of the forward current slows down considerably (c, d) due to the effect of series resistance

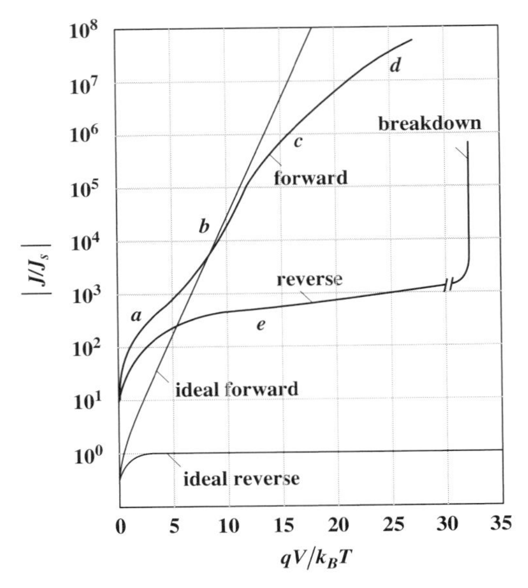

Figure 19.11 The general diode I–V curve, consisting of ideal forward and reverse currents and recombination, generation, and breakdown currents. (taken from Sze, S. M. and Ng, K. K. (2006) Physics of Semiconductor Devices, 3rd edn., Wiley Interscienece.)

and other factors. Also, the generation current considerably enhances the saturation current I_S, as expected (*e*).

The explosive increase of the reverse current I_R beyond a certain range of the reverse voltage, called the breakdown voltage V_{BR}, is caused by a few basic physical processes. The avalanche multiplication, for example, is an important factor for the breakdown. Figure 19.12 graphically illustrates the multiplication mechanism. The electron–hole pairs, when generated in the junction interface region via the band-to-band or trap-assisted excitations are subject to a strong electric field, namely, the built-in space charge field, which is further reinforced by the reverse bias. Under such strong local field, the carriers can gain in between collisions kinetic energies sufficient to generate e–h pairs via the impact ionization of the host atoms. The impact ionization process occurring in cascade gives rise to the avalanche breakdown.

The Fowler–Nordheim (F–N) tunneling provides additional mechanism responsible for the breakdown. Figure 19.12 also illustrates the F–N tunneling occurring in the reverse-biased junction interface. Because the electrons cannot reside in the bandgap, the gap acts as the potential barrier for electrons in the valence bands. The barrier assumes a triangular shape under reverse bias with height E_G and the width narrowed by strong local electric field E present therein. Thus, the electrons in the valence band can tunnel through this barrier to the conduction band with the probability given from F–N tunneling (6.12) by

$$T \approx \exp-\frac{4(2m_n)^{1/2}}{3q\mathrm{E}\hbar}E_G^{3/2}.$$

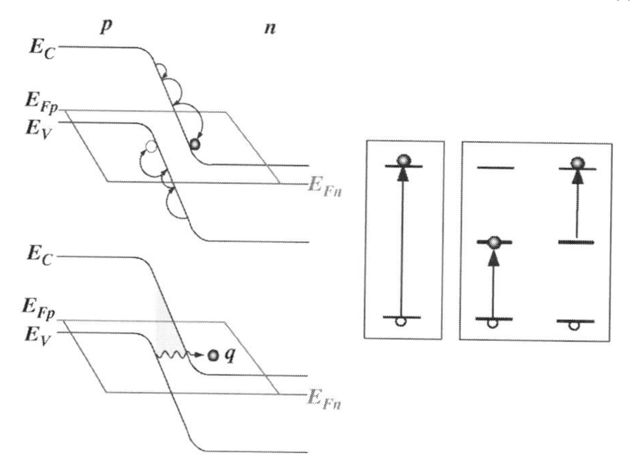

Figure 19.12 The schematic illustration of diode breakdown: the band-to-band or trap-assisted generation of e–h pairs (right); avalanche multiplication due to impact ionizations occurring in cascade (top left); Zener breakdown due to Fowler–Nordheim tunneling of valence band electrons to conduction band (bottom left).

Indeed, the probability of tunneling increases exponentially with increasing E and/or reverse voltage V_R. This tunneling-induced breakdown is known as Zener breakdown.

19.4 Applications of p–n Junction Diodes

19.4.1 Optical Absorption in Semiconductors

As mentioned, a variety of innovative applications of the p–n junction interface has been devised and some of these applications are based upon the interaction of radiation with semiconductors. Thus, the optical absorption in a semiconducting medium is first considered and that in a direct bandgap semiconductor, for simplicity of discussion. Figure 19.13 shows the conduction and valence bands in a direct bandgap semiconductor, together with $E-k$ dispersion curves of electrons and holes in conduction and valence bands, respectively. In a direct bandgap material, the minimum E_C and the maximum E_V in the dispersion curves coincide at $k = 0$, as shown in the figure. Near the bottom of E_C and the top of E_V, the dispersion curves are generally characterized by $E \propto k^2$ and are well approximated by those of a free particle, as detailed in Chapter 7.

The quantum states of electrons above E_F in the conduction band are mostly empty, while those below E_F in the valence band are nearly filled up. Also, an electron, when promoted into the conduction band, moves as a free particle with an effective mass m_n and kinetic energy $E = \hbar^2 k^2/2m_n$. The hole left behind in the valence band also moves as a free particle with energy $E = \hbar^2 k^2/2m_p$. With increasing kinetic energy,

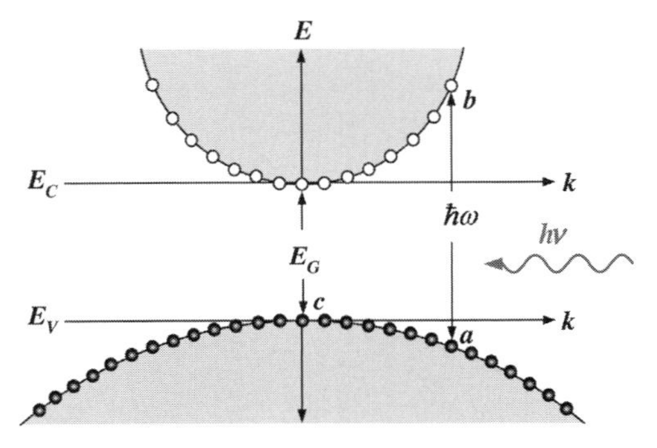

Figure 19.13 The dispersion curves of electrons and holes in conduction and valence bands, respectively, in a direct bandgap semiconductor and the band to band generation of e–h pair via the absorption of a photon.

the electron in the conduction band moves up the dispersion curve. Likewise, with increasing kinetic energy, the hole moves down the dispersion curve in the valence band. This is because it takes a certain amount of energy for an electron to move from (a) to (c) in Figure 19.13 in the valence band, for example. This is equivalent to hole moving from (c) to (a) and gaining the same amount of kinetic energy.

Consider an electron making a band-to-band transition by absorbing a photon, as sketched in Figure 19.13. The interaction Hamiltonian between the light and matter was discussed in Chapter 14 and is given from (14.42) by

$$\hat{H}' = -\hat{\mu}\frac{\mathrm{E}_0}{2}\left[e^{i(\omega t - \boldsymbol{k}_{\mathrm{opt}}\cdot\boldsymbol{r})} + e^{-i(\omega t - \boldsymbol{k}_{\mathrm{opt}}\cdot\boldsymbol{r})}\right], \quad \hat{\mu} \equiv q(\hat{e}_f\cdot\boldsymbol{r}), \tag{19.39}$$

where $\hat{e}_f$ is the polarization vector of the light wave and the time and position dependence of the electric field is explicitly accounted for in (19.39).

With this interaction Hamiltonian, the transition rate of the electron from valence to conduction band can be described by the Fermi's golden rule. The rate is given from (14.43) by

$$W_{vc} = \frac{2\pi}{\hbar}|\hat{H}'|^2\delta(E_b - E_a - \hbar\omega), \quad |\hat{H}'|^2 = \frac{\tilde{\mu}^2\mathrm{E}_0^2}{4}, \tag{19.40}$$

where the transition matrix element reads in terms of the Bloch wavefunction (7.4) and (7.11) as

$$\tilde{\mu} = q\int d\boldsymbol{r}u_c^*(\boldsymbol{r})u_v(\boldsymbol{r})(\hat{e}_f\cdot\boldsymbol{r})e^{-i(\boldsymbol{k}_c - \boldsymbol{k}_v \pm \boldsymbol{k}_{\mathrm{opt}})\cdot\boldsymbol{r}}. \tag{19.41}$$

Here, u_c and u_v denote the periodic modulation functions in conduction and valence bands, respectively, and $\boldsymbol{k}_c$ and $\boldsymbol{k}_v$ are the corresponding crystal wave vectors.

Inasmuch as the Bloch wavefunction is extended over the entire volume of the crystal, the spatial integration of (19.41) has to be carried out over the same crystal volume. In this case, the rapidly varying phase factor in the integrand will make the transition matrix vanishingly small unless the condition is satisfied, namely,

$$\boldsymbol{k}_c - \boldsymbol{k}_v \pm \boldsymbol{k}_{\text{opt}} = 0. \tag{19.42}$$

In the optical wavelength regime, $k \approx 2\pi/\lambda \approx 10^5\ \text{cm}^{-1}$ at $\lambda = 500$ nm, while $k_c \approx k_v \approx 2\pi/d \approx 10^8\ \text{cm}^{-1}$ for the lattice spacing of $d \approx 0.5$ nm. Hence, k_{opt} is much smaller than k_c and k_v and can therefore be neglected in (19.42), so that one can put

$$\boldsymbol{k}_c = \boldsymbol{k}_v = \boldsymbol{k}. \tag{19.43}$$

It thus follows from (19.43) that the optical transitions occur vertically in k space with the magnitude of the Bloch wave vector remaining intact (Figure 19.13).

Thus, the evaluation of the transition matrix element has been reduced to simply calculating the inner product of the modulation functions in the conduction and valence bands, namely,

$$\tilde{\mu} = q\langle u_c(\boldsymbol{r})|(\hat{e}_f \cdot \boldsymbol{r})|u_v(\boldsymbol{r})\rangle.$$

Next, the difference in energy between the initial and final states has to be analyzed to examine the transition rate in (19.40). The difference in energy consists of the sum of the kinetic energies of the electron and hole plus the bandgap (Figure 19.13):

$$E_f - E_i = \frac{\hbar^2 k^2}{2}\left(\frac{1}{m_c} + \frac{1}{m_v}\right) + E_G = \frac{\hbar^2 k^2}{2\mu_{\text{eff}}} + E_G, \tag{19.44}$$

where $1/\mu_{\text{eff}} = 1/m_c + 1/m_v$, and (19.43) has been used. The effective mass of electron m_n and that of hole m_p are also denoted by m_c and m_v, respectively.

Hence, the total number of transitions N per unit time in volume V is obtained by multiplying the transition rate W_{vc} in (19.40) by the density of states:

$$N = \frac{2\pi}{\hbar}\frac{\tilde{\mu}^2 \text{E}_0^2 V}{4}\int \delta\left(\frac{\hbar^2 k^2}{2\mu_{\text{eff}}} + E_G - \hbar\omega\right)\frac{k^2 dk}{\pi^2}, \quad g(k) = \frac{k^2}{\pi^2}, \tag{19.45}$$

where $g(k)$ is the density of states in k space derived in (4.25). Implicit in (19.45) is an assumption that the quantum states of electrons in valence band are occupied by electrons, while those in conduction band are not. This condition is certainly valid at $T = 0$. The integral in (19.45) can be done in a simple manner by introducing a new variable of integration,

$$\xi = \frac{\hbar^2 k^2}{2\mu_{\text{eff}}} + E_G - \hbar\omega,$$

in which case (19.45) is transformed into a form amenable to a simple δ-function integration:

$$\begin{aligned} N/V &= \frac{\tilde{\mu}^2 \text{E}_0^2 \mu_{\text{eff}}^{3/2}}{\sqrt{2}\pi\hbar^4}\int \delta(\xi)(\xi + \hbar\omega - E_G)^{1/2} d\xi \\ &= \frac{\tilde{\mu}^2 \text{E}_0^2 \mu_{\text{eff}}^{3/2}}{\sqrt{2}\pi\hbar^4}(\hbar\omega - E_G)^{1/2}. \end{aligned} \tag{19.46}$$

The absorption coefficient is therefore obtained by dividing the power absorbed per unit volume, $\hbar\omega \times (N/V)$, by the incident power crossing the unit area, namely, the Poynting vector:

$$\alpha(\omega) \equiv \frac{\hbar\omega N/V}{c\varepsilon E_0^2}$$
$$= A^*(\hbar\omega - E_G)^{1/2}, \quad A^* = \frac{\omega\tilde{\mu}^2\mu_{\text{eff}}^{3/2}}{\sqrt{2}\pi\hbar^3 c\varepsilon}. \tag{19.47}$$

Here, c and ε are the velocity of light and permittivity of the medium, respectively. It is therefore clear from (19.47) that the photon energy should be larger than the bandgap for the optical absorption to occur and further that the absorption coefficient increases with increasing photon energy as more electron states participate in the absorption process.

19.4.2 Photodiodes

The photodiode is the p–n junction diode used for detecting the optical signal. The detector has high sensitivity and speed and is utilized for various optoelectronic applications. When used as photodetector, the diode is operated in the reverse bias mode to increase the speed of response. Thus, consider a p–n junction reverse biased and illuminated by photons with energy larger than the bandgap, as sketched in Figure 19.14. The absorption of photons in the junction interface generates electron–hole pairs by the band-to-band excitation. The e–h pairs thus generated are propelled by the strong local field, namely, the built-in field, reinforced by the reverse bias, in opposite directions, holes rolling up the potential barrier to p region and electrons rolling down the barrier to n region (Figures 19.14 and 19.5). On

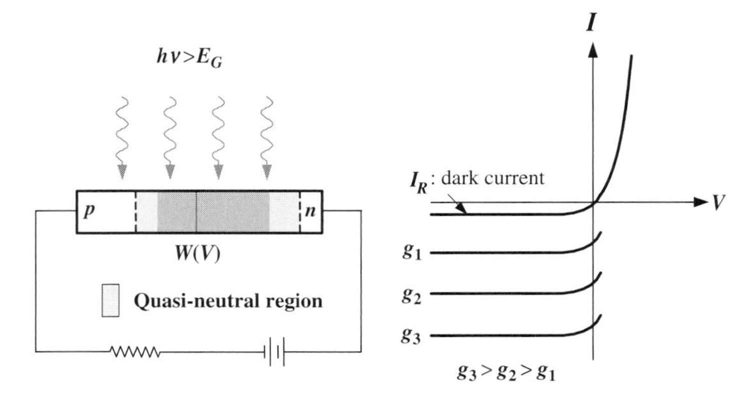

Figure 19.14 The photoresponse of the p–n junction: electrons and holes are photogenerated and transported to n and p regions, respectively, and give rise to photocurrent on top of the reverse dark current of the diode.

reaching n and p bulk regions, these excess carriers join the respective majority carriers and are endowed with a long lifetime.

The output current for the incoming light signal is analyzed as flows. The generation rate of e–h pairs at y from the surface,

$$g(y) = g_0 e^{-\alpha y}, \quad g_0 = \alpha[I_0(1-R)/h\nu], \tag{19.48}$$

is given by the linear absorption coefficient α of the medium, the reflection coefficient of light, R, at the surface, and the incident photon flux $I_0/h\nu$. The drift current as contributed by e–h pairs photo-generated in the illuminated area specified by the depletion depth W and width w namely Ww and swept across the depletion region in opposite directions are given by,

$$I_{\mathrm{drift}} = -qWw \int_0^{T_j} dy g(y) = -qA\tilde{g}_0 W, \quad A = wT_j \tag{19.49a}$$

where A is the cross-sectional area of photocurrent with T_j denoting the diode thickness and

$$\tilde{g}_0 = g_0[(1-e^{-\alpha T_j})/\alpha T_j] \tag{19.49b}$$

Thus, for $T_j \ll 1/\alpha, \tilde{g}_0 = g_0$ while $\tilde{g}_0 = g_0/\alpha T_j$ for $T_j \gg 1/\alpha$.

Concomitant with the absorption of light in W, photons are also absorbed in the two quasi-neutral regions, generating e–h pairs therein. For instance, consider the e–h pair generated in the quasi-neutral region on the n side. The photogenerated electrons drift to the n region driven by the reverse bias to join the majority carriers therein. Simultaneously the holes are diffused in the opposite direction toward the junction region where the holes are depleted due to the reverse bias. (see Figures 19.7 and 19.8). On reaching the junction edge, the holes are swept across the depletion depth to the p region, propelled by the strong electric field present therein. The resulting diffusion current can be analyzed by considering a generalized diffusion equation, namely,

$$p''_n - \frac{p_n - p_{n0}}{L_p^2} + \frac{\tilde{g}_0}{D_p} = 0, \quad L_p \equiv (D_p \tau_p)^{1/2} \quad \text{for } x \geq x_n, \tag{19.50}$$

where the photogeneration rate has been embedded into the usual diffusion equation (19.23). The solution can be obtained in a straightforward manner, using two boundary conditions: (i) $p_n(x = x_n) = 0$ under reverse bias and (ii) $p_n(x \to \infty) = p_{n0} + g\tau_p$ in n bulk. Thus, the solution reads as

$$p_n(x) = (p_{n0} + \tilde{g}_0 \tau_p)[1-e^{-(x-x_n)/L_p}]. \tag{19.51}$$

The hole diffusion current is therefore given by

$$\begin{aligned} I_{p,\mathrm{diff}} &\equiv -qAD_p \frac{\partial p_n(x = x_n)}{\partial x} \\ &= -qAD_p \frac{p_{n0} + \tilde{g}_0 \tau_p}{L_p} = -qA\left(\frac{p_{n0} D_p}{L_p} + \tilde{g}_0 L_p\right), \quad D_p \tau_p = L_p^2. \end{aligned} \tag{19.52}$$

The first term of (19.52) is precisely the saturation current I_S contributed by thermally generated holes (see (19.29)). The second term is the photocurrent, contributed by the holes generated within the diffusion length from the junction edge x_n. This is consistent with (19.30), since holes as minority carriers are diffused on the average by the diffusion length L_p. Therefore, only those holes generated within L_p distance from x_n could reach on the average the junction edge to contribute to the photocurrent. In practice, the second term is larger than the first term and one can put

$$I_{p,\text{diff}} = -qA\left(\frac{p_{n0}D_p}{L_p} + \tilde{g}_0 L_p\right) \approx -qA\tilde{g}_0 L_p \tag{19.53}$$

One can similarly obtain the photo current in the quasi-neutral region on p side as

$$I_{n,\text{diff}} = -qA\left(\frac{n_{p0}D_n}{L_n} + \tilde{g}_0 L_n\right) \approx -qA\tilde{g}_0 L_n. \tag{19.54}$$

The total photocurrent therefore consists of the three components, (19.49), (19.53), and (19.54), namely,

$$I_{\text{ph}} = I_{\text{drfit}} + I_{n,\text{diff}} + I_{p,\text{diff}} \equiv -I_l, \tag{19.55a}$$

where it is important to notice that the photocurrent flows from n to p regions, opposite to the junction forward current at a level given by

$$I_l = qA\tilde{g}_0(W + L_p + L_n). \tag{19.55b}$$

The characteristics of the photodiode operating in the third quadrant of the diode I–V plane are shown in Figure 19.14. Here, the dark current due to the saturation current I_S constitutes the background noise and the output photocurrents are linear with respect to the input light intensity, as expected, and are also flat with the operating voltage V_R. This is because the photogenerated carriers are swept across the junction depletion depth by the built-in electric field already present regardless of V_R.

19.4.3 Photovoltaic Effect and Solar Cell

The photovoltaic effect generally refers to the physical processes where an incident light generates a voltage across a certain portion of the illuminated medium. The p–n junction provides a convenient platform supporting such effect and carries one of the most important applications, namely, the solar cell. The physical processes involved in the solar cell operation are substantially the same as those for photodetectors but the operating bias regimes are different.

The photovoltaic effect is again triggered by incident photons, generating e–h pairs via the band-to-band excitation. The excess carriers thus generated in the junction interface subsequently undergo the usual drift under an open circuit condition, propelled by the built-in space charge field. As a consequence, photogenerated excess holes roll up the potential hill and end up in the p region and excess electrons roll down the barrier to join the n region, as sketched in Figure 19.15.

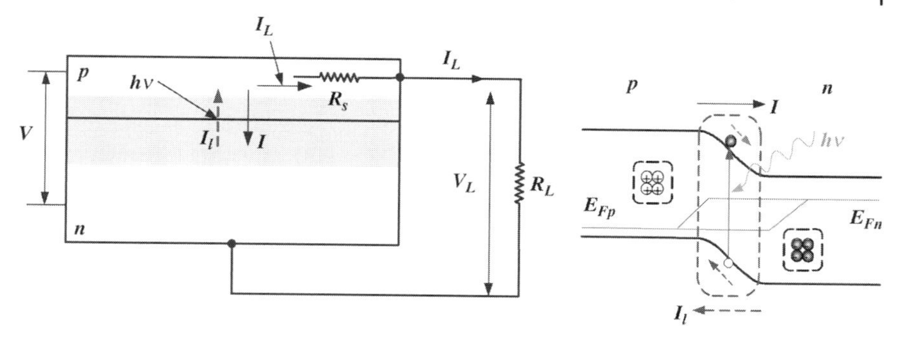

Figure 19.15 The junction solar cell, its equivalent circuit and energy band: e–h pairs generated by solar radiation induce simultaneously the forward voltage and photocurrent in reverse direction, thereby enabling the power extraction, namely $IV < 0$.

These excess charges in n and p regions in turn set up the forward bias V for the p–n junction and induce the forward current from p to n, thereby reducing the photocurrent flowing in the opposite direction. Hence, if an external circuit is connected to the illuminated p–n junction, both the photovoltage and photocurrent can be measured, while the junction serves as a battery. The current in the illuminated junction is thus given by

$$I = I_S(e^{qV/k_BT} - 1) - I_l, \tag{19.56}$$

where the first component is the usual diode forward current (19.29), taken ideal for simplicity of discussion, while the second component is the photocurrent (19.55). The open-circuit photovoltage is therefore found by setting $I = 0$ in (19.56), namely,

$$V_{\text{oc}} = \frac{k_BT}{q}\ln\left(\frac{I_l}{I_S} + 1\right) \approx \frac{k_BT}{q}\ln\frac{I_l}{I_S}, \tag{19.57a}$$

while the short-circuit current for $V = 0$ is given by

$$I_{\text{sc}} = -I_l. \tag{19.57b}$$

When a finite load resistance is connected to the junction, as shown in Figure 19.15, a positive voltage V is set up, while the net current flows against V in the same direction as the photocurrent, namely, in the negative direction from n to p. Therefore, $IV < 0$ and the power is extracted from the device. This constitutes the working principle of the solar cell.

The I–V behavior of (19.56) is shown in Figure 19.16 with the light intensity as parameter. Clearly, the I–V curve passes through the fourth quadrant on the diode I–V plane, indicating that $IV < 0$. This I–V curve is plotted in the same figure by taking the photocurrent as positive. Clearly, the I–V curve intersects with the voltage and current axes at V_{oc} and I_l, respectively, as it should. Also, the shaded region inside the I–V curve represents the maximum power rectangle. The figure also shows V_{oc} and I_l plotted versus incident light intensity. Indeed, I_l increases linearly with the

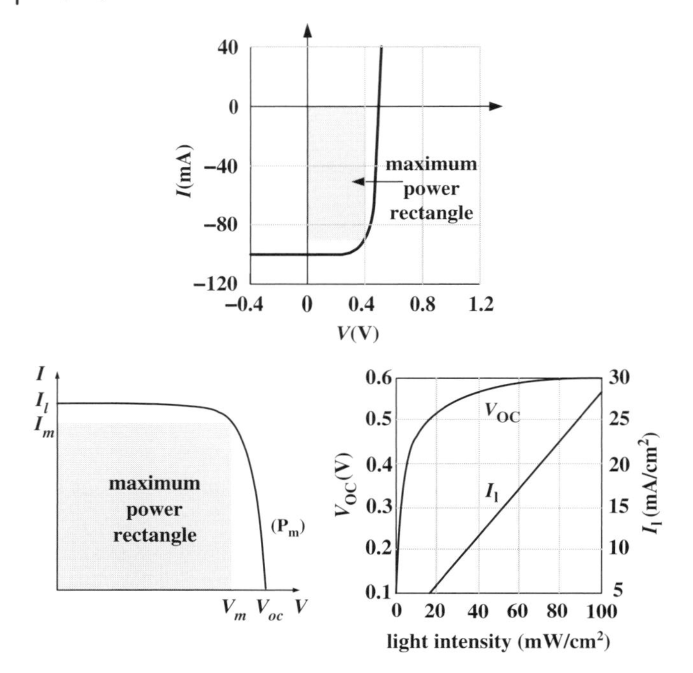

Figure 19.16 The solar cell *I–V* (top) and the enlarged version of *I–V* curve with the direction of photocurrent I_l taken positive (bottom left). Also shown are I_l, V_{oc} versus light intensity (bottom right). (taken from Pankove, J. I. (1971) Optical Processes in Semiconductors, Dover)

light intensity, as it should, while V_{oc} rises logarithmically, approaching the built-in voltage asymptotically.

In this manner, the p–n junction converts the solar radiation energy into electrical energy and is used as the solar cell. Naturally, a key parameter of the cell is the efficiency of the energy conversion. The parameters pertinent for enhancing the efficiency are the light absorption coefficient α and others such as V_{oc}, I_l. Ideally, the cell should absorb the entire spectrum of the solar radiation. The smaller the cell bandgap, the larger the fraction of the solar spectrum absorbed, as is clear from (19.47). However, a larger bandgap can induce larger photovoltage and smaller leakage current via reduced saturation current level I_S (see (19.29) and (19.57a)).

The factors pertinent to the performance of the solar cell can be more explicitly brought out by revisiting the equivalent circuit of the cell, as shown in Figure 19.15. Here, R_L and R_S are the load and series resistances and I_L the output load photocurrent. As already discussed, when photons are absorbed in the junction interface, thereby inducing the photocurrent I_l, the diode leakage current is also generated, and these two currents flow in the direction opposite to the junction forward current. In the presence of R_S, the voltage across the load V_L is smaller than the photogenerated voltage V across the junction, that is,

$$V_L = V - I_L R_S. \tag{19.58}$$

Obviously, V is determined by the incident light intensity and the output load current I_L can be expressed from (19.56) as

$$I_L = I_l - I_S(e^{qV/k_BT} - 1), \tag{19.59}$$

where the direction of I_l has been taken positive for convenience. It thus follows from (19.58) and (19.59) that

$$V_L = \frac{k_BT}{q}\ln\left(\frac{I_l - I_L}{I_S} + 1\right) - I_L R_S. \tag{19.60}$$

The power extracted is therefore given from (19.59) and (19.60) by

$$P \equiv V_L I_L = V_L[I_l - I_S(e^{qV/k_BT} - 1)]. \tag{19.61}$$

The general features for the power extraction can be brought out by considering the simple case where R_S is neglected, so that $V_L = V$. One can then estimate the maximum power from (19.61) by putting $V_L = V$ and imposing the condition

$$\partial P/\partial V_L = 0,$$

obtaining

$$e^{qV_{\mathrm{Lm}}/k_BT} = (1 + I_l/I_S)/(1 + qV_{\mathrm{Lm}}/k_BT), \tag{19.62a}$$

where V_{Lm} is the load voltage corresponding to the maximum power extraction. The V_{Lm} in (19.62) can be recast into more transparent form using (19.57) as

$$V_{\mathrm{Lm}} = V_{\mathrm{oc}} - \frac{k_BT}{q}\ln[1 + V_{\mathrm{Lm}}/(k_BT/q)] \tag{19.62b}$$

and is shown to be dictated primarily by V_{oc}. This open-circuit voltage in turn depends on I_S, hence the bandgap of the material (19.57a). Once V_{Lm} is found, the load current is obtained from (19.59) and (19.62) as

$$I_{\mathrm{Lm}} = I_l - I_S(e^{qV_{\mathrm{Lm}}/k_BT} - 1) \approx I_l\left(1 - \frac{k_BT/q}{V_{\mathrm{Lm}}}\right), \tag{19.62c}$$

where an approximation has been made of, namely, $I_l/I_S > V_m/(k_BT/q) \gg 1$. Indeed, I_{Lm} is commensurate with I_l, indicating that the cell efficiency is contingent upon the maximal absorption of the solar radiation. Needless to say, devising a high-efficiency solar cell would require an innovative bandgap engineering of the cell material to optimally compromise the absorption versus the photovoltage.

19.4.4
LD and LED

One of the key technologies developed in the communication area is the optical fiber communication and the overwhelming advantages of the fiber communication consist of (i) the low signal loss, (ii) a wide bandwidth, and (iii) the small diameter of silica fibers. A key component enabling the pervasive utilization of the fiber communication is the laser diode and light-emitting diode (LED). These photonic devices are pumped by an electrical current and can therefore be incorporated readily into the communication circuitry. Moreover, light-emitting diodes are fast becoming the mainstream light source with long lifetime and low power usage. The p–n junction again provides the convenient platforms on which to implement these photonic devices. The physics underlying the device operation is discussed next.

A factor crucial for these devices is to convert the electrical power into light and to achieve the optical gain. The gain generally requires the population inversion between two lasing levels, as discussed in Chapter 14. For p–n junctions used as laser diodes, the conduction and valence bands play the role of two lasing levels and the pumping for the population inversion is done by injecting carriers into the junction interface.

Thus, consider a p–n junction fabricated in a direct bandgap semiconductor such as GaAs. The semiconductor is doped heavily with donor and acceptor atoms, so that both n and p bulk regions are degenerate. In this case, the Fermi level at equilibrium penetrates deep into both the conduction and valence bands and these two bands overlap with each other, as shown in Figure 19.17. Under a forward bias, the junction band bending is reduced as usual and the electron and hole concentrations, n and p, are described separately by two quasi-Fermi levels, E_{Fn} ($\equiv E_{Fc}$) and E_{Fp} ($\equiv E_{Fv}$), whose splitting in the depletion depth $E_{Fc}-E_{Fv}$ is a measure of excess electrons and holes being injected therein for either radiative or nonradiative recombinations.

To analyze the optical gain or loss on a general ground at arbitrary temperature, the transition rate (19.45) should incorporate the probability factors to ensure that the

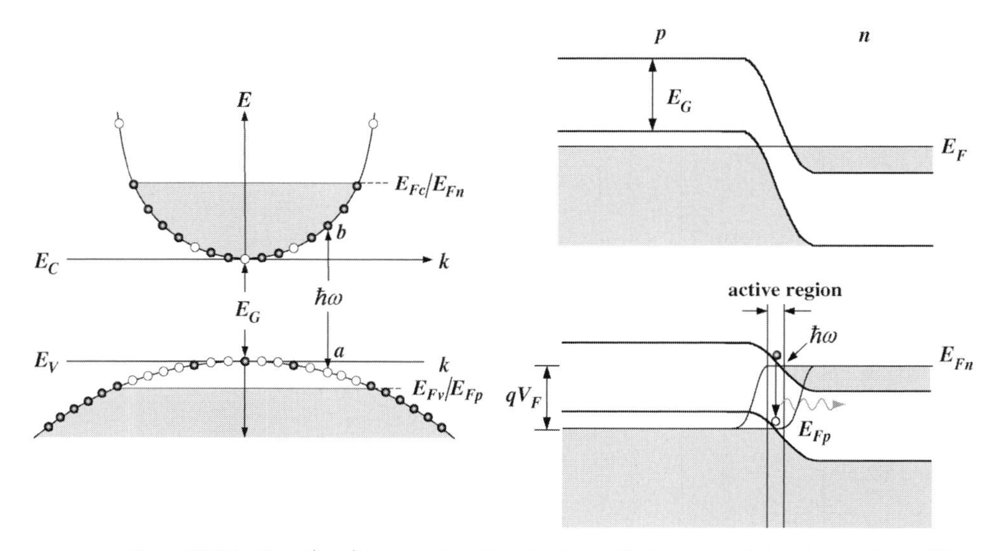

Figure 19.17 The p$^+$–n$^+$ junction band bending in equilibrium (top right) and under forward bias (bottom right). Under bias, excess electrons and holes are injected into the junction region to recombine radiatively. Also shown is the band to band excitation or recombination of e–h pair (left).

electrons are present in the initial state and absent in the final state of the transition in accordance with the exclusion principle. Thus, (19.45) should be generalized to read as

$$N_{ab}-N_{ba} = \frac{2\pi}{\hbar}\frac{\tilde{\mu}^2 E_0^2 V}{4}\int \delta\left(\frac{\hbar^2 k^2}{2\mu_{\text{eff}}}+E_G-\hbar\omega\right) \times \frac{k^2 dk}{\pi^2}[f_v(E_a)(1-f_c(E_b))-f_c(E_b)(1-f_v(E_a))], \tag{19.63}$$

where the Fermi–Dirac distribution function or the occupation factor

$$f_j(E_\gamma) = \frac{1}{1+e^{(E_\gamma - E_{Fj})/k_B T}}, \quad \gamma = a, b, j = c/n, v/p, \tag{19.64}$$

is specified by the respective quasi-Fermi levels (c/n) for electrons in the conduction band and for holes (v/p) in the valence band and the two energy levels involved in transition (a, b) (Figure 19.17). Under bias, the junction is pushed away from the equilibrium and the distribution function f departs from the equilibrium Fermi–Dirac distribution function. However, it is reasonable to still represent f by the Fermi–Dirac distribution function. Inserting (19.64) into (19.63) and repeating the same algebra as was done in (19.45)–(19.47), one finds

$$\tilde{\alpha}(\omega) = \alpha(\omega)[f_v(E_a)-f_c(E_b)], \tag{19.65}$$

where $\alpha(\omega)$ is the absorption coefficient derived in (19.47) for $T = 0$.

It is therefore clear from (19.65) that a net absorption

$$\tilde{\alpha}(\omega) = \alpha(\omega)[f_v(E_a)-f_c(E_b)] \approx \alpha(\omega)$$

ensues if

$$f_v(E_a) > f_c(E_b). \tag{19.66a}$$

This condition can be recast into a more transparent form using (19.64) and the identification $E_b - E_a = \hbar\omega$ as

$$\hbar\omega > E_{Fc}-E_{Fv}. \tag{19.66b}$$

On the other hand, there is a net gain, namely,

$$\gamma(\omega) = -\tilde{\alpha}(\omega) = \alpha(\omega)[f_c(E_b)-f_v(E_a)], \tag{19.67}$$

provided

$$f_c(E_b) > f_v(E_a) \tag{19.68a}$$

or equivalently

$$\hbar\omega < E_{Fc}-E_{Fv}. \tag{19.68b}$$

The two criteria (19.66) and (19.68) for attenuation and gain, as given by Dumke, can be intuitively understood, using the energy level diagram in Figure 19.17. The condition (19.66) states that the probability of an electron being in the valence band is higher than that in the conduction band. In this case, there should be more upward transitions of electrons from the valence band into the empty states in conduction band, resulting in net absorption of light.

On the other hand, the condition (19.68) states the reverse case in which the probability of electrons in the conduction band is higher than that in the valence band. In this case, there should be more downward transitions of electrons from the conduction band to the empty states in valence band, emitting more photons than absorption for the optical gain.

The p–n junction again provides a convenient platform for achieving the optical gain. The gain is achieved by forward biasing the degenerately doped p–n junction and the splitting of two quasi-Fermi levels. The splitting of two Fermi levels should be large enough to ensure a sufficient injection of electrons and holes in the junction interface for the radiative recombination. For this purpose, E_{Fn} and/or E_{Fc} should be well above E_C, so that the sufficient amount of electrons are made available in the junction region. At the same time, E_{Fp} and/or E_{Fv} should be considerably below E_V to maintain sufficient amount of holes in the junction region, as is clearly shown in Figure 19.17. This gain condition is to be met with a modest value of the forward bias in the p^+–n^+ junction where there is considerable overlap between the conduction and valence bands to begin with. The gain ensues when the emitted photon energy $\hbar\omega$ is less than the splitting of two quasi-Fermi levels.

An additional parameter of importance is the luminescence efficiency, which specifies the fraction of the radiative recombination *vis-à-vis* the total recombination, namely,

$$\eta = R_{\text{rad}}/R_{\text{total}} = \frac{1/\tau_r}{1/\tau_r + 1/\tau_{nr}}, \tag{19.69}$$

where τ_r and τ_{nr} denote the radiative and nonradiative lifetimes, respectively. The high luminescence efficiency is generally attained in direct bandgap semiconductors in which the optical transitions are the first-order process and therefore have higher probabilities. In indirect semiconductors on the other hand, the optical transitions are of the second-order process and the efficiency is generally low.

Naturally, this optical conversion of the electron–hole recombination constitutes the physical basis for light-emitting diodes. A variety of LEDs have thus far been realized in various kinds of semiconductors and emitted radiations span a wide range of spectrum from the infrared to visible wavelengths. Moreover, LED-based applications are fast increasing in fiber communications, displays, energy-saving lamps, and so on. A typical p–n junction structure used for LED is sketched in Figure 19.18.

Furthermore, the optical gain can also be attained in which case the diode can be turned into a coherent source of light, namely, the laser diode. Figure 19.19 shows the cross-sectional view of the laser diode. Here, the cavity is of the Fabry–Perot type, composed of a pair of parallel, cleaved planes perpendicular to the plane of current injection. The lasing frequencies can be varied and controlled using the bandgap engineering, combined with the appropriate choice of base semiconducting materials.

Once enough gain is attained above the threshold value, the operation of the laser diode can be understood in the same context as detailed in Chapter 14. There are however a few features unique to laser diodes. Here, the lasing is triggered by passing

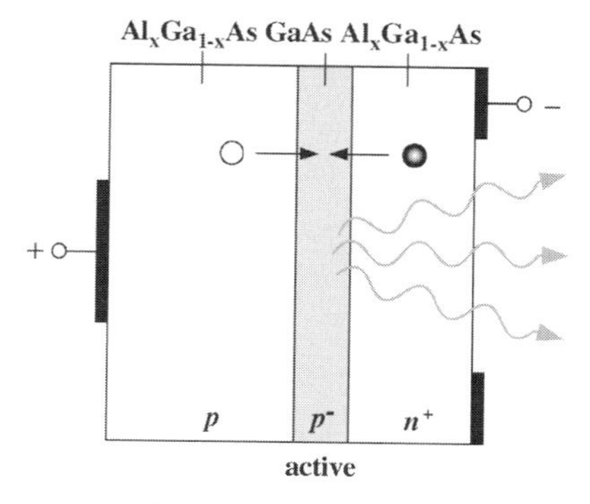

Figure 19.18 The cross-sectional view of light-emitting diode.

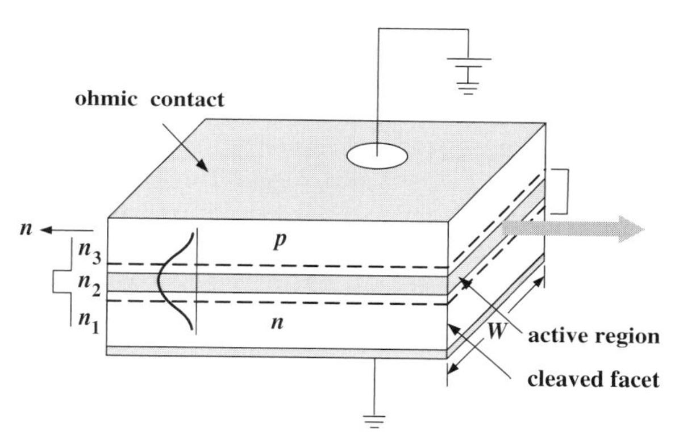

Figure 19.19 The cross-sectional view of laser diode: Fabry–Perot cavity, active lasing layer, and optical waveguide arrangement for confining lasing field within the active region.

the forward current through the diode interface, which makes the device simple to operate. Also, the light intensity can be readily modulated even at high frequencies.

A factor crucial for minimizing the cavity loss and for lowering the threshold current density for lasing is the optimal waveguiding, that is, an efficient confinement of the laser field intensity within the active region. Otherwise, a substantial fraction of lasing intensity would be wasted via the tailing of field amplitude out of the lasing layer, thereby increasing the net cavity loss. To this end, either single or double heterostructures can be incorporated, so that the improved profiling of refractive index entailed in the structure better confines the field intensity within the active layer (Figure 19.19).

In addition, laser diodes are fabricated in superlattice heterostructures with built-in quantum wells, as shown in Figure 19.20. When the thickness of the superlattice active layer is reduced to about 20 nm range, electrons and holes injected therein form two-dimensional gases. Furthermore, these excess carriers reside in the subbands with constant density of states, independent of energy, as discussed in

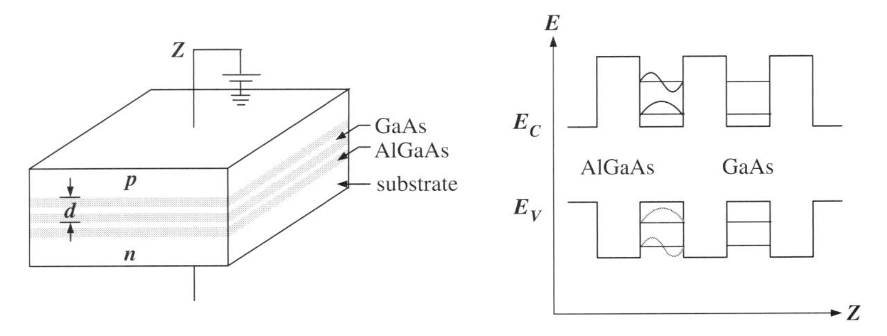

Figure 19.20 The cross-sectional view of the quantum well laser diode. The active region is made up of superlattice structure and electrons and holes are injected into respective subbands in quantum wells and recombine rediatively.

Chapter 4. A major advantage of this kind of laser diodes made of quantum wells is the reduced threshold current density. This is because the transparency density of carriers in the active region is roughly commensurate with the thickness of the active region and the active layers in superlattice structures are thinner than those of conventional laser diodes. Moreover, the electron–hole pairs can now recombine between the two subbands, well confined in narrow spatial region. This is in contrast with the conventional laser diodes in which the excess electrons and holes are distributed in energy over 3D density of states and are swept fast out of the spatial region of recombination. Clearly, these improvements are made possible with the innovative utilization of quantum wells.

19.4.5 Tunnel Diodes

The p^+–n^+ junction was shown to provide the working basis for LEDs or laser diodes. The same degenerate junctions can also be utilized as high-speed electronic switches or logic devices, capable of providing a negative resistance regime in the static I–V characteristics. The device is known as tunnel diode or Esaki diode and is based upon the controlled tunneling of electrons.

Figure 19.21 shows the energy band diagram of p^+–n^+ diode at thermodynamic equilibrium (b), in which the Fermi level E_F is located in both conduction and valence bands simultaneously and is lined up and flat, as it should. Moreover, the valence and conduction bands are overlapped in energy. Inasmuch as the Fermi level is flat across the junction, there is no net flux of electrons or holes, as was shown on a general ground in Chapter 18 (see (18.41) and (18.45)).

Under the reverse bias (a), however, the p side is lifted with respect to n side, and E_F splits into two quasi-Fermi levels and E_{Fp} is lifted above E_{Fn}. As a result, the valence band electrons can tunnel from p to n regions, giving rise to current from n to p (Figure 19.21). The potential barrier encountered by these tunneling electrons is well approximated by the triangular shape and therefore the tunneling is of the F–N type. Because the junction is heavily doped, the built-in space charge field is high to begin

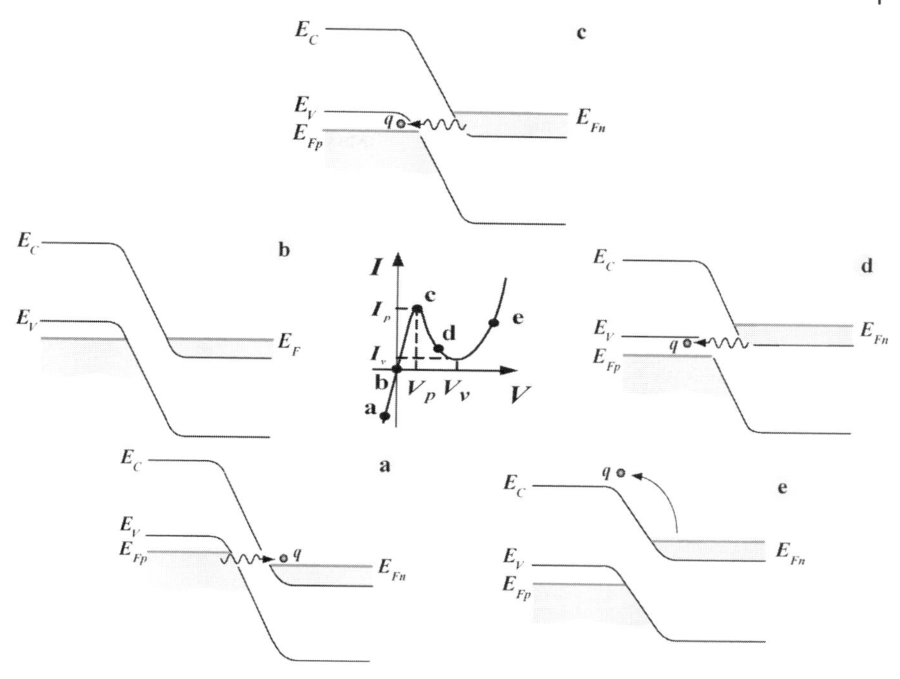

Figure 19.21 The *I–V* behavior of tunnel diode and energy band diagrams corresponding to different operating voltage regimes.

with and this built-in field is further reinforced by the reverse bias. Hence, there exists a strong local field in the interface, which drastically enhances the F–N tunneling probability. As a result, the magnitude of the reverse current rapidly increases with reverse bias, as clearly seen from Figure 19.21.

Under the forward bias, the p side is lowered instead and the energy level of electrons in the conduction band on the n side begins to overlap with that of the holes in the valence band on the p side. In this case, the conduction band electrons can tunnel into the empty states in the valence band, giving rise to current from p to n. This forward current reaches the peak level when the energy level of tunneling electrons maximally overlaps with that of holes, so that the maximum number of electrons and holes can take part in tunneling (c). With further increase in the forward voltage, the degree of overlap between the electron and hole energies reduces and the current is decreased, thereby providing a negative resistance regime (d).

With further increase in the forward bias, the overlap of electron and hole energies disappears and no final states are available for conduction band electrons to tunnel from n to p regions, hence no tunneling current. However, for this voltage range, the usual diode forward current takes over adding to the current (e). This static *I–V* behavior is shown in Figure 19.21, together with the energy level diagram corresponding to each regime. The peak (I_p) and valley (I_v) values of the current occurring at V_p and V_f, respectively, determine the magnitude of the negative resistance and the current ratio I_p/I_v is used as the figure of merit of the tunnel device. Also, the corresponding voltage ratio V_p/V_f represents a measure of spread of voltage for the dynamic negative resistance. Because of the extremely short tunneling time, the tunnel diode holds up the promise for high-speed device.

19.5 Problems

19.1. The operational principle of p–n junction diode consists in the breaking of the detailed balancing between drift and diffusion of electrons and holes in equilibrium by applying voltages and using the resulting restoring force.
(a) Discuss the reasons why the balance is broken under forward and reverse biases.
(b) Compare the drift and diffusion fluxes of electrons and holes before and after the forward and reverse biases.

19.2. The dipole space charge ϱ used in the text is based on the depletion approximation. Check the validity of this approximation by estimating the width of the region near x_n and $-x_p$, in which the carrier concentrations are not negligible, and by comparing the widths with typical x_n and x_p values.

19.3. A linearly graded junctions are often processed by thermally diffusing in acceptor atoms in n substrate. The space charge can thus be modeled by

$$\varrho = \begin{cases} q[N_A(x) - N_D] = ax, & -W/2 \leq x \leq W/2 \\ 0, & \text{otherwise.} \end{cases}$$

(a) Find the space charge, field, potential, and built-in potential.
(b) Discuss the band bending and resulting I–V characteristics.

19.4. The quasi-Fermi levels in the junction region under both forward and reverse biases are taken flat in quasi-equilibrium approximation. Check the accuracy of this approximation by (i) taking typical doping levels of p–n junction, (ii) estimating the electron and hole fluxes under forward and reverse biases, and (iii) relating these fluxes to the gradient of the respective quasi-Fermi levels (18.51).

19.5. (a) Is it possible to achieve the junction band bending by an amount greater than the energy gap of the semiconductor in contact, namely, $q\varphi_{bi} > E_G$?
(b) Estimate the donor and acceptor doping levels that will make $q\varphi_{bi} \simeq E_G$ in Si and Ge.

19.6. For fabricating laser diode, p and n regions are degenerately doped
(a) Estimate the donor and acceptor doping levels, for which the conduction and valence bands are overlapped by an amount 0.2 eV in silicon and gallium arsenide.
(b) Estimate electron and hole fluxes under forward bias in the two junctions in (a).

19.7. Consider p^+–n step junctions in silicon in which $N_A = 2 \times 10^{18}\ \text{cm}^{-3}$, but the donor levels vary from $1 \times 10^{15}\ \text{cm}^{-3}$ to $2 \times 10^{17}\ \text{cm}^{-3}$.

(a) Find x_n, x_p, E_{max}, W, and φ_{bi} versus N_D doping level.

(b) At which reverse biases will these p–n junctions undergo breakdown if the maximum field for breakdown is taken at 3×10^5 V/cm?

19.8. In which semiconductor is the Zener breakdown more likely to occur among silicon, germanium, and gallium arsenide if the same breakdown field of 3×10^5 V/cm is assumed. Estimate the reverse biases at which the breakdown sets in these semiconductors.

19.9. Consider a one-sided step p–n^+ junction in which the quasi-neutral region between x_n and the metal contact is shorter than the hole diffusion length L_p.

(a) Solve the diffusion equation of the excess holes using the boundary condition that $\delta p_n = p_n - p_{n0} = 0$ at the metallic contact.

(b) Derive the ideal diode I–V for this junction.

19.10. (a) The steady-state diffusion of minority carriers under illumination is an important element of modeling photodiode and solar cells. Derive (19.51) in the text, filling in algebraic steps.

(b) Derive V_{Lm}, I_{Lm} in (19.62) in the text, filling in algebraic steps.

19.11. (a) In analyzing the power extraction from the solar cell, the series resistance has been neglected. Examine the effect of the series resistance in the power extraction using an iteration scheme or a numerical method and discuss the results obtained.

(b) For optimal power extraction from the solar cell, the effects of various device and material parameters have to be carefully assessed. Discuss the roles of these variables and their combined effect for maximal power extraction.

Suggested Reading

1 Streetman, B.G. and Banerjee, S. (2005) *Solid State Electronic Devices*, 6th edn, Prentice Hall.

2 Muller, R.S., Kamins, T.I., and Chan, M. (2002) *Device Electronics for Integrated Circuits*, 3rd edn, John Wiley & Sons, Inc.

3 Sze, S.M. and Ng, K.K. (2006) *Physics of Semiconductor Devices*, 3rd edn, Wiley-Interscience.

4 Neudeck, G.W. (1989) *The Bipolar Junction Transistor, Modular Series on Solid State Devices*, 2nd edn, vol. III, Prentice Hall.

5 Grove, A.S. (1967) *Physics and Technology of Semiconductor Devices*, John Wiley & Sons, Inc.

6 Yariv, A. (1982) *An Introduction to Theory and Applications of Quantum Mechanics*, John Wiley & Sons, Inc.

7 Bhattacharya, P. (1996) *Semiconductor Optoelectronic Devices*, 2nd edn, Prentice Hall.

8 Pankove, J.I. (1971) *Optical Processes in Semiconductor*, Dover.

20
The Bipolar Junction Transistor: Device Physics and Technology

The invention of transistors is one of the towering intellectual achievements made in the twentieth century. These semiconductor devices are deeply rooted in the basic concepts of quantum mechanics and statistical physics and are based on the ingenious applications of the concepts. As such these devices have played a major role in bringing about information technology. In this chapter, the bipolar junction transistor is singled out for discussion as a direct consequence of two interacting diodes. The discussion is focused on the basic physical concepts underlying the device operation rather than the device modeling per se.

As well known, BJT is one of the earliest transistors ever realized and still remains one of the fastest devices, capable of delivering large amount of current. Because of the overwhelming advantages of the device, namely, the capability for batch processing *en masse* and the miniaturization, it served as a driving force for opening up the semiconductor and integrated circuit industry.

20.1
Bipolar Junction Transistor: Overview

The idea of the bipolar junction transistor (BJT) was conceived by W. Shockley and is based upon the innovative utilization of two interacting p–n junction diodes. BJT was one of the earliest devices produced in mass and miniaturized for integrated circuit applications. BJT is the three-terminal device, used as "switch and amplifier" and has played key roles in integrated circuits. The device consists of two interacting p–n junctions, as sketched in Figure 20.1. Here, two electrodes are attached to the emitter and collector junctions and an additional third terminal is introduced in the base, which electrically separates the two p–n junctions by providing neutral bulk in between. There are two kinds of BJTs, namely, npn and pnp, depending on how emitter, collector, and base regions are doped (Figure 20.1). Also shown in the figure are typical output currents, namely, the collector currents versus emitter–collector and base–emitter voltages, respectively. The former is the usual transistor I–V curve, while the latter represents the transfer characteristics, namely,

Introductory Quantum Mechanics for Semiconductor Nanotechnology. Dae Mann Kim

ISBN: 978-3-527-40975-4

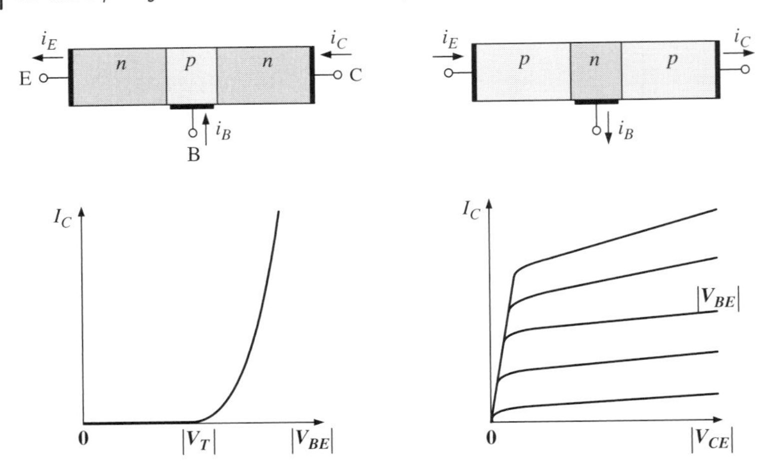

Figure 20.1 The npn and pnp bipolar junction transistors, consisting of emitter, base, and collector. Also shown are collector currents I_C versus V_{BE} at fixed V_{CE} and I_C versus V_{CE} at fixed V_{BE}.

the base-voltage-controlled output current behavior. The design of optimal device structures is directed toward tailoring these output currents, as desired.

In BJT, two diodes are interacting with the minority carriers acting as the mediator. Specifically, the minority carriers are injected from emitter junction and collected at the collector junction without suffering appreciable recombination in the base. That is, the two junctions are electrically linked by making the base width thinner than the minority carrier diffusion length and by controlling the charge transfer via the terminal voltages.

A key factor for understanding the device operation is to comprehend the band bending in equilibrium and nonequilibrium under bias. Thus, the band bending and the corresponding carrier concentrations are first discussed on a general ground. Figure 20.2 presents the energy band diagrams for both npn and pnp transistors in equilibrium and typical carrier concentration profiles therein. Here, the emitter doping level is higher than that of the collector doping for the reasons to become clear later.

Naturally, the physics of band bending in the junction interface is identical to what has already been covered in p–n junction diodes. The device operation is again based on controlling and harnessing the junction interface phenomena. It is interesting and important to notice at the outset from Figure 20.2 that in pnp transistor, the band bending in two junctions provides a quantum well for electrons in the base region. This means that electrons as majority carriers therein tend to be confined in the well. In contrast, holes as minority carriers in the base can roll up the potential hill near the junction edges to be easily pushed out of the base region.

By the same token, in npn transistor, an opposite situation prevails, namely, that quantum well is formed in the base for holes as majority carriers, while electrons as minority carriers simply roll down the potential hill in the junction edges to be

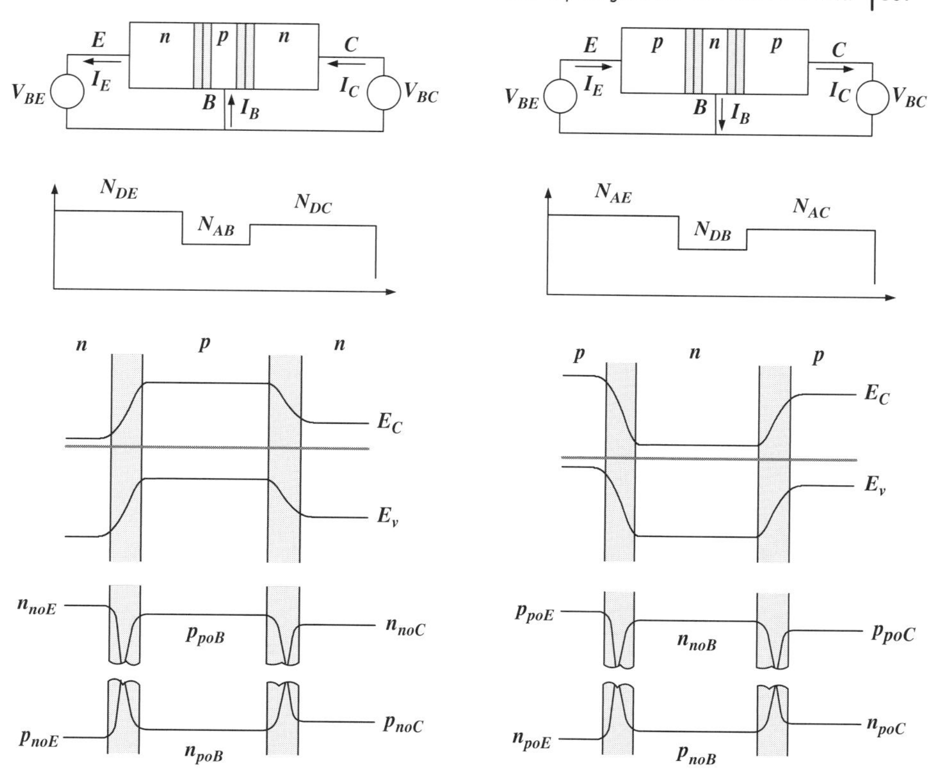

Figure 20.2 The energy band diagrams, carrier concentrations, and doping profiles in npn and pnp transistors in equilibrium.

pushed out of the base. This provides one of the key features of the bipolar junction transistor and it is the minority carriers that provide both the input and output currents. Also, both electrons and holes inextricably take part in the device operation, hence the name bipolar transistor.

20.1.1
npn Transistor

Operating Bias Regimes The discussion is now focused on the npn transistor, but it also applies to pnp transistors by properly interchanging the roles of electrons and holes. Consider first the band bending under the bias. The biasing regimes of the two interacting diodes are shown in Figure 20.3, with the base–emitter (V_{BE}) and base–collector (V_{BC}) voltages forming the horizontal and vertical axes, respectively.

In the first quadrant of Figure 20.3, both junctions are forward biased, namely, $V_{BE} > 0$ and $V_{BC} > 0$, and the band bendings in these two junctions are reduced, as detailed in Chapter 19. In this case, electrons spill over these reduced potential barriers into the base region from both junctions to form a pool of excess minority

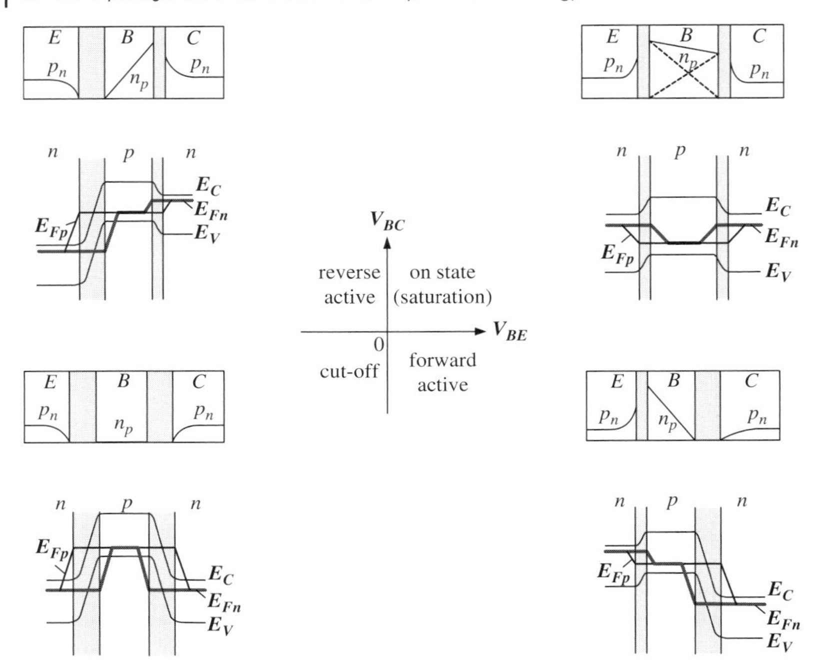

Figure 20.3 The four bias regimes of npn transistor and the corresponding junction band bending and carrier concentration profiles.

carriers ready for conduction, hence the name "on" state or saturation. In the third quadrant, both junctions are reverse biased and the opposite situation prevails. That is, the band bendings of the two junctions are enhanced and as a consequence the base is depleted of the minority carriers, hence the name cut-off or simply "off" state.

In the fourth quadrant, the base–emitter junction is forward biased while the base–collector junction is reverse biased. In this case, electrons spill into base from the emitter and are drained at the collector junction. In this way, the two junctions are electrically connected and the transistor action ensues, hence the name "forward active" regime. The second quadrant is the mirror reflection of the fourth quadrant, hence the name "reverse active" regime.

The band bending under bias corresponding to each quadrant is shown in Figure 20.3 and the accompanying minority carrier profiles can be understood based on the same junction interface physics as in junction diodes. In forward active region $V_{BE} > 0$, the band bending in the base–emitter junction is decreased. The concomitant splitting of two imrefs to accommodate V_{BE} ensures that excess electrons spill into the base and at the same time excess holes are injected into the emitter. Since $V_{BC} < 0$ on the other hand, the band bending in the base–collector junction is enhanced, and the minority carriers (electrons in this case) near the collector–base junction edge are nearly depleted. The resulting profile of electrons in the base region indicates that electrons injected from emitter are diffused to the collector junction to be drained out. In reverse active regime, electrons diffuse in the opposite direction.

For the “on” state, the electron profile in the base can be understood in terms of the linear superposition of two minority carrier injections from both sides and for the cut-off state, the electron profile can likewise be understood in terms of two minority carrier extractions induced by two reverse-biased junctions. It is therefore clear that both on and off states are characterized by the minority carrier concentrations in the base. The switch between the two states is made by redistributing the minority carriers via forward or reverse active biases. Also, in the cut-off regime, there should exist the usual leakage current arising from the generation of electron–hole pairs in the junction interface regions.

20.2 The Physics of Transistor Action

Gross Features of Carrier Transport in the Base As shown in Figure 20.2, the band bendings in n–p and p–n junctions form a quantum well for holes in the base. As a consequence, holes are confined in the well. When a hole approaches the junction edges on both sides of the base, it encounters downward bending of E_V and is therefore pushed back toward the bulk of the base region. In contrast, when an electron approaches the same junction edges, it simply rolls down the potential hill and is pushed out of the base into emitter or collector.

With this general picture in mind, one may put

$$J_p = 0 = q\mu_p p\mathrm{E} - qD_p \frac{dp}{dx} \tag{20.1}$$

and find the built-in electric field in the base in terms of the hole concentration and its spatial gradient:

$$\mathrm{E} = \frac{D_p}{\mu_p}\frac{1}{p}\frac{dp}{dx} = \frac{k_B T}{q}\frac{1}{p}\frac{dp}{dx}, \tag{20.2}$$

where the Einstein relation between μ_p and D_p has been used (see (18.36)).

Using this expression of the electric field E, the electron current density in the base (20.1) reads as

$$\begin{aligned} J_n &= q\mu_n n\mathrm{E} + qD_n \frac{dn}{dx} \\ &= q\left(\mu_n n \frac{k_B T}{q}\frac{1}{p}\frac{dp}{dx} + D_n \frac{dn}{dx}\right) \\ &= qD_n \frac{1}{p}\left(n\frac{dp}{dx} + p\frac{dn}{dx}\right) = qD_n \frac{1}{p}\frac{d}{dx}(np), \end{aligned} \tag{20.3}$$

or equivalently

$$\frac{d}{dx}(np) = \frac{J_n qp}{q^2 D_n}. \tag{20.4}$$

Hence, performing the x-integration from the base–emitter junction edge ($x = 0$) to the base–collector junction edge ($x = W_B$), one obtains from (20.4)

$$np|_{W_B} - np|_0 = \frac{J_n}{q^2} \int_0^{W_B} dx \frac{qp}{D_n} = \frac{J_n}{q^2} \frac{Q_B}{\tilde{D}_n}, \tag{20.5}$$

where

$$Q_B \equiv q \int_0^{W_B} dx p_p(x) = q \int_0^{W_B} dx N_{AB}(x) \tag{20.6}$$

represents the hole charge per unit width as the majority carrier in the base and is called the Gummel number with N_{AB} denoting the acceptor doping level in the base. Also, a few facts are worth noting at this point. In performing the x-integration in (20.5), J_n was taken out of the integral since it should be constant, independent of x. If J_n varies with x, the charge can be constantly accumulated or depleted in some points along the current path, which would make it impossible to have the steady-state current. In addition, D_n can weakly depend on x but it can be taken into account by introducing effective diffusion coefficient $\tilde{D}_n$ and taking it out of the integral via the mean value theorem.

Now the carrier concentrations at the two edges of the base are obviously controlled by V_{BE} and V_{BC}, as detailed in Chapter 19 (see (19.19) and Figure 19.7):

$$np|_{W_B} - np|_0 = n_i^2 (e^{qV_{BC}/k_B T} - e^{qV_{BE}/k_B T}). \tag{20.7}$$

Hence, combining (20.5) and (20.7) and multiplying J_n by the effective cross-sectional area of the emitter A_E, one finds the electron current in the base as

$$I_n = I_S (e^{qV_{BC}/k_B T} - e^{qV_{BE}/k_B T}), \quad I_S = \frac{A_E q^2 n_i^2 \tilde{D}_n}{Q_B}. \tag{20.8}$$

Equation 20.8 represents essential features of the bipolar junction transistor. The current contributed by electrons as minority carrier in the base, namely, I_n, mediates the interaction between the two junctions and also provides the output collector current. In addition, the transistor is switched on and off by V_{BE} and V_{BC}. Thus, it is this voltage-controlled minority carrier current linking the two junctions that the operation of the bipolar junction transistor is based upon.

Forward Active Bias Regime As mentioned, the transistor action is triggered by the junction forward current resulting from the electrons being injected from the emitter into the base and reaching the output collector terminal. Specifically, under forward active bias, the band bending in the base–emitter junction is reduced, thereby inducing excess electrons spilling over this reduced barrier into the base. The amount of excess electrons injected into the base from the emitter junction is given from (19.24) by

$$n(x = 0) = n_{p0} e^{qV_{BE}/k_B T}, \quad V_{BE} > 0, \tag{20.9a}$$

whereas electrons are nearly depleted at the edge of reverse-biased base–collector junction, that is,

$$n(x=W_B)=n_{p0}e^{qV_{BC}/k_BT}\approx 0,\quad V_{BC}<0. \tag{20.9b}$$

Now, the excess electrons in the base are governed by the diffusion equation, as detailed in the p–n junction modeling:

$$\frac{\partial^2 n_p}{\partial x^2}-\frac{n_p-n_{p0}}{L_n^2}=0,\quad L_n=(D_n\tau_n)^{1/2}$$

with n_{p0} denoting the equilibrium electron concentration (see (19.23)) and are subject to the boundary conditions (20.9). In the limit where the base width is much shorter than the diffusion length, that is, $W_B \ll L_n$ one can put $(n_p-n_{p0})/L_n^2\approx 0$, in which case the solution is approximately given by

$$n(x)=n_{p0}e^{qV_{BE}/k_BT}\left(1-\frac{x}{W_B}\right)+n_{p0}e^{qV_{BC}/k_BT}\frac{x}{W_B} \tag{20.10}$$

as can be readily verified by substitution. The resulting diffusion current provides the output current, namely,

$$I_C\equiv I_n=A_EqD_n\frac{\partial n(x)}{\partial x}=-I_S\left(e^{qV_{BE}/k_BT}-e^{qV_{BC}/k_BT}\right),\quad I_S=\frac{A_EqD_nn_i^2}{N_{AB}W_B},\ n_{p0}=\frac{n_i^2}{N_{AB}}. \tag{20.11}$$

with N_{AB} denoting the acceptor doping level in the base. Note that I_S differs from the corresponding quantity in (19.29) in that the electron diffusion length L_n is replaced by the base width W_B. The output current (20.11) as analyzed from the diode I–V point of view is in agreement with (20.8), with $\tilde{D}_n\approx D_n$ and the identification

$$Q_B\equiv q\int_0^{W_B}dx\,p(x)=qN_{AB}W_B.$$

Current Gain The transfer characteristics of the device, namely, I_C versus V_{BE} curve are shown in Figure 20.4, together with the base current versus V_{BE}, for comparison. From this figure, one can observe a few general features regarding the I–V behavior. First, I_C increases exponentially according to the diode factor, $[\exp(qV_{BE}/k_BT)-1]$. However, I_C level is considerably higher than that of the diode forward current, obviously due to the narrow base width, that is, $W_B<L_n$. It is particularly noteworthy that with V_{BE} swing of mere 60 mV, I_C is increased by one decade. It is also interesting that the base current I_B exhibits two distinct regimes, namely, the recombination-dominated regime for small V and the diffusion-dominated regime for large V, as expected from the diode I–V theory. However, I_C increases consistently with the diode factor. This is understandable, since the collector current is contributed solely by the minority carrier diffusion, and is therefore independent of the recombination process in the junction depletion region.

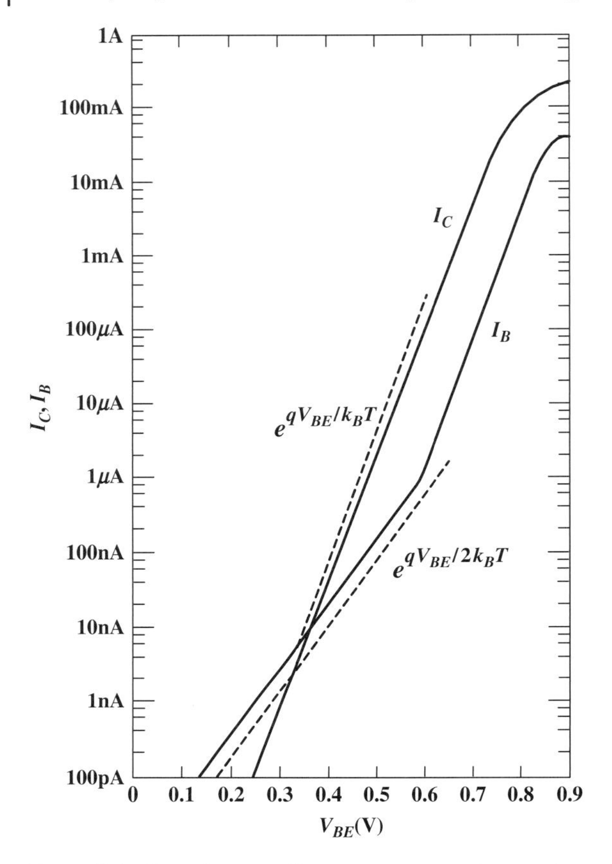

Figure 20.4 The collector and base currents versus base–emitter voltage. (taken from Grove, A. S. (1967) *Physics and Technology of Semiconductor Devices*, John Wiley and Sons, Inc.)

Figure 20.5 graphically depicts the essential physics entailed in transistor action. For the output current I_C at collector terminal, the base and emitter terminals are the control electrodes. Specifically, the electron injection from emitter to base region for inducing I_C is inseparably intermingled with hole injection from base to emitter region, thereby inducing the input current I_B. The smaller the input current, the more effective amplifier the transistor. Now, the recombination current resulting from the excess electrons recombining in the base can be calculated using (20.10):

$$I_{rB} \equiv qA_E \int_0^{W_B} \frac{n-n_{p0}}{\tau_n} dx \simeq \frac{qA_E n_i^2 W_B}{2N_{AB}\tau_n} e^{qV_{BE}/k_BT}, \qquad n_{p0} = \frac{n_i^2}{N_{AB}} \tag{20.12}$$

where the excess term $\propto n_{p0} \exp qV_{BE}/k_BT$ has been taken much larger than n_{p0}. The loss of electrons due to the recombination in the base is incorporated into the transport factor, defined as

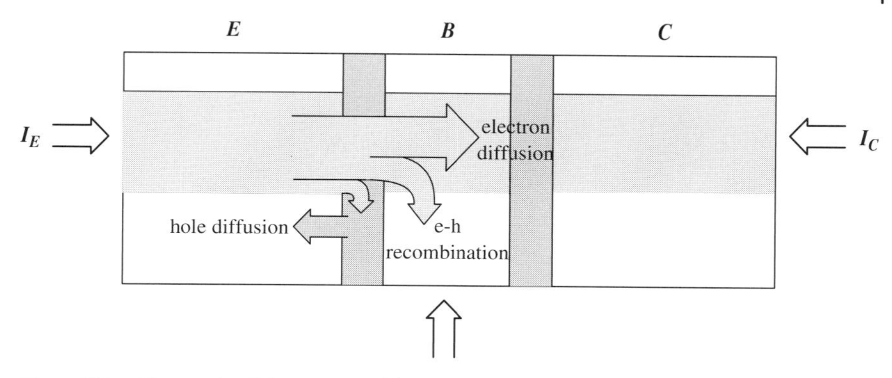

Figure 20.5 The graphical description of the transistor action: under forward active bias the base–emitter junction current consists of hole and electron diffusion currents. The latter is consumed in part by the recombination in the base, and contributes to the output collector current, while the former constitutes the input current.

$$\alpha_T \equiv \frac{|I_C|}{|I_{nE}|} = \frac{|I_{nE}|-|I_{rB}|}{|I_{nE}|} = 1-\left|\frac{I_{rB}}{I_{nE}}\right| = 1-\frac{W_B^2}{2L_{nB}^2}, \quad L_{nB}^2 = \tilde{D}_n\tau_n, \tag{20.13}$$

where

$$\begin{aligned} I_{nE} &= -I_S(e^{qV_{BE}/k_BT} - e^{qV_{BC}/k_BT}) \\ &\simeq -I_S e^{qV_{BE}/k_BT}, \quad I_S = \frac{A_E q D_{nB} n_i^2}{N_{AB} W_B}, \end{aligned} \tag{20.14}$$

is the forward current of the emitter junction, contributed by electrons injected into the base (see (20.11)).

Simultaneously with the electron injection into the base for inducing the output current via I_{nE}, the holes are injected from base to emitter, inducing the input current

$$\begin{aligned} I_{pE} &= \tilde{I}_{pE}(e^{qV_{BE}/k_BT} - 1) \\ &\simeq \tilde{I}_{pE} e^{qV_{BE}/k_BT}, \quad \tilde{I}_{pE} \equiv \frac{-qA_E n_i^2 D_{pE}}{N_{DE} l_{pE}}, \quad l_{pE} = W_E, L_p, \end{aligned} \tag{20.15}$$

where l_{pE} stands for the lesser of the emitter width W_E and hole diffusion length L_p and N_{DE} is the donor doping level in the emitter. Thus, an additional parameter gauging the current gain is the emitter injection efficiency, defined as

$$\gamma \equiv \frac{I_{nE}}{I_{nE}+I_{pE}} = \frac{1}{1+\frac{W_B N_{AB} D_{pE}}{W_E N_{DE} D_{nB}}} = \frac{1}{1+\frac{GN_B \tilde{D}_{pE}}{GN_E \tilde{D}_{nB}}}, \tag{20.16}$$

where GN_B and GN_E are the Gummel numbers of holes and electrons in base and emitter regions, respectively (see (20.6)).

Hence, the net current gain is given by

$$\alpha_F \equiv \frac{|I_C|}{|I_E|} = \frac{|I_{nE}|}{|I_{nE}|+|I_{pE}|}\frac{|I_C|}{|I_{nE}|} = \gamma\alpha_T. \tag{20.17}$$

And using Kirchhoff's circuit law,

$$I_B + I_E + I_C = 0$$

and replacing I_E by I_C via (20.17), namely,

$$I_B - \frac{I_C}{\alpha_F} + I_C = 0,$$

one obtains the gain factor β_F of the output collector current with respect to the input base current:

$$\beta_F \equiv \frac{I_C}{I_B} = \frac{\alpha_F}{1-\alpha_F}. \tag{20.18}$$

It is therefore clear from (20.18) that the gain factor can be made large by reducing the electron recombination in the base and also by suppressing the hole injection from base to emitter, thereby rendering $\alpha_F \approx 1$.

The former necessitates a narrow base width, while the latter requires heavier emitter doping, N_{DE}, so that the equilibrium hole concentration in the emitter is reduced (see (19.29)). More important, the base–emitter junction can be made of a heterojunction in which the emitter has larger bandgap, so that an additional potential barrier is provided as an effective suppressor of the hole injection. The heterojunction BJT thus designed and implemented, based purely on simple yet fundamental quantum mechanical considerations, has yielded one of the largest gain factors.

Although a large gain factor can be attained by making $\alpha_F \approx 1$, it is difficult to control tightly the fluctuations in β-values. For example, for $\alpha_F \simeq 0.995$, a large current gain ensues, namely, $\beta \simeq 200$. However, small fluctuation of α_F centered at 0.995 obviously induces large fluctuations of the gain factor.

20.3 Ebers–Moll Equations

Inasmuch as the bipolar junction transistor is comprised of two p–n junction diodes in voltage-controlled interaction, the two coupled diode equations can provide the working model of the device. The Ebers–Moll equations provide such a model and are briefly discussed in this section. Thus, consider the base–emitter junction current, contributed by both the electron and hole currents. The former is given by (20.11) or (20.14), while the latter is specified by (20.15), so that the total current flowing into the emitter lead is given by

$$I_E = I_S(e^{qV_{BC}/k_BT} - e^{qV_{BE}/k_BT}) - \tilde{I}_{pE}(e^{qV_{BE}/k_BT} - 1). \tag{20.19}$$

Also, the hole current in the base–collector junction can be specified in similar way as in (20.15), and one can write

$$I_{pC} = \tilde{I}_{pC}(e^{qV_{BC}/k_BT} - 1), \quad \tilde{I}_{pC} \equiv \frac{-qA_C n_i^2 D_{pC}}{N_{DC} l_{pC}}, \quad l_{pC} = W_C, L_p, \tag{20.20}$$

with l_{pC} denoting the lesser of either the collector width W_C or the hole diffusion length L_p. Hence, the total current flowing into the collector lead consists of (20.11) and (20.20), namely,

$$I_C = I_S(e^{qV_{BE}/k_BT} - e^{qV_{BC}/k_BT}) - \tilde{I}_{pC}(e^{qV_{BC}/k_BT} - 1). \tag{20.21}$$

It is thus a simple matter to regroup the terms in emitter and collector currents, (20.19) and (20.21), as

$$I_E = -I_{ES}(e^{qV_{BE}/k_BT} - 1) + \alpha_R I_{CS}(e^{qV_{BC}/k_BT} - 1), \tag{20.22a}$$

$$I_C = -I_{CS}(e^{qV_{BC}/k_BT} - 1) + \alpha_F I_{ES}(e^{qV_{BE}/k_BT} - 1), \tag{20.22b}$$

with

$$I_{ES} \equiv I_S + \tilde{I}_{pE}; \quad I_{CS} \equiv I_S + \tilde{I}_{pC}. \tag{20.23a}$$

In this notation, the current gain factors are given by

$$\alpha_F \equiv \frac{I_S}{I_S + \tilde{I}_{pE}}; \quad \alpha_R \equiv \frac{I_S}{I_S + \tilde{I}_{pC}}. \tag{20.23b}$$

And the specification of the three terminal currents can be completed with the base current expressed in terms of I_E and I_C via the Kirchhoff circuit law, that is,

$$I_B = -I_E - I_C. \tag{20.24}$$

Equations 20.22 are the Ebers–Moll (E–M) equations for npn transistors. The corresponding equations for pnp transistors can be readily transcribed from (20.22) by interchanging the roles of electrons and holes. Clearly, four parameters enter in Ebers–Moll equations, that is, I_{ES}, I_{CS}, α_F, and α_R. However, there is one constraint, namely, the reciprocity relation,

$$\alpha_F I_{ES} \equiv \alpha_R I_{CS} \equiv I_S, \tag{20.25}$$

which leaves only three independent parameters in E–M equations.

The E–M equation (20.22) is often expressed in terms of the forward active (I_F) and reverse active (I_R) currents in the form

$$I_E = -I_F + \alpha_R I_R, \tag{20.26a}$$

$$I_C = -I_R + \alpha_F I_F, \tag{20.26b}$$

where

$$I_F \equiv I_{ES}(e^{qV_{BE}/k_BT} - 1), \tag{20.27a}$$

$$I_R \equiv I_{CS}(e^{qV_{BC}/k_BT} - 1). \tag{20.27b}$$

In this notation, the base current reads as

$$I_B = -(I_E + I_C) = I_F(1 - \alpha_F) + I_R(1 - \alpha_R). \tag{20.27c}$$

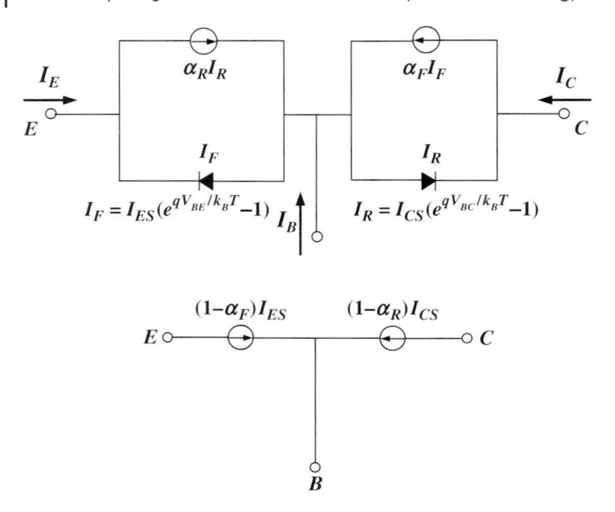

Figure 20.6 The equivalent circuit for Ebers–Moll model: two coupled diodes (top) and cut-off state (bottom).

The equivalent circuit representing (20.26) is shown in Figure 20.6. As expected, the circuit consists of two coupled diodes and two current sources. The current source term, connected in parallel with each diode, accounts for the current contributed by other coupled diode. With the three independent parameters specified from the device I–V data, the E–M equations well describe the I–V characteristics of the bipolar junction transistor, based on two coupled diodes.

20.4
Base Transit Time and Charge Control Model

Base Transit Time The speed with which the device responds to changing terminal voltages is one of the key parameters. For devices banking on the charge control such as BJT, the speed is obviously dictated by the charging or discharging time entailed in the changing terminal voltages. For example, the switching time from "on" to "off" states is the time required to drain the excess charge from both the base and the junction depletion regions and to attain the off-state charge configuration. A key parameter in this context is the average time an electron spends for transiting the quasi-neutral base region as the minority carrier.

To analyze this base transit time, first consider the total excess charge of electrons that are injected into the base under forward active bias. The excess charge stored in the base is given by

$$Q_{nB} = -\int_0^{W_B} qA_E[n_p(x) - n_{p0}]dx \tag{20.28}$$

and is dictated by the terminal voltages used. Once the charging is completed, electrons are still constantly drained out to the collector and at the same time replenished from the emitter. Naturally, these balanced influx and outflux of electrons keep Q_{nB} constant in time.

Therefore, if τ_{trB} represents the average base transit time of electrons, the output current can also be expressed, by definition, as

$$|I_C| \equiv \frac{|Q_{nB}|}{\tau_{trB}}. \tag{20.29}$$

This operational definition of I_C represents the basic fact that the collector current is contributed by $|Q_{nB}|/q$ electrons crossing the collector terminal every τ_{trB} second. Moreover, (20.29) can also be used for estimating the charging time required for the forward active bias.

Now, one can find Q_{nB} by inserting the excess electrons in (20.10) into (20.28), obtaining for $V_{BC} < 0$,

$$Q_{nB} = A_E \frac{1}{2} q n_{p0} (e^{qV_{BE}/k_BT} - 1) W_B \simeq A_E \frac{1}{2} q n_{p0} e^{qV_{BE}/k_BT} W_B. \tag{20.30}$$

When Q_{nB} in (20.30) is combined with I_C in (20.11), the transit time of electron in the base is obtained as

$$\tau_{trB} \equiv \frac{|Q_{nB}|}{|I_C|} = \frac{W_B^2}{2D_n}. \tag{20.31a}$$

In terms of the mobility, τ_{trB} can be recast as

$$\tau_{trB} = \frac{W_B^2}{2(k_BT/q)\mu_n} = \frac{W_B}{2[(k_BT/q)/W_B]\mu_n} = \frac{W_B}{2v_{dn}} \tag{20.31b}$$

and is therefore correlated to the drift velocity, driven by the thermal voltage k_BT/q.

As pointed out, the transit time is intricately correlated with the charging or discharging time and the output current. This can be further elucidated by revisiting the distributed carrier profiles under the forward active bias, as shown in Figure 20.3 and further amplified in Figure 20.7. Here, V_{BE} induces the excess charge of electrons in the base and that of holes in the emitter and the total magnitude of the induced charge is therefore the sum of these two,

$$Q_F = |Q_{nB}| + |Q_{pE}|. \tag{20.32a}$$

Moreover, these two charge components are controlled simultaneously by the diode factor, $\exp(qV_{BE}/k_BT) - 1$ (see (20.30)). Hence, one can write

$$Q_F = Q_{F0}(e^{qV_{BE}/k_BT} - 1). \tag{20.32b}$$

Clearly, Q_F acts as the common source, supporting both collector and base currents and one can therefore introduce two time constants for I_C and I_B, namely,

$$I_C = \frac{Q_F}{\tau_F}, \tag{20.33}$$

$$I_B = \frac{Q_F}{\tau_{BF}}. \tag{20.34}$$

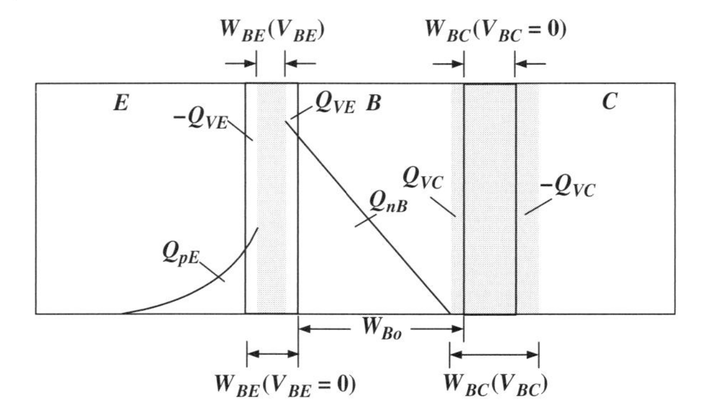

Figure 20.7 Uncompensated donor or acceptor charges resulting from electrons or holes injected or extracted under forward active bias. Q_{VE}, Q_{VC} are the charges resulting from electrons and holes being injected into emitter base junction and extracted from base collector junction, respectively.

The three time constants, τ_{trB}, τ_F, τ_{BF}, can be better appreciated when correlated with each other in conjunction with the current gain factor. Thus, under the forward active bias and in the limit of high injection efficiency, for example, in which $I_{nE} \gg I_{pE}$ and $\gamma \approx 1$ (see (20.16)), one can put $Q_F \approx Q_{nB}$. In this case, the electron–hole recombination in the base becomes the dominant factor for determining I_C and I_B (see Figure 20.5). It then follows from (20.34) that τ_{BF} is well approximated by the recombination time of electrons, namely, $\tau_{BF} \approx \tau_n$. Moreover, it also follows from (20.29), (20.31a), and (20.33) that $\tau_F \approx W_B^2/2D_n$. Hence, one may write

$$\beta_F \equiv \frac{I_C}{I_B} = \frac{\tau_{BF}}{\tau_F} = \frac{\tau_n}{W_B^2/2\tilde{D}_n} = \frac{2L_n^2}{W_B^2}, \quad L_n^2 = \tilde{D}_n\tau_n, \tag{20.35}$$

in agreement with the previously derived result of β_F in (20.18), since $\alpha_F = \gamma\alpha_T \approx \alpha_T = 1 - W_B^2/2L_n^2$ and $\beta_F \approx 1/(1-\alpha_F) = 2L_n^2/W_B^2$.

In this manner, the currents flowing in through terminal leads for the given bias can be expressed via the time variation of the stored charges. For the forward active bias shown in Figure 20.7, for example, the base current can be represented as usual via Q_F, τ_{BE} and also by the time rate of change in Q_F, τ_{BE} itself. Moreover, the stored charges at two junctions also change in time due to the voltages applied. Thus, one can generally express the base current as

$$I_B = \frac{Q_F}{\tau_{BF}} + \frac{dQ_F}{dt} + \frac{dQ_{VE}}{dt} + \frac{dQ_{VC}}{dt}, \tag{20.36}$$

where the charge stored in base–emitter junction Q_{VE} should decrease in time due to reduced depletion depth resulting from the forward voltage applied, while the stored charge in base–collector junction Q_{VC} increases in time as the depletion depth is increased with the reverse voltage applied.

Likewise, one can represent the collector current as

$$I_C = \frac{Q_F}{\tau_F} - \frac{dQ_{VC}}{dt} \tag{20.37}$$

and from the Kirchhoff's law, the emitter current reads in terms of I_C and I_B as

$$\begin{aligned} I_E &= -I_C - I_B \\ &= -Q_F\left(\frac{1}{\tau_F} + \frac{1}{\tau_{BF}}\right) - \frac{dQ_F}{dt} - \frac{dQ_{VE}}{dt} \end{aligned} \tag{20.38}$$

Equations 20.36, 20.37, and 20.38 represent the charge control model under the forward active bias and describe the transient behavior of the current as well. The model is based upon the concept of the voltage controlled charging and discharging, coupled with the transit time of carriers, as dictated by transport coefficients and recombination and/or generation time. The model can also be applied to other bias regimes in similar manner.

20.5 Problems

20.1. The quantum well in the base region plays a key role in the operation of the bipolar junction transistor. The role of quantum well for electrons in pnp transistor is obvious from Figure 20.2. Discuss in detail why the mirror image of the electron quantum well acts as the quantum well for holes.

20.2. Discuss the pnp bipolar junction transistor, starting from the basic energy band diagrams operative therein.

(a) Derive the linking current corresponding to (20.8) which is contributed by holes as minority carrier in the base.

(b) Write down the output hole current under forward active bias corresponding to (20.11) and compare with the result obtained in (a).

(c) Discuss the current gain based upon the transport factor, emitter injection efficiency.

20.3. In both npn and pnp junction transistors, the collector current is always well described by the ideal p–n junction forward current, namely, $I_C \propto \exp q|V_{BE}|/k_BT$, in spite of the usual generation or recombination processes occurring in the junction depletion region. Discuss the precise reasons for this observed behavior of the collector current.

20.4. Set up the Ebers–Moll equations for pnp transistor, together with its corresponding equivalent circuit.

20.5. (a) Re-express the collector and base currents in pnp transistors under forward active bias in terms of the charges stored and the time constants involved (see (20.31), (20.33), and (20.34)).

(b) Discuss the gain factor in terms of these time constants under the same bias (see (20.35)).

(c) Set up the charge control equations under forward active bias and discuss the rearrangements occurring in the stored charges from those of equilibrium.

20.6. Discuss the I–V behaviors in n^{++}–p^{+}–n and p^{++}–n^{+}–p junction transistors in comparison with *npn* and *pnp* transistors, respectively, and compare the current gain and amplification factors.

Suggested Reading

1 Muller, R.S., Kamins, T.I., and Chan, M. (2002) *Device Electronics for Integrated Circuits*, 3rd edn, John Wiley & Sons, Inc.
2 Sze, S.M. and Ng, K.K. (2006) *Physics of Semiconductor Devices*, 3rd edn, Wiley-Interscience.
3 Neudeck, G.W. (1989) *The Bipolar Junction Transistor, Modular Series on Solid State Devices*, 2nd edn, vol. III, Prentice Hall.
4 Streetman, B.G. and Banerjee, S. (2005) *Solid State Electronic Devices*, 6th edn, Prentice Hall.
5 Grove, A.S. (1967) *Physics and Technology of Semiconductor Devices*, John Wiley & Sons, Inc.
6 Yariv, A. (1982) *An Introduction to Theory and Applications of Quantum Mechanics*, John Wiley & Sons, Inc.
7 Bhattacharya, P. (1996) *Semiconductor Optoelectronic Devices*, 2nd edn, Prentice Hall.

21
Metal Oxide Silicon Field Effect Transistors I: Overview of Device Behavior and Applications

The idea of the field effect transistor is rather simple and was conceived as early as 1930s but it took approximately three decades to actually realize the device. Once its feasibility was demonstrated in 1960s, however, MOSFET soon emerged as the leading technology driver for digitalization and information technology. The device has retained the role ever since and may perhaps exert the same role for some more years to come in this era of nanotechnology.

MOSFET is simple in structure, capable of batch processing, and lends to very large-scale integration for the multifunctional system on chip (SOC) applications. More important, there has been the relentless drive for downsizing the device for higher integration and higher performance. The fundamental limitations encountered in such drive via the top–down approach, based on etch, lithography, and self-oxidation, have prompted the idea of the bottom–up approach.

The latter approach is understandably centered around the self-assembly and self-organization of atoms and molecules, which constitute the core area of the quantum mechanics. The top–down and bottom–up approaches could complement with each other for pushing back the frontiers of electronic devices. It is thus fitting to single out MOSFETs for discussion as a mainstream technology and also as a technological background for devising more novel devices. The discussion is again focused on highlighting the quantum mechanical foundations upon which the device operation is based, rather than the in-depth I–V modeling.

21.1
MOSFET: Overview

MOSFET is the acronym for the metal oxide silicon field effect transistor. The device was also known in the past as the metal oxide semiconductor transistor (MOST) or more functionally as insulated gate field effect transistor (IGFET).

It is a three-terminal and unipolar device, as opposed to bipolar, and is used as switch and amplifier. The overwhelming advantage of the device consists of the simplicity of structure and the low cost of fabrication, coupled with the capability of

Introductory Quantum Mechanics for Semiconductor Nanotechnology. Dae Mann Kim

ISBN: 978-3-527-40975-4

large-scale integration for multifunctional IC applications. In addition, the device has the inherent capability for downscaling into the very midst of nanoregime and can also provide convenient platforms for a number of useful applications such as memory cells and sensors.

From the quantum mechanical point of view, MOSFET has been one of the first devices in which the quantum wells were induced and the two-dimensional electrons were utilized for generating currents.

MOSFET is the normally off device and requires the gate voltage to sustain currents, in contrast with the normally on device such as GaAs MESFET. There are three types of MOSFETs, namely, the n-type (NMOS), p-type (PMOS), and complementary FET (CMOS). In this chapter, NMOS is singled out for discussion, but the results can be readily transcribed into PMOS by interchanging the roles of electrons by those of holes and adjusting the device parameters accordingly.

21.1.1 NMOS

Thus, consider the n-channel MOSFET, which is composed of n^+ source and drain plus the third terminal called gate (Figure 21.1). The gate is electrically isolated from the semiconductor via an insulator, for example, the silicon dioxide. The source and drain are separated by p-type silicon substrate, so that n^+–p and p–n^+ junctions are formed back-to-back.

With the gate voltage off ($V_G = 0$) and the drain voltage on ($V_D > 0$), the reverse-biased p–n^+ junction at the drain end cuts off the current (off state). With the gate voltage on at a value above the threshold voltage, that is, $V_G > V_T$, the channel is inverted at the oxide–silicon interface. In which case, the n^+–p junction barriers at the source and drain are reduced to the n^+–n barriers and, consequently, electrons spill over from the source to the bias-induced quantum well and these inverted electrons contribute to the output drain current with V_D on (on state). The on and off states are separated by the subthreshold regime, in which $0 < V_G < V_T$.

Figure 21.1 also shows the device output current, namely, the drain current I_D versus V_D with V_G as parameter. The resulting family of I–V curves is called the transistor I–V. Here, each I_D–V_D curve is comprised of two regimes, namely, the linear or triode region, in which I_D grows linearly or sublinearly with V_D. The triode region merges with the saturation region in which I_D is approximately pinned at a constant level for the given V_G. The drain conductance

$$g_d \equiv \left.\frac{\partial I_D}{\partial V_D}\right|_{V_G}$$

decreases practically linearly in triode region to become nearly zero in saturation, as evident from the transistor I–V curves.

Figure 21.1 also shows I_D versus V_G at fixed V_D and this I_D–V_G curve represents the transfer characteristics. Here, I_D takes off essentially from the threshold voltage V_T and increases with V_G. It is customary to also plot I_D in a log scale to amplify on the subthreshold current, connecting the off current at $V_G = 0$ to on current at $V_G > V_T$.

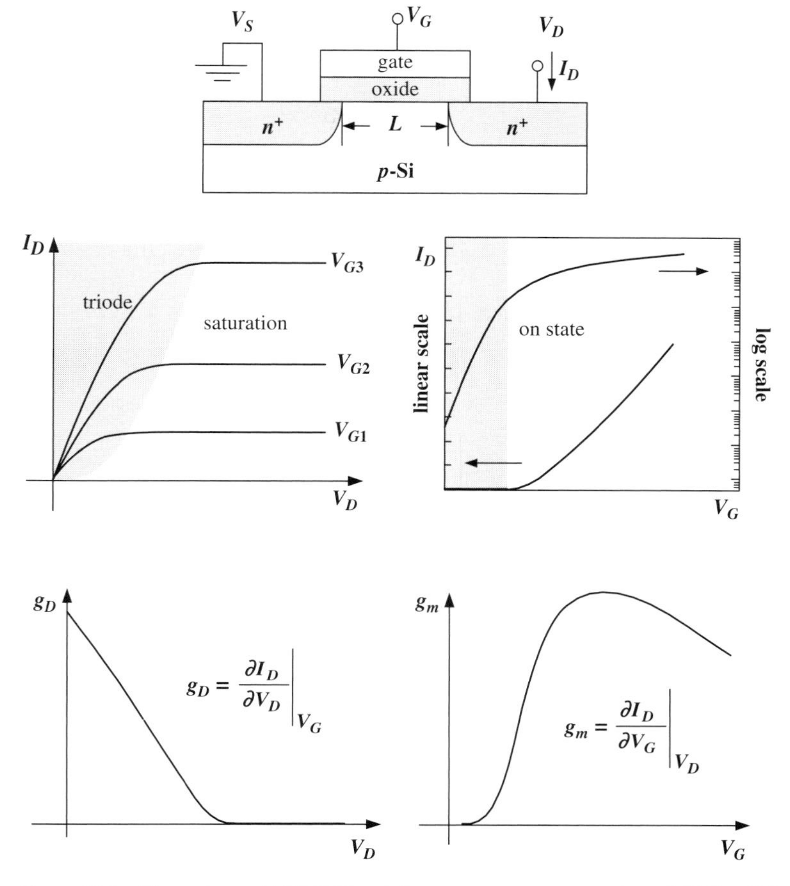

Figure 21.1 The n-channel MOSFET and the output currents: transistor I–V ($I_D - V_D$), drain conductance (g_d), transfer characteristics ($I_D - V_G$), and transconductance (g_m).

The on to off current ratio is understandably a key parameter and typically takes up the value $I_{\text{on}}/I_{\text{off}} \geq 10^6$. Also, the transconductance

$$g_m \equiv \left.\frac{\partial I_D}{\partial V_G}\right|_{V_D}$$

is a key device parameter, determining the device performance, including the speed.

Gross Features of Drain Current The modeling of the device consists of quantifying I_D as a function of both V_D and V_G and also in terms of pertinent device parameters. It is thus useful to have at the outset an overview of the gross features of the drain current, based on a simple model. Thus, consider the well-known simulation program with integrated circuit emphasis (SPICE), level 1, in which the drain current is described by

$$I_D = \begin{cases} \dfrac{W}{L} C_{OX}\mu_n \left(V_G - V_T - \dfrac{1}{2} V_D \right) V_D, & V_D \leq V_{DSAT} \equiv V_G - V_T \\ \dfrac{W}{2L} C_{OX}\mu_n (V_G - V_T)^2, & V_D \geq V_{DSAT}. \end{cases} \quad (21.1)$$

Here, the ratio between the width and length of the channel W/L is known as the aspect ratio; C_{OX} ($\equiv \varepsilon_{ox}/t_{ox}$) given in terms of the permittivity, ε_{ox} of the silicon dioxide and the thickness, t_{ox} is the oxide capacitance per unit area; and μ_n is the channel mobility of the electron.

In this simple model, the on current I_D sets in for V_G greater than the threshold voltage V_T. Further, the triode and saturation regimes are demarcated by $V_{DSAT}(\equiv V_G - V_T)$ below which I_D increases with V_D, while it is pinned at fixed level for $V_D > V_{DSAT}$. The higher level SPICE models consist, in essence, of incorporating necessary physical refinements by taking the various device parameters as appropriate functions of V_D and V_G. Although (21.1) represents the simplest I–V model, it nevertheless provides the essential features of I_D and lends to a simple interpretation.

To interpret the drain current, multiply both the denominator and numerator of (21.1) by the channel length L and regroup the device parameters as

$$\begin{aligned} I_D &= WLC_{OX}\left(V_G - V_T - \frac{1}{2} V_D \right) \frac{\mu_n V_D}{L^2} \\ &\equiv Q_{nINV}/\tau_{tr}. \end{aligned} \quad (21.2a)$$

In this format, the quantity

$$Q_{nINV} \equiv WLC_{OX}\left(V_G - V_T - \frac{1}{2} V_D \right) \quad (21.2b)$$

obviously represents the electron charge inverted under the gate electrode via the gate voltage V_G. Q_{nINV} is understandably specified by the product of the total capacitance under the gate electrode, $C_{ox}WL$, and the gate voltage approximately at the midpoint of the channel minus V_T. Thus, Q_{nINV} should, by definition, represent the charge, which is capacitively coupled to V_G on silicon substrate, and V_T enters the expression for singling out the mobile electron charge from the fixed charge, arising from uncompensated acceptor atoms, as will become clear subsequently. The remaining quantities are lumped into

$$\frac{\mu_n V_D}{L^2} = \mu_n (V_D/L) \frac{1}{L} = v_D \frac{1}{L} = \frac{1}{L/v_D} = \frac{1}{\tau_{tr}} \quad (21.2c)$$

and represents the inverse transit time of the electron from source to drain through the channel, as dictated by the drift velocity driven by the longitudinal channel field, $v_D = \mu_n(V_D/L)$.

Thus, the drain current in (21.2a) is shown to carry a clear physical meaning. That is, I_D is contributed by Q_{nINV}/q electrons transiting across the channel via the drain voltage induced drift every τ_{tr} second. Further the steady-state I_D is maintained by Q_{nINV}/q electrons being constantly injected from the source and drained out at the drain at the same time, so that Q_{nINV} is pinned at a level determined by given V_G. Therefore, the output current is specified by two factors, namely, the mobile electron charge in the channel, capacitively coupled to V_G, and the carrier drift driven by V_D. The discussion to follow is addressed to these basic quantities.

21.2
Charge Control and Metal Oxide Silicon System

21.2.1
The Channel Inversion: Classical Theory

The Surface Charge and Surface Potential In this section, the channel inversion is discussed, taking NMOS as an example. First, the classical theory is presented and the modifications arising from quantum effects are discussed later. Needless to say, electrons are taken in classical theory as point charges occupying an infinitesimal spatial extent. Figure 21.2 shows the metal oxide silicon (MOS) system, consisting of the metal and/or n^+ polysilicon (gate electrode), the silicon dioxide (insulator), and the silicon (p-substrate). In a semiconductor, the work function is defined as the sum of the affinity factor $q\chi$ and the difference between the bottom of the conduction band E_C and E_F, namely, $E_C - E_{Fp}$ in this case. The affinity factor is the energy required to excite an electron from the bottom of the conduction band to the vacuum level. The work function of the metal is generally different from that of the silicon substrate.

The silicon dioxide has a large bandgap of about 9 eV and the affinity factor of 0.95 eV. The work function of aluminum $q\varphi_m$ chosen here for consideration is 4.05 eV, which is approximately the same as that of n^+ polysilicon. The work function of the p-type silicon substrate, $q\varphi_s \equiv q\chi + (E_C - E_{Fp}) > q\chi + E_G/2$, should be larger than

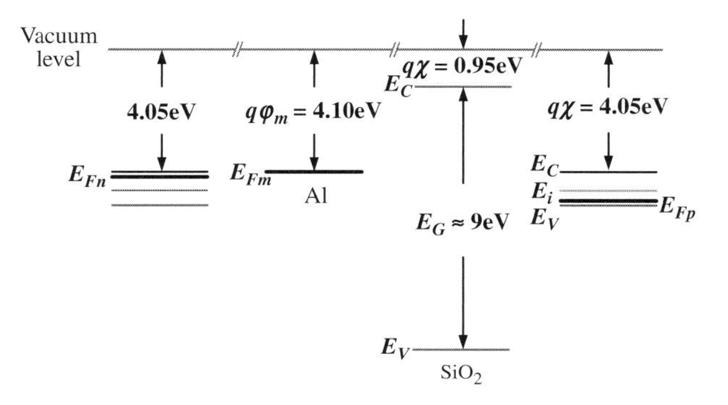

Figure 21.2 NMOS system, consisting of metal and/or n^+ polysilicon, gate insulator (silicon dioxide), and silicon p-substrate. Also shown are metal work function, affinity factor, $q\chi$, and Fermi levels.

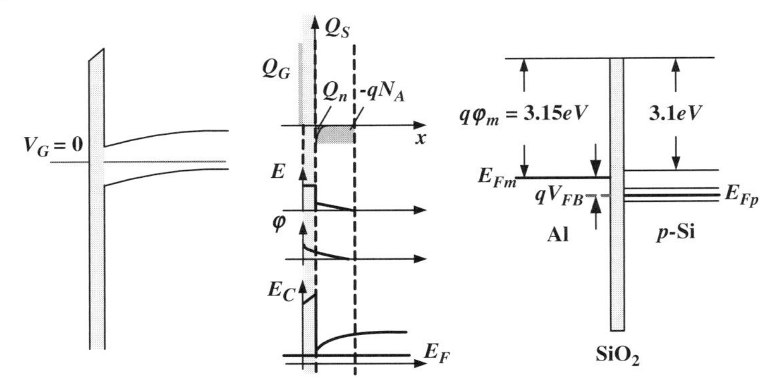

Figure 21.3 NMOS system in equilibrium contact and the space charge, field, potential, and the electron potential energy supporting the band bending. Also shown is the flat band voltage at which the band is flattened out.

4.6 eV (Figure 21.2). When these three subcomponents are brought into equilibrium contact, the Fermi level should line up and be flat, as detailed in Chapter 18 and as also shown in Figure 21.3. This condition for equilibrium again necessitates the band bending, just as in the case of the p–n junction.

Equivalently, the equilibrium contact of these subsystems and the accompanying band bending come about due to electrons being exchanged between the metal gate and semiconductor substrate. Specifically, for the case under consideration, E_{Fm} lies above E_{Fp}, and electrons are therefore transferred from the metal to the silicon substrate. This leaves behind the positive charge sheet at the metal surface, which in turn induces the dipole charge in the interface by pushing holes away from the surface, thereby exposing acceptor ions uncompensated. The resulting space charge, field, potential, and the band bending are shown in Figure 21.3.

It is important to notice that, the difference in Fermi energies, $E_{Fm} - E_{Fp}$, and/or work function difference $q\varphi_m - q\varphi_s$ is accommodated via the band bending in both silicon dioxide and substrate. This partitioning of the band bending comes about due to the fact that a bulk charge can be sustained in the semiconductor. This is in contrast with the case of metal in which the space charge can reside only at the surface, thereby precluding any band bending inside the metal.

It is also important to notice that the equilibrium band bending can be made flat by applying the voltage at the gate by an amount

$$qV_{FB} \equiv q\varphi_m - q\varphi_s \tag{21.3}$$

The voltage V_{FB} thus introduced is called the flat band voltage. Since $q\varphi_m < q\varphi_s$ for the case under consideration, V_{FB} is negative and applying V_{FB} at the gate is tantamount to cancelling the positive charge sheet on the metal surface, so that the dipole charge disappears, together with the band bending (Figure 21.3).

Next, to analyze the channel inversion, it is convenient to introduce the charging voltage

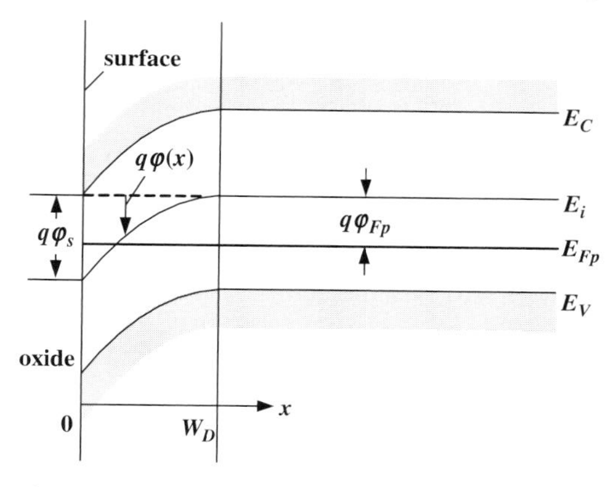

Figure 21.4 Cross-sectional view of the surface potential and band bending.

$$V'_G \equiv V_G - V_{FB} \tag{21.4}$$

and examine the partitioning of V'_G in the gate dielectric and the substrate, as shown in Figure 21.4, namely,

$$V'_G = V_{OX} + \varphi_S. \tag{21.5}$$

Here, φ_S is the space charge potential at the surface, namely, $\varphi_S \equiv \varphi(x=0)$ and plays a key role in analyzing the channel inversion. The potential $\varphi(x)$ obeys the Poisson equation in the depletion region near the interface:

$$\frac{d^2\varphi(x)}{dx^2} = -\frac{1}{\varepsilon_S}\varrho(x), \tag{21.6}$$

where the space charge is generally specified by

$$\varrho(x) = q(N_D^+ - N_A^- + p_p(x) - n_p(x)). \tag{21.7}$$

Deep in the substrate the charge neutrality prevails at every point, and one can therefore write

$$N_D^+ - N_A^- = n_{p0} - p_{p0} \tag{21.8}$$

with n_{p0} and p_{p0} denoting respective minority and majority equilibrium carrier concentrations in p-substrate. But near the surface where the band bending occurs, the electron and hole concentrations are different from the respective equilibrium values in the bulk and become position dependent:

$$p_p(x) = p_{p0}e^{-\beta\varphi(x)}, \quad \beta \equiv q/k_BT, \tag{21.9a}$$

$$n_p(x) = n_{p0}e^{\beta\varphi(x)} \tag{21.9b}$$

(see Figure 21.4) and $\varphi(x)$ should obviously be zero at the edge of the bulk substrate. If $\varphi(x)$ increases as the interface is approached from the bulk substrate, so that the band $-q\varphi(x)$ bends down, $p_p(x)$ decreases while $n_p(x)$ increases, as is clear from Figure 21.4. On the other hand, if $\varphi(x)$ decreases, the band bends up and the opposite situation prevails. Inserting (21.7)–(21.9) into (21.6), the Poisson equation reads as

$$\frac{d^2\varphi(x)}{dx^2} = -\frac{q}{\varepsilon_S}\left[p_{p0}(e^{-\beta\varphi}-1)-n_{p0}(e^{\beta\varphi}-1)\right] \tag{21.10}$$

and is shown a strongly nonlinear differential equation.

However, it is always possible to perform the first integration in an exact fashion by multiplying both sides of (21.10) with $d\varphi$ and carrying out the integration. Thus, from the left-hand side of (21.10), there results

$$\int_0^{\varphi} d\varphi \frac{d^2\varphi}{dx^2} \equiv \int_0^{d\varphi/dx} \frac{d\varphi}{dx} d\left(\frac{d\varphi}{dx}\right) = \int_0^{-E} (-\mathrm{E})d(-\mathrm{E}) = \frac{1}{2}\mathrm{E}^2, \tag{21.11a}$$

where the variable of integration has been transformed from φ to the space charge field $\mathrm{E}(= -\partial\varphi/\partial x)$.One can likewise find from the right-hand side

$$\begin{aligned} &\frac{-q}{\varepsilon_S}\int_0^{\varphi} d\varphi[p_{p0}(e^{-\beta\varphi}-1)-n_{p0}(e^{\beta\varphi}-1)], \qquad \beta = \frac{q}{k_B T}, \\ &= \mathrm{E}_0^2\left[(e^{-\beta\varphi}+\beta\varphi-1)+\frac{n_{p0}}{p_{p0}}(e^{\beta\varphi}-\beta\varphi-1)\right], \quad \mathrm{E}_0^2 = \left(\frac{qp_{p0}}{\beta\varepsilon_S}\right), \end{aligned} \tag{21.11b}$$

where E_0 can be recast as

$$\mathrm{E}_0^2 = \frac{1}{\beta}\left(\frac{qp_{p0}}{\varepsilon_S}\right) = \frac{1}{\beta^2}\frac{1}{L_D^2} = \left(\frac{k_B T}{q}\right)^2\frac{1}{L_D^2}, \tag{21.11c}$$

where

$$L_D \equiv \left(\frac{k_B T\varepsilon_S}{q^2 p_{p0}}\right)^{1/2} \tag{21.11d}$$

is the well-known Debye screening length.

Hence, with the introduction of the function

$$F(\beta\varphi, n_{p0}/p_{p0}) \equiv \left[(e^{-\beta\varphi}+\beta\varphi-1)+\frac{n_{p0}}{p_{p0}}(e^{\beta\varphi}-\beta\varphi-1)\right]^{1/2}, \tag{21.12}$$

the space charge field is specified from (21.11) and (21.12) as

$$\mathrm{E}(x) = -\frac{d\varphi}{dx} = \pm\sqrt{2}\frac{k_B T}{q}\frac{1}{L_D}F(\beta\varphi, n_{p0}/p_{p0}), \tag{21.13}$$

with the polarity of the field determined by that of the gate voltage. Given the surface field $\mathrm{E}_S(\equiv \mathrm{E}(x=0))$, the surface charge Q_S is quantified by the well-known basic boundary condition at the oxide–substrate interface, namely, $Q_S \equiv -\varepsilon_S \mathrm{E}_S$. Now that

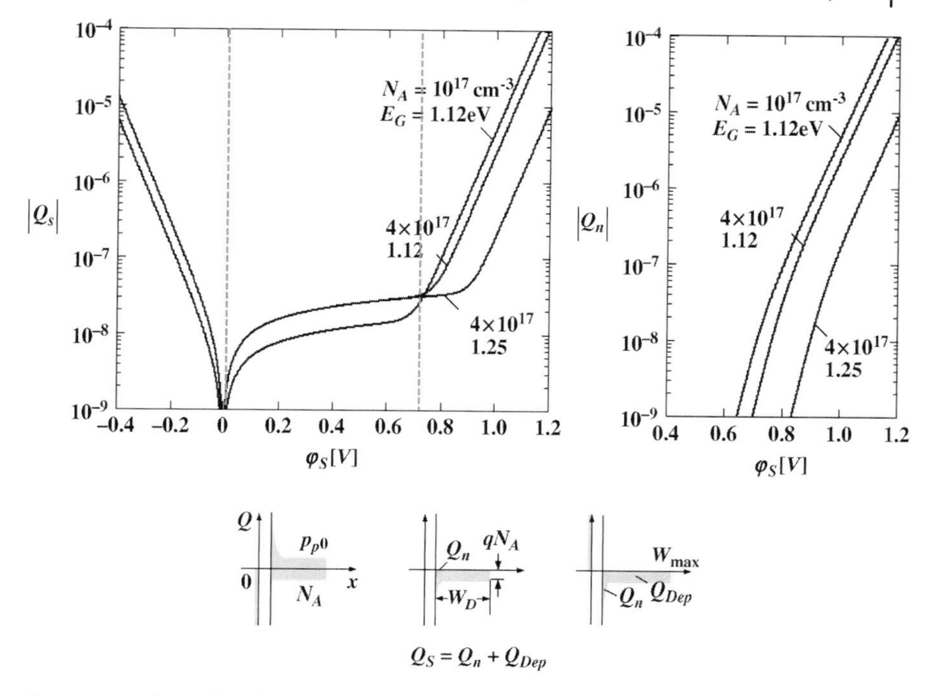

Figure 21.5 The total surface charge Q_S versus surface potential φ_S at different doping levels and bandgaps. Solid lines show the results obtained by using silicon and typical doping levels, and the effect of different bandgaps is also shown by considering E_G larger than that of silicon. The surface charge consisting of uncompensated N_A and n is also indicated and the electron surface charge $|Q_n|$ is singled out and plotted versus φ_S.

E_S has been specified in terms of φ_S, one can generally write

$$Q_S \equiv -\varepsilon_S E_S = \mp\varepsilon_S\sqrt{2}\frac{k_B T}{q}\frac{1}{L_D}F(\beta\varphi_s, n_{p0}/p_{p0}) \tag{21.14}$$

Equation 21.14 explicitly correlates the surface charge in terms of the surface potential and provides a basic relation by which to quantify the channel inversion in terms of doping level, temperature, and other parameters.

Figure 21.5 shows the surface charge Q_S versus the surface potential φ_S at different doping levels and bandgaps. Also shown in the figure are the space charges associated with each regime. As noted, for $V_G = V_{FB}$, there is no band bending and therefore $\varphi_S = 0$ and $Q_S = 0$. For $\varphi_S < 0$, the band bends up and consequently holes as the majority carrier in the substrate are accumulated at the surface. This is described by

$$F(\beta\varphi_S, n_{p0}/p_{p0}) \approx e^{\beta|\varphi_S|/2}, \quad Q_S \propto e^{\beta|\varphi_S|/2}. \tag{21.15}$$

In this accumulation regime, Q_S consists of the excess holes. In the depletion region, $0 < \varphi_S < \varphi_{Fp}$, the band bends down, supported by the fixed dopant charge $-qN_A$ near the interface. In this regime, Q_S consists primarily of fixed acceptor ions $-qN_A W_D$

with W_D denoting the surface depletion depth of holes. In weak inversion, namely, $\varphi_{Fp} < \varphi_S < 2\varphi_{Fp}$, electrons begin to populate appreciably the interface region, but Q_S still consists primarily of the dopant charge. One can thus write for both depletion and weak inversion regions, namely, $0 < \varphi_S < 2\varphi_{Fp}$,

$$F(\beta\varphi_S, n_{p0}/p_{p0}) \approx (\beta\varphi_S)^{1/2}, \quad Q_S \propto (\beta\varphi_S)^{1/2}. \tag{21.16}$$

When $\varphi_S \approx 2\varphi_{Fp}$, it clearly follows from (21.12) that the electron concentration as minority carrier has reached at the surface, the level comparable to the majority carrier concentration p_{p0}, namely,

$$n_S = n_{p0}\exp(\beta 2\varphi_{Fp}) \approx p_{p0}.$$

Once this condition is attained, any further increase of φ_S beyond $2\varphi_{Fp}$ mainly results in increasing the mobile electrons at the interface and one can write

$$F(\beta\varphi_S, n_{p0}/p_{p0}) \approx e^{\beta\varphi_S/2}, \quad Q_S \propto e^{\beta\varphi_S/2}. \tag{21.17a}$$

In this regime, the electrons are induced practically at the surface and do not contribute significantly to the band bending, and φ_S is therefore pinned nearly at $2\varphi_{Fp}$. In this way, the channel is inverted and the condition

$$\varphi_S = 2\varphi_{Fp} \tag{21.17b}$$

represents the onset of the strong inversion, with φ_{Fp} denoting the hole Fermi potential in p substrate.

The Charge Control and Channel Inversion Now that the channel inversion has been explicitly specified by the surface potential φ_S, the charge control by the gate can be specified by correlating φ_S to the gate voltage V_G. For this purpose, consider again the charging voltage partitioned into the gate oxide and the substrate, namely (21.5),

$$\begin{aligned} V_G - V_{FB} &\equiv V_{OX} + \varphi_S \equiv \frac{|Q_S|}{C_{OX}} + \varphi_S \\ &= \frac{1}{C_{OX}}\sqrt{2}\frac{k_B T}{q}\frac{\varepsilon_S}{L_D}F(\beta\varphi_S, n_{p0}/p_{p0}) + \varphi_S, \end{aligned} \tag{21.18}$$

where the voltage drop across the oxide is given by definition by $V_{OX} \equiv |Q_S|/C_{OX}$ and (21.5) and (21.14) have been combined. Evidently, (21.18) correlates the surface charge Q_S to V_G via φ_S.

A typical φ_S–V_G curve obtained from (21.18) is shown in Figure 21.6. In depletion and weak inversion regimes, φ_S increases nearly linearly with charging voltage V'_G, indicating that the surface electron concentration n_S is enhanced exponentially with V'_G from n_{p0} to p_{p0}. However, near the onset of strong inversion, namely, $\varphi_S \approx 2\varphi_{Fp}$, the rate of increase slows down drastically. In fact φ_S tends to be pinned at about $2\varphi_{Fp}$, as any further increase of the gate voltage can now be accommodated by the inverted electrons at the surface with a minimal change in φ_S.

MOS Capacitor The channel inversion and the control of the mobile electron charge therein can be conveniently quantified by considering the MOS capacitor C.

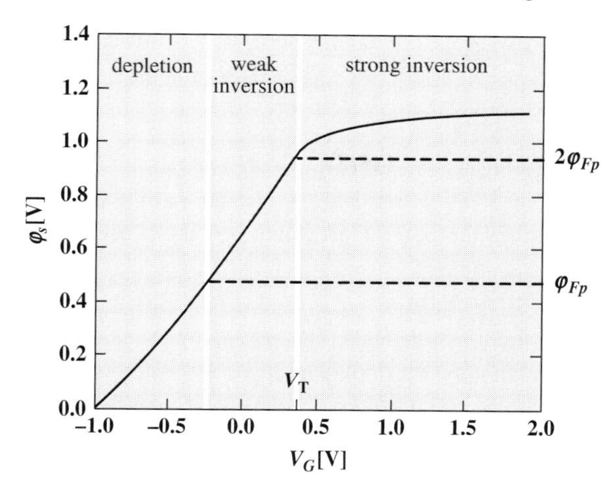

Figure 21.6 Surface potential (φ_S) versus gate voltage, $V_G - V_{FB}$.

Obviously, C consists of the oxide (C_{ox}) and the surface (C_S) capacitors connected in series, as shown in Figure 21.7. Thus, one can write

$$\frac{1}{C} = \frac{1}{C_{OX}} + \frac{1}{C_S}. \tag{21.19}$$

Here, $C_{OX} = \varepsilon_{ox}/t_{ox}$ and C_S representing the change in the surface charge with respect to φ_S is given from (21.14) by

$$C_S \equiv \frac{\partial |Q_S|}{\partial \varphi_S} = \frac{\varepsilon_S}{\sqrt{2}L_D} \frac{\left|(1-e^{-\beta\varphi_S}) + \frac{n_{p0}}{p_{p0}}(e^{\beta\varphi_S}-1)\right|}{F(\beta\varphi_s, n_{p0}/p_{p0})}. \tag{21.20}$$

Inasmuch as φ_S and W_D vary with V_G (Figure 21.5), C_S is a variable capacitance.

It follows from (21.20) that in accumulation where $\varphi_S < 0$,

$$C_S \approx \frac{\varepsilon_S}{\sqrt{2}L_D} \frac{e^{-\beta\varphi_S}}{e^{-\beta\phi_S/2}} \approx \frac{\varepsilon_S}{\sqrt{2}L_D} e^{-\beta\phi_S/2} \tag{21.21a}$$

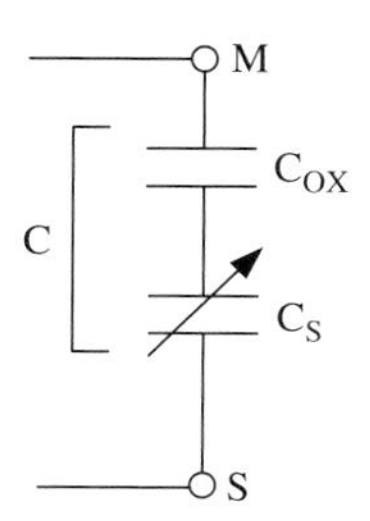

Figure 21.7 The MOS capacitor, consisting of C_{OX} and the variable surface capacitor C_S connected in series.

and grows exponentially with φ_S, while at the flat band, where $\varphi_S = 0$, one can write by letting $\varphi_S \to 0$ in (21.20),

$$C_S = \frac{\varepsilon_S}{\sqrt{2}L_D} \frac{\beta\varphi_S}{(\beta\varphi_S/\sqrt{2})}\bigg|_{\varphi_S \to 0} = \frac{\varepsilon_S}{L_D}. \tag{21.21b}$$

Similarly, in depletion and weak inversion regions, in which $0 < \varphi_S < 2\varphi_{Fp}$,

$$C_S \approx \frac{\varepsilon_S}{\sqrt{2}L_D\sqrt{\beta\varphi_S}} = \frac{\varepsilon_S}{\sqrt{2\varepsilon_S\varphi_S/qN_A}} \approx \frac{\varepsilon_S}{W_D} \tag{21.21c}$$

and C_S is primarily determined by the dopant depletion depth W_D at the surface. In (21.21), the expression of L_D in (21.11d) has been used together with $\varphi_S \simeq qN_A W_D^2/\varepsilon_S$. Finally, for $\varphi_S > 2\varphi_{Fp}$,

$$C_S \approx \frac{\varepsilon_S}{\sqrt{2}L_D}\sqrt{\frac{n_{p0}}{p_{p0}}}e^{\beta\varphi_S/2} \tag{21.21d}$$

and again grows exponentially with φ_S.

Figure 21.8 shows a typical C–V curve, obtained from (21.19) and (21.20). In accumulation and inversion, $C_S \gg C_{OX}$, so that $C \approx C_{OX}$. However, in depletion and weak inversion, $C_S \ll C_{OX}$ and C is therefore mainly determined by C_S. The overall behavior of the coupling of surface charge to V_G is contained in this C–V curve.

A comment is due at this point. Ideally, only the mobile electron charge should be coupled to the gate voltage. However, in the bulk MOSFET structure, the channel inversion requires the surface band bending, which has to be supported by the dopant charge. Hence, the fixed dopant charge unavoidably enters into the channel inversion process.

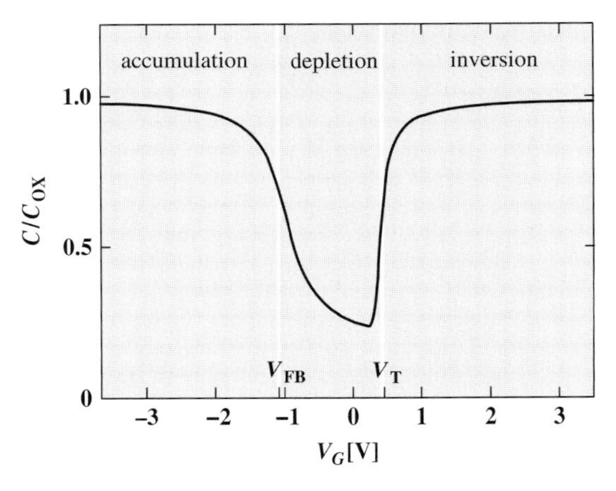

Figure 21.8 C–V characteristics of MOS system, exhibiting accumulation, depletion, and inversion regimes versus $V_G - V_{FB}$.

21.2.2
Quantum Description of Channel Inversion

Subbands It is clear from the classical treatment of the channel inversion that the charge control at the gate consists of forming the quantum well at the interface and inducing mobile 2D electrons therein. The inverted electrons are spatially confined by the quantum well in the direction normal to the surface, while freely moving in the interface plane, as shown in Figure 21.9. The quantum well is formed by the potential barrier provided by the large bandgap of the gate oxide and the surface band bending.

The quantum analysis of the channel inversion therefore requires a self-consistent numerical computation for solving the coupled Schrödinger and Poisson equations. However, the gross features of the channel inversion can be brought out using the triangular potential well approximation. In this approximation, the quantum well is taken to consist of infinite potential barrier at the gate–dielectric interface and straight line band bending in the p-substrate, with the surface field E_S providing the slope (Figure 21.9).

The Schrödinger equation in the direction normal to the surface is then given by

$$\left[-\frac{\hbar^2}{2m_n}\frac{d^2}{dz^2}-q\varphi(z)\right]\psi(z) = E\psi(z), \tag{21.22a}$$

where m_n is the effective mass of the electron for the given crystallographic direction and the electrostatic potential $\varphi(z)$ reads as

$$\varphi(z) = \begin{cases} E_S z, & z > 0, \\ \infty, & z \leq 0. \end{cases} \tag{21.22b}$$

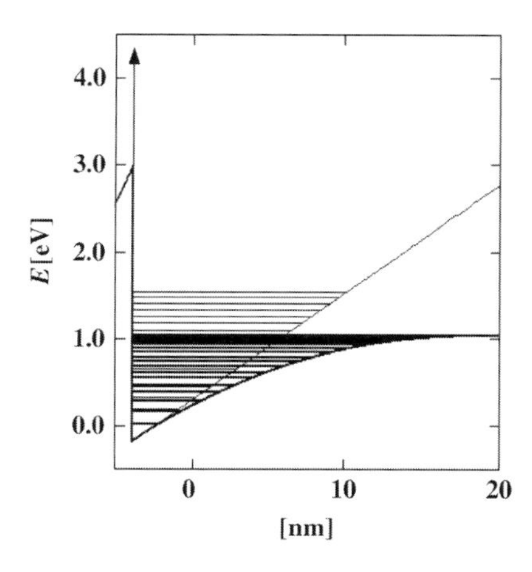

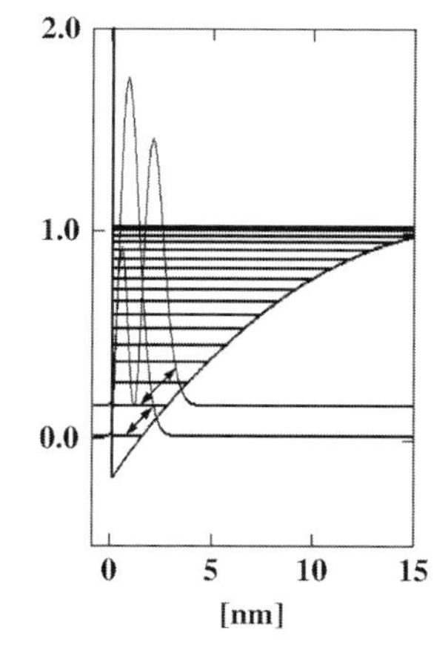

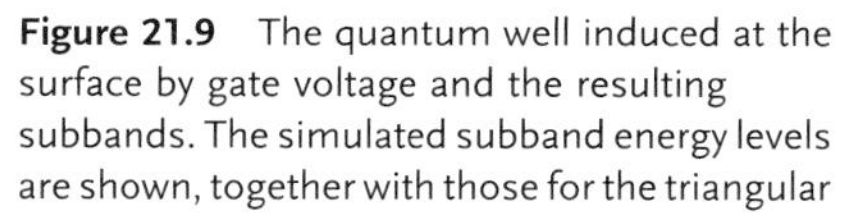

Figure 21.9 The quantum well induced at the surface by gate voltage and the resulting subbands. The simulated subband energy levels are shown, together with those for the triangular potential well, for comparison. The simulated wavefunctions of the electrons in ground and first excited subbands are also shown.

The solution of the energy eigenequation (21.22) for the triangular potential well is given by the well-known Airy function $A(\xi_j)$:

$$\psi(\xi) = \left(\frac{2m_n q E_S}{\hbar^2}\right)^{1/6} \frac{A(\xi)}{|A'(\xi_0)|}, \tag{21.23a}$$

where the prime denotes the differentiation with respect to the dimensionless distance variable ξ defined as

$$\xi = z\left(\frac{2m_n q E_S}{\hbar^2}\right)^{1/3} - \xi_0, \quad \xi_0 \equiv \frac{E}{qE_S}\left(\frac{2m_n q E_S}{\hbar^2}\right)^{1/3}. \tag{21.23b}$$

The Airy function has the asymptotic behavior,

$$A(\xi) \sim \frac{1}{2}(\xi)^{-1/4} \exp-\left(\frac{2\xi^{3/2}}{3}\right), \quad \xi > 0, \tag{21.24a}$$

while

$$A(\xi) \sim |\xi|^{-1/4} \sin\left(\frac{2|\xi^{3/2}|}{3} + \frac{1}{4}\right), \quad \xi \leq 0. \tag{21.24b}$$

It is clear from (21.24a) that the probability density of electron penetrating into the classically forbidden region decays exponentially as expected (Figure 21.9). Also, at the interface where $z = 0$ and $\xi = -\xi_0$, the wavefunction should vanish, that is, $\psi(|\xi_0|) = 0$, because of the infinite barrier height assumed at the interface. Hence, one can write

$$\sin\left(\frac{2|\xi_0|^{3/2}}{3} + \frac{\pi}{4}\right) = 0, \tag{21.25}$$

or

$$\frac{2|\xi_0|^{3/2}}{3} + \frac{\pi}{4} = j\pi, \quad j = 1, 2, 3, \ldots, \tag{21.26}$$

and the quantized energy levels, called the subbands, naturally follow from (21.23b) and (21.26) as

$$E_j = \beta_j E_S^{2/3}, \quad \beta_j \equiv \left(\frac{\hbar^2}{2m_n}\right)^{1/3} \left[\frac{3}{2}\pi q\left(j - \frac{1}{4}\right)\right]^{2/3}, \quad j = 1, 2, 3 \ldots. \tag{21.27}$$

These subbands are plotted in Figure 21.9. Also shown in the figure are the subbands obtained by solving numerically the coupled Poisson and Schrödinger equations in Hartree approximation. In this approximation, the single-electron wavefunction is used, while treating the rest of electrons in the atom as the core background charge. In the simulation, the band bending is accounted for in terms of the electron charge cloud and uncompensated dopant charge near the surface.

Clearly, the subbands are located well above the conduction band in both cases and the level spacing between the subbands decreases progressively with increasing energy eigenvalues. The general features of these two sets of subbands are similar, except that the subbands in the triangular potential well are consistently higher than those obtained by simulation. This is expected since the triangular potential overestimates the steepness of the quantum well.

Also shown in Figure 21.9 are simulated ground and first excited-state wavefunctions. It is important to notice that the peak of the wavefunction is located away from the interface. This indicates that the 2D electron charge is induced at a distance appreciably away from the interface. This is in contrast with the classical theory of channel inversion, in which electrons are induced practically at the interface. As a result, in the quantum analysis of channel inversion, the effective oxide thickness becomes larger than the deposited thickness t_{OX}. This together with the effective broadening of the bandgap due to the formation of subbands considerably affects the channel inversion.

Figure 21.10 shows the simulated classical and quantum mechanical C–V curves obtained, using the same device parameters. Also shown in the figure are the density profiles of inverted electrons resulting from both classical and quantum descriptions of channel inversion. As is clearly seen from this figure, the concentration profiles of inverted electrons in classical theory are sharply peaked at the surface, while the corresponding profiles of 2D electrons are peaked at a distance appreciably away from the surface. Although small, this distance can lead to substantially different device performances, especially for small oxide thickness. Specifically, because of the difference in effective oxide thickness the classical C–V curve lies consistently above the corresponding quantum C–V curve for 2D electrons. Since $I_D \propto C_{OX}$, the quantum modification is therefore shown to yield smaller output current for the given process parameters and bias conditions.

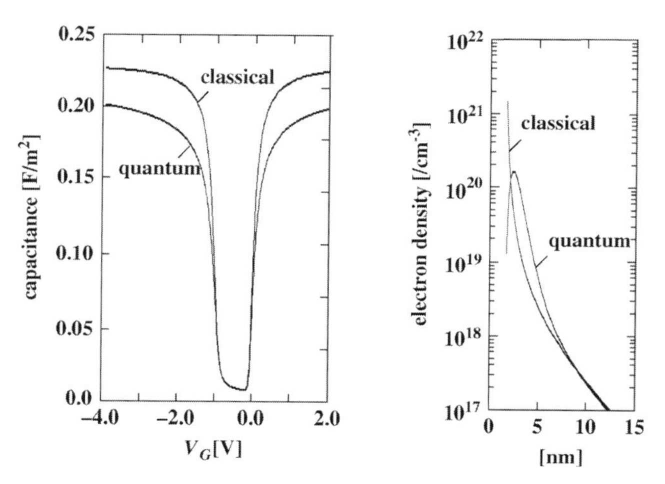

Figure 21.10 Simulated classical and quantum mechanical C–V characteristics and the corresponding inverted electron density profiles versus $V_G - V_{FB}$.

21.3 NMOS I–V

21.3.1 On Current and Variable Depletion Approximation

Variable Depletion Depth W_D In Section 21.1, the lumped view of "on" current was presented, based on the mobile charge induced under the gate electrode and the carrier transit time across the channel. In Section 21.2, the quantum modifications in the channel inversion have been discussed. With this modification in mind, the device on current is modeled next, using the variable depletion approximation.

A key to modeling the I–V behavior is to quantify the mobile charges, coupled into the gate voltage V_G. Thus, a convenient starting point of discussion is (21.14) in which the total surface charge is partitioned into the mobile electron and fixed dopant charges, namely,

$$Q_S \equiv -C_{OX}V_{OX} = Q_n + Q_{DEP} \tag{21.28a}$$

Clearly, (21.28a) reiterates a simple but basic fact that the field lines emanating from the positive surface charge sheet in the gate electrode due to V_G should be terminated by both electron and dopant charges induced on the semiconductor surface (see (Figure 21.5)).

Now, because electrons are induced practically at the surface, the depletion charge $Q_{DEP} = -qN_A W_D$, is responsible for the band bending and is correlated to φ_S via the relation $\varphi_S = qN_A W_D^2/2\varepsilon_S$. Hence, one can write

$$Q_{DEP} = -qN_A W_D = -(2\varepsilon_S qN_A\varphi_S)^{1/2}. \tag{21.28b}$$

Also, the charging voltage introduced in (21.4) can be generalized to include the bulk bias V_B and can again be divided into the voltage drop across the oxide and the surface potential, namely,

$$V_G' \equiv V_G - V_B - V_{FB} = V_{OX} + \varphi_S. \tag{21.29}$$

Hence, combining (21.28) and (21.29), the mobile charge is singled out as

$$\begin{aligned} Q_n &= -C_{OX}V_{OX} - Q_{DEP} = -C_{OX}(V_G - V_B - V_{FB} - \varphi_S) - Q_{DEP} \\ &= -C_{OX}(V_G - V_B - V_{FB} - \varphi_S - \gamma\varphi_S{}^{1/2}), \quad \gamma \equiv \frac{(2\varepsilon_S qN_A)^{1/2}}{C_{OX}}, \end{aligned} \tag{21.30}$$

and γ thus introduced is called the body effect coefficient.

When V_D is turned on, it is distributed in the channel. Hence, at a channel position y with the channel voltage V $(0 \le V \le V_D)$, the onset of strong inversion is specified by

$$\varphi_S = 2\varphi_{Fp} + V - V_B. \tag{21.31}$$

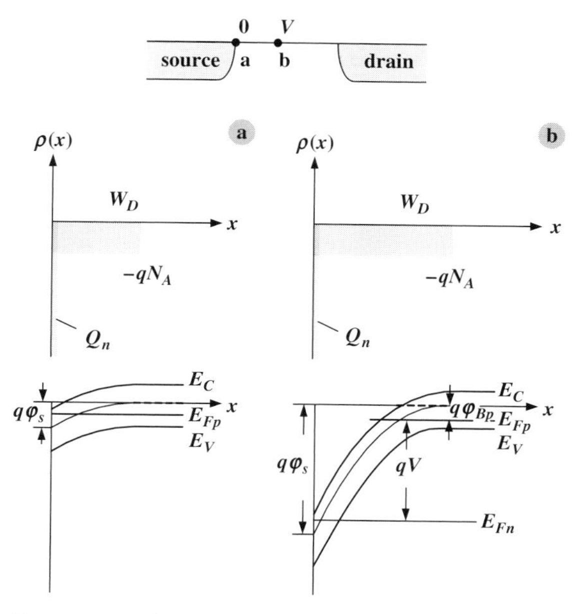

Figure 21.11 The surface band bending at source, where $V = 0$ (a), and at a channel position y with channel voltage V (b).

Here, the first term simply accounts for the condition for the strong inversion, that is, the electron concentration at the surface being equal to the bulk majority hole concentration, that is, $n_S = p_{p0}$, as detailed in the previous section for $V = 0$ (see (21.17b)). In the presence of V and V_B, the quasi-Fermi level of electrons E_{Fn} in the channel and that of holes E_{Fp} in the bulk should split to accommodate V and V_B, as shown in Figure 21.11. This is analogous to the reverse-biased p–n junction discussed in Chapter 19. Therefore, the onset of strong inversion at a channel position y with channel voltage V necessitates larger band bending, hence increased depletion depth W_D to support the bending. This means that with V_D and the distribution of the channel voltage, there should be concomitant distribution of depletion depths along the channel. This is the essence of the variable depletion approximation. Inserting (21.31) into (21.30) results in

$$Q_n = -C_{OX}[V_G - V_{FB} - 2\varphi_{Fp} - V - \gamma(2\varphi_{Fp} + V - V_B)^{1/2}]. \tag{21.32}$$

Now given the channel element y to $y + dy$ with V to $V + dV$, the voltage drop therein is given by

$$dV \equiv I_D dR = I_D \varrho \frac{dy}{W t_{ch}}, \tag{21.33a}$$

where ϱ is the resistivity and t_{ch} the effective thickness of the channel. Expressing ϱ in terms of the conductivity, one can also express (21.33a) as

$$dV \equiv I_D \frac{dy}{q\mu_n \tilde{n} W t_{ch}} = I_D \frac{dy}{W\mu_n |Q_n|}, \quad |Q_n| \equiv q\tilde{n}t_{ch}, \tag{21.33b}$$

with $\tilde{n}$ denoting the average electron concentration in the channel. Combining (21.32) and (21.33), one can write

$$\begin{aligned} \int_0^L I_D dy &= \int_0^{V_D} dV \mu_n W |Q_n| \\ &= \int_0^{V_D} dV \mu_n W C_{OX} [V_G - V_{FB} - 2\varphi_{Fp} - V - \gamma(2\varphi_{Fp} + V - V_B)^{1/2}] \end{aligned} \tag{21.34}$$

and carry out the integration using the fact that I_D is constant throughout, obtaining

$$I_D = \frac{W}{L}\mu_n C_{OX}\left\{\left[(V_G - V_{FB} - 2\varphi_{Fp})V_D - \frac{1}{2}V_D^2\right] - \Gamma\right\}, \tag{21.35a}$$

with

$$\Gamma \equiv \frac{2}{3}\gamma\left[(2\varphi_{Fp} + V_D - V_B)^{3/2} - (2\varphi_{Fp} - V_B)^{3/2}\right]. \tag{21.35b}$$

Equation 21.35 is the well-known expression for the "on" current, resulting from the variable depletion approximation. The equation can be simplified using a nonessential simplification for $V_B < 0$, namely,

$$\Gamma \equiv \frac{2}{3}\gamma(2\varphi_{Fp} - V_B)^{3/2}\left[\left(1 + \frac{V_D}{2\varphi_{Fp} - V_B}\right)^{3/2} - 1\right] \approx \gamma(2\varphi_{Fp} - V_B)^{1/2} V_D. \tag{21.35c}$$

Hence, inserting (21.35c) into (21.35a) and regrouping the terms, one finds

$$I_D = \frac{W}{L}\mu_n C_{OX}\left(V_G - V_T - \frac{1}{2}V_D\right)V_D, \tag{21.36}$$

where the threshold voltage

$$V_T \equiv V_{FB} + 2\varphi_{Fp} + \gamma(2\varphi_{Fp} - V_B)^{1/2} \tag{21.37}$$

naturally enters for singling out the mobile electron charge from the dopant charge in the channel.

Clearly, the I–V model thus derived is valid as long as $Q_n > 0$. When the channel is pinched off at the drain end, that is, $Q_n(y = L) = 0$, however, the model ceases to be operative. The pinch-off voltage V_{DSAT} is found from (21.32) by putting $Q_n = 0$ and replacing V by V_{DSAT}, that is,

$$Q_n = -C_{OX}[V_G - V_{FB} - 2\varphi_{Fp} - V_{DSAT} - \gamma(2\varphi_{Fp} + V_{DSAT} - V_B)^{1/2}] = 0. \tag{21.38}$$

The resulting quadratic equation for V_{DSAT} readily yields

$$V_{DSAT} = V_G - V_{FB} - 2\varphi_{Fp} - \frac{\varepsilon_S q N_A}{C_{OX}^2}\left\{\left[1 + \frac{2C_{OX}^2}{\varepsilon_S q N_A}(V_G - V_{FB} - V_B)\right]^{1/2} - 1\right\}. \tag{21.39}$$

where γ introduced in (21.30) has been used.

For simplicity, however, V_{DSAT} is often taken to be the drain voltage at which I_D reaches the maximum value at given V_G, that is, $\partial I_D/\partial V_D = 0$. This condition when applied to (21.36) yields

$$V_{DSAT} = V_G - V_T. \tag{21.40}$$

Hence, the on current is specified by (21.36) in the linear or triode region and is taken pinned at the maximum level, called the saturation current for $V > V_{DSAT}$:

$$I_{DSAT} = \frac{W}{2L}\mu_n C_{OX}(V_G - V_T)^2. \tag{21.41}$$

21.3.2
The Subthreshold Current

As mentioned, the subthreshold current I_{SUB} connects "off" current to "on" current and the subthreshold region is specified by $0 < V_G < V_T$ or equivalently by $0 < \varphi_S < 2\varphi_{Fp}$. In this regime, the surface charge can be approximately partitioned from (21.14) and (21.12) as

$$\begin{aligned} Q_S \equiv Q_{DEP} + Q_n &= -\varepsilon_S\sqrt{2}\frac{k_B T}{q}\frac{1}{L_D}F(\beta\varphi_s, n_{p0}/p_{p0}), \quad \beta \equiv q/k_B T, \\ &\approx -(2N_A\varepsilon_S k_B T)^{1/2}\sqrt{\beta\varphi_S}\left[1 + \frac{e^{\beta(\varphi_S - 2\varphi_{Fp})}}{\beta\varphi_S}\right]^{1/2}, \quad \frac{n_{p0}}{p_{p0}} = e^{-2\beta\varphi_{Fp}}, \\ &\approx -(2qN_A\varepsilon_S\varphi_S)^{1/2}\left(1 + \frac{1}{2}\frac{e^{\beta(\varphi_S - 2\varphi_{Fp})}}{\beta\varphi_S}\right), \end{aligned} \tag{21.42}$$

where the Debye length L_D defined in (21.11d) has been used and the bulk and channel voltage V_B, V have been taken zero for simplicity of discussion. Clearly, the first term on the right-hand side of (21.42) represents the depletion charge (see (21.28b)), while the second term denotes the electron charge. Thus, one can write using (21.11d) for Debye length,

$$Q_n = -\frac{1}{\beta}\left(\frac{qN_A\varepsilon_S}{2\varphi_S}\right)^{1/2} e^{\beta(\varphi_S - 2\varphi_{Fp})} = -qN_A L_D\left(\frac{1}{2\beta\varphi_S}\right)^{1/2} e^{\beta(\varphi_S - 2\varphi_{Fp})}. \tag{21.43}$$

It has therefore become clear from (21.43) that Q_n grows exponentially with the surface potential φ_S in the subthreshold regime. Furthermore, since $Q_{DEP} \gg Q_n$ in

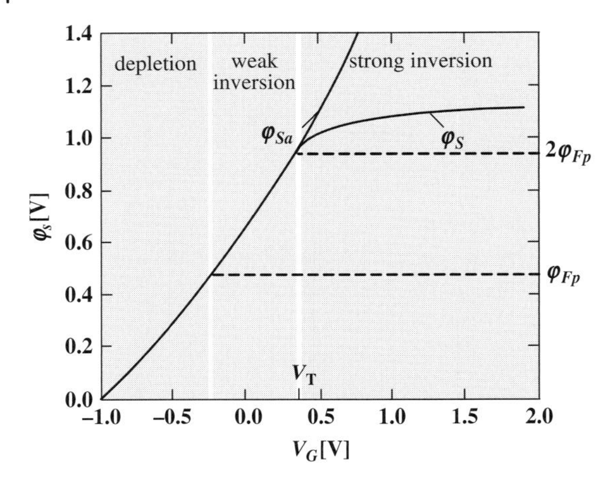

Figure 21.12 The simulated surface potential versus the gate voltage $V_G - V_{FB}$. The analytical result (φ_{Sa}) in (21.45) is plotted together with the simulated curve (φ_S) for comparison.

this region, one can write from (21.18) and (21.42),

$$V_G' = V_{OX} + \varphi_S \approx -\frac{Q_{DEP}}{C_{OX}} + \varphi_S = \gamma\varphi_S^{1/2} + \varphi_S, \tag{21.44}$$

where (21.28b) was used for Q_{DEP} together with the body coefficient defined in (21.30). Hence, one can solve this quadratic equation for $\varphi_S^{1/2}$, obtaining

$$\varphi_S = \left[-\frac{\gamma}{2} + \left(\frac{\gamma^2}{4} + V_G'\right)^{1/2}\right]^2. \tag{21.45}$$

The surface potential thus found is plotted in Figure 21.12 for $V_B = 0$, for simplicity, together with the exact $\varphi_S - V_G$ curve plotted in Figure 21.6. Indeed, (21.45) is shown an excellent approximation in the subthreshold regime, indicating that Q_n increases exponentially with V_G (see (21.43) and (21.45)).

In the presence of V_B and channel voltage V, one can generalize (21.43) using (21.31) and write

$$Q_n(0) = -qN_A L_D \left(\frac{1}{2\beta\varphi_S}\right)^{1/2} e^{\beta(\varphi_S - 2\varphi_{Fp} - V_{SB})}, \quad V_{SB} = V_S - V_B, \tag{21.46a}$$

at the source and

$$Q_n(L) = -qN_A L_D \left(\frac{1}{2\beta\varphi_S}\right)^{1/2} e^{\beta(\varphi_S - 2\varphi_{Fp} - V_{DS} - V_{SB})} \tag{21.46b}$$

at the drain. It has therefore become clear from (21.46) that Q_n decreases exponentially from source to drain as a function of the channel voltage. Hence, the subthreshold current is diffusion dominated and is well approximated by

$$|I_{SUB}| \approx WD_n \frac{Q_n(0)-Q_n(L)}{L}$$
$$= \frac{W}{L} D_n q N_A L_D \left(\frac{1}{2\beta\varphi_S}\right)^{1/2} e^{\beta(\varphi_S - 2\varphi_{Fp} - V_{SB})}(1-e^{-\beta V_{DS}}). \quad (21.47)$$

The subthreshold current thus derived connects the large gap existing between the on and off current levels, namely, $I_{on}/I_{off} > 10^6$.

21.4 Applications of Metal Oxide Silicon Field Effect Transistor

21.4.1 Dynamic Random Access Memory and Electrically Erasable and Programmable Read Only Memory Cells

In addition to the use of MOSFETs as logic devices in integrated circuits, the simple structure of MOSFET itself serves as a convenient platform for a variety of other useful device applications. For example, the dynamic random access memory (DRAM) cell is based on the simple MOSFET structure itself with a storage capacitor attached at the drain end (Figure 21.13). The programming and erase simply consists of injecting and extracting electrons into and out of the storage capacitor by opening the channel via VG. The reading is done likewise by monitoring the voltage difference at the source end caused by the presence and absence of stored electrons.

The scalability of the DRAM cell and the advantages inherent therein prompted continued downscaling of the cell. The cell size has been reduced deep into nano regime and memory densities have been enhanced correspondingly. However, the cell off is not low enough, so that the excess electrons have to be constantly refreshed. The stored electrons could tunnel across gate oxide, leading to gate leakage. The subthreshold conduction also adds to the leakage path. With stored excess electrons raising the energy band at the drain electrode, the stored electrons could leak into the channel over the lowered p–n$^+$ barrier, in a manner similar to the conduction process for subthreshold currents. These physical processes are further discussed in connection with the hot carrier effects and the nonvolatile memory cell operation.

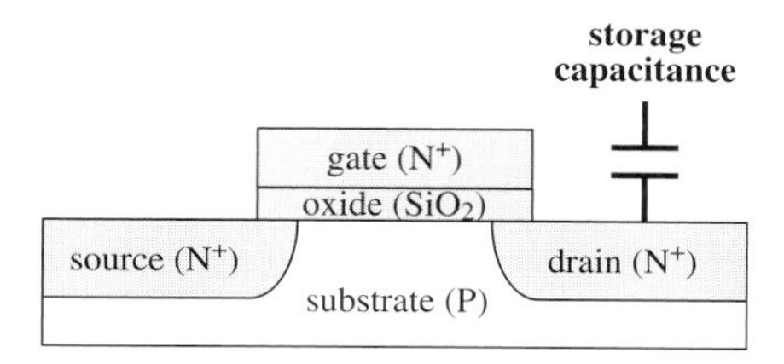

Figure 21.13 The cross-sectional view of dynamic random access memory cell, with the storage capacitance at the drain end.

21.4.2
Nonvolatile Electrically Erasable and Programmable Read Only Memory Cell

Hot Carrier Effect The nonvolatile memory cell without the need to refresh the stored charge has also been devised, based again on the simple structure of MOSFET. Before discussing this memory cell, however, the fundamental physical processes affecting the performance of the device are further discussed, in connection with the hot carrier effect. Figure 21.14 illustrates the various effects accompanying the channel electron becoming hot. The hot electrons are generated in general in device saturation in which $V_D > V_{DSAT}$. As detailed in the channel inversion, the mobile electron charge Q_n decreases monotonously from source to drain in the presence of the channel voltage distribution (see (21.32)), so that the channel pinch off sets in at the drain end at $V_D = V_{DSAT}$.

When the channel is pinched off, there ensues a drastic redistribution of the channel voltage. This is seen as follows. First, the drain current is driven mainly by the drift of electrons or holes and is therefore commensurate with the longitudinal channel field, namely, $I_D \propto \mu_n Q_n \mathrm{E}(y)$. Since Q_n decreases progressively from source to drain, the channel field $\mathrm{E}(y)$ should increase accordingly to keep $I_D \propto Q_n \mathrm{E}(y)$ constant throughout the channel. This means that with Q_n becoming practically zero near the drain end in device saturation, $\mathrm{E}(y)$ therein should grow large enough to keep the current constant throughout the channel. Hence, for $V_D > V_{DSAT}$, most of the excess voltage $V_D - V_{DSAT}$ should be concentrated in the narrow pinch-off region near the drain.

Now, in the presence of an electric field, the electrons generally absorb energy from the field by an amount proportional to the field intensity, that is, $\Delta E \propto \mathrm{E}^2$. Normally, the same amount of energy gained by the electron is deposited to the lattice, so that a steady state is reached at an elevated lattice temperature. However, in the presence of a large field, the energy gained by electrons could exceed the energy lost to the lattice. Consequently, the electron temperature is raised higher than the lattice temperature

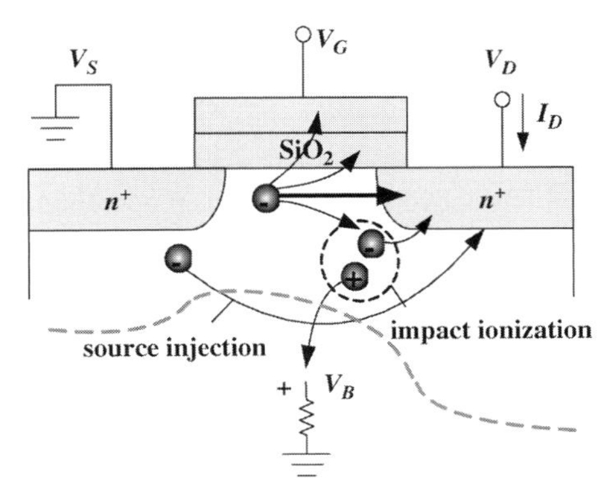

Figure 21.14 The schematic illustration of channel hot electron effects: (i) the generation of e–h pairs via impact ionization, (ii) the hot electron injection into gate electrode or gate oxide, and (iii) the electron injection at the forward-biased n^+–p junction at the source end.

and the electrons could become hot in the presence of large electric field. The resulting effective temperature of electrons could reach tens of thousands of degrees Kelvin and this phenomenon is known as the hot carrier effect.

The hot electrons once generated give rise to a series of effects, as sketched in Figure 21.14. First, the hot electrons could generate electron–hole pairs via the impact ionization of the host atoms. The electrons thus generated induce in turn the impact ionizations in cascade and add to the drain current in uncontrolled manner. The holes thus generated are collected at the bulk terminal, yielding the leakage path and at the same time raising the potential in the substrate. The substrate voltage thus elevated is transferred to the source end and acts as a forward bias in the n^+–p source junction, thereby inducing the injection of electrons from the source to the substrate. The injected electrons again add to the drain current in an uncontrolled manner.

In addition, a fraction of hot electrons in the upper tail of the Gaussian energy distribution can have enough energy to be injected from the channel to the oxide, thereby inducing the gate leakage. In so doing, some of those injected hot electrons can be trapped in the oxide, causing time-dependent or permanent oxide damage and increasing the threshold voltage. This tunneling-induced gate leakage is a fundamental limiting factor for device scaling. A comprehensive drain engineering has thus been devised to reduce the hot carrier effects, in particular to reduce the gate current. It is rather interesting, however, to notice that the unwanted gate current was also exploited instead as the driving force for devising the nonvolatile memory cell, which is briefly discussed next.

Electrically Erasable and Programmable Read Only Memory Cell The cross-sectional view of the nonvolatile flash electrically erasable and programmable read only memory (EEPROM) cell is shown in Figure 21.15. As is clear from the figure, the cell consists of the simple MOS structure, however, with an additional floating gate inserted as the storage node in between the control gate and the substrate. The programming is done by injecting electrons from the channel to the floating gate and the required electron injection is based on two basic physical processes, namely, (i) the thermionic emission by channel hot electrons over the oxide barrier (in NOR-type cell) or (ii) the F–N tunneling of channel electrons across the potential barrier (in NAND-type cell).

The energy level diagram during the programming is also shown in Figure 21.15, and its typical bias conditions are obvious from the band bending. Here, the channel hot electrons are intentionally induced rather than suppressed by applying large V_D. Acquiring enough energy, the hot electrons can fly over the barrier or undergo thermally assisted tunneling for programming. Alternatively, the electrons are transported across the gate dielectric from the channel via F–N tunneling. Once the excess electrons are injected into the floating gate, they are confined therein by the quantum well, which is formed by surrounding the n^+ polysilicon floating gate with silicon dioxide layers (Figure 21.15).

The erase is done by the reverse process of extracting the excess electrons from the floating gate into the channel via F–N tunneling through the thin oxide, as shown in

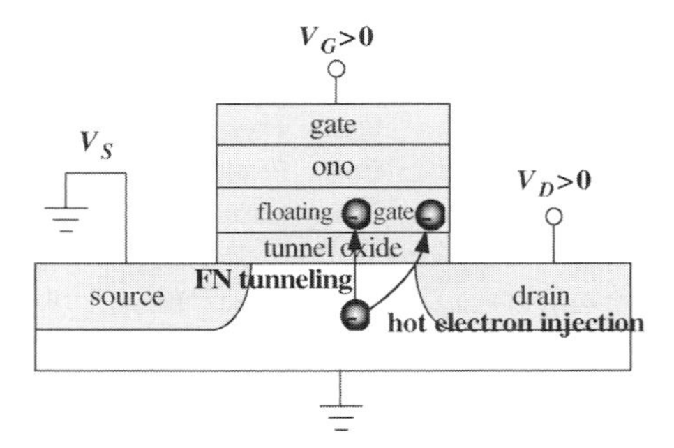

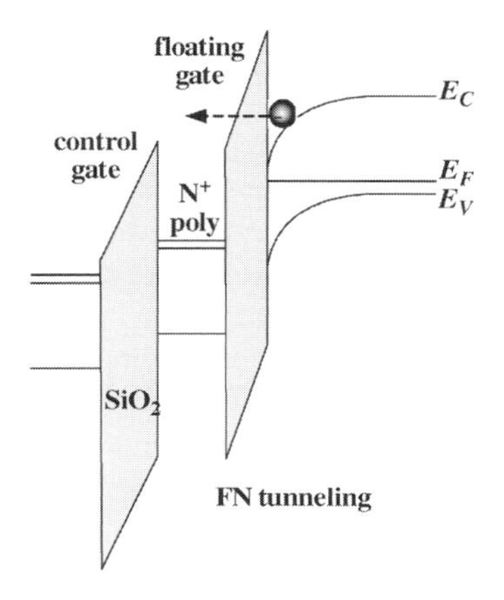

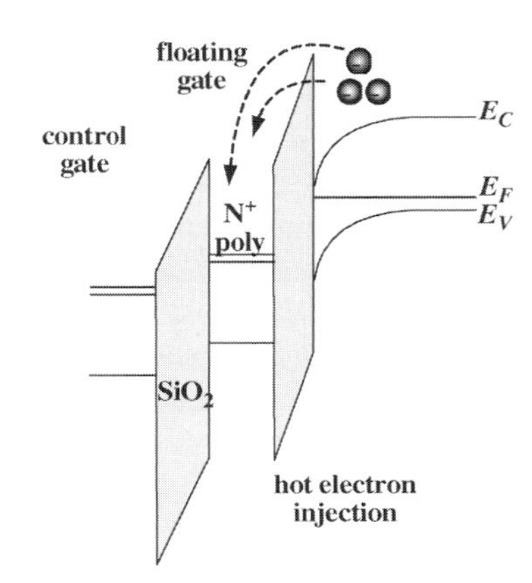

Figure 21.15 The flash EEPROM cell and its band bending during programming. The cell consists of standard MOS structure with additional floating gate as storage node. Under a positive bias at the control gate, hence at the floating gate, inverted electrons in the channel are injected into the floating gate via F–N tunneling throughout the channel or energetically near the drain end after becoming hot.

Figure 21.16. The typical bias conditions for inducing F–N tunneling for erase are again obvious from the band bendings involed. It is also noted that the dielectric deposited on top of the floating gate consists of the composite oxide/nitride/oxide layers, thick enough to ensure that practically no electrons may tunnel into the control gate.

Also without the erase bias applied for enhancing the tunneling probability via FN tunneling electrons are well confined in the quantum well of the floating gate. Thus there is no need for refreshing stored electrons and the memory cell operates based on controlled variation of the tunneling barrier potential.

Memory Operation The electrons are transported across the tunnel oxide for programming or erase via a simple capacitive coupling of applied voltages. The equivalent capacitance of the cell is shown in Figure 21.17. Here, the capacitance at the storage node of the floating gate consists of control gate, source, drain, and bulk terminal capacitances connected in parallel. Hence, when V_{CG} is applied at the control gate while grounding all other terminals, V_{CG} is partitioned from the basic

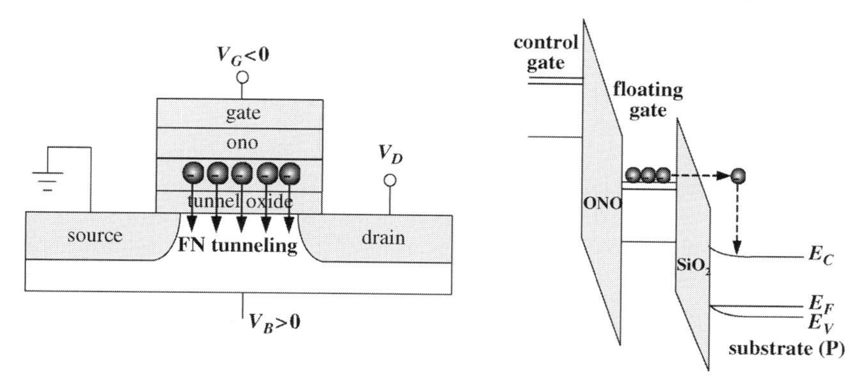

Figure 21.16 The flash EEPROM cell and its band bending during erase. Under a negative bias at the control gate, hence at the floating gate, electrons are tunneled out from the floating gate into the channel via F–N tunneling.

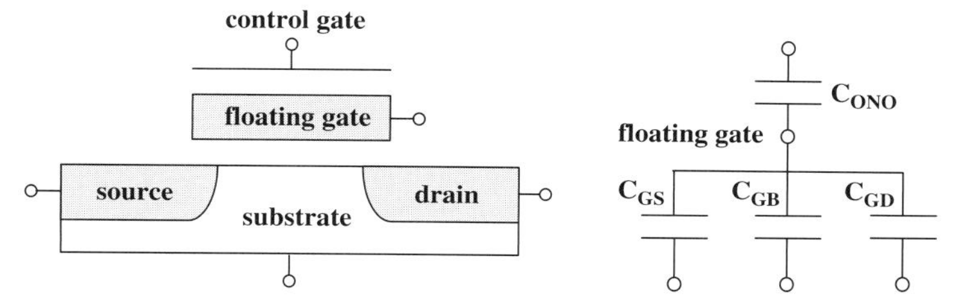

Figure 21.17 The equivalent capacitance at the floating gate. The storage node is connected in parallel to each terminal via respective capacitances.

definition of the capacitance as

$$C_{\mathrm{ONO}}(V_{\mathrm{CG}}-V_{\mathrm{FG}}) = (C_{GS}+C_{GB}+C_{GD})V_{\mathrm{FG}}, \tag{21.48a}$$

where V_{FG} is the voltage at the floating gate and the capacitances appearing in (21.48a) connect the storage node with respective terminals (Figure 21.17). Rearranging (21.48a) results in

$$V_{\mathrm{FG}} = \alpha_{\mathrm{CG}} V_{\mathrm{CG}}, \tag{21.48b}$$

where the coupling coefficient α_{CG} is defined in terms of the total capacitance as

$$\alpha_{\mathrm{CG}} = \frac{C_{\mathrm{ONO}}}{C_{GS}+C_{GB}+C_{GD}+C_{\mathrm{ONO}}} \equiv \frac{C_{\mathrm{ONO}}}{C_T}. \tag{21.48c}$$

One can repeat the same procedure by applying the bias at each terminal, while grounding the rest. One thus obtains from the principle of linear superposition,

$$V_{FG} = \alpha_{\mathrm{CG}} V_{\mathrm{CG}} + \alpha_S V_S + \alpha_B V_B + \alpha_D V_D + \frac{Q_{\mathrm{FG}}}{C_T}, \tag{21.49}$$

where the coupling coefficients are given by $\alpha_j = C_j/C_T$, with j denoting the control gate, source, bulk, and drain, respectively. The last term in (21.49) accounts for the excess charge present in the floating gate and contributing to the voltage therein. The electrically isolated floating gate acts as the storage node, but it also acts as the gate controlling the channel. However, its voltage V_{FG} is controlled at the control gate (see Figure 21.17).

For programming, the channel is turned on by applying an appropriate positive voltage at the control gate and the electrons inverted in the channel are injected into the storage node either via thermionic or field emission. Likewise, the erase is done by applying negative gate voltage and tunneling out the stored excess electrons from the floating gate into the channel via F–N tunneling.

The programmed or erased state is monitored simply by detecting the shift in threshold voltages in programmed and erased cells. This is illustrated in Figure 21.18. The threshold voltage, V_{TCG}, at the control gate is to be defined operationally as the value of V_{CG} at which the drain current of 1 μA at $V_D = 0.1$ V, for example, is attained. Thus, in the erased cell in which $Q_{FG} = 0$, the threshold voltage V_{TCGE} at the control gate is the voltage at which the voltage transferred to the floating gate V_{FG} induces the channel inversion to attain the specified drain current. In the presence of Q_{FG} in the programmed cell, however, it is clear from the figure that an additional voltage is needed at the control gate to compensate Q_{FG}, namely, $\Delta V = |Q_{FG}|/C_{ONO}$ to invert the channel and therefore the threshold voltage of the programmed cell is given by

$$V_{TCGP} = V_{TCGE} + |Q_{FG}|/C_{ONO}.$$

Therefore, reading consists of probing the cell at a gate voltage in between these two threshold voltages and detecting the drain current. Specifically, depending on whether $I_D \geq 1$ μA or $I_D \approx 0$, the cell is read as erased or programmed. Figure 21.18 also sketches the distribution in both programmed and erased threshold voltages. Obviously, making these two distributions of programmed and erased threshold voltages well separated to meet the given specifications constitutes one of the key process challenges, especially for high-density memory applications. In this manner,

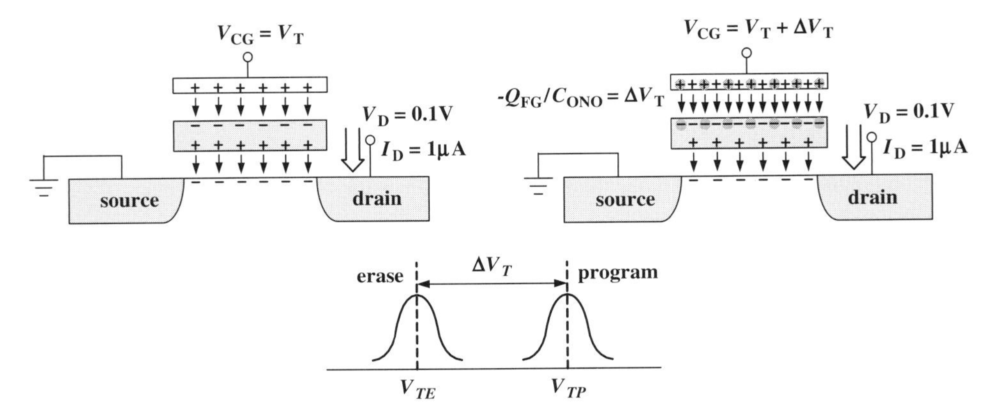

Figure 21.18 A graphical illustration of the shift in threshold voltages in programmed and erased cells. The presence or absence of excess electrons in the floating gate is responsible for the threshold voltage shift. Also shown are the respective V_T-distributions.

the basic physical mechanism such as tunneling, thermionic emission, or hot carrier effects are harnessed for useful applications.

21.4.3 CMOS Image Sensors

With the incorporation of storage nodes, the simple MOSFET structure has thus been shown to provide efficient means for realizing high-density memory cells. Moreover, an efficient image sensor can be made out of a few MOS devices working together. Figure 21.19 shows the simplified configuration of a pixel in CMOS image sensor. As well known, an image to be sensed is generally decomposed into dots of information, including the color information as selected by color filters. These basic units of information are transferred via strings of photons from the object to the pixels in the sensor.

As shown in this figure, a string of photons emanating from a dot in the object is impinging on the n^+–p photodiode, generating e–h pairs therein. The electrons and holes thus generated are transported to n^+ and p regions, respectively, driven by the built-in junction field, as detailed in Chapter 19. It is interesting to notice that the same n^+ region in the photodiode also serves as the source diffusion for NMOS device, called the transfer transistor. With the transfer gate (TG) voltage turned on, the barrier potential at the n^+ diffusion and p-substrate is lowered and the photogenerated

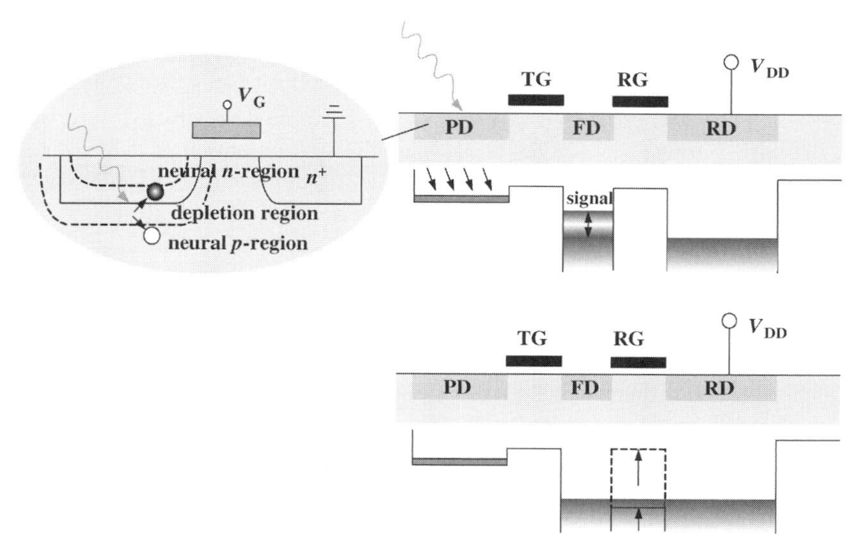

Figure 21.19 A simplified pixel configuration of the CMOS image sensor and energy band diagram in operation. The photogenerated electrons spill into the transfer gate, from n^+–p photodiode and/or source diffusion of TG transistor, and are stored in floating diffusion (top). After detection, the reset gate (RG) is opened to discharge the electrons (solid line at bottom figure) and is closed to initialize the potential at FD to repeat the detection process (broken line at the bottom).

electrons spill over into the channel and are transported to the drain to be stored therein.

To store these electrons, a capacitance is added to the drain or a stray capacitance naturally formed therein can be used as the storage node. In this way, the image information carried in by photons is converted into an electrical signal in the form of stored electron charge in floating diffusion (FD). This stored electrical signal is to be routinely monitored with the use of a source follower, which is a well-known circuitry for detection and consists essentially of two transistors connected in series. After monitoring the input information turned into electrical signal, the reset gate (RG) is turned on to discharge the stored electrons and simultaneously to initialize the voltage at the storage node, so that the next cycle of detection can be repeated.

The image information converted into electrical signal on a pixel level is in turn processed back into the original image by image processing techniques. The pixel itself consists of only four transistors, namely, the transfer and reset transistors for detecting and converting optical signal into electrical signal and two additional transistors used for monitoring the stored electrical signal. A few important physical parameters are involved in the performance of the CMOS image sensor. The quantum efficiency specifying the ratio of photogenerated electrons with respect to the incoming photon numbers, namely, $\mathrm{QE}(\lambda) = N_{\mathrm{ele}}/N_{\mathrm{photon}}$, is one of the key parameters. Equally important is the conversion efficiency representing the change in voltage in storage node induced by stored electrons, namely, $\Delta V_{FD} = qN_{\mathrm{ele}}/C_{FD}$ with C_{FD} denoting the diffusion capacitance.

In summary, the CMOS image sensor is based on innovative coupling of detection scheme for incoming photons and control scheme of electrons. The control of these electrons is in turn based on the utilizations of quantum wells via the voltages applied.

21.5 Problems

21.1. Consider the p-channel MOSFET on n-type silicon substrate.

(a) Starting from the general Poisson equation (21.10), derive the expression of the surface charge Q_S as a function of the surface potential φ_s for various donor doping levels (see (21.14)).

(b) Discuss the accumulation, depletion, weak, and strong inversion regions in PMOS system.

(c) Discuss the resulting C–V behavior.

(d) Derive the I–V characteristics of the p-channel MOSFET and compare the threshold voltage with that of n-channel MOSFET.

21.2. The drain current in MOSFETs increases linearly with the gate voltage, $I_D \propto V_G - V_T$, while the collector current in bipolar junction transistors grows exponentially with the base emitter voltage, that is, $I_C \propto \exp qV_{BE}/k_BT$. Discuss the physical reasons for this difference in output currents versus the input voltage.

21.3. Sketch the energy band diagrams at both thermal equilibrium and flat band for ideal MOS system consisting of (i) n^+ polysilicon as gate electrode and $1\,\Omega\,\text{cm}$ p-type silicon and (ii) p^+ polysilicon as gate electrode and $1\,\Omega\,\text{cm}$ n-type silicon. Take $\mu_n = 700\,\text{cm}^2/\text{V}\,s$ and $\mu_n = 400\,\text{cm}^2/\text{V}\,s$.

21.4. An n-channel MOSFET has the aspect ratio $W/L = 5$, the gate oxide thickness 50 nm, and the channel electron mobility $\mu_n = 600\,\text{cm}^2/\text{V}\,\text{s}$.

(a) Find the inverted electron density that is required for the MOSFET to be used as variable resistor of 5, 2.5, 1 kΩ at $V_{DS} = 0.1$ V.

(b) Find the required gate voltage $V_G - V_T$.

21.5. The drain current in NMOS is given in linear region by

$$I_D = \frac{W}{L} C_{\text{OX}}\mu_n\left(V_G - V_T - \frac{1}{2}V_D\right)V_D, \quad V_D \leq V_{\text{DSAT}} \equiv V_G - V_T. \tag{21.50}$$

The I_D can be formally expressed in terms of the channel voltage V at y from the source with the identification $V_D \rightarrow V(y)$ and $L \rightarrow y$, namely,

$$I_D = \frac{W}{y} C_{\text{OX}}\mu_n\left(V_G - V_T - \frac{1}{2}V(y)\right)V(y). \tag{21.51}$$

(a) Find the profile of the channel voltage by finding $V(y)$ from (21.51) in terms of the drain current I_D in (21.50).

(b) Find the channel field $E(y) = -\partial V(y)/\partial y$.

(c) Using the result of (b), find the transit time of the electron from source to drain,

$$\tau_{\text{tr}} = \int_0^L \frac{dy}{v_d} = \int_0^L \frac{dy}{\mu_n \text{E}(y)}.$$

21.6. (a) Find the threshold voltage in NMOS with n^+ polysilicon gate and in PMOS with p^+ polysilicon gate versus dopant concentrations in the substrate ranging from 10^{16} to $5 \times 10^{17}\,\text{cm}^{-3}$ for ideal MOS system where there is no surface state. Take the oxide thickness of 10 nm.

(b) If there is a fixed positive charge density of $10^{11}\,\text{cm}^{-3}$ at the oxide interface, how does it affect the threshold voltages?

Repeat the calculations in (a) in the presence of this positive charge density.

21.7. The floating gate in the flash memory cell is in essence a two-dimensional quantum well, as sketched in Figure 21.20.

(a) Find the lifetime of an electron confined in the well and moving with thermal velocity at room temperature as a function of well width W, ranging from 80–150 nm.

(b) Calculate the voltage applied at the floating gate with respect to the channel at which the lifetime of this electron is reduced to 1 μs.

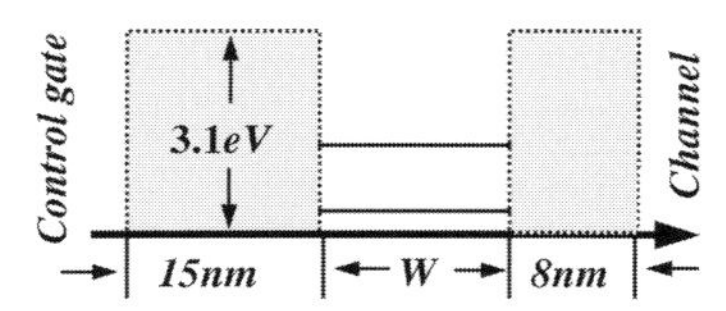

Figure 21.20 Cross-sectional view of the floating gate quantum well.

(c) Find the velocity v_z in the ground state of the well as a function of the well width. Approximate the quantum well by infinite square well potential of varying width W for simplicity of analysis.

(d) Calculate the width of the well at which $v_z = v_{zT}$ where v_{zT} is the z-component of the thermal velocity at room temperature.

(e) Calculate the lifetime of an electron in the ground state as a function of the well width ranging from 10 to 100 nm.

The lifetime can be defined by the condition $T\tilde{N} = 1$ where T is the tunneling probability of a particle encountering a potential barrier and $\tilde{N}$ is the total number of encountering the barrier during the lifetime. For simplicity of analysis, the subbands of the quantum well can be found analytically by taking the barrier height to be infinite and then using the actual barrier given in the figure for estimating the lifetime.

21.8. Consider the same floating gate as sketched in Figure 21.20. The cell is programmed by injecting excess electrons into the floating gate and the presence of excess electrons is monitored by measuring the shift in threshold voltage at the control gate. Calculate the number of excess electrons injected if the measured shift of threshold voltage is 2.5 and 5 V, respectively.

Suggested Reading

1 Muller, R.S., Kamins, T.I., and Chan, M. (2002) *Device Electronics for Integrated Circuits*, 3rd edn, John Wiley & Sons, Inc.

2 Sze, S.M. and Ng, K.K. (2006) *Physics of Semiconductor Devices*, 3rd edn, Wiley-Interscience.

3 Pierret, R.F. (1990) *Field Effect Devices, Modular Series on Solid State Devices*, 2nd edn, vol. IV, Prentice Hall.

4 Schroder, D.K. (1987) *Advanced MOS Devices, Modular Series on Solid State Devices*, vol. VII, Addison-Wesley Publishing Company.

5 Taur, Y. and Ning, T. H. (1998) *Fundamentals of Modern VLSI Devices*, Cambridge University Press.

22
Metal Oxide Silicon Field Effect Transistors II: Device Scaling and Schottky Contact

The MOS transistor has been continually downscaled for several decades to enhance both the device performance and integration. The scaling has been carried out, based upon the top–down approach, namely, etch, lithography, and self-oxidation. As a consequence, the channel length of the device has been successfully reduced deep into nanoregime, down to about 10 nm. At the same time, the limitations encountered in such scaling have prompted the bottom–up approach, utilizing the self-assembly and self-organization of atoms and molecules. Naturally, these constitute the essential core of quantum mechanics and of nanoscience and technology. The basic limiting factors for device downscaling are discussed, together with process innovations devised for overcoming those limitations. In addition, the device structures evolving from 3D bulk to 1D wire are discussed from the standpoint of device downscaling. With 1D and molecular devices rapidly advancing, the Schottky contact is becoming increasingly important and is singled out for discussion.

22.1
Device Scaling: Physical Issues and Limitations

22.1.1
Constant Field Scaling

As pointed out, the overwhelming advantage of MOSFET is its simple structure and scalability. When successfully scaled, the performance of the device is improved, and with it the functionality of the integrated circuits. Also, the cost of fabrication is lowered. This virtuous circle has been kept up and manifested itself in the celebrated Moor's law as a landmark achievement of semiconductor technology in the latter half of the twentieth century. However, the scaling process has encountered a number of quantum mechanical and other basic limitations and these limitations are briefly discussed to provide a proper perspective of nanoscale devices.

A useful guideline for device downsizing has been the constant field scaling, as indicated in Figure 22.1. In this approach, both the operating voltage $V(V_G, V_D)$ and

Introductory Quantum Mechanics for Semiconductor Nanotechnology. Dae Mann Kim

ISBN: 978-3-527-40975-4

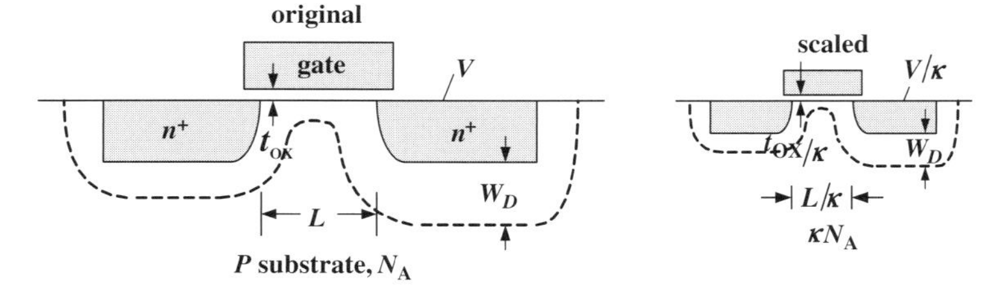

Figure 22.1 The schematic illustration of the constant field scaling of MOSFETs.

the channel length (L) are scaled simultaneously by a factor $\varkappa$, so that the longitudinal channel field $\propto V_D/L$ remains constant. Other parameters scaled are the width (W) of the channel, the oxide thickness (t_{OX}), the depth (x_j) of source and drain implants, and dopant depletion depth (W_D) therein. On the other hand, the substrate doping, N_A or N_D are increased by $\varkappa$ instead (Figure 22.1).

The scaling of these device parameters leads to interesting consequences. For example, the current drivability

$$I_D \approx (W/L)C_{OX}(V_G - V_T)V_D, \quad C_{OX} = \varepsilon_S/t_{OX},$$

is, in effect, reduced by $\varkappa$. The total capacitance C ($\equiv WLC_{OX}$) under the gate is also decreased by $\varkappa$, leading to an important advantage, namely, the reduced circuit delay time τ ($\approx CV/I_D$) by $\varkappa$. In addition, the power dissipation P ($= VI$) drops by $\varkappa^2$, and the power delay product $P\tau$ decreases by $\varkappa^3$, while the power density, namely, the power dissipated per area remains the same. On the other hand, the scaling of these device parameters leads to reduced dynamic range due to scaled supply voltage and more seriously the short channel effects.

Short Channel and Reverse Short Channel Effects The shortened channel length L brings in a number of adverse effects, one of which is the threshold voltage roll-off. The physical mechanism responsible for decreased V_T is illustrated in Figure 22.2 for $V_D = 0$, for simplicity. The V_T roll-off is caused essentially by the edge effect, which is amplified by shortened L. The channel inversion necessitates, as detailed in Chapter 21, a surface band bending that is supported by the uncompensated dopant ions near the surface. The same dopant ions at the channel edge should also support the junction band bending, inherent in n^+–p junctions at source and drain.

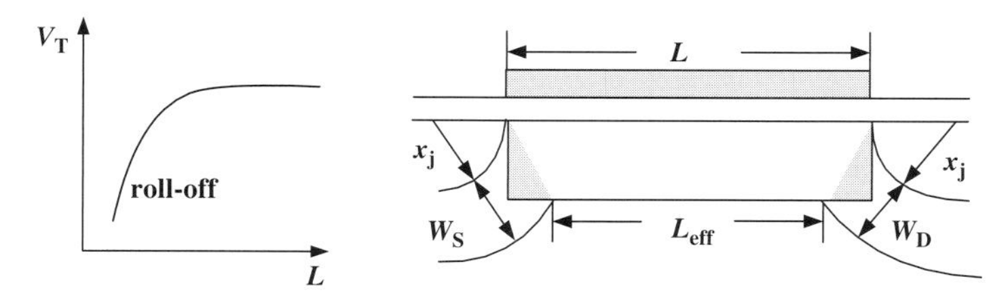

Figure 22.2 The threshold voltage roll-off and sharing of the depletion charge for both junction band bending at source and drain and the surface band bending for channel inversion.

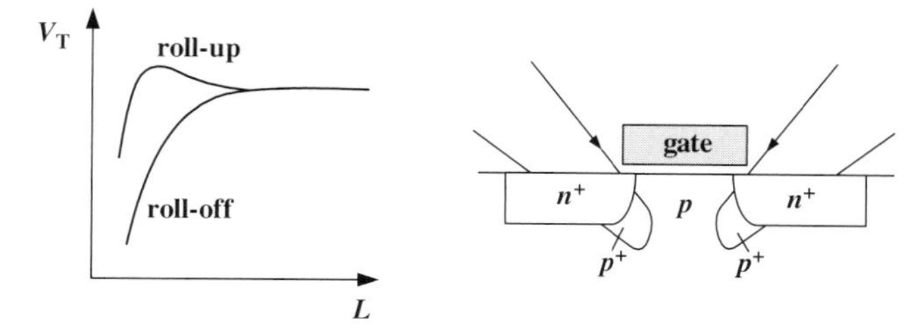

Figure 22.3 The threshold voltage roll up and p^+ pocket implants.

Therefore, those dopant charges at the channel edge and within the depletion depth of the source and drain junctions are to be shared for two kinds of band bending. This means that the gate field lines emanating from V_G are terminated only by those dopant ions in the trapezoidal area under the gate, as shown in Figure 22.2. The reduced number of dopant ions involved in the channel inversion points to reduced V_T. Naturally, this edge effect induced V_T roll-off would be negligible when the channel length is long enough. However, with decreased L, it becomes a tangible effect to contend with.

It is also worth noting that with further reduction of L, V_T could roll up instead, as indicated in Figure 22.3. The V_T roll-up is, in essence, an artifact arising from the channel engineering. An important part of the channel engineering is the pocket p^+ implants (Figure 22.3), which have been devised for preserving the neutral substrate bulk and for sustaining the device operation with shortened L. The pocket implant effectively raises the substrate doping, resulting in increased V_T, as detailed in Chapter 21 (See Figure 21.5). The resulting increase in V_T is called the reverse short channel effect.

Punchthrough Effect One of the most fundamental and obvious limitations for device scaling originates from the fact that there should be a channel to be controlled. Equivalently, a neutral substrate bulk should exist under the gate electrode, so that the source and drain are kept electrically isolated. A most critical concern in this regard is the substrate punchthrough effect, as illustrated in Figure 22.4. With V_D on and the drain p–n^+ junction reverse biased, the junction depletion region therein is broadened and the two junction boundaries at source and drain could touch with each other, especially with L shortened. In this case, the neutral bulk under the gate disappears. Moreover, the electrons may punchthrough from source to drain, adding to off currents and degrading the on to off current ratio. The effect can, however, be

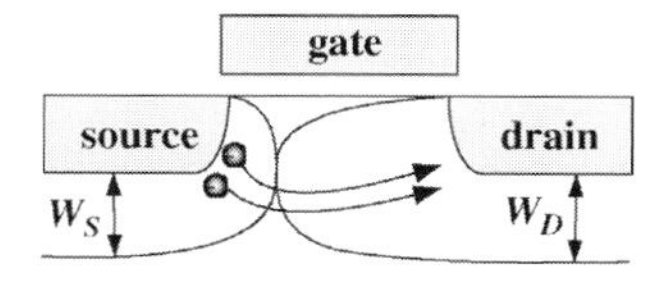

Figure 22.4 The punchthrough effect: the joining of source and drain depletion boundaries, the disappearance of bulk substrate, and the electron emission from the source through the depleted substrate.

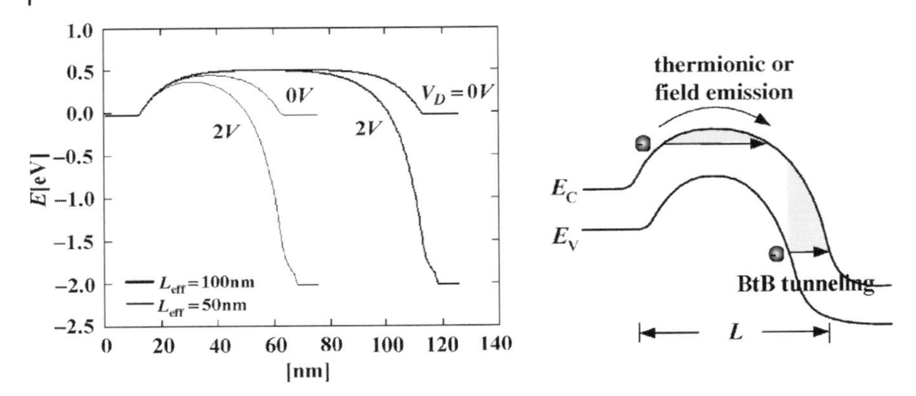

Figure 22.5 The drain voltage induced barrier lowering at source junction in short channel device (left) and the resulting thermionic and field emission of electrons from the source and band-to-band tunneling near the drain (right).

circumvented in part via higher substrate doping but the higher doping could in turn lead to reverse short channel effect, as discussed, and degradation of carrier mobility due to enhanced impurity scattering.

Drain-Induced Barrier Lowering With L shortened, V_D applied at the drain could pull down the source $p\!-\!n^+$ junction barrier height, as shown in Figure 22.5. This is known as the drain-induced barrier lowering (DIBL) and could affect the threshold voltage. More seriously, the electrons in the source terminal may tunnel through or thermally emit across the lowered barrier to the drain. Or the body to drain emission of electrons could ensue via the band-to-band tunneling across the energy gap near the drain (Figure 22.5). All of these processes are uncontrolled and add to off currents, degrading the device performance.

Moreover, the hot carrier effect discussed in Chapter 21 is further enhanced with the shortening of the channel length. In addition, with decreased L, the carrier mobility could become field dependent. Inasmuch as the drift velocity is the output of the input field with the mobility constituting the response function, as detailed in Chapter 1, that is,

$$\mathbf{v}_D = \mu_n(\mathrm{E})\mathbf{E}, \quad \mathbf{v}_D = \mu_p(\mathrm{E})\mathbf{E}$$

the mobility should in general depend on the input field strength. This is explicitly shown in the data presented in Figure 22.6, where the measured drift velocity of electrons is plotted versus the applied field in different semiconductors. The field-dependent mobility could obviously affect the on current.

Power Consumption and Gate Leakage Additional fundamental limitation for scaling the device is the issue of power consumption, active as well as standby. The standby power is increasing faster than the active one with the reduction of device dimensions and is contributed primarily by the gate leakage through the thin gate oxide. Although the scaling of t_{OX} is essential for controlling the channel by

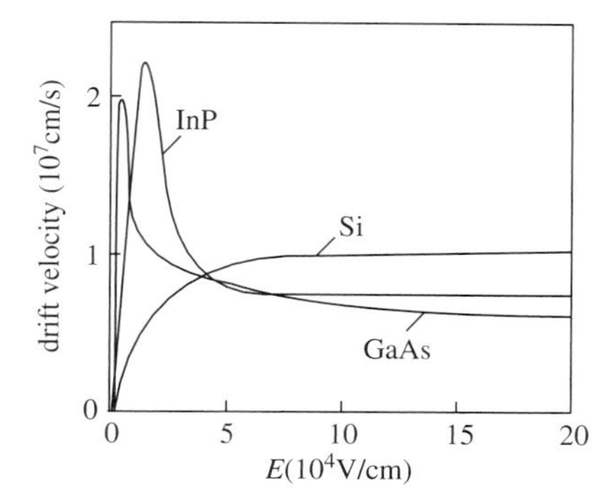

Figure 22.6 The drift velocity of electrons versus applied electric field measured from Si, GaAs, and InP. ($N_D = 10^{17}$ cm^{-3}, $T = 300$ K) (taken from Sze, S.M and Ng, K.K. (2006) *Physics of Semiconductor Devices*, Wiley-Interscience.).

the gate via reduced V_G, thinner oxide results in enhanced gate current. The physical mechanism for such gate leakage I_{GL} is the well-known Fowler–Nordheim or direct tunneling as detailed in Chapter 6:

$$I_{\mathrm{GL}} \propto \exp\left(-\frac{\eta\varphi_B^{3/2}}{\mathrm{E_{OX}}}\right), \quad \eta = 4(2m_n)^{1/2}/3q\hbar,$$

where $\mathrm{E_{OX}}$ is the electric field in the oxide and φ_B the oxide barrier encountered by tunneling electrons, as shown in Figure 22.7 (see (6.11) and (6.12)).

The scaling of t_{OX} leads to increased gate capacitance $C_{\mathrm{OX}}(\propto 1/t_{\mathrm{OX}})$, which in turn reduces the fraction of V_G partitioned into V_{OX}. However, $\mathrm{E_{OX}}(= V_{\mathrm{OX}}/t_{\mathrm{OX}})$ still makes the gate leakage grow exponentially with decreased t_{OX}. Thus, the search for and utilization of new dielectric materials with high dielectric constant $K(= \varepsilon/\varepsilon_0)$ is necessitated for continued device scaling. With high K material, the drain current

$$I_D \propto C_{\mathrm{OX}} \propto K/t_{\mathrm{OX}}$$

can be controlled via scaled V_G, while keeping t_{OX} to a reasonable thickness for suppressing the gate leakage.

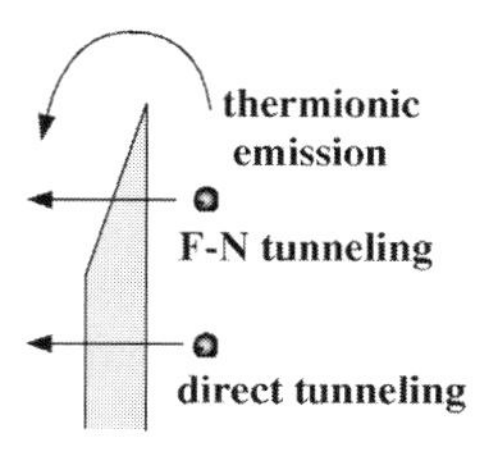

Figure 22.7 The sketch of thermionic emission over the barrier and F–N or direct tunneling of electrons through triangular and trapezoidal potential barriers.

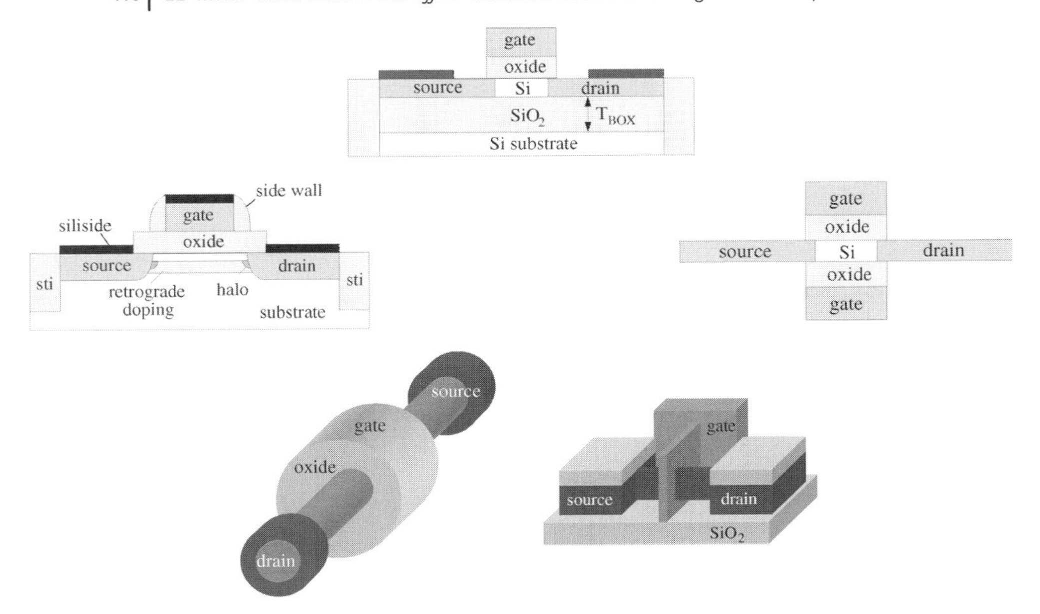

Figure 22.8 The evolving device structures from planar to single-gate SOI to double-gate SOI to FinFET to quantum wire. Also sketched are the process innovations devised for the planar structure.

Process Innovations and Device Structure Evolution There have been devised extensive innovations in process technologies for circumventing the limitations in device scaling. Concomitantly, the device structures have also evolved from 3D bulk to 1D quantum wire. This is summarized in Figure 22.8. It is clear from the figure that the channel engineering has been devised extensively to cope with scaling limitations and for minimizing off currents, while maximizing on currents. For instance, the super steep retrograde well was introduced for suppressing the substrate punchthrough and minimizing the off current via reduced depletion depths on both sides of the channel. Also, the halo implant and the source drain extension have been devised to minimize the short channel effects. However, the channel engineering also leads to the degradation of carrier mobility via enhanced impurity scattering entailed in higher doping and it also results in higher junction capacitance and leakages.

Concomitant with these process innovations, novel device structures have been designed, as sketched in Figure 22.8. A feature common to all of these device structures is centered around reducing the substrate bulk, which acts as the source for various adverse effects in downsizing the device dimension. In silicon on oxide (SOI) technology, for example, a thin silicon substrate is placed on top of the oxide layer and in so doing, the substrate punchthrough and junction leakages are effectively eliminated. This is because there is no current path away from the gate, so that the stray or leakage current is limited. Moreover, the channel can be made of near intrinsic silicon, thereby attaining higher carrier mobility via reduced impurity scattering. More important, the adverse effects of the dopant fluctuation can be suppressed and the uniformity of device performance can be enhanced.

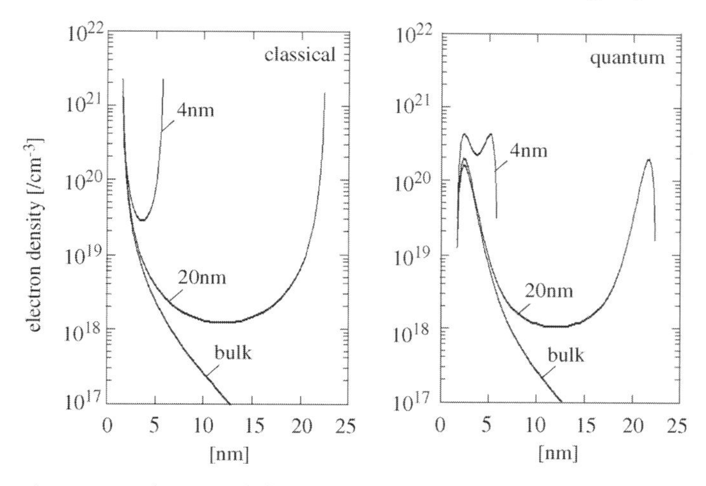

Figure 22.9 The inverted electron density profiles resulting from classical (left) and quantum mechanical (right) simulations in 3D bulk and 1D quantum wires with diameters of 4 and 20 nm, respectively ($N_A = 3 \times 10^{16}\,\mathrm{cm}^{-3}$; $t_{OX} = 1.5\,\mathrm{nm}$).

The near intrinsic silicon can be utilized in SOI technology, since the electrons can be directly coupled into the channel via the gate voltage in this geometry without the need for the dopant charge supporting the band bending for channel inversion. Moreover, the tighter capacitive coupling leads to steeper subthreshold slope and better control of short channel effects. As shown in the figure, the SOI device structure has evolved from the single gate to double gate to triple gate FinFET and ultimately to 1D quantum wire structures.

Figure 22.9 shows the electron density profiles, obtained from the classical and quantum mechanical simulations, which have been performed using the same process parameters for both 3D bulk and 1D quantum wire structures. Clearly, the classical simulation results reveal the planar nature of channel inversion sharply peaking at the surface, regardless of the device structures. However, the quantum results clearly indicate the transformation taking place from the planar inversion in bulk structure to the volume inversion in quantum wire structures. This transformation is shown particularly pronounced for smaller diameter quantum wires. Naturally, the striking differences existing between classical and quantum profiles of inverted electrons are due to the wave nature of electrons the wavefunctions of which yield the peak probability away from the interface.

The corresponding C–V curves are shown in Figure 22.10. In all of these C–V curves, the accumulation, depletion, weak, and strong inversion regimes are clearly manifested. Also, for the given device parameters and bias conditions, the capacitance values resulting from the classical simulations are consistently higher than the corresponding values obtained with quantum simulations. This is understandable, since the total capacitance consists of the oxide and surface capacitances, connected in series, namely,

$$\frac{1}{C} = \frac{1}{C_{OX}} + \frac{1}{C_S}$$

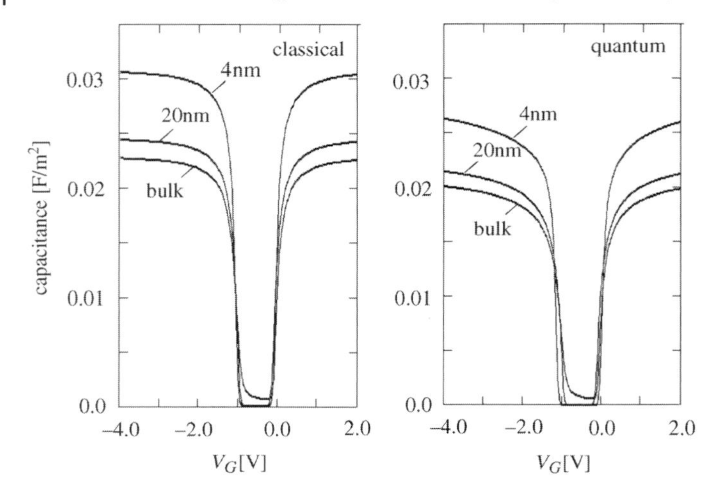

Figure 22.10 The *C–V* characteristics resulting from the classical (left) and quantum mechanical (right) simulations in 3D bulk and 1D quantum wires with diameters of 4 and 20 nm, respectively ($N_A = 3 \times 10^{16}\ \mathrm{cm}^{-3}$; $t_{OX} = 1.5$ nm).

(see (21.19)), and C_S for classical charge inversion is larger than that of quantum inversion. This can be clearly seen from the respective inverted electron profiles in Figure 22.9. In fact, the *C–V* behaviors shown in Figure 22.10 are entirely consistent with the corresponding profiles of the inverted electrons in Figure 22.9.

The discussions made in connection with Figures 22.9 and 22.10 can be summarized as follows. First, the difference between quantum and classical channel inversions becomes more pronounced as the device structure is changed from 3D bulk to 1D wire, in particular to smaller diameter quantum wires. Also, the efficiency of channel inversion is significantly improved as the structure evolves from 3D bulk to 1D wire. Finally, the volume channel inversion in smaller diameter quantum wires suggests that the inverted electrons therein are relatively free of surface scattering and the resulting degradation in mobility.

22.2 Metal–Semiconductor Contacts

22.2.1 The Schottky Contact

Traditionally, the metal–semiconductor contact has played an important role in microelectronics and it is expected to play an equally important role in nanoelectronics as well. The semiconductor devices processed in batch and assembled together in integrated circuits are connected via such contacts and the integrated circuit itself communicates with rest of the system through such contacts. A feature unique to this metal–semiconductor contact is its ability to exhibit both active diode and passive ohmic behaviors. Inasmuch as the nanodevices have nanoscale channel lengths, and correspondingly small channel resistance, it is important to come up with a contact resistance much smaller than the channel resistance. In this regard, the metal–semiconductor contact is an essential element of nanoelectronic systems and is discussed in this section.

The Equilibrium Contact Without Surface States The metal–semiconductor contact and the contact potential inherent therein were analyzed by Schottky, based on the space charge formed at the interface. The contacts are thus called Schottky contacts and are discussed in this section both in equilibrium and nonequilibrium under bias and also with and without surface states.

Thus, consider a metal, for example, Au in equilibrium contact with semiconductor, for example, n-type silicon, in the absence of interface states. In this example, the metal work function is greater than that of n-type silicon by about 0.25 eV or more, as sketched in Figure 22.11. Upon equilibrium contact, the two Fermi levels E_{Fm} and E_{FS} should be lined up and flat throughout, as detailed in Chapter 19. This basic equilibrium condition together with those conditions for preserving the bulk properties of gold and silicon away from the interface necessitates the band bending. The resulting band bending gives rise to two barriers, one encountered by electrons in the metal at the Fermi level, $q\varphi_{Bn}$ called Schottky barrier and the other encountered by electrons in the silicon conduction band, namely, the usual built-in barrier potential $q\varphi_{\text{bi}}$. From the geometry of the energy level diagram shown in Figure 22.11, one can readily specify these two barriers in terms of E_{Fm} and E_{FS}, the metal work function $q\varphi_m$, and the affinity factor $q\chi$ as

$$q\varphi_{Bn} \equiv q\varphi_m - q\chi, \tag{22.1}$$

$$q\varphi_{\text{bi}} \equiv E_{FS} - E_{Fm} = q\varphi_m - q(\chi + V_n), \tag{22.2}$$

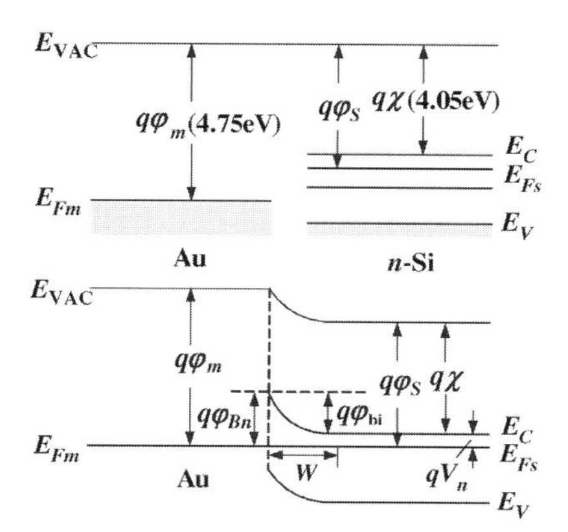

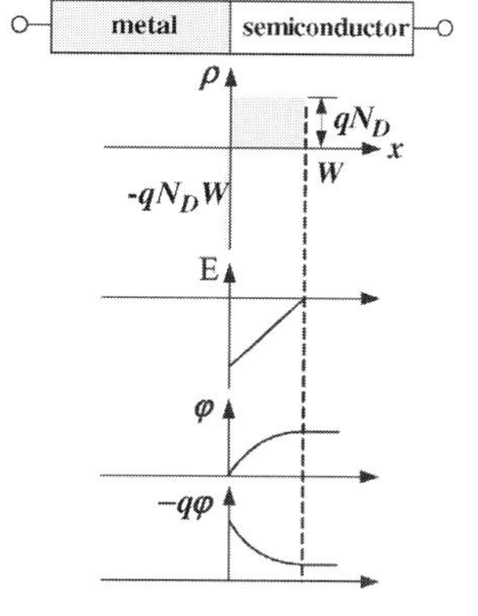

Figure 22.11 The energy band diagram of the metal (Au)–semiconductor (n-type Si) system before and after equilibrium contact (left). Also shown are Schottky ($q\varphi_{Bn}$) and built-in ($q\varphi_{\text{bi}}$) potential barriers for electrons and the space charge, field, potential, and electron potential energy responsible for band bending (right).

where the quantity $qV_n = E_C - E_{FS}$ specifies the position of E_{FS} with respect to E_C in silicon n-bulk and is determined strictly by the doping level N_D.

Naturally, the required band bending is supported by the space charge at the interface, resulting from the exchange of electrons between the two subsystems. For the case under consideration, E_{FS} lies above E_{Fm} before contact, and electrons are therefore transferred from silicon to Au, leaving behind donor ions uncompensated. Once transferred into the metal, the excess electrons reside at the metal surface. Thus, the space charge ϱ consists of the surface charge sheet of electrons on the metal side and the positive charge of ions in the semiconductor in the surface depletion depth W.

Once ϱ is known, the space charge field E, potential φ, and the electron potential energy $-q\varphi$ can be routinely obtained for quantifying the band bending. For example, the built-in barrier $q\varphi_{bi}$ represents the total area under the E-curve, as discussed in Chapter 19. In fact, the analysis is identical to that of p^+-n step junction with heavily doped p-type silicon replaced by the metal. The surface depletion depth thus reads as

$$W = (2\varepsilon_S \varphi_{bi}/qN_D)^{1/2}, \tag{22.3}$$

since the case corresponds to $N_A \gg N_D$. (see (19.8b)).

One can likewise consider the equilibrium contact of Au with a p-type silicon by interchanging the role of electrons with that of holes. As shown in Figure 22.12, E_{Fm} lies above E_{FS} in this case before contact and electrons are therefore transferred from Au to silicon, leaving behind surface hole charge sheet. This is equivalent to holes being transferred from silicon to Au, leaving behind the acceptor ions uncompensated. Once in the metal, holes reside only at the surface of the metal. As a consequence, the space charge ϱ has the same dipolar shape as in Figure 22.11, however, with opposite polarity and the band bends down. The two barriers for holes become operative, one on the metal side at E_{Fm} and the other on the semiconductor

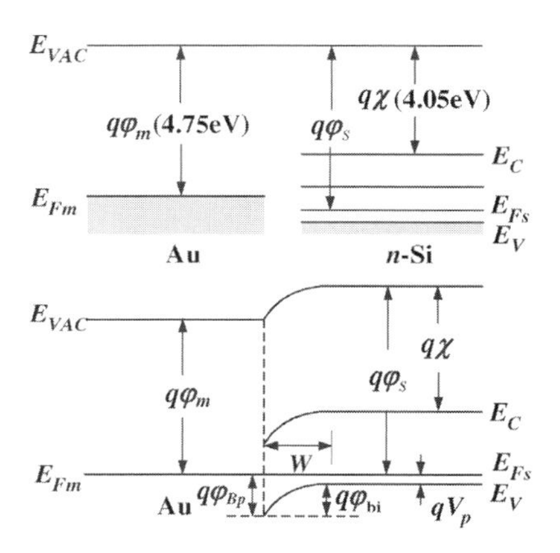

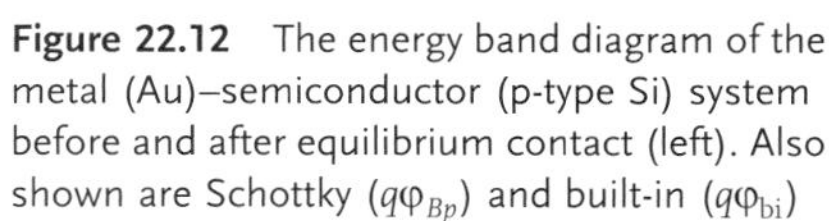

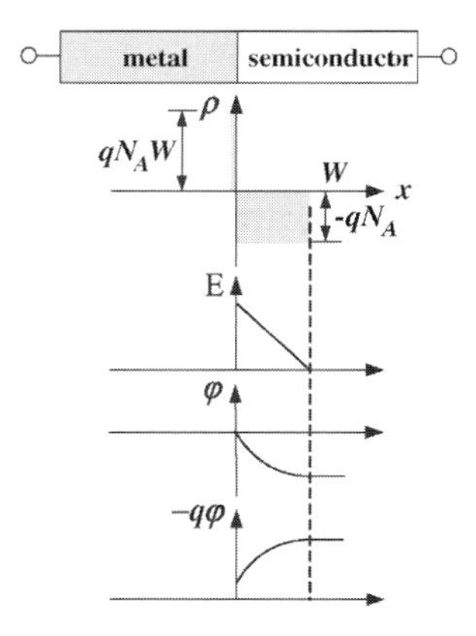

Figure 22.12 The energy band diagram of the metal (Au)–semiconductor (p-type Si) system before and after equilibrium contact (left). Also shown are Schottky ($q\varphi_{Bp}$) and built-in ($q\varphi_{bi}$) potential barriers for holes and the space charge, field, potential, and electron potential energy responsible for band bending (right).

side at E_V, as can be clearly seen from Figure 22.12:

$$q\varphi_{Bp} = q\chi + E_G - q\varphi_m \tag{22.4}$$

$$q\varphi_{bi} = E_{FS} - E_{Fm} = q\chi + E_G - qV_p - q\varphi_m, \tag{22.5}$$

with $qV_p = E_{FS} - E_V$ denoting this time the position of the Fermi level with respect to E_V in silicon p-substrate and determined by N_A. The depletion depth corresponds in this case to n^+–p step junction in which $N_D \gg N_A$, and one can write

$$W = (2\varepsilon_S \varphi_{bi}/qN_A)^{1/2}. \tag{22.6}$$

The Equilibrium Contact with Surface States At the metal–semiconductor interface, localized surface states are inherently present in the bandgap due to the broken periodicity of the semiconductor. That is, the electrons at the surface are bonded only from the bulk side and the electronic states at such sites should be different, in general, from those in the bulk. This kind of surface states are called Tam or Shockley states. In this section, only acceptor-like surface states are considered, for simplicity, in which the state is neutral when empty and negatively charged when filled with electrons. In addition, a thin surface layer of atomic dimension is formed in between the metal and the semiconductor and the effect of the surface states in the presence of such surface layer is discussed.

The roles of the surface states are illustrated in Figure 22.13 for the case of Au-contacted n-type silicon, the same example which has been considered in the absence of surface states (Figure 22.11). The electrons now fill up the surface states, even before the contact, from the bottom of the bandgap up to E_{FS}, in accordance with the Fermi occupation factor (see (16.38)). Also, the electrons when trapped in surface states leave behind the donor ions near the surface uncompensated, giving rise to the

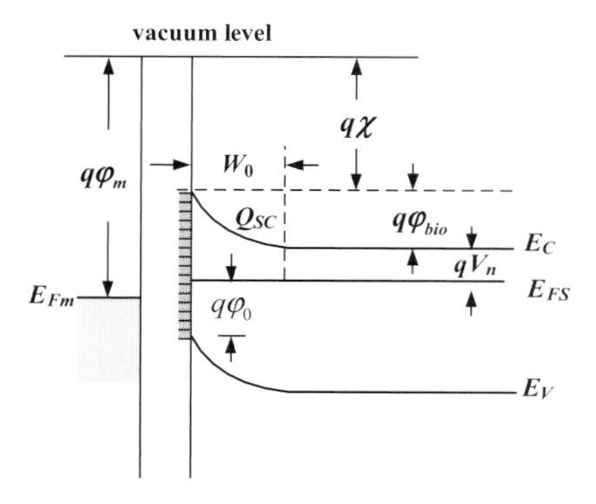

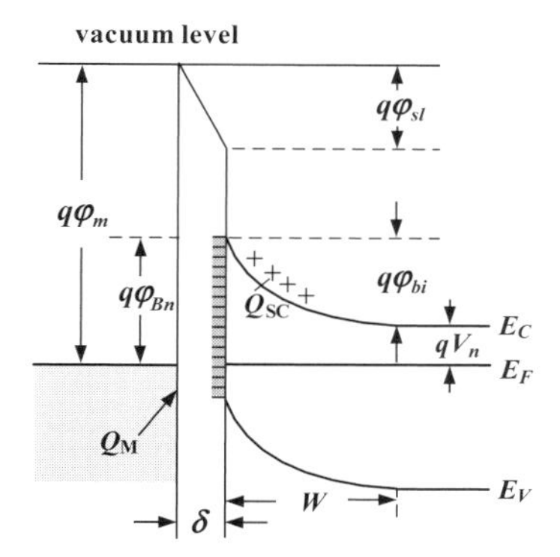

Figure 22.13 The energy band diagram of the metal (Au)–semiconductor (n-type Si) system before and after equilibrium contact in the presence of surface states. The intrinsic band bending before contact due to electrons trapped at surface states (left), and the band bending after contact due to the difference in Fermi levels or two work functions (right). The voltage drop in the surface layer is also shown.

space charge and ultimately the band bending (Figure 22.13). The resulting space charge in the depletion depth W_0 yields the usual band bending and built-in potential φ_{bi0}:

$$\varphi_{\text{bi0}} = qN_D W_0^2/2\varepsilon_S. \tag{22.7}$$

Thus, before the contact, the surface states below E_F are filled up with electrons supplied from the depletion region, that is,

$$qN_D W_0 = qD_S(E_G - q\varphi_{\text{bi0}} - qV_n), \tag{22.8}$$

where D_S is the density of surface states per area (cm^2) per energy (eV) and is taken constant, for simplicity of discussion. Obviously, the quantity in the parenthesis on the right-hand side of (22.8) represents the portion of the energy gap lying below E_{FS} (Figure 22.13).

Prior to contact, both W_0 and φ_{bi0} are determined self-consistently from (22.7) and (22.8) in terms of N_D and D_S and the filled surface states range from the bottom of the energy gap to the level

$$q\varphi_0 \equiv N_D W_0/D_S = E_G - q\varphi_{\text{bi0}} - qV_n \tag{22.9}$$

and $q\varphi_0$ is determined solely by N_D and D_S (Figure 22.13).

After the contact, the difference between E_{FS} and E_{Fm} induces the usual transfer of electrons from silicon to Au. Consequently, the band bending is adjusted to make two Fermi levels line up (Fig. 22.13). The accompanying charge redistribution leads to a new voltage drop across the surface layer and/or new built-in potential and the depletion depth, that is

$$\varphi_{\text{bi}} = qN_D W^2/2\varepsilon_S. \tag{22.10}$$

In the process the filled surface states lying below E_{FS} are readjusted and the difference in electron surface charge before and after the contact is thus given by

$$Q_{SS} = -qD_S[E_G - q(\varphi_{\text{bi}} + V_n) - q\varphi_0] \tag{22.11}$$

(see Fig. 22.13), while electrons transferred to the metal end up at the surface of the metal. Combining (22.9) and (22.11) the electron surface charge can also be expressed in terms of the difference in built-in potentials before and after the contact:

$$Q_{SS} = q^2 D_S(\varphi_{\text{bi}} - \varphi_{\text{bi0}}). \tag{22.12}$$

Also, the uncompensated dopant charge in the surface depletion depth is now given from (22.10) by

$$Q_{SC} = qN_D W = (2q\varepsilon_S N_D \varphi_{\text{bi}})^{1/2}. \tag{22.13}$$

The charge on the metal surface can then be specified from the charge neutrality condition, namely,

$$Q_m = -(Q_{SS} + Q_{SC}) \tag{22.14}$$

Now, the difference between the two Fermi levels or equivalently between two corresponding work functions is partitioned into the interface layer and the silicon, namely,

$$\varphi_m - (\chi + V_n) \equiv \Lambda = \varphi_{sl} + \varphi_{bi}, \tag{22.15}$$

where the potential drop across the surface layer φ_{sl} is given from the well-known electrostatic theorem by

$$\varphi_{sl} = -(Q_m/\varepsilon_i)\delta, \tag{22.16}$$

with ε_i and δ denoting the permittivity and layer thickness.

Hence, rewriting (22.15) using (22.12)–(22.16) results in a quadratic equation for $\varphi_{bi}^{1/2}$:

$$\left(1 + \frac{\delta q^2 D_S}{\varepsilon_i}\right)\varphi_{bi} + \frac{\delta}{\varepsilon_i}(2q\varepsilon_S N_D)^{1/2}\varphi_{bi}^{1/2} - \left(\Lambda + \frac{\delta q^2 D_S \varphi_{bi0}}{\varepsilon_i}\right) = 0. \tag{22.17}$$

Fermi Level Pinning One can readily solve (22.17) and find φ_{bi} explicitly to complete the analysis of the surface states affecting the equilibrium contact. But the essential role of the surface states can be brought out directly without solving the equation by taking a nonessential simplification, namely, by letting $\delta \to 0$. In this case, the built-in potential simply reads as

$$\varphi_{bi} \simeq \frac{\Lambda + (\delta q^2 D_S/\varepsilon_i)\varphi_{bi0}}{1 + \delta q^2 D_S/\varepsilon_i}. \tag{22.18}$$

It is therefore clear from (22.18) that in the limit of small surface state density, that is, $D_S \to 0$, $\varphi_{bi} \simeq \Lambda$ and the band bending or the built-in potential is thus strictly dictated by the difference between two Fermi levels, as it should. The barrier potentials φ_{Bn} and φ_{bi} then sensitively depend on the metal work function. On the other hand, for large surface state density, that is, $D_S \to \infty$, $\varphi_{bi} \simeq \varphi_{bi0}$. This indicates that the Fermi level is in this case pinned at the level set before the contact, regardless of the metal work function. This effect is known as the pinning of Fermi level.

22.3 Metal–Semiconductor *I–V* Behavior

22.3.1 Schottky Diode *I–V*

Under bias, the metal–semiconductor contact exhibits a rectifying behavior just like the p–n junction diode. To discuss this *I–V* behavior, consider again the Au-contacted n-type silicon as an example. When a positive or negative voltage V is applied to Au with respect to silicon, as shown in Figure 22.14, E_{Fm} therein is lowered or raised with respect to E_{FS}. But the two work functions remain unchanged, and the Schottky barrier $q\varphi_{Bn}$ also remains intact. To meet these conditions, the built-in potential $q\varphi_{bi}$ should decrease or increase, depending on the polarity of V (Figure 22.14).

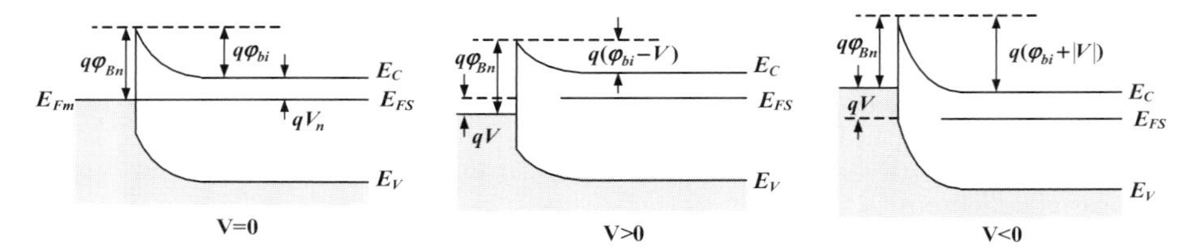

Figure 22.14 The energy band diagrams of Schottky contact both in equilibrium (left) and nonequilibrium under positive (middle) and negative (right) biases for Au-contacted n-type silicon, in which $E_{FS} > E_{Fm}$.

Consequently, the detailed balancing of the electron transfer between the two subsystems is broken and the current flows.

Thermionic Emission Theory There are a few different approaches for describing Schottky diode I–V behavior. In the thermionic emission theory by Bethe, the Schottky barrier $q\varphi_{Bn}$ is taken larger than the thermal energy k_BT. Also, the carriers are taken free of collisions in the interface depletion region. Under these conditions, the electrons are to be emitted from semiconductor to metal by overcoming the built-in potential barrier energetically. To analyze the current, consider first the simple case of equilibrium, that is, for $V = 0$. The current density resulting from the emission of electrons from semiconductor to metal is given by

$$J_{sm}(V=0) = q\int_{E_B}^{\infty} dEg_{3D}(E)f(E)v_x, \quad E_B = E_{FS} + qV_n + q\varphi_{\text{bi}}, \tag{22.19}$$

where

$$g_{3D} = \frac{\sqrt{2}m_n^{3/2}}{\pi^2\hbar^3}(E-E_C)^{1/2} \tag{22.20a}$$

is the 3D density of states derived in (4.27), with the electron rest mass replaced by its effective mass m_n in the semiconductor and the kinetic energy of the electron is taken from the bottom of the conduction band, that is,

$$E-E_C = \frac{1}{2}m_n v^2. \tag{22.20b}$$

For the nondegenerate case to which the analysis is confined, the Fermi occupation factor simply reduces to the Boltzmann probability factor (see (16.38)), namely,

$$f(E) \simeq \exp-\frac{E-E_F}{k_BT} = \exp-\frac{E-(E_C-qV_n)}{k_BT}. \tag{22.20c}$$

Also, E_B represents the minimum energy required for overcoming the built-in potential barrier $q\varphi_{\text{bi}}$ (Figure 22.14).

Inserting (22.20) into (22.19), transforming the variable of integration from $E-E_C$ to v via (22.20b), and finally converting the volume element in velocity space from the spherical to Cartesian frame of references, namely, $4\pi v^2 dv = dv_x dv_y dv_z$, one can write

$$J_{sm}(V=0) = \frac{qm_n^3}{4\pi^3\hbar^3} e^{-qV_n/k_BT} \int_{-\infty}^{\infty} dv_y e^{-\beta v_y^2} \int_{-\infty}^{\infty} dv_z e^{-\beta v_z^2} \int_{v_B}^{\infty} dv_x v_x e^{-\beta v_x^2}, \quad \beta \equiv \frac{m_n}{2k_BT}, \tag{22.21a}$$

where v_B represents the minimum velocity, required for overcoming $q\varphi_{bi}$, namely,

$$\frac{1}{2} m_n v_B^2 = q\varphi_{bi}. \tag{22.21b}$$

One can readily perform the integrations in (22.20a) using (1.24), obtaining

$$\begin{aligned} J_{sm}(V=0) &= A_n^* T^2 e^{-qV_n/k_BT} e^{-q\varphi_{bi}/k_BT}, \; A_n^* = \frac{qm_n k_B^2}{2\pi^2\hbar^3}, \\ &= A_n^* T^2 e^{-q\varphi_{Bn}/k_BT}, \quad \varphi_{Bn} = \varphi_{bi} + V_n, \\ &\equiv J_{ms}(V=0). \end{aligned} \tag{22.22}$$

The constant A^* thus introduced is called the Richardson constant and the detailed balancing at equilibrium assures that $J_{sm}(V=0) = J_{ms}(V=0)$.

Next, when an external voltage is applied, the two Fermi levels E_{FS}, E_{Fm} should split to accommodate the bias, while the Schottky barrier $q\varphi_{Bn}$ determined solely by the difference between the two work functions remains unchanged, as shown in Figure 22.14. These boundary conditions clearly necessitate the change in built-in potential $q\varphi_{bi}$ under bias, namely,

$$q\varphi_{Bn} = qV + qV_n + q\varphi_{bi}(V). \tag{22.23}$$

That is, $\varphi_{bi}(V)$ decreases or increases in the presence of the positive or negative bias at the metal:

$$\begin{aligned} q\varphi_{bi}(V) &= q\varphi_{Bn} - qV_n - qV \\ &= q\varphi_{bi} - qV, \quad q\varphi_{bi} \equiv q\varphi_{Bn} - qV_n, \end{aligned} \tag{22.24}$$

where φ_{bi} is the built-in potential in equilibrium.

In the thermionic emission theory, it is further assumed that under bias the quasi-equilibrium condition still prevails, so that the equilibrium distribution function can be used. Under this assumption, the analysis of J_{sm} under bias should follow exactly the same steps as those used for obtaining equilibrium current density, and one can therefore use (22.22), with minor modification, namely, φ_{bi} should now be replaced by $\varphi_{bi}(V)$ in (22.24). Hence, one can write

$$\begin{aligned} J_{sm}(V) &= A^* T^2 e^{-qV_n/k_BT} e^{-[q\varphi_{bi} - qV]/k_BT} \\ &= A^* T^2 e^{-q\varphi_{Bn}/k_BT} e^{qV/k_BT}, \quad \varphi_{Bn} \equiv \varphi_{bi} + V_n. \end{aligned} \tag{22.25}$$

The reverse current density J_{ms} from metal to semiconductor must remain the same as in equilibrium, since the Schottky barrier remains unchanged under bias. Therefore, the detailed balancing of current flow is broken and the current

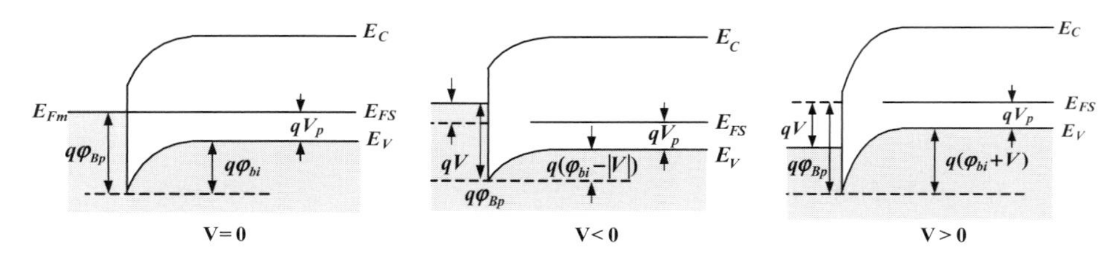

Figure 22.15 The energy band diagrams of Schottky contact both in equilibrium (left) and nonequilibrium under negative (middle) and positive (right) biases for Au-contacted p-type silicon, in which $E_{FS} < E_{Fm}$.

flows, namely,

$$\begin{aligned} J_n &= J_{sm}(V) - J_{ms}(0) \\ &= A^* T^2 e^{-q\varphi_{Bn}/k_BT}[e^{qV/k_BT} - 1]. \end{aligned} \tag{22.26}$$

For Au-contacted p-type silicon, one can carry out similar analysis. Here, the roles of electrons are replaced by those of holes, as indicated in Figure 22.15. In this case, the barrier potential for holes is lowered or raised with negative or positive voltage applied at the metal and one can write

$$\begin{aligned} J_p &= J_{sm}(V) - J_{ms}(0) \\ &= A^* T^2 e^{-q\varphi_{Bp}/k_BT}[e^{-qV/k_BT} - 1], \end{aligned} \tag{22.27a}$$

where the Schottky barrier for holes is given by

$$q\varphi_{Bp} = q\varphi_{\mathrm{bi}} + qV_p. \tag{22.27b}$$

Obviously, the diode forward current is induced in this case with negative voltage at the metal. It is clear that the I–V behavior of Schottky diodes is substantially same as that of p–n junction diodes. However, there is an important difference in the behavior of the two diodes. As evident from the I–V analysis, Schottky diode is a unipolar device, whose current is contributed either by electrons or holes, depending on the work function difference between the metal and semiconductor. This is in marked contrast with the bipolar behavior of p–n junction diodes in which the current is always contributed by both electrons and holes.

Drift Diffusion Theory The I–V behavior in the same gold-contacted n-type silicon is again considered, based on the drift and diffusion theory by Schottky. In this theory, the collisions suffered by electrons in the surface depletion region are accounted for by starting from the general expression of electron current density, namely,

$$J_n = qn\mu_n \mathrm{E} + qD_n \frac{dn}{dx} \tag{22.28}$$

(see (18.6)). Using the Einstein relation (18.33) and the definition of the electrostatic potential, namely,

$$D_n/\mu_n = k_BT/q; \quad -\frac{\partial\varphi}{\partial x} = \mathrm{E},$$

(22.28) can be recast into

$$J_n e^{-\beta\varphi(x)} = qD_n \frac{\partial}{\partial x}\left[e^{-\beta\varphi(x)} n(x)\right], \quad \beta \equiv \frac{q}{k_B T}. \tag{22.29}$$

Clearly, (22.29) is identical to (22.28) aside from the fact that both sides of (22.28) are multiplied by the integration factor exp $-\beta\varphi(x)$, and the identity between (22.28) and (22.29) can be easily verified by carrying out the differentiations in (22.29).

Performing the integrations on both sides of (22.29) and using the fact that J_n is constant throughout results in

$$J_n = \frac{qD_n[e^{-\beta\varphi(x)} n(x)]_0^W}{\int_0^W dx e^{-\beta\varphi(x)}}. \tag{22.30}$$

To proceed further, $\varphi(x)$ has to be specified explicitly. For this purpose, recall that the space charge field E in this case is identical to that of p^+–n junction for which $N_A \gg N_D$, so that one can write

$$\mathrm{E}(x) = -\mathrm{E}_{\max}(V)[1-x/W(V)], \quad \mathrm{E}_{\max}(V) = qN_D W(V)/\varepsilon_S \tag{22.31a}$$

(see Figures 22.11 and 22.13 and (19.2) and (19.7)). Integrating (22.31a) with respect to x results in

$$\begin{aligned}\varphi(x) &= \mathrm{E}_{\max}(V)\left[x - \frac{x^2}{2W(V)}\right] - \varphi_{\mathrm{Bn}} \\ &= \frac{qN_D}{\varepsilon_S}\left[W(V)x - \frac{x^2}{2}\right] - \varphi_{\mathrm{Bn}}\end{aligned} \tag{22.31b}$$

In (22.31b) the Schottky barrier φ_{Bn} has been introduced as the boundary condition for $\varphi(x)$ to satisfy. Thus, $\varphi(0) = -\varphi_{\mathrm{Bn}}$ so that the conduction band, $E_C(0) = -q\varphi(0)$ bends up at the surface from $\mathrm{E_F}$ by an amount, $q\varphi_{\mathrm{Bn}}$ regardless of the bias V (see Figures 22.11 and 22.13). Also, the built-in potential under bias, $\varphi_{\mathrm{bi}}(V) = \varphi_{\mathrm{bi}} - V$ increases or decreases from the equilibrium value, depending on the polarity of V (see (22.23).

With $\varphi(x)$ thus specified,

$$n(0)e^{-\beta\varphi(0)} = N_C e^{-(E_C(0)-E_F)/k_B T} e^{\beta\varphi_{\mathrm{Bn}}} = N_C \tag{22.32a}$$

$$n(W)e^{-\beta\varphi(W)} = N_C e^{-(E_C(W)-E_F)/k_B T} e^{\beta(V_n+V)} = N_C e^{\beta V} \tag{22.32b}$$

where $E_C(0) - E_F = q\varphi_{\mathrm{Bn}}$, $E_C(W) - E_F = qV_n$ (see Figures 22.11 and 22.13).

Hence, the numerator of (22.30) is to be explicitly evaluated as

$$qD_n[e^{-\beta\varphi(x)} n(x)]_0^W = qD_n N_C(e^{\beta V} - 1), \ \beta = q/k_B T \tag{22.33}$$

The integration in the denominator of (22.30) can be carried out using (22.31b), but first by completing the square in the exponent, obtaining

$$\begin{aligned}\int_0^{W(V)} dx e^{-\beta\varphi(x)} &= e^{\beta\varphi_{Bn}} \int_0^{W(V)} dx e^{(\beta q N_D/\varepsilon_S)\{[x-W(V)]^2 - W(V)^2\}} \\ &= e^{\beta\varphi_{Bn}} \left(\frac{2\varepsilon_S}{q N_D \beta}\right)^{1/2} D(\varsigma),\end{aligned} \tag{22.34a}$$

where the function

$$D(\varsigma) \equiv e^{-\varsigma^2} \int_0^{\varsigma} ds e^{s^2}, \quad \varsigma \equiv (\beta q N_D/2\varepsilon_S)^{1/2} W(V), \tag{22.34b}$$

is the well-tabulated Dawson integral and is defined in terms of dimensionless variable $s \equiv (\beta q N_D/2\varepsilon_S)^{1/2} x$. In the asymptotic limit, the Dawson integral is well approximated by

$$D(\varsigma) \simeq 1/2\varsigma. \tag{22.34c}$$

Hence, in this limit, (22.34a) is compacted as

$$\int_0^{W(V)} dx e^{-\beta\varphi(x)} = e^{\beta\varphi_{Bn}} \frac{\varepsilon_S}{q N_D \beta W(V)}. \tag{22.35}$$

Hence, inserting (22.33) and (22.35) into (22.30) results in

$$J_n = J_{SD} e^{-\beta\varphi_{Bn}} (e^{\beta V} - 1), \ \beta = q/k_B T \tag{22.36a}$$

where φ_{Bn} is the Schottky barrier (see (22.23)) and

$$J_{SD} = \frac{q^2 D_n N_C N_D \beta W(V)}{\varepsilon_S} = \frac{q^2 D_n N_C}{k_B T} \left[\frac{2 q N_D \varphi_{bi}(V)}{\varepsilon_S}\right]^{1/2}, \quad \varphi_{bi}(V) = (\varphi_{bi} - V). \tag{22.36b}$$

In (22.36b), $W(V)$ has been specified in terms of $\varphi_{bi}(V)$ (see (22.10)) and $\beta = q/k_B T$ has been used. One can analyze in a similar fashion Au-contacted p-type silicon by interchanging the role of electrons with that of holes.

Schottky Barrier Lowering In discussing the diode I–V behavior, the changes in Schottky barrier height induced by external field have not been accounted for and are now briefly discussed. This change in barrier height arises specifically from the image force, operative on a charged particle near any metallic surface. Figure 22.16 illustrates the potential acting on an electron resulting from both external field and image force. The same kinds of potentials are also operative for an electron near the surface of Au-contacted n-type silicon, as considered in Figure 22.13. The electron near the surface of

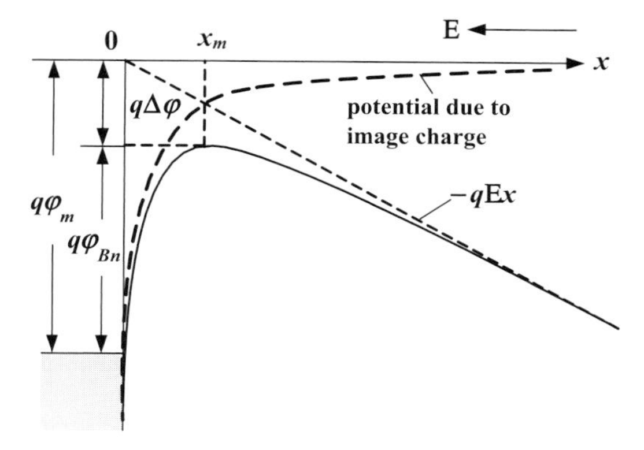

Figure 22.16 The Schottky barrier lowering due to external electric field and image force induced at the metal surface.

metal–semiconductor contact is to be viewed as a charged particle with effective mass m_n moving in a medium with permittivity ε_S. Hence, the electron should induce the image charge on the metal surface. Also, the space charge induced near the surface gives rise to the space charge field that effectively acts as external field on the electron.

With this fact in mind, consider Figure 22.16. An electron at x distance away from the surface induces an image charge of opposite polarity at $-x$ distance away. The resulting Coulomb force $\text{E}(x) = -q^2/16\pi\varepsilon_S x^2$ gives rise to the image force potential $\propto 1/x$. Thus, the total potential acting on the electron in the presence of external field $\mathbf{E} = -\hat{x}\text{E}$ is given by

$$q\varphi(x) = -\frac{q^2}{16\pi\varepsilon_S x} - q\text{E}x. \tag{22.37}$$

Hence, the maximum value of $\varphi(x)$ occurs at x_m, at which

$$0 = \frac{\partial\varphi(x_m)}{\partial x} = \frac{q}{16\pi\varepsilon_S x_m^2} - \text{E}.$$

Inserting x_m thus determined into (22.37), one finds the effective lowering of the Schottky barrier as

$$q\Delta\varphi = \left(\frac{q^3\text{E}}{4\pi\varepsilon_S}\right)^{1/2}. \tag{22.38}$$

Obviously, this barrier lowering has an important bearing on the diode I–V behavior (see (22.27) or (22.36)).

22.3.2
Ohmic Contact

In addition to exhibiting active diode behavior, Schottky contacts can also be used as ohmic contacts. As noted, ohmic contacts are essential elements in semiconductor devices, connecting the devices to the outside with negligible resistance. There are two different ways by which to make ohmic contacts out of Schottky contacts.

Tunnel-Based Ohmic Contact In previous two examples of Au-contacted n- and p-type silicon diodes, the electrons or holes as the majority carrier are always depleted near the surface (Figures 22.11 and 22.12). This leaves the surface donor or acceptor ions uncompensated, thereby leading to the band bending and concomitant formation of built-in barrier potential φ_{bi}. The control of φ_{bi} via the applied voltage has been shown the essential ingredient of diode operation.

Now, the depletion depth of Au-contacted n-type silicon, for example, develops so as to support the built-in potential, namely,

$$W = \left(\frac{2\varepsilon_S \varphi_{bi}}{qN_D}\right)^{1/2}$$

(see (22.10)). It is therefore apparent that W can be made extremely narrow by increasing the doping level for the given φ_{bi} or for that matter for the given Schottky barrier φ_{Bn}. In fact, the depletion depth can be narrowed down to a few nanometers in the limit of heavy doping. In such cases, electrons can be transported freely from metal to semiconductor or vice versa via tunneling with the barrier becoming virtually transparent.

Thus, when negative voltage is applied to metal, the Fermi level E_{Fm} therein is raised with respect to E_{FS} and electrons on top of E_{Fm} are emitted with negligible voltage drop across the thin tunnel barrier, as illustrated in Figure 22.17. Likewise, when positive bias is applied, E_{Fm} is lowered and there ensues free reverse flow of electrons from semiconductor to metal via tunneling. For the case of Au-contacted

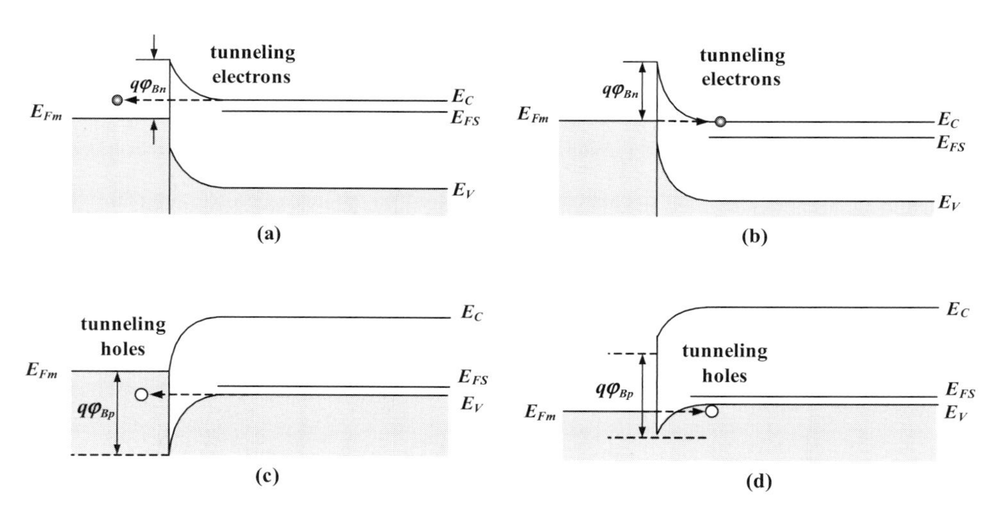

Figure 22.17 The energy band diagrams of tunnel-based ohmic contact: for electrons under positive (a) and negative (b) biases at the metal; for holes under negative (c) and positive (d) biases at the metal.

p-type silicon, the ohmic contact for holes can likewise be realized by heavily doping the substrate with acceptor atoms. The energy band diagrams therein and operational principle are also shown in Figure 22.17.

Schottky Ohmic Contact There is an alternative method by which to attain the ohmic contact. In this approach, the majority carriers in the substrate are accumulated rather than depleted near the surface so that the resistivity therein is made much lower than the device resistivity. The majority carriers can be accumulated at the surface by choosing the work function of the metal smaller than that of the semiconductor. This is illustrated in Figure 22.18 in which the metal work function is smaller than that of n-type semiconductor, namely, $q\varphi_m < q\chi + qV_n$.

Upon contact, the electrons are then transferred from metal to semiconductor, giving rise to a dipole charge, which consists of the hole charge sheet at the metal surface and excess electrons in the semiconductor near the surface. As a result, the space charge field and potential peak at the surface decrease exponentially away from the surface into the bulk, as indicated in Figure 22.18. This brings down the conduction band near the surface, thereby inducing the electron accumulation at the surface and the ohmic contact therein.

A key parameter in this kind of ohmic contact is its spatial extent from the contacting surface. To examine this, consider the Poisson equation near the surface, namely,

$$\frac{d^2\varphi}{dx^2} = -\frac{\varrho}{\varepsilon_S}, \quad \varrho = -q(n_{n0}e^{\beta\varphi} - n_{n0}), \quad \beta = \frac{q}{k_B T}. \tag{22.39}$$

Here, the source term represents the excess electron charge accumulated near the surface with n_{n0} denoting the electron concentration as the majority carrier in the bulk substrate. The potential φ ranges from the maximum value φ_S at the surface for $x = 0$ to zero in the bulk (Figure 22.18).

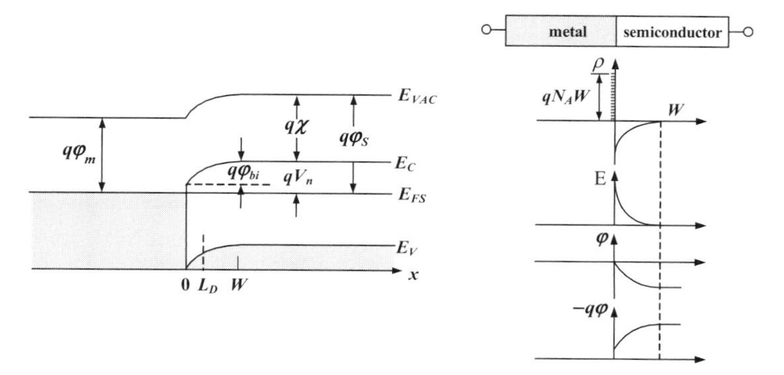

Figure 22.18 The energy band diagram of Schottky ohmic contact, made up of n-type semiconductor contacting the metal, in which $E_{Fn} < E_{Fm}$ (left). Also sketched are the space charge, field, potential, electron potential energy (right). The depletion depth W is scaled by Debye screening distance L_D.

The first integration of the Poisson equation can be carried out, as detailed in Chapter 21, by multiplying both sides of (22.39) by $d\varphi$ and carrying out the integration with respect to φ. From the left-hand side, one finds

$$\int_0^{\varphi} d\varphi \frac{d^2\varphi}{dx^2} = \int_0^{d\varphi/dx} \frac{d\varphi}{dx} d\left(\frac{d\varphi}{dx}\right) = \frac{1}{2}\mathrm{E}^2, \quad \mathrm{E} \equiv -\frac{d\varphi}{dx}, \tag{22.40a}$$

while from the right-hand side of (22.39), one obtains

$$\int_0^{\varphi} d\varphi \frac{-\varrho}{\varepsilon_S} = \frac{qn_{n0}}{\beta\varepsilon_S}[e^{\beta\varphi}-\beta\varphi]_0^{\varphi} \simeq \frac{qn_{n0}}{\beta\varepsilon_S}e^{\beta\varphi} \tag{22.40b}$$

Equating (22.40a) and (22.40b), one can write

$$\mathrm{E} \equiv -\frac{d\varphi}{dx} = \left(\frac{2qn_{n0}}{\beta\varepsilon_S}\right)^{1/2} e^{\beta\varphi/2}. \tag{22.41}$$

One can recast (22.41) into the integral form

$$-\int_{\varphi_S}^{\varphi} d\varphi e^{-\beta\varphi/2} = \left(\frac{2qn_{n0}}{\beta\varepsilon_S}\right)^{1/2} \int_0^x dx \tag{22.42}$$

and carry out the integrations in (22.42), obtaining

$$e^{\beta\Delta\varphi/2}-1 = \frac{x}{\sqrt{2}L_D}, \tag{22.43}$$

where

$$\Delta\varphi \equiv \varphi_S-\varphi(x) \tag{22.44a}$$

represents the variation of $\varphi(x)$ from its maximum value at the surface and

$$L_D = \left(\frac{\varepsilon_S}{q\beta n_{n0}e^{\beta\varphi_S}}\right)^{1/2} = \left(\frac{\varepsilon_S k_B T}{q^2 n_s}\right)^{1/2}, \quad n_s = n_{n0}e^{\beta\varphi_S}, \quad \beta = q/k_B T \tag{22.44b}$$

is the Debye screening distance given in terms of the surface electron concentration n_s.

Hence, the space charge potential φ can be obtained as a function of x from (22.43) as

$$\varphi(x) = \varphi_S-\frac{2}{\beta}\ln\left(1+\frac{x}{\sqrt{2}L_D}\right) \tag{22.45}$$

and is shown to decrease monotonously from the surface into the bulk, as expected. Once $\varphi(x)$ is found as a function of x, the space charge field and the profile of excess electrons can be obtained using (22.45) as

$$E = -\frac{d\varphi}{dx} = \frac{\sqrt{2}k_B T}{qL_D}\frac{1}{1+x/\sqrt{2}L_D}, \qquad (22.46)$$

$$\varrho(x) \equiv -\varepsilon_S\frac{d^2\varphi(x)}{dx^2} = -\frac{qn_s}{(1+x/\sqrt{2}L_D)^2}. \qquad (22.47)$$

Finally, the depth in which the excess electrons extend from the surface to the bulk can be readily obtained from (22.45) in conjunction with the boundary condition, that is, $\varphi(x = W) = 0$, namely,

$$W = \sqrt{2}L_D(e^{\beta\varphi_S/2} - 1), \qquad (22.48)$$

where the surface band bending is specified by the difference between the two work functions $\varphi_S = \chi + V_n$ and φ_m. As expected, the spatial profiles of E, $\varrho(x)$ and W, are all scaled by the Debye screening length, as it should, and L_D in turn is specified by the difference in work functions and doping level (see (22.44b)).

22.4 Problems

22.1. Discuss the merits and demerits of the quantum wire FETs, as compared with planar MOSFETs, from scaling, performance, and fabrication points of view.

22.2. Consider the quantum wire FET, made of intrinsic silicon, surrounded by 10 nm thick silicon dioxide, as sketched in Figure 22.19.
- (a) Find the wavefunctions and energy levels of the ground and first few excited states as a function of cross-sectional area W^2.
- (b) Calculate the electron concentrations in the conduction band versus W^2.
- (c) Discuss the inverted electrons versus the gate voltage and W^2, based on the classical theory.
- (d) Discuss the modifications in (c) arising from the wave nature of electrons. For simplicity of discussion, the barrier height of the wire may be taken to be infinite.

22.3. Consider the Au-contacted p-type silicon.
- (a) Find Schottky and built-in barrier potentials for holes as a function of acceptor doping level in the range 10^{17}–10^{20} cm^{-3}.

Figure 22.19 A sketch of the quantum wire FET made of intrinsic silicon surrounded by SiO_2.

(b) Draw the energy level diagram at a typical doping level and both in the absence and the presence of surface states Q_{SS}.

(c) Specify the built-in potential barrier for holes in terms of surface state density and work function difference, corresponding to (22.18) for electrons.

22.4. Derive the *I–V* behavior of Schottky diode consisting of Au-contacted p-type silicon based on both thermionic emission and drift and diffusion of holes.

22.5. Consider n- and p-type silicon in contact with copper with the work function 4.5 eV.

(a) Draw the energy band diagrams both in the absence and presence of the surface states Q_{SS}.

(b) If the e–h pairs are generated by the incident light, indicate the direction of the photocurrent in these two junctions when connected to the circuits.

(c) Find the open-circuit voltage in these two junctions versus the doping levels of donor and acceptor atoms.

22.6. Consider the case in which Schottky ohmic contacts are made of intrinsic silicon for both electrons and holes.

(a) Specify the respective metal work functions required.

(b) Estimate the respective metal work functions if Debye screening distance of 10 and 100 nm are required.

Suggested Reading

1 Taur, Y. and Ning, T.H. (2009) *Fundamentals of Modern VLSI Devices*, 2nd edn, Cambridge University Press.

2 Muller, R.S., Kamins, T.I., and Chan, M. (2002) *Device Electronics for Integrated Circuits*, 3rd edn, John Wiley & Sons, Inc.

3 Sze, S.M. and Ng, K.K. (2006) *Physics of Semiconductor Devices*, 3rd edn, Wiley-Interscience.

Index

Introductory Quantum Mechanics for Semiconductor Nanotechnology. Dae Mann Kim

ISBN: 978-3-527-40975-4

p

Important Physical Numbers and Quantities

$1\,\text{cm} = 10^4\,\mu\text{m} = 10^8\,\text{Å} = 10^7\,\text{nm}$
Electron volt: $1\text{eV} = 1.602 \times 10^{-19}\,\text{J}$
Electron charge: $q = 1.602 \times 10^{-19}\,\text{C}$
Coulomb constant: $1/(4\pi\varepsilon_0) = 8.988 \times 10^9\,\text{N·m}^2/\text{C}^2$
Planck's constant: $h = 6.626 \times 10^{-34}\,\text{J·s} = 4.136 \times 10^{-15}\,\text{eV·s}$
$\hbar = h/2\pi = 1.055 \times 10^{-34}\,\text{J·s} = 6.582 \times 10^{-16}\,\text{eV·s}$
Boltzmann constant: $k_B = 1.381 \times 10^{-23}\,\text{J/K} = 8.617 \times 10^{-5}\,\text{eV/K}$
Bohr radius: $a = 0.529\,\text{Å} = 0.0529\,\text{nm}$
Avogadro's number: $N = 6.022 \times 10^{23}$ particles/mole
Electron mass in free space: $m_0 = 9.109 \times 10^{-31}\,\text{kg}$
Proton mass in free space: $m_p = 1.673 \times 10^{-27}\,\text{kg}$
Permeability of free space: $\mu_0 = 1.256 \times 10^{-8}\,\text{H/cm}$
Permittivity of free space: $\varepsilon_0 = 8.854 \times 10^{-14}\,\text{F/cm}$
Speed of light in free space: $c = 2.998 \times 10^8\,\text{m/s}$
Thermal voltage at room temperature (300K): $k_B T/q = 0.0259\,\text{V}$
Wavelength of 1eV photons: $1.24\,\mu\text{m}$

Important Electronic Properties of Semiconductors at Room Temperature

	Ge	Si	GaAs
Atoms/cm^3	4.42×10^{22}	5.0×10^{22}	4.42×10^{22}
Breakdown field (V/cm)	$\sim 10^5$	$\sim 3 \times 10^5$	$\sim 4 \times 10^5$
Dielectric constant	16.0	11.9	13.1
Effective density of states (cm^{-3})			
conduction band N_c	1.04×10^{19}	2.8×10^{19}	4.7×10^{17}
valance band N_v	6.0×10^{18}	1.04×10^{19}	7.0×10^{18}
Electron affinity, χ (V)	4.0	4.05	4.07
Energy gap (eV)	0.66	1.12	1.424
Intrinsic carrier concentration (cm^{-3})	2.4×10^{13}	1.45×10^{10}	1.79×10^6
Intrinsic Debye length (μm)	0.68	24	2250
Lattice constant (Å)	5.646	5.430	5.653
Lattice (intrinsic) mobility (cm^2/V·s)			
electrons	3900	1500	8500
holes	1900	450	400